Physicochemical, Sensory and Nutritional Properties of Foods Affected by Processing and Storage

Physicochemical, Sensory and Nutritional Properties of Foods Affected by Processing and Storage

Editors

Sidonia Martínez Suárez
Francisco Javier Carballo García

MDPI • Basel • Beijing • Wuhan • Barcelona • Belgrade • Manchester • Tokyo • Cluj • Tianjin

Editors
Sidonia Martínez Suárez   Francisco Javier Carballo García
University of Vigo   University of Vigo
Spain   Spain

Editorial Office
MDPI
St. Alban-Anlage 66
4052 Basel, Switzerland

This is a reprint of articles from the Special Issue published online in the open access journal *Foods* (ISSN 2304-8158) (available at: https://www.mdpi.com/journal/foods/special_issues/Physicochemical_Sensory_and_Nutritional_Properties_of_Foods_Affected_by_Processing_and_Storage).

For citation purposes, cite each article independently as indicated on the article page online and as indicated below:

LastName, A.A.; LastName, B.B.; LastName, C.C. Article Title. *Journal Name* **Year**, *Volume Number*, Page Range.

ISBN 978-3-0365-2732-1 (Hbk)
ISBN 978-3-0365-2733-8 (PDF)

Cover image courtesy of Sidonia Martínez Suárez

Contents

About the Editors . . . . **ix**

Preface to "Physicochemical, Sensory and Nutritional Properties of Foods Affected by Processing and Storage" . . . . **xi**

Sidonia Martínez and Javier Carballo
Physicochemical, Sensory and Nutritional Properties of Foods Affected by Processing and Storage
Reprinted from: *Foods* **2021**, *10*, 2970, doi:10.3390/foods10122970 . . . . **1**

Santiago P. Aubourg, Marcos Trigo, Beatriz Martínez and Alicia Rodríguez
Effect of Prior Chilling Period and Alga-Extract Packaging on the Quality of a Canned Underutilised Fish Species
Reprinted from: *Foods* **2020**, *9*, 1333, doi:10.3390/foods9091333 . . . . **9**

Lucía Gómez-Limia, Roxana Cutillas, Javier Carballo, Inmaculada Franco and Sidonia Martínez
Free Amino Acids and Biogenic Amines in Canned European Eels: Influence of Processing Step, Filling Medium and Storage Time
Reprinted from: *Foods* **2020**, *9*, 1377, doi:10.3390/foods9101377 . . . . **23**

Lucía Gómez-Limia, Nicolás Moya Sanmartín, Javier Carballo, Rubén Domínguez, José M. Lorenzo and Sidonia Martínez
Oxidative Stability and Antioxidant Activity in Canned Eels: Effect of Processing and Filling Medium
Reprinted from: *Foods* **2021**, *10*, 790, doi:10.3390/foods10040790 . . . . **43**

Amanda N. Dainton, Hulya Dogan and Charles Gregory Aldrich
The Effects of Select Hydrocolloids on the Processing of Pâté-Style Canned Pet Food
Reprinted from: *Foods* **2021**, *10*, 2506, doi:10.3390/foods10102506 . . . . **63**

Aarón García-Blázquez, Natalia Moratalla-López, Cándida Lorenzo, M. Rosario Salinas and Gonzalo L. Alonso
Effect of *Crocus sativus* L. Stigmas Microwave Dehydration on Picrocrocin, Safranal and Crocetin Esters
Reprinted from: *Foods* **2021**, *10*, 404, doi:10.3390/foods10020404 . . . . **77**

Jun-Jie Xing, Dong-Hui Jiang, Zhen Yang, Xiao-Na Guo and Ke-Xue Zhu
Effect of Humidity-Controlled Dehydration on Microbial Growth and Quality Characteristics of Fresh Wet Noodles
Reprinted from: *Foods* **2021**, *10*, 844, doi:10.3390/foods10040844 . . . . **91**

Begüm Önal, Giuseppina Adiletta, Marisa Di Matteo, Paola Russo, Inês N. Ramos and Cristina L. M. Silva
Microwave and Ultrasound Pre-Treatments for Drying of the "Rocha" Pear: Impact on Phytochemical Parameters, Color Changes and Drying Kinetics
Reprinted from: *Foods* **2021**, *10*, 853, doi:10.3390/foods10040853 . . . . **105**

Alberto González-Mohino, Trinidad Pérez-Palacios, Teresa Antequera, Jorge Ruiz-Carrascal, Lary Souza Olegario and Silvia Grassi
Monitoring the Processing of Dry Fermented Sausages with a Portable NIRS Device
Reprinted from: *Foods* **2020**, *9*, 1294, doi:10.3390/foods9091294 . . . . **123**

Lihong Pan, Jiali Xing, Xiaohu Luo, Yanan Li, Dongling Sun, Yuheng Zhai, Kai Yang and Zhengxing Chen
Influence of Electron Beam Irradiation on the Moisture and Properties of Freshly Harvested and Sun-Dried Rice
Reprinted from: *Foods* **2020**, *9*, 1139, doi:10.3390/foods9091139 . . . . **135**

Giuseppe Natrella, Graziana Difonzo, Maria Calasso, Giuseppe Costantino, Francesco Caponio and Michele Faccia
Evolution of VOC and Sensory Characteristics of Stracciatella Cheese as Affected by Different Preservatives
Reprinted from: *Foods* **2020**, *9*, 1446, doi:10.3390/foods9101446 . . . . **149**

Chrystalla Antoniou, Angelos Kyratzis, Youssef Rouphael, Stelios Stylianou and Marios C. Kyriacou
Heat- and Ultrasound-Assisted Aqueous Extraction of Soluble Carbohydrates and Phenolics from Carob Kibbles of Variable Size and Source Material
Reprinted from: *Foods* **2020**, *9*, 1364, doi:10.3390/foods9101364 . . . . **169**

Diana I. Santos, Maria João Fraqueza, Hugo Pissarra, Jorge A. Saraiva, António A. Vicente and Margarida Moldão-Martins
Optimization of the Effect of Pineapple By-Products Enhanced in Bromelain by Hydrostatic Pressure on the Texture and Overall Quality of Silverside Beef Cut
Reprinted from: *Foods* **2020**, *9*, 1752, doi:10.3390/foods9121752 . . . . **179**

Woo-Hee Cho and Jae-Suk Choi
Sensory Quality Evaluation of Superheated Steam-Treated Chicken Leg and Breast Meats with a Combination of Marination and Hot Smoking
Reprinted from: *Foods* **2021**, *10*, 1924, doi:10.3390/foods10081924 . . . . **201**

Asad Nawaz, Enpeng Li, Ibrahim Khalifa, Noman Walayat, Jianhua Liu, Sana Irshad, Anam Zahra, Shakeel Ahmed, Mario Juan Simirgiotis, Mirian Pateiro and José M. Lorenzo
Effect of Different Processing Methods on Quality, Structure, Oxidative Properties and Water Distribution Properties of Fish Meat-Based Snacks
Reprinted from: *Foods* **2021**, *10*, 2467, doi:10.3390/foods10102467 . . . . **221**

Soraya Mousavi, Roberto Mariotti, Vitale Stanzione, Saverio Pandolfi, Valerio Mastio, Luciana Baldoni and Nicolò G. M. Cultrera
Evolution of Extra Virgin Olive Oil Quality under Different Storage Conditions
Reprinted from: *Foods* **2021**, *10*, 1945, doi:10.3390/foods10081945 . . . . **235**

Xiaoxue Yan, Jun Yan, Siyi Pan and Fang Yuan
Changes of the Aroma Composition and Other Quality Traits of Blueberry 'Garden Blue' during the Cold Storage and Subsequent Shelf Life
Reprinted from: *Foods* **2020**, *9*, 1223, doi:10.3390/foods9091223 . . . . **255**

Lin Zhu, Xianrui Liang, Yushuang Lu, Shiyi Tian, Jie Chen, Fubin Lin and Sheng Fang
Effect of Freeze-Thaw Cycles on Juice Properties, Volatile Compounds and Hot-Air Drying Kinetics of Blueberry
Reprinted from: *Foods* **2021**, *10*, 2362, doi:10.3390/foods10102362 . . . . **267**

Khadijeh Yasaminshirazi, Jens Hartung, Michael Fleck and Simone Graeff-Hönninger
Impact of Cold Storage on Bioactive Compounds and Their Stability of 36 Organically Grown Beetroot Genotypes
Reprinted from: *Foods* **2021**, *10*, 1281, doi:10.3390/foods10061281 . . . . **279**

Rong Lin, Shasha Cheng, Siqi Wang, Mingqian Tan and Beiwei Zhu
Influence of Refrigerated Storage on Water Status, Protein Oxidation, Microstructure, and
Physicochemical Qualities of Atlantic Mackerel (*Scomber scombrus*)
Reprinted from: *Foods* **2021**, *10*, 214, doi:10.3390/foods10020214 . . . . **299**

About the Editors

Sidonia Martínez Suárez has been an Associate Professor of Food Technology at the Faculty of Sciences of the University of Vigo, Spain since November 2007. She received her B.S. (1996) and Ph.D. (2002) in Veterinary Medicine from the University of León (Spain). Later, she completed a research stage at the Faculty of Biotechnology in Porto (Portugal), working to identify natural antioxidants in different vegetable matrixes. In recent years, she developed her research interests in the study and improvement of traditional foods (meat and meat products, milk and dairy products, starter cultures, fishery products, vegetables, antioxidant components). In this scientific field, she is co-author of 43 articles published in different journals indexed in the JCR, most of them in prestigious international journals; six peer-reviewed research articles in prestigious national journals not indexed in the JCR; and co-author of two teaching publications. She is also the co-author of one book, five book chapters, thirty long communications and seventy-nine short communications presented at conferences/events, both nationally and internationally. She is a regular reviewer for the most reputed journals in Food Science and Technology. She has collaborated in 17 projects funded by public institutions/administrations (Principal Investigator in two of them) and 5 contracts signed with companies (Principal Investigator in three of them).

Francisco Javier Carballo García is a Professor of Food Technology at the Faculty of Sciences of the University of Vigo, Spain. He received his B.S. (1984) and obtained his Ph.D. (1989) in Veterinary Medicine from the University of León (Spain), both with Special End of Degree Awards. Later, he completed a research stage at the Station de Recherches Laitières (INRA), Jouy-en-Josas, France. In the last 30 years, he has developed his research interest in the study and improvement of traditional foods, both of animal and vegetable origin. In this scientific field, he published more than 200 research articles in the most reputed international journals, and wrote more than 25 book chapters. Once the former President (from 2008 to 2016) of the Food Microbiology Section at the Spanish Society of Microbiology, he is now an Associate Editor of the journals Food Microbiology and Frontiers in Microbiology and is part of the editorial boards of several reputable international journals. He is also an assiduous translator to Spanish of scientific books in the field of food science and technology.

Preface to "Physicochemical, Sensory and Nutritional Properties of Foods Affected by Processing and Storage"

Due to the perishable nature of foods, possibility of transmitting pathogens, seasonality of the production of many of them, and need to obtain a wide variety of products to satisfy the demands, preferences and needs of consumers, food processing and storage are very common activities with great economic and commercial importance.

The effects of processing produce can change or convert foods into more stable, safer, more edible, enjoyable, and/or palatable products, and can increase the digestibility and/or bioavailability of some nutrients and bioactive compounds—in some cases, their concentrations.

Food processing and storage, however, can cause modifications and interactions in food components that have negative effects on nutritional value, organoleptic characteristics or even food safety.

In recent years, intense research was carried out, which focused on understanding the changes and effects during the processing and storage of foods, in order to promote the positive consequences and minimize the negative effects to achieve safe processed and stored foods with the highest nutritional and sensory quality.

It is evident that much progress has been made in generating knowledge in this field. However, there are many factors to consider, such as the chemical diversity of food components and complexity of the interactions, as well as the effects that occur in such components as a consequence of the environmental conditions established during the different technological processes applied before and during storage. Especially, the requirements of consumers who increasingly demand safer food with a higher organoleptic and nutritional quality mean that more research is necessary to achieve more efficient, economical and sustainable processes.

This volume aims to gather some of the latest advances in this field and become a useful tool for students, researchers and professionals in this scientific area.

Sidonia Martínez Suárez, Francisco Javier Carballo García

Editors

Editorial

Physicochemical, Sensory and Nutritional Properties of Foods Affected by Processing and Storage

Sidonia Martínez * and Javier Carballo

Food Technology Area, Faculty of Sciences, University of Vigo, 32004 Ourense, Spain; carbatec@uvigo.es
* Correspondence: sidonia@uvigo.es

Citation: Martínez, S.; Carballo, J. Physicochemical, Sensory and Nutritional Properties of Foods Affected by Processing and Storage. *Foods* **2021**, *10*, 2970. https://doi.org/10.3390/foods10122970

Received: 18 November 2021
Accepted: 29 November 2021
Published: 2 December 2021

Publisher's Note: MDPI stays neutral with regard to jurisdictional claims in published maps and institutional affiliations.

Due to their chemical composition and physicochemical characteristics, most foods are very perishable and are easily altered by physical, chemical and biological agents. To increase their shelf-life and thus enable their transport and storage, it is necessary to carry out specific technological treatments that protect them from deterioration agents. In other cases, technological treatments are applied with a sanitizing purpose, destroying pathogens, or with the aim of diversification, seeking to obtain from a concrete raw material a wide range of different products in order to satisfy the diverse needs and demands of consumers. In any case, sanitizing or diversification treatments are almost always accompanied by an increase in shelf life. Treatments can also be intended to condition food, making it more suitable for consumption. Food processing is therefore a common and necessary activity in order to obtain the wide range of foods that we know, safe and with their diverse sensory and nutritional profiles. However, sometimes these processes have negative effects that result in a decrease in nutritional value or in the alteration of sensory properties. Even sometimes compounds negative for the health of the consumer are generated. It is therefore necessary to gather as much knowledge as possible about the effects of technological treatments on food components in order to optimize these treatments and to minimize their negative effects and achieve food of the highest quality and wholesomeness.

The articles included in the present special issue show important and interesting advances and new approaches in this field.

Canning is one of the most important procedures of food preservation. From the initial works of Nicolas Appert [1], extensive and ongoing research has been carried out aimed at improving the process and focused on the development of new materials for packaging, on optimizing heat treatments and their adaptation to the characteristics of the food to be preserved and on the study of the effect of the filling medium to select the most appropriate in each case.

Marine foods are traditionally preserved by canning, with canned fish and seafood being products of proven quality that are very popular and well-established among consumers. Constituents and organoleptic properties of marine foods have different thermal sensitivity and according to this, different detrimental effects were reported in the literature. In this special issue, three articles make interesting contributions within this research topic. Aubourg et al. [2] studied the effect of the duration of the previous chilling period (0, 4 and 9 days) and of the use of an aqueous extract of the macroalga *Fucus spiralis* in the brine-packaging medium (final contents of 0.00, 3.50, 10.50 and 21.00 mg extracted alga/mL packaging medium) on the chemical changes related to quality registered after 3 months of storage in a canned underutilized mackerel species (*Scomber colias*). Increased chilling times increase the free fatty acid content, but the use of the alga extract reduced this content. An increased chilling time led to the increase of the values of the lipid oxidation indexes and the presence of the alga extract had an uneven effect on these parameters. Trimethylamine content markedly increased after the sterilization process and no effects of the chilling time or of the alga extract addition were observed.

Gómez-Limia et al. [3] evaluated the effects of the steps of the canning process (raw, after frying, after sterilization and after 2 and 12 months of storage) and of different filling media (sunflower oil, olive oil, or spiced olive oil) on the free amino acid and biogenic amine content of the European eel (*Anguilla anguilla*). Relative to raw samples, all steps in the manufacture and storage significantly affected the free amino acid content of eels, this content being also affected by the type of oil used in the frying process and as the filling medium. The step of the canning process, the type of oil used in the frying process or as filling medium and the storage time also affected the contents of the different biogenic amines. The authors reported very low biogenic amine contents and an absence of histamine in all samples. The low biogenic amine index observed indicate the good quality of canned eels.

Gómez-Limia et al. [4] also studied the effects of the steps of the canning process and of different filling media on the oxidative stability and antioxidant activity in canned eels. After the same steps as in the previous study and testing the same oils as filling medium, the authors assessed the oxidation parameters (acidity, peroxide value and thiobarbituric acid reactive substances), antioxidant capacity and total phenolic and vitamin E contents. Despite the losses, the canning process and subsequent storage preserved a great part of the antioxidant capacity and vitamin E content of the filling medium, which is of great interest to the consumer.

Pet food can also appear in the markets canned. Most of these goods are products consisting of meat and water containing binding/structural ingredients, commercially sterilized and similar to stews and pâtés in appearance. Hydrocolloids are commonly used in canned pet food for firmness control, but their functional effects have not been quantified in this food format. Dainton et al. [5] determined the effects of select hydrocolloids (1% dextrose and 0.5% guar gum with 0.5% of either dextrose, kappa carrageenan, locust bean gum, or xanthan gum) on batter consistency, heat penetration and the texture of canned pet food. The authors showed that hydrocolloids influenced heat penetration, probably due to differences in batter consistency and affected finished product texture.

Dehydration is one of the most ancient and effective method for food preservation. The reduction of the water content hinders the activity of the agents of deterioration, both of the microorganisms and enzymes. At the same time, the elimination of water reduces the weight and volume of the food and facilitates its package, transport and storage. Additionally, dehydration provides particular organoleptic characteristics, creating new foods in some cases [6]. In this special issue, three different articles reported the effects of dehydration achieved by different procedures on several food constituents and characteristics.

Saffron is the dried stigmas of *Crocus sativus*, its high value being due to the color, flavor and aroma. The dehydration procedure used affects the main metabolites of saffron that are responsible for these organoleptic characteristics and that define the quality of the product. García-Blázquez et al. [7] studied the effect of microwave dehydration on saffron main metabolites (picrocrocin, safranal and crocetin esters) from *Crocus sativus* L. stigmas at three determinate powers and different time lapses. A control of saffron dehydrated using the traditional dehydration called "toasting" in which stigmas are put on a sieve with a silk bottom placed over a heating source was obtained and analyzed. The results reported suggest that microwave dehydration is a suitable process for obtaining high quality saffron, 800 W with 6 lapses of 20 s being the best conditions among those studied.

Fresh wet noodles (FWN) have high water and nutrient contents that favor microbial growth. Therefore, a preservation treatment is necessary in order to extend the shelf-life without decreasing the organoleptic and nutritional quality. Several treatments such as pasteurization, pH decrease, natural or chemical preservatives, heat, irradiation, non-thermal processing technologies, ozone water treatment, modified atmosphere packaging, hurdle technology, etc., have been tried for fresh wet noodle preservation with unequal results. Drying is of course another valid alternative. The industrial drying process of noodle products was often at low temperature with divided steps, and the humidity of the hot air was usually taken into consideration for fine dried noodles. The relative humidity, and the temperature of the hot air, should be controlled during the drying process for both semi-dried noodles and fresh wet noodles because heat treatment with excessive temperature and humidity may result in the over gelatinization or melting of starch and the generation of cracks or chaps in fresh noodles, thus deteriorating the noodle quality. The humidity-controlled dehydration (HCD) at moderate temperatures could be a successful technique to control the microbial growth in fresh wet noodles. Xing et al. [8] investigated the effects of HCD treatment using different temperatures, relative humidity and treatment times on the total plate count, the shelf-life and qualities (color, degree of gelatinization, cooking and textural properties and sensory attributes) of FWN. Data reported by these authors showed that HCD reduced the initial microbial load on the fresh noodles and extended their shelf-life by up to 14–15 days under refrigerated (10 °C) storage. The lightness (L*) values, the apparent stickiness and the cooking properties of the noodle body were improved by HCD while good sensory and texture quality of noodles were still maintained after the dehydration process.

Pear (*Pyrus communis* L.) can be maintained for long periods of time (close to a year) under controlled atmosphere. However, long-term storage in these conditions can promote deterioration processes such as internal browning and superficial scald, negatively influencing the consumer acceptability. Drying can be an alternative to increase the shelf-life of pears with the added benefit of reducing transportation and storage costs, also offering new ways of presentation for consumers. Conventional hot air drying is widely used in the food industry, but it can promote the partial degradation of nutrients and undesirable changes in color, appearance and structural properties. Pre-treatments with modern technologies such as microwaves or ultrasound can accelerate the drying process, reducing drying time, enhancing mass transfer phenomena, preserving functional components, inactivating enzymes and improving rehydration characteristics. However, these pre-treatments can also have negative effects on the quality of the final product if the application conditions are not suitable. Therefore, more research is needed to stablish the optimal conditions for the application of the pre-treatments. In this context, Önal et al. [9] evaluated the effect of drying temperature and innovative pre-treatments (microwave and ultrasound) on drying behavior and quality characteristics, such as color, total phenolic content and antioxidant activity of "Rocha" pear, a traditional Portuguese cultivar. The authors concluded that ultrasound application is a promising technology to obtain healthy/nutritious dried "Rocha" pear snacks.

Fermentation followed by drying is a common combined procedure for obtaining the most reputed meat and dairy products. Dry-fermented sausages are widely manufactured and consumed meat products, deeply rooted in the culture and tradition of several regions and countries and with outstanding social and economic relevance. Most of the dry-fermented sausage varieties are manufactured in small-scale production plants following traditional procedures with scarce degree of mechanization and with limited control of the quality and safety of the final product. As a result, products are very heterogeneous in their organoleptic characteristics which hinders in many cases the loyalty of consumers and expansion to markets far from their areas of origin. The implementation of systems of control for the whole dry-fermented sausage process, ensuring the safety and quality of the product, would be of use to overcome these issues. However, due to the particularities of the producers, such a control system should be simple, cheap and easily implemented. Exploring the possibilities of one of these control systems, González-Mohino et al. [10] studied the ability of a MicroNIR device to monitor the dry fermented sausage process with the use of multivariate data analysis. Thirty sausages were manufactured, subjected to dry fermentation and studied into four main stages of the production process (raw material, end of fermentation, end of intense drying and final product). The authors performed physicochemical (weight lost, pH, moisture content, water activity, color, hardness and thiobarbiruric reactive substances analysis) and sensory (quantitative descriptive analysis) analysis of samples on different steps of the ripening process. Near-infrared (NIR) spectra were taken throughout the process at three points of the samples. The results of multivariate data analysis showed the ability to monitor and classify the different stages of ripening process (mainly the fermentation and drying steps). Based on their results, the authors concluded that a portable NIR device is a nondestructive, simple, noninvasive, fast and cost-effective tool with the ability to monitor the dry fermented sausage processing and to classify samples as a function of the stage. This device constitutes a feasible decision method for sausages to progress to the following processing stage.

Irradiation is one of the classical procedures for food preservation [11] based on the ability of ionizing radiation to destroy the biological agents of food spoilage, microorganisms and enzymes. Despite their limited penetration, electron beam irradiation (E-beam) has a widely demonstrated effectiveness for food preservation or sanitation. Rice (*Oryza sativa* L.) storage is largely affected by the moisture content. Rice with high moisture (HM) content has better taste but is difficult to store. Farmers generally use sunlight drying to reduce the moisture content of rice and to extend storage time. However, although moisture reduction extends rice shelf-life, it will also reduce the rice taste by increasing hardness. Therefore, another procedure for rice preservation will be of use. E-beam irradiation could be used for this purpose, but before exploiting this possibility, it would be interesting to see the effect of E-beam irradiation on the quality properties of rice. In a study included in this special issue, Pan et al. [12] applied low-dose electron beam irradiation (EBI) in newly harvested HM rice and in dried rice and studied the effect of the irradiation on the cooking quality and moisture migration of irradiated rice and on the thermodynamics and digestion properties of the isolated starches. Overall, low-dose EBI had little effect on the properties of rice. High moisture rice showed superior quality and taste, whereas low moisture rice exhibited superior nutritional quality. Therefore, low-dose EBI seems to be more effective than other techniques for rice preservation.

Food spoilage can also be prevented or slowed by using preservatives that delay microbial growth or slow spoilage reactions such as oxidation of food components. However, consumers are increasingly reluctant to chemical additives, which makes it necessary to replace traditional preservatives with natural substances that have similar functions. Natrella et al. [13] essayed the use of different preservatives to avoid the formation of undesirable volatile organic compounds (VOC) in stracciatella, a traditional Italian cream cheese. Samples of cheese were prepared by adding two different preservatives (sorbic acid and an olive leaf extract) alone or combined. Their effect on flavor preservation during the refrigerated storage of cheese was investigated by chemical, microbiological and sensory analyses. The best chemical and sensory results were obtained in the samples containing sorbic acid, alone or in combination with leaf extract, demonstrating a significant shelf-life extension. According to the results, the use of the olive leaf extract, at the concentration tested (400 mg/kg), seemed to be interesting only in the presence of sorbic acid.

Some compounds present in foods have particular (i.e., functional, bioactive, etc.) properties and it is desired to obtain them in purity for their inclusion in the formulation of other foods. In other cases, some compounds have negative effects and should be removed without further ado. Sometimes, therefore, food processing is aimed at the extraction of molecules present in its composition. Extraction processes have different degree of complexity and involve different techniques and are based on the chemical and physical characteristics of the components to be extracted. Aqueous extraction of carob kibbles is the most important step in the production of carob juice and carob molasses. Improving the theoretical yield in sugars during organic solvent-free aqueous extraction is of prime interest to the food industry. However, collateral extraction of phenolic compounds must be monitored as it influences the functional and sensory profile of carob juice. In a study present in this special issue, Antoniou et al. [14] examined the impact of source material, kibble size, temperature and duration on the efficiency of aqueous extraction of sugars and phenolics by conventional heat-assisted (HAE) and ultrasound-assisted (UAE) methods. Authors observed that source material was the most influential factor determining the concentration of phenolics extracted by either method. Disproportionate extraction of phenolics over sugars limits the use of heat-assisted extraction to improve sugar yield in carob juice production and may shift the sensory profile of the product toward astringency.

However, prolonged extraction at near ambient temperature can improve sugar yield, keeping collateral extraction of phenolic compounds low. Authors concluded that ultrasound agitation constitutes an effective procedure of extracting sugars from powder-size kibbles and that the industrial application of both methodologies depends on the targeted functional and sensory properties of carob juice.

The term "marinade" is mainly used to refer a mixture of ingredients, in a liquid solution or powder form that is applied to raw foods to improve their organoleptic characteristics [15]. Marinating by food immersion or injection using solutions of different compounds (i.e., sauces, herbs, spices, organic acids, etc.) is a common practice for food conditioning before cooking or direct consumption.

Marinating can be used for meat tenderization by including a tenderizing agent in the liquid that is injected or in which the meat is submerged. Bromelain is a proteolytic enzyme widely essayed and used for meat tenderization. With the aim of meat tenderization by marinating, Santos et al. [16] added in marinades dehydrated pineapple by-products enriched in bromelain using a hydrostatic pressure treatment. Steaks from the silverside beef cut characterized as harder and cheaper were immersed in marinades containing dehydrated and pressurized pineapple (*Ananas comosus* L.) by-products that corresponded to a bromelain concentration of 0–20 mg tyrosine/100 g meat, 0–24 h time, according to the central composite factorial design matrix. After marinating, samples were characterized in terms of marination yield, pH, color and histology. Next, samples were cooked in a water-bath, stabilized and analyzed for cooking loss, pH, color, hardness and histology. Marinating times of 12–24 h and bromelain concentrations of 10–20 mg tyrosine/100 g meat reduced pH and hardness, increased marination yield and resulted in a lighter color. Meat hardness decreased by 41%, although refrigeration was not an optimal temperature for bromelain activity. Authors concluded that the use of pineapple by-products in brine allowed for the tenderization of lower commercial value steak cuts.

Marinating may be combined with other seasoning/conditioning methods for optimal results. In order to select the best processing conditions to obtain chicken meat with the highest sensory quality, Cho and Choi [17] after defrosting using different procedures (room temperature, running tap water, or high-frequency defroster) marinated chicken meat (leg and breast) with herbal extract solutions (bay leaves, coriander powder, fennel whole, thyme whole, cumin seeds, basil whole, basil powder, or star anise). Next, marinated meats were treated with superheated steam and then smoked with wood sawdust from different species (oak, apple, chestnut, walnut or cherry). Sensory evaluation was performed after each processing step. The products were also analyzed for fatty acids and nutrients (moisture, ash, salinity, calories, sodium, carbohydrates, sugars, dietary fiber, crude fat, trans fat, saturated fat, cholesterol, crude protein, calcium, iron, potassium and vitamin D), along with storage tests under different conditions monitored by microbiological, chemical and sensory analyses. High-frequency defrosting was the best method in terms of drip loss and thawing time. Bay leaves for marinating and oak wood for smoking were selected as the best materials for higher sensory scores. Optimal superheated steam conditions that showed the highest overall acceptance were 225 °C during 12 min and 20 s for leg meat and 223 °C during 8 min and 40 s for breast meat. The final meat products possessed good nutritional composition and no severe sensory spoilages were detected during storage.

Cooking is the most widely used procedure for food conditioning before consumption. Cooking can have positive or negative impact on the organoleptic characteristics and nutritional value of foods. Therefore, cooking method and parameters should be adapted to each raw material for optimal quality of cooked products. Nawaz et al. [18] evaluated the effect of various cooking methods (frying, baking and microwave cooking) on quality, structure, pasting, water distribution and protein oxidation of fish meat-based snacks. The findings suggest the endorsement of baking and microwave cooking for obtaining quality, safe and healthy snacks.

Some foods have a seasonal production but their demand and consumption is produced throughout the year. In other cases, it is very difficult to adapt the rhythm of production to that of consumption. Therefore, food storage is a necessary operation. Some methods of food preservation (canning, dehydration, irradiation) destroy or inhibit biological spoilage agents and allow long-term storage of food. However, such preservation procedures modify the organoleptic properties of foods, significantly distancing them from the sensory attributes that the same foods have in the fresh state. Some foods therefore should be stored in the fresh state and it is necessary to adopt other strategies to minimize their alteration during storage. In such cases, it is extremely important to know the changes that occur during storage to try to avoid or minimize them. Low temperatures retard microbial growth to a different degree depending on the nature of the microorganisms and slow down enzymatic reactions. That is why the storage of food at low temperatures is a useful and widely adopted procedure for prolonging the shelf-life of foods by minimally altering their organoleptic characteristics and nutritional value. In this special issue, several works studied the physical and chemical changes experienced by foods during storage.

The extra virgin olive oil (EVOO) is a typical food in which the extent and conditions of storage may affect its stability and quality. Mousavi et al. [19] evaluated the effects of different conditions of storage (ambient, 4 °C and −18 °C temperatures, and argon headspace) on three EVOOs (with low, medium and high phenol contents) over 18 and 36 months, through the analyses of the main metabolites at six time points. The organoleptic attributes of the oil samples were evaluated at all time points by a panel of tasters. The results showed that low temperatures are able to maintain all three EVOOs within the legal limits established by the current EU regulations for most compounds up to 36 months. The best temperature for conservation during 36 months was 4 °C, but −18 °C represented the optimum temperature to preserve the organoleptic properties. This study provided new insights that should guide EVOO manufacturers and traders to apply the most efficient storage methods to maintain the characteristics of the freshly extracted oils for a long conservation time.

The blueberry is widely cultivated throughout the world and its production has increased considerably in recent years. Yan et al. [20] studied the changes of volatile composition and other quality traits (weight loss, decay index, firmness and physicochemical parameters) of blueberry "garden blue" during postharvest storage. Odor was also evaluated by a sensory panel. Blueberries were packaged in vented clam-shell containers and stored at 0 °C for 0, 15 and 60 days, followed by storage at room temperature (25 °C) for up to 8 days for quality evaluation. The results of this work proved that cold storage was a dependable way to maintain the quality of blueberry. Nevertheless, a flavor deterioration was observed during subsequent shelf life.

Due to seasonality and short shelf life, approximately 50% of blueberries produced worldwide are processed, transforming them mainly into juice and dry fruits. Most of the blueberries used for juice production or drying are frozen fruits. During frozen storage and transportation, blueberries might be subjected to several freeze–thaw (FT) cycles. Understanding the changes in frozen blueberries during repeated FT cycles is essential for the processing of blueberry products because FT treatment will affect the qualities and flavors of the final food products. With this aim, Zhu et al. [21] studied the effects of FT cycles on the juice properties and aroma profiles and the hot-air drying kinetics of frozen blueberry. The authors reported that after FT treatment, the juice yield increased while pH and total soluble solids of the juice keep unchanged. The total anthocyanins contents and DPPH antioxidant activities of the juice decreased by FT treatments. The electronic nose showed that FT treatments significantly change the aroma profiles of the juice. The authors observed that one FT treatment can shorten the drying time by about 30% to achieve the same water content. Undoubtedly, the results of this work will be of use for the processing of frozen blueberry into juice or dried fruits.

In order to exploit the functional properties of fresh beetroot throughout the year, it is essential to maintain the health-benefiting compounds. With this purpose, Yasaminshirazi et al. [22] studied the impact of cold storage on bioactive compounds and their stability of beetroot. In their work, thirty-six beetroot genotypes collected from two field experiments, which were conducted under organic conditions, were evaluated regarding their content of total dry matter, total phenolic compounds, betalain, nitrate and total soluble sugars. Samples were analyzed directly after harvest and after cold storage periods of one and four months. The outcome of this study revealed a significant influence of genotype on all measured compounds. Furthermore, significant impacts were shown for storage period on total dry matter content, nitrate and total phenolic compounds. Therefore, the authors concluded that selection of the suitable genotype based on the intended final use is recommended to retain the quality of the beetroot for an extended time after harvest.

Finally, in this special issue, Lin et al. [23] studied the moisture migration, protein oxidation, microstructure and the physicochemical characteristics of Atlantic mackerel (*Scomber scombrus*) during storage at 4 °C and 0 °C. A slightly continuous decrease in the content of water and a certain degree of protein oxidation were observed over the course of storage. The storage process also caused changes in the secondary structure of proteins, the contraction and fracture of myofibrils and the granulation of endolysin protein. In addition, the drip loss, total volatile basic nitrogen content, thiobarbituric acid-reactive substances value and yellowness (b*) value significantly increased with the storage time.

It is evident that in recent years, much progress has been made in generating knowledge in this field. However, the chemical diversity of food components, the complexity of the interactions and effects that occur in such components as a consequence of the environmental conditions established during the different technological processes applied and during storage, and the requirements of consumers who increasingly demand safer food with higher organoleptic and nutritional quality make even more research necessary in order to achieve more efficient, economical and sustainable processes.

Author Contributions: S.M. and J.C. Conceptualization, data curation, writing—original draft preparation, writing—review and editing; visualization, supervision, project administration. This manuscript has not been submitted to, nor is under review at, another journal or other publishing venue. All authors have read and agreed to the published version of the manuscript.

Funding: This research received no external funding.

Conflicts of Interest: The authors declare no conflict of interest.

References

1. Appert, N. *L'Art de Conserver, Pendant Plusieurs AnnéES, Toutes Les Substances Animales ET Végétales*; Patris et Cie; Imprimeurs-Libraires: Paris, France, 1810; p. 116.
2. Aubourg, S.P.; Trigo, M.; Martínez, B.; Rodríguez, A. Effect of prior chilling period and alga-extract packaging on the quality of a canned underutilised fish species. *Foods* **2020**, *9*, 1333. [CrossRef] [PubMed]
3. Gómez-Limia, L.; Cutillas, R.; Carballo, J.; Franco, I.; Martínez, S. Free amino acids and biogenic amines in canned European eels: Influence of processing step, filling medium and storage time. *Foods* **2020**, *9*, 1377. [CrossRef] [PubMed]
4. Gómez-Limia, L.; Moya San Martín, N.; Carballo, J.; Domínguez, R.; Lorenzo, J.M.; Martínez, S. Oxidative stability and antioxidant activity in canned eels: Effect of processing and filling medium. *Foods* **2021**, *10*, 790. [CrossRef] [PubMed]
5. Dainton, A.N.; Dogan, H.; Aldrich, C.G. The effects of selected hydrocolloids on the processing of pâté-style canned pet food. *Foods* **2021**, *10*, 2506. [CrossRef] [PubMed]
6. Berk, Z. Dehydration. In *Food Process Engineering and Technology*, 3rd ed.; Berk, Z., Ed.; Academic Press: London, UK, 2018; pp. 513–566. ISBN 9780128120187.
7. García-Blázquez, A.; Moratalla-López, N.; Lorenzo, C.; Salinas, M.R.; Alonso, G.L. Effect of Crocus sativus L. stigmas microwave dehydration on picrocrocin, safranal and crocetin esters. *Foods* **2021**, *10*, 404. [CrossRef] [PubMed]
8. Xing, J.-J.; Jiang, D.-H.; Yang, Z.; Guo, X.-N.; Zhu, K.-X. Effect of humidity-controlled dehydration on microbial growth and quality characteristics of fresh wet noodles. *Foods* **2021**, *10*, 844. [CrossRef] [PubMed]
9. Önal, B.; Adiletta, G.; Di Matteo, M.; Russo, P.; Ramos, I.N.; Silva, C.L.M. Microwave and ultrasound pre-treatments for drying of the "Rocha" pear: Impact on phytochemical parameters, color changes and drying kinetics. *Foods* **2021**, *10*, 853. [CrossRef]
10. González-Mohino, A.; Pérez-Palacios, T.; Antequera, T.; Ruíz-Carrascal, J.; Olegario, L.S.; Grassi, S. Monitoring the processing of dry-fermented sausages with a portable NIRS device. *Foods* **2020**, *9*, 1294. [CrossRef]
11. Farkas, J.; Mohácsi-Farkas, C. History and future of food irradiation. *Trends Food Sci. Technol.* **2011**, *22*, 121–126. [CrossRef]
12. Pan, L.; Xing, J.; Luo, X.; Li, Y.; Sun, D.; Zhai, Y.; Yang, K.; Chen, Z. Influence of electron beam irradiation on the moisture and properties of freshly harvested and sun-dried rice. *Foods* **2020**, *9*, 1139. [CrossRef] [PubMed]

13. Natrella, G.; Difonzo, G.; Calasso, M.; Costantino, G.; Caponio, G.; Facia, M. Evolution of VOC and sensory characteristics of stracciatella cheese as affected by different preservatives. *Foods* **2020**, *9*, 1446. [CrossRef] [PubMed]
14. Antoniou, C.; Kyratzis, A.; Rouphael, Y.; Stylianou, S.; Kyriacou, M. Heat- and ultrasound-assisted aqueous extraction of soluble carbohydrates and phenolics from carob kibbles of variable size and source material. *Foods* **2020**, *9*, 1364. [CrossRef] [PubMed]
15. Yusop, S.M.; O'Sullivan, M.G.; Kerry, J.P. Marinating and enhancement of the nutritional content of processed meat products. In *Processed Meats. Improving Safety, Nutrition and Quality*; Kerry, J.P., Kerry, J.F., Eds.; Woodhead Publishing Series in Food Science, Technology and Nutrition; Elsevier Ltd.: Amsterdam, The Netherlands, 2011; pp. 421–449. ISBN 9781845694661.
16. Santos, D.I.; Fraqueza, M.J.; Pissarra, H.; Saraiva, J.A.; Vicente, A.A.; Moldao-Martins, M. Optimization of the effect of pineapple by-products enhanced in bromelain by hydrostatic pressure on the texture and overall quality of silverside beef cut. *Foods* **2020**, *9*, 1752. [CrossRef] [PubMed]
17. Cho, W.-H.; Choi, J.-S. Sensory quality evaluation of superheated steam-treated chicken leg and breast meats with combination of marination and hot smoking. *Foods* **2021**, *10*, 1924. [CrossRef] [PubMed]
18. Nawaz, A.; Li, E.; Khalifa, I.; Walayat, N.; Liu, J.; Irshad, S.; Zahra, A.; Ahmed, S.; Simirgiotis, M.J.; Pateiro, M.; et al. Effect of different processing methods on quality, structure, oxidative properties and water distribution properties of fish meat-based snacks. *Foods* **2021**, *10*, 2467. [CrossRef] [PubMed]
19. Mousavi, S.; Mariotti, R.; Stanzione, V.; Pandolfi, S.; Mastio, V.; Valdoni, L.; Cultrera, N.G.M. Evolution of extra virgin olive oil quality under different storage conditions. *Foods* **2021**, *10*, 1945. [CrossRef] [PubMed]
20. Yan, X.; Yan, J.; Pan, S.; Yuan, F. Changes of the aroma composition and other quality traits of blueberry 'garden blue' during the cold storage and subsequent shelf life. *Foods* **2020**, *9*, 1223. [CrossRef] [PubMed]
21. Zhu, L.; Liang, X.; Lu, Y.; Tian, S.; Chen, J.; Lin, F.; Fang, S. Effect of freeze-thaw cycles on juice properties, volatiles compounds and hot-air drying kinetics of blueberry. *Foods* **2021**, *10*, 2362. [CrossRef] [PubMed]
22. Yasaminshirazi, K.; Hartung, J.; Fleck, M.; Graeff-Hönninger, S. Impact of cold storage on bioactive compounds and their stability of 36 organically grown beetroot genotypes. *Foods* **2021**, *10*, 1281. [CrossRef] [PubMed]
23. Lin, R.; Cheng, S.; Wang, S.; Tan, M.; Zhu, B. Influence of refrigerated storage on water status, protein oxidation, microstructure, and physicochemical qualities of Atlantic mackerel (Scomber scombrus). *Foods* **2021**, *10*, 214. [CrossRef] [PubMed]

Article

Effect of Prior Chilling Period and Alga-Extract Packaging on the Quality of a Canned Underutilised Fish Species

Santiago P. Aubourg [1],*, **Marcos Trigo** [1], **Beatriz Martínez** [2] **and Alicia Rodríguez** [3]

1   Department of Food Technology, Marine Research Institute (CSIC), C/Eduardo Cabello, 6. 36208 Vigo, Spain; mtrigo@iim.csic.es
2   Department of Food Technologies, CIFP Coroso, Avda. da Coruña, 174, 15960 Ribeira, Spain; bmartinezr@edu.xunta.gal
3   Department of Food Science and Chemical Technology, Faculty of Chemical and Pharmaceutical Sciences, University of Chile, C/Santos Dumont 964, Santiago 8380000, Chile; arodrigm@uchile.cl
*   Correspondence: saubourg@iim.csic.es

Received: 10 August 2020; Accepted: 18 September 2020; Published: 21 September 2020

Abstract: The effect of a prior chilling period and an alga extract packaging on the quality of a canned underutilised mackerel species (*Scomber colias*) was investigated. For this different chilling times (0, 4 and 9 days) were taken into account and three concentrations of aqueous extracts of the macroalga *Fucus spiralis* were tested in a brine-packaging medium. Chemical changes related to quality were analysed after 3 months of canned storage. A substantial increase ($p < 0.05$) in free fatty acid content was observed in canned fish by increasing the chilling time; however, alga extract presence in the packaging medium led to decreased mean values. Concerning lipid oxidation development, an increased chilling time led to higher values ($p < 0.05$) of thiobarbituric acid index and fluorescent compounds formation; remarkably, an increased presence of alga extract led to a higher ($p < 0.05$) peroxide retention and lower ($p < 0.05$) fluorescent compounds content. Average colour L^* and a^* values showed a decrease and an increase, respectively, with chilling time; however, such changes were minimised with the alga extract content in the packaging system. Trimethylamine content revealed a marked increase as a result of the sterilisation step, but no influence ($p > 0.05$) of the chilling time or the alga-packaging medium could be implied.

Keywords: *Scomber colias*; prior chilling; *Fucus spiralis*; packaging medium; canning; lipid damage; colour; trimethylamine; quality

---

1. Introduction

Canning represents one of the most important and traditional means of marine species preservation [1,2]. According to the thermal sensitivity of a broad number of constituents included in marine species, different kinds of detrimental effects have been reported, especially if over processing is carried out [3,4]. Among them, oxidation of lipids and vitamins and leaching of water-soluble vitamins and minerals can be mentioned. Contrary to other muscle food, marine species are generally caught or harvested in distant locations, so that the time elapsed till arrival to cannery can be a decisive period for the quality of the final product. Consequently, canneries are in the need of storing the raw material before it is canned or just transported to factory. With this purpose, two strategies have been employed abundantly, namely frozen and chilled storage. As a result, quality of canned marine products will strongly depend on the adequacy of storage times and temperatures employed to hold the raw material [5,6].

For centuries, marine algae have been included in the Asian diet, especially in countries like China, Japan and Korea. They have revealed to be an important source of beneficial constituents such as vitamins, trace minerals, dietary fibre, amino acids and unsaturated lipids [7]. Remarkably, algae are known to be exposed to a combination of high oxygen concentration and light. The lack of structural damage in their organs has led to the consideration that their protection against damage arises from their content on preservative substances [8]. Consequently, marine algae (brown, red and green) are attracting a great interest as a source of bioactive molecules such as polyphenols, alkaloids, terpenes, phycocyannins and carotenoids, all of them showing antioxidant activity [9,10]. According to the European Council regulation [11], algae are considered food or food ingredients, so that their use in food technology in general should not constitute any hazard to health. In spite of such advantages, algae use in foods as preservatives can be limited because of flavour, odour and colour considerations since effective preservative doses may exceed sensorial acceptable limits.

Among brown macroalgae, *Fucus spiralis* is an abundant species living in the Atlantic coasts of North America and Europe. This macroalga has attracted a great attention because of its valuable nutrition content [12,13] and the presence of different kinds of bioactive constituents [14–16], with reported antioxidant and antimicrobial activity during fish refrigeration [17], chilling [18] and canning [19].

Availability of traditional species is being constantly reduced, so that the search for unconventional sources has turned necessary for the fish industry [20,21]. Thus, small pelagic fish species represent a promising valuable and economic choice in different geographic areas. One such under-valued species is Atlantic chub mackerel (*Scomber colias*) [22,23], abundantly found in the Mediterranean Sea, the Atlantic Ocean and the Black Sea. Concerning its technological aptitude, its quality loss evolution was studied as a result of freezing [24,25], refrigerated storage under modified atmosphere and vacuum packaging [26], chilling storage [27] and canning [19].

In this study, the effect of a prior chilling period and the presence of a *F. spiralis* extract in the packaging medium employed for canning was investigated in canned chub mackerel quality. Quality analyses (lipid damage, colour changes and trimethylamine content) were carried out in mackerel muscle after 3 months of canned storage. These indices were chosen as being important markers of fish quality changes during the technological processes encountered (chilling storage and canning).

2. Materials and Methods

2.1. Initial Raw Fish and Chilling Storage

Specimens (130 fish) of fresh mackerel (length and weight ranges: 24.5–28.0 cm and 157–175 g, respectively) were obtained at Vigo harbour (North-Western Spain) in November 2017. Once at the laboratory, 10 fish individuals were taken and divided into five groups (two individuals per group). This fish (initial raw fish) were beheaded, eviscerated and filleted. Then, the white muscle was separated, pooled together within each group, minced and analysed independently ($n = 5$).

From the remaining whole fish individuals, 40 of them were immediately taken for the canning process (0 days of chilling time; day-0 samples). On the other side, 80 whole fish specimens were surrounded by ice at a 1:1 fish-to-ice ratio and placed in a small (2 m × 2 m, 2.5 m height) refrigerated room (4 °C). After 4 and 9 days of chilling storage, specimens (40 at each sampling time) were taken for the canning analysis. Storage temperature of fish specimens was +0.5 °C throughout the storage period. Boxes employed allowed draining, ice being renewed when required.

2.2. Preparation of Alga Extract

Lyophilised alga *F. spiralis* was obtained from Porto-Muiños (Cerceda, A Coruña, Spain). A mixture of 14 g of alga and 140 mL of distilled water was submitted to stirring for 30 s, sonication for 30 s and centrifugation at 3500× *g* for 30 min at 4 °C. The supernatant was then recovered and the extraction process was repeated. Supernatants were finally pooled together, made up to 250 mL with distilled water and then employed in the preparation of the packaging medium during the canning process, as expressed in the following sub-section.

2.3. Canning Process and Sampling Procedure

As previously indicated, the canning process was carried out on fish that was previously chilled for 0, 4 and 9 days. At each canning time, fish (40 whole fish individuals) were divided into 5 groups (8 specimens in each group). Then, fish were beheaded, eviscerated, filleted and 45 g pieces of mackerel fillets were introduced in flat rectangular cans (105 × 60 × 25 mm; 150 mL). Such fish portions included skin and whole muscle (i.e., white and dark muscles). Each can was prepared from one fish individual, so that eight cans were prepared in each group. For the packaging medium, 0, 5, 15 or 30 mL of the above-mentioned alga extract (corresponding to 0.00, 0.28, 0.84 and 1.68 g of lyophilised alga, respectively) were added to the cans and labelled as C-CT (canned control), C-F1 (low *F. spiralis* concentration), C-F2 (medium *F. spiralis* concentration) and C-F3 (high *F. spiralis* concentration) batches, respectively, followed by the addition of 40, 35, 25 or 10 mL of distilled water, respectively. Then, a brine solution (40 mL; 4% *w/v*) was added to each can, so that a final content of 0.00, 3.50, 10.50 and 21.00 mg extracted alga mL^{-1} packaging system was attained, respectively. At each canning time and within each group, two cans were prepared having the same alga extract content.

Contents of alga extract employed were based on preliminary trials. A concentration of 21.00 mg mL^{-1} packaging medium showed to be the highest concentration without modifying the flesh odour, colour and flavour of canned fish. Thus, this concentration was considered in the C-F3 batch, together with two less concentrated batches (C-F1 and CF-2).

After vacuum sealing, all cans were subjected to heat sterilisation treatment in a steam retort at 115 °C for 45 min (F_0 = 7 min) at the CIFP Coroso (Ribeira, A Coruña, Spain). After completing the heating time, the steam was cut off, the remaining steam was flushed away by air employment, and cans were cooled at reduced pressure. After a 3-month storage at 20 °C, the cans were opened and the liquid part was carefully drained off gravimetrically. Then, the fish white muscle was separated, wrapped in filter paper, and used for analysis. At each canning time, the fish white muscle of two cans with the same alga extract content was pooled together, minced and employed to carry out the different quality analyses. Each batch (C-CT, C-F1, C-F2 or C-F3) was analysed by means of five replicates (n = 5).

Solvents and chemical reagents used were in all cases reagent grade (Merck, Darmstadt, Germany).

2.4. Determination of Lipid Damage

Lipids were obtained by extraction of the mackerel white muscle by applying the Bligh and Dyer [28] method, which employs a chloroform-methanol (1:1) mixture. Lipid content is expressed as g lipid·kg^{-1} muscle.

Assessment of the free fatty acids (FFA) content was carried out on the muscle lipid extract according to Lowry and Tinsley [29]; this method is based on a complex formation with cupric acetate-pyridine followed by spectrophotometric determination at 715 nm (Beckman Coulter DU 640 spectrophotometer, Beckman Coulter Inc., Brea, CA, USA). Results are expressed as g FFA·kg^{-1} muscle.

Peroxide value (PV) was determined spectrophotometrically (520 nm) on the lipid extract by peroxide reduction with ferric thiocyanate [30]. Results are expressed as meq. active oxygen·kg^{-1} lipids.

Thiobarbituric acid index (TBA-i) was assessed according to the Vyncke [31] procedure. This method is based on the reaction between a trichloracetic acid extract of the fish white muscle and thiobarbituric acid. The content of thiobarbituric acid reactive substances (TBARS) is spectrophotometrically measured (532 nm) and calculated from a standard curve prepared from commercial 1,1,3,3-Tetraethoxypropane (TEP). Results are expressed as mg malondialdehyde·kg^{-1} muscle.

The fluorescent compounds formation (LS 45 fluorimeter; Perkin Elmer España; Tres Cantos, Madrid, Spain) was determined in the lipid extract of the fish white muscle as described by Aubourg and Medina [32]. The relative fluorescence (RF) was calculated as follows: RF = F/F_{st}, where F is the fluorescence measured at each excitation/emission wavelength pair (namely 393/463 nm and 327/415 nm) and F_{st} is the fluorescence intensity of a quinine sulphate solution (1 µg·mL^{-1} in 0.05 M

H_2SO_4) at the corresponding wavelength pair. Results are expressed as the fluorescence ratio (FR), which was calculated as the ratio between the two RF values: $FR = RF_{393/463\ nm}/RF_{327/415\ nm}$.

2.5. Determination of Colour Changes and Trimethylamine Content

A tristimulus HunterLab Labscan 2.0/45 colorimeter (HunterLab, Reston, VA, USA) was applied in order to carry out the instrumental colour analysis (CIE 1976). Colour scores corresponding to each sample were averaged over four measurements, which were taken by rotating the measuring head 90° among triplicate measurements per position.

Determination of trimethylamine (TMA)-nitrogen (TMA-N) values was carried out by employing the picrate spectrophotometric (410 nm) method [33]. For it, a 5%-trichloracetic acid extract of the fish white muscle (10 g 25 mL^{-1}) was prepared. Results are expressed as mg TMA-N·kg^{-1} muscle.

2.6. Statistical Analysis

Chemical and physical values were subjected to the ANOVA method to explore differences resulting from the effect of the prior chilling storage period and alga concentration in the packaging medium. As expressed above, five replicates ($n = 5$) were considered throughout the study. The least-squares difference (LSD) method was used to perform the comparison of means. Analyses were carried out using the PASW Statistics v.18 software for Windows (SPSS Inc., Chicago, IL, USA); differences were considered significant for a confidence interval at the 95% level ($p < 0.05$).

3. Results and Discussion

3.1. Determination of Lipid Hydrolysis Development

Compared with the initial raw fish, all kinds of canned samples corresponding to a 0-day chilling time showed a substantial increase ($p < 0.05$) of FFA content (Figure 1).

Furthermore, an additional general increase ($p < 0.05$) was also observed in canned fish by increasing the chilling time up to 4 days; contrary, a chilling storage extension up to 9 days did not lead to significant differences ($p > 0.05$) in the FFA content. Concerning the effect of the alga extract in the packaging medium, average FFA values showed a decreasing effect of *F. spiralis* extract content when considering samples corresponding to day-0 chilling time. If a 4 or 9-day chilling period is considered, average values corresponding to control canned fish (C-CT batch) were higher than their counterparts from the alga- packaging batches (C-F1, C-F2 and C-F3 batches); however, average values did not show a decreasing effect by increasing the alga extract presence.

FFA are considered to be the result of hydrolysis of high-molecular-weight lipid compounds such as triacylglycerols (TG) and phospholipids (PL). In the current research, FFA content can be considered the result of different factors. First, their formation during chilling storage should increase with chilling time by action of endogenous and microbial enzymes (phospholipases and lipases in general) [5,34]. Furthermore, the sterilization process can lead to hydrolysis of lipid classes such as TG and PL [6]. Interestingly, FFA are known to be rapidly oxidised by heating according to the fact that they provide a greater accessibility to oxygen and other oxidants in general when compared with TG and PL [35]. Finally, preservative compounds present in the alga extract can protect FFA from their breakdown during the heating process. On the basis of the data obtained, it can be concluded that an important effect of prior chilling time was implied, while the possible preserving effect of alga extract was not especially important.

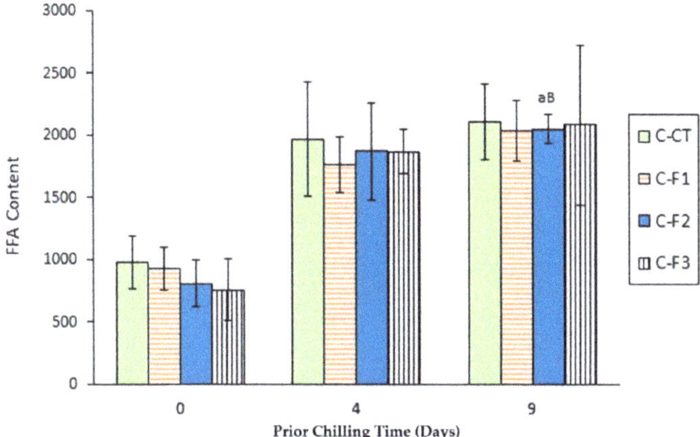

Figure 1. Assessment of free fatty acids (FFA) value (g·kg^{-1} muscle) in canned mackerel previously subjected to different chilling times and packaged with different alga extract concentrations. Mean values of five replicates (n = 5). Standard deviations are indicated by bars. Initial raw fish value: 37.15 ± 11.24. For each chilling time, the same low-case letter (a) denotes that no significant differences ($p > 0.05$) were obtained related to the alga extract presence. For each alga extract concentration, different capital letters (A, B) denote significant differences ($p < 0.05$) by effect of the chilling time. Alga extract concentrations in packaging media: C-CT (control), C-F1 (low alga content), C-F2 (medium alga content) and C-F3 (high alga content), in agreement with the Material and Methods section.

In agreement with the current results, sunflower- or brine-packaged sprat (*Clupeonella cultriventris*) [36], as well as sunflower oil-packaged canned albacore tuna (*Thunnus alalunga*) subjected to different sterilisation conditions [37] revealed a marked FFA formation when compared with the starting raw material. Contrary to the present study, increasing the prior chilling period in brine-canned sardine (*Sardina pilchardus*) [32] and in sunflower oil-canned salmon (*Oncorhynchus kisutch*) [38] led to marked increases of the FFA content.

According to the present study, no effect of the alga extract concentration was produced on the FFA content of canned Atlantic mackerel (*Scomber scombrus*) packaged with an aqueous extract of macroalga *Bifurcaria bifurcata* [39]. Contrary, an increased formation of FFA was implied in canned Atlantic chub mackerel (*S. colias*) by increasing the presence of macroalgae *Ulva lactuca* and *F. spiralis* extracts in the packaging system [19].

3.2. Determination of Lipid Oxidation Development

Lipid oxidation development was measured by assessing different and complementary indices in order to obtain a satisfactory overview on the advance of this damage mechanism.

Canned samples corresponding to all chilling conditions showed a marked PV increase ($p < 0.05$) when compared with the initial raw fish (Table 1). Canned fish corresponding to control batch showed a progressive increase ($p < 0.05$) with chilling time, while alga-packaged fish did not provide differences ($p > 0.05$) as a result of the chilling time period. For all chilling times considered, peroxides content showed to be higher ($p < 0.05$) by increasing the alga extract content in the packaging medium. Thus, canned fish corresponding to C-F2 and C-F3 batches showed significant differences ($p < 0.05$) when compared with their counterpart control.

Table 1. Determination of peroxide value (PV), thiobarbitruric acid index (TBA-i) and fluorescence ratio (FR) in canned mackerel packaged with different alga extract concentrations and previously subjected to different chilling times.

Quality Index	Alga Extract Concentration	Prior Chilling Time (Days)		
		0	4	9
PV (meq. active oxygen·kg^{-1} lipids)	C-CT	a 1.01 ± 0.24 A	a 1.31 ± 0.81 AB	a 2.14 ± 0.55 B
	C-F1	b 3.08 ± 0.91 A	ab 2.08 ± 0.72 A	ab 2.89 ± 0.57 A
	C-F2	b 3.68 ± 0.61 A	bc 3.01 ± 0.55 A	b 3.63 ± 0.53 A
	C-F3	c 6.88 ± 1.60 A	d 6.84 ± 1.47 A	c 6.50 ± 1.24 A
TBA-i (mg malondialdehyde·kg^{-1} muscle)	C-CT	a 0.73 ± 0.22 A	a 1.22 ± 0.40 AB	a 1.39 ± 0.40 B
	C-F1	a 0.51 ± 0.20 A	a 1.77 ± 0.65 B	a 1.28 ± 0.33 B
	C-F2	a 0.44 ± 0.07 A	a 1.69 ± 0.35 B	a 1.27 ± 0.45 B
	C-F3	a 0.48 ± 0.09 A	a 1.36 ± 0.39 B	a 1.02 ± 0.14 B
FR	C-CT	c 4.34 ± 0.57 A	a 5.27 ± 0.55 A	b 6.08 ± 0.14 B
	C-F1	ab 3.30 ± 0.16 A	a 5.06 ± 0.50 B	b 6.28 ± 0.31 C
	C-F2	b 3.56 ± 0.14 A	a 4.73 ± 0.10 B	b 6.03 ± 0.25 B
	C-F3	a 2.96 ± 0.33 A	a 4.99 ± 0.12 B	a 5.01 ± 0.36 B

Average values ± standard deviations of five replicates (*n* = 5). Initial raw fish values: 0.43 ± 0.31 (PV), 0.52 ± 0.24 (TBA-i) and 2.00 ± 0.47 (FR). For each chilling time, values preceded by different low-case letters (a, b, c, d) denote significant differences ($p < 0.05$) by effect of the alga extract concentration. For each alga extract concentration, values followed by different capital letters (A, B, C) denote significant differences ($p < 0.05$) related to the chilling time. Alga extract concentrations in packaging media: C-CT (control), C-F1 (low alga content), C-F2 (medium alga content) and C-F3 (high alga content), according to the Material and Methods section.

A significant formation of TBARS was not detected ($p > 0.05$) in canned fish subjected previously to a 0-day chilling in any of the batches under study (Table 1). However, if a 4-day storage was applied, a general increase was obtained in all batches, differences being significant ($p < 0.05$) in all alga-treated batches. Contrary, no significant differences ($p > 0.05$) were implied by increasing the chilling time up to 9 days; in this case, average values provided an increase in control batch, while a decrease was detected in canned batches including any alga extract concentration. A definite trend could not be concluded related to the alga extract presence in the packaging system. Thus, control canned fish showed higher average values after 0 and 9 days of chilling storage, while this batch showed the lowest average value when a 4-day storage is taken into account.

Comparison between initial raw fish and canned fish corresponding to day-0 storage showed an important fluorescent compounds formation ($p < 0.05$) as a result of the sterilisation step in all batches (Table 1). Furthermore, a progressive FR increase ($p < 0.05$) was produced in all batches by increasing the chilling time. This FR increase would be the result of interaction between primary and secondary lipid oxidation compounds (electrophilic behaviour) and food constituents possessing nucleophilic functions [32,40]. Remarkably, an inhibitory effect on fluorescent compounds formation was observed by the alga extract presence in the packaging medium. Thus, considering the canned fish corresponding to the 0-day time, all alga concentrations tested led to lower ($p < 0.05$) levels than the control; additionally, lower levels ($p < 0.05$) were observed in C-F3-batch canned fish when compared with their counterparts if a 9-day storage is taken into account.

Lipid oxidation is considered a complex deteriorative mechanism since it involves the formation of a wide range of molecules, most of them unstable, and consequently, able to breakdown and give rise to lower-weight compounds susceptible to react with nucleophilic-type molecules (proteins, peptides, free amino acids, etc.) present in fish muscle. As expressed above, this would be the case of peroxides and TBARS, widely reported to breakdown and give rise to tertiary (or interaction compounds) lipid oxidation compounds [32,40]. Since endogenous enzymes and microbial development are inactivated by heat, most attention in canned fish has been accorded to lipid oxidation and further interaction of oxidised lipids with other constituents, proteins especially, in agreement with their heat denaturation and consequently turning into more reactive molecules. Lipid damage development may be especially important if a fatty fish species is encountered as in the current study (lipid content: 74.3 ± 14.5 g kg^{-1} white muscle).

On the basis of the strong processing included in canning (i.e., sterilisation step), previous research has shown that assessment of primary and secondary lipid oxidation products does not lead to a definite trend about lipid quality changes in different kinds of canned fish such as oil- packaged canned albacore tuna (*T. alalunga*) sterilised under different conditions [37], as well as in other pelagic fish species such as olive-oil packaged bluefin tuna (*Thunnus thynnus*) and tomato sauce-packaged sardine (*Sardina pilchardus*) [41]. Indeed, the TBA-i led to a decreasing value in brine-canned sardine (*S. pilchardus*) by increasing the prior chilling time [32]. Remarkably, and in agreement with the current data, fluorescent compounds formation has shown to be a valuable quality index in canned fish. Thus, a progressive formation of fluorescent compounds was implied by increasing the chilling time in brine-canned sardine (*S. pilchardus*) [32] as well as in fish muscle and packaging system in sunflower-packaged canned sardine (*S. pilchardus*) [42].

In agreement with the current study, previous research has shown an increasing effect on peroxides content in canned fish by increasing the *B. bifurcata* water-extract [39] or the *U. lactuca* and *F. spiralis* brine-extract [19] contents in the packaging media employed for mackerel canning. In both studies, it was considered that peroxides breakdown was partially avoided as a result of the algae extract presence. Also in agreement with the current study, the formation of fluorescent compounds during canning was partially inhibited by increasing the macroalga extract presence both in water- [39] and brine-packaged [19] canned fish.

The antioxidant effect of alga *F. spiralis* extracts has already been proved in in vitro tests [12], showing a marked content on different kinds of antioxidant molecules such as polyphenols [15], alpha-tocopherol [13] and phlorotannins [14]. Antioxidant compounds, polyphenols especially, have been described as being able to stabilise free radical molecules so that lipid oxidation development would be inhibited and damage to other muscle constituents be decreased. Furthermore, water extracts obtained from macroalgae have shown to include different kinds of preserving phenolic acids (i.e., caffeic, chlorogenic, vanillic, etc.) [43], as well as sulphate polysaccharides, proteins, peptides, glycosides, low-molecular organic acids and salts) having potential preserving properties [44,45].

3.3. Determination of Colour Changes

Colour is considered as an important property in the appearance and acceptability of seafood by the consumer. Current data on colour assessment are depicted in Figures 2 and 3. Taking into account the L^* value of the initial raw fish, a general increase was observed as a result of canning, independently of the chilling period applied (Figure 2). Remarkably, no significant differences ($p > 0.05$) were practically obtained as a result of the chilling period, although the highest mean values were obtained in canned fish corresponding to the 0-day period in all batches. For all chilling times considered, decreasing L^* values were obtained by increasing the alga extract concentration in the packaging system. Compared with control, significant differences ($p < 0.05$) were implied for canned fish including the most concentrated alga extract in canned samples corresponding to all storage times (0, 4 and 9 days).

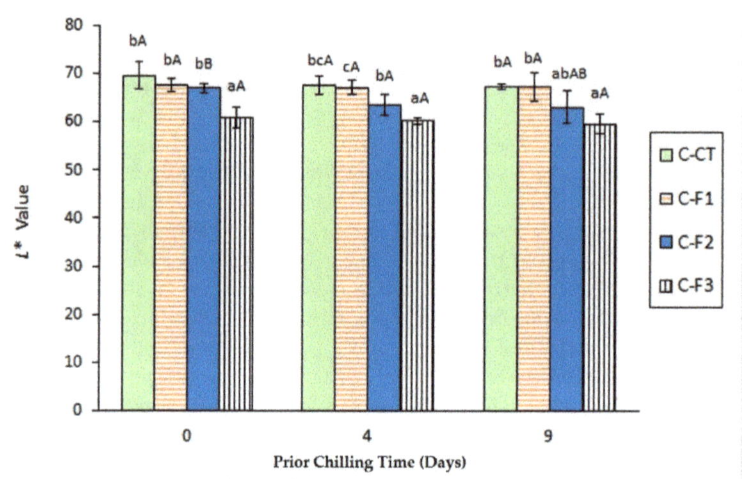

Figure 2. Determination of *L** colour value in canned mackerel previously subjected to different chilling times and packaged with different alga extract concentrations. Mean values of five replicates ($n = 5$). Standard deviations are indicated by bars. Initial raw fish value: 38.30 ± 2.41. For each chilling time, different low-case letters (a, b, c) denote significant differences ($p < 0.05$) as a result of the alga extract concentration. For each alga extract concentration, different capital letters (A, B) denote significant differences ($p < 0.05$) by effect of the chilling time. Alga extract concentrations in packaging media as expressed in Figure 1.

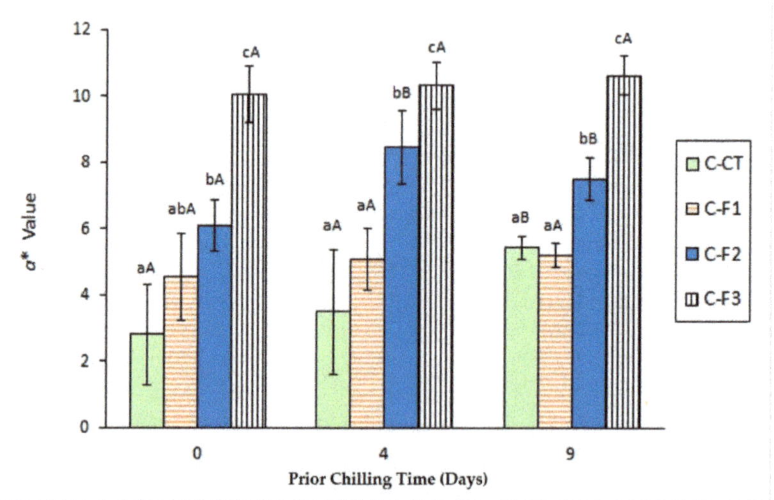

Figure 3. Determination of *a** colour value in canned mackerel previously subjected to different chilling times and packaged with different alga extract concentrations. Mean values of five replicates ($n = 5$). Standard deviations are indicated by bars. Initial raw fish value: 15.56 ± 3.36. For each chilling time, different low-case letters (a, b, c) denote significant differences ($p < 0.05$) as a result of the alga extract concentration. For each alga extract concentration, different capital letters (A, B) denote significant differences ($p < 0.05$) related to the chilling time. Alga extract concentrations in packaging media as indicated in Figure 1.

Canning process led to a substantial decrease ($p < 0.05$) of a^* value in all kinds of canned samples (Figure 3).

However, this decrease was partially inhibited ($p < 0.05$) by the presence of the alga extract in the packaging medium. Remarkably, an increase of the chilling time led to a progressive increase of average a^* values in most batches, this increase being significant ($p < 0.05$) in fish corresponding to batches C-CT (9-day period) and C-F2 (4-day period). Concerning the effect of the alga-extract packaging, an increased value was detected for this colour parameter with the alga extract content in canned fish corresponding to all chilling times. It is concluded that the presence in the packaging system of preservative compounds (antioxidants, especially) from the alga extract has led to an inhibitory effect on the breakdown of molecules responsible for the a^* value (carotenoids and pigments in general) in the mackerel muscle during the canning process.

Valuable results were not obtained by determination of the b* colour score; consequently, a definite effect of the chilling time and the alga extract presence in the packaging medium could not be observed for this parameter.

Previous related research has addressed the colour parameters assessment. Thus, Barbosa et al. [39] showed a marked increase for L^* value in canned Atlantic mackerel (*S. scombrus*) by comparison with the initial raw fish. Remarkably, this increase was partially inhibited by increasing the concentration of an aqueous extract of macroalga *B. bifurcata* in the packaging system. Furthermore, increasing the previous storage temperature and time led to an increase in L^* value and a decrease in a^* value in canned skipjack tuna (*Katsuwonus pelamis*) [46] and coho salmon (*O. kisutch*) [38]. Indeed, an a^* value decrease has been pointed out as being correlated with haemoglobin-mediated lipid oxidation in fish and to show an inverse relationship with the content on secondary lipid oxidation compounds [47,48].

3.4. Trimethylamine Content Assessment

Contents on TMA values are depicted in Table 2. Prior chilling storage did not lead to a definite effect on TMA content in canned fish. Thus, highest average values ($p < 0.05$) were obtained after 4 days in samples corresponding to C-CT and C-F3 batches, while a 9-day chilling period showed that canned fish from C-F1 and C-F2 batches revealed the highest levels ($p < 0.05$). Furthermore, a strong formation ($p < 0.05$) of TMA was observed after the sterilisation process in all kinds of canned samples. Concerning the effect of the alga extract in the packaging medium, some decreasing effect could be observed in TMA content in canned samples corresponding to a 0-day storage; however, canned samples related to 4 and 9 days of storage did not provide a definite trend ($p > 0.05$). Thus, higher mean TMA-N values were detected in canned control fish than in any other counterpart fish when a 0- and 4-day chilling period is considered. However, if the longest chilling time (i.e., 9 days) is taken into account, a higher mean value was obtained in C-F1-batch canned fish.

Table 2. Determination of trimethylamine-nitrogen (TMA-N) content in canned mackerel packaged with different alga extract concentrations and previously subjected to different chilling times.

Quality Index	Alga Extract Concentration	Prior Chilling Time (Days)		
		0	4	9
TMA-N (g·kg^{-1} muscle)	C-CT	a 25.41 ± 2.64 A	b 31.40 ± 3.79 A	ab 28.12 ± 6.10 A
	C-F1	a 23.78 ± 5.65 A	a 18.06 ± 3.26 A	b 33.58 ± 2.77 B
	C-F2	a 21.21 ± 4.22 A	a 18.55 ± 3.32 A	a 24.96 ± 5.11 A
	C-F3	a 21.70 ± 1.87 A	b 28.71 ± 2.05 B	a 26.40 ± 2.64 B

Mean values ± standard deviations of five replicates ($n = 5$). Initial raw fish value: 1.04 ± 0.12 (TMA-N). For each chilling time, values preceded by different low-case letters (a, b) denote significant differences ($p < 0.05$) related to the alga extract concentration. For each alga extract concentration, values followed by different capital letters (A, B) denote significant differences ($p < 0.05$) by effect of the chilling time. Alga extract concentrations in packaging media as indicated in Table 1.

TMA formation in the present research can be justified as a result of two different pathways. One side, trimethylamine oxide (TMAO) can be broken down by bacterial hydrolysis during chilled storage [34]. On the other side, TMA can be produced by breakdown of TMAO and other muscle constituents during the sterilisation step [6]. Since a great difference between raw and day-0 canned samples was observed, the second effect has shown to be largely more important than the first one. Interestingly, as being a tertiary amine, TMA would not be leached into the brine-packaging system as not being a water-soluble compound. Since no effect of the alga extract presence in the packaging medium was perceived, it is concluded that alga preservative compounds did not alter the TMA formation during the sterilisation process.

Although including different packaging media than the current one, similar results on TMA values have been obtained previously in canned fish. Thus, Gallardo et al. [49] observed a marked formation of TMA as a result of cooking and sterilisation in oil-packaged canned albacore tuna (*T. alalunga*), while a marked loss of TMAO was detected in sunflower-packaged canned sardine (*S. pilchardus*) by increasing the prior chilling time [42]. A substantial formation of TMA was also detected in olive-oil packaged bluefin tuna (*T. thynnus*), tomato sauce-packaged sardine (*S. pilchardus*) muscle [41] and in sunflower oil-packaged canned salmon (*O. kisutch*) that was previously stored under traditional and slurry icing conditions [38]. Concerning the effect of the alga extract content in the packaging system, previous research has shown, as in the current research, no definite effect on the TMA content in canned mackerel species when a water- [39] or a brine-packaged [19] medium was employed.

4. Conclusions

This study focused on the quality of a canned fish product prepared from a fatty under-valued species. In it, the effect of a prior chilling period and the employment of a packaging system including an alga extract on different physico-chemical properties of canned chub mackerel were investigated. As a result, an increased chilling time led to a quality loss ($p < 0.05$) that was observed in lipid hydrolysis (FFA formation) and oxidation (TBA-i and FR determinations) development; remarkably, no effect ($p > 0.05$) was detected in TMA levels and colour parameters (L^* and a^*). On the other side, packaging including *F. spiralis* extracts medium led to an average decrease of lipid hydrolysis development (FFA content) and to a significant ($p < 0.05$) quality loss inhibition according to values obtained for the FR and colour parameters (L^* and a^*); furthermore, no effect ($p > 0.05$) of alga-packaging was detected in TMA content, while a higher retention ($p < 0.05$) of primary oxidation molecules (i.e., peroxides) was concluded by increasing the alga concentration.

Current results show the strong dependence of canned fish quality on the holding conditions of the raw material employed. Remarkably, a packaging system including a macroalga extract has been tested and found profitable to enhance the quality of a canned under-valued fish species. It is considered that the development of optimised conditions of this alga-packaging system may open the way to its application on all kinds of fish species, these including high-value fish such as tuna, bonito or salmon.

Author Contributions: Conceptualization—S.P.A., M.T., B.M. and A.R.; methodology—M.T. and B.M.; draft preparation—S.P.A., review and editing—S.P.A. and A.R. All authors have read and agreed to the published version of the manuscript.

Funding: This work was supported by Consejo Superior de Investigaciones Científicas (CSIC) Spain, Research Project 2013-70E001.

Acknowledgments: The authors thank Porto-Muiños (Cerceda, A Coruña, Spain) for providing the lyophilised alga.

References

1.   Horner, W. Canning fish and fish products. In *Fish Processing Technology*, 2nd ed.; Hall, G., Ed.; Blackie Academic & Professional An Imprint of Chapman & Hall: London, UK, 1997; pp. 119–159.

2. Lukoshkina, M.; Odoeva, G. Kinetics of chemical reactions for prediction of quality of canned fish during storage. *App. Biochem. Microb.* **2003**, *39*, 321–327. [CrossRef]

3. Pigott, G.M.; Tucker, B.W. *Seafood. Effects of Technology on Nutrition*; Marcel Dekker, Inc.: New York, NY, USA; Basel, Switzerland, 1990.

4. García-Arias, T.; Sánchez-Muniz, J.; Castrillón, A.; Navarro, P. White tuna canning, total fat, and fatty acid changes during processing and storage. *J. Food Comp. Anal.* **1994**, *7*, 119–130. [CrossRef]

5. Sikorski, Z.; Kolakowski, E. Endogenous enzyme activity and seafood quality: Influence of chilling, freezing, and other environmental factors. In *Seafood Enzymes*; Haard, N., Simpson, B., Eds.; Marcel Dekker: New York, NY, USA, 2000; pp. 451–487.

6. Aubourg, S.P. Loss of quality during the manufacture of canned fish products. *Food Sci. Technol. Int.* **2001**, *7*, 199–215. [CrossRef]

7. Díaz-Rubio, MªE.; Pérez-Jiménez, J.; Saura-Calixto, F. Dietary fiber and antioxidant capacity in *Fucus vesiculosus* products. *Int. J. Food Sci. Technol.* **2009**, *60*, 23–34.

8. Smit, A. Medicinal and pharmaceutical uses of seaweed natural products: A Review. *J. App. Phycol. 1* **2004**, *16*, 245–262. [CrossRef]

9. Zubia, M.; Fabre, M.S.; Kerjean, V.; Le Lann, K.; Stiger-Pouvreau, V.; Fauchon, M.; Deslandes, E. Antioxidant and antitumoral activities of some *Phaeophyta* from Britanny coasts. *Food Chem.* **2009**, *116*, 693–701. [CrossRef]

10. Gupta, S.; Abu-Ghannam, N. Bioactive potential and possible health effects of edible brown seaweeds. *Trends Food Sci. Technol.* **2011**, *22*, 315–326. [CrossRef]

11. European Council Regulation (14/02/1997). European Community (EC), No 258/97, 27 January 1997. Concerning novel foods and novel food ingredients. *CELEX-EUR Off. J.* **1997**, *L-43*, 1–7.

12. Andrade, P.; Barbosa, M.; Pedro Matos, R.; Lopes, G.; Vinholes, J.; Mouga, T.; Valentão, P. Valuable compounds in macroalgae extracts. *Food Chem.* **2013**, *138*, 1819–1828. [CrossRef] [PubMed]

13. Paiva, L.; Lima, E.; Ferreira Patarra, R.; Neto, A.; Baptista, J. Edible Azorean macroalgae as source of rich nutrients with impact on human health. *Food Chem.* **2014**, *164*, 128–135. [CrossRef] [PubMed]

14. Cérantola, S.; Breton, F.; Gall, E.; Deslandes, E. Co-occurrence and antioxidant activities of fucol and fucophlorethol classes of polymeric phenols in *Fucus spiralis*. *Botanica Marina* **2006**, *49*, 347–351. [CrossRef]

15. Tierney, M.; Smyth, T.; Hayes, M.; Soler-Vila, A.; Croft, A.; Brunton, N. Influence of pressurised liquid extraction and solid-liquid extraction methods on the phenolic content and antioxidant activities of Irish macroalgae. *Int. J. Food Sci. Technol.* **2013**, *48*, 860–869. [CrossRef]

16. Peinado, I.; Girón, J.; Koutsidis, G.; Ames, J.M. Chemical composition, antioxidant activity and sensory evaluation of five different species of brown edible seaweeds. *Food Res. Int.* **2014**, *66*, 36–44. [CrossRef]

17. García-Soto, B.; Miranda, J.; Rodríguez-Bernaldo de Quirós, A.; Sendón, R.; Rodríguez-Martínez, A.; Barros-Velázquez, J.; Aubourg, S.P. Effect of biodegradable film (lyophilised alga *Fucus spiralis* and sorbic acid) on quality properties of refrigerated megrim (*Lepidorhombus whiffiagonis*). *Int. J. Food Sci. Technol.* **2015**, *50*, 1891–1900. [CrossRef]

18. Miranda, J.; Trigo, M.; Barros-Velázquez, J.; Aubourg, S.P. Effect of an icing medium containing the alga *Fucus spiralis* on the microbiological activity and lipid oxidation in chilled megrim (*Lepidorhombus whiffiagonis*). *Food Cont.* **2016**, *59*, 290–297. [CrossRef]

19. Barbosa, R.G.; Trigo, M.; Campos, C.A.; Aubourg, S.P. Preservative effect of algae extracts on lipid composition and rancidity development in brine-canned Atlantic Chub mackerel (*Scomber colias*). *Eur. J. Lipid Sci. Technol.* **2019**, *121*, 1900129. [CrossRef]

20. Venugopal, V.; Shahidi, F. Traditional methods to process underutilized fish species for human consumption. *Food Rev. Int.* **1998**, *14*, 35–97. [CrossRef]

21. Blanco, M.; Sotelo, C.G.; Chapela, M.J.; Pérez-Martín, R.I. Towards sustainable and efficient use of fishery resources: Present and future trends. *Trends Food Sci. Technol.* **2007**, *18*, 29–36. [CrossRef]

22. Wituszynska, B. Determination of some group B vitamins in fresh and canned fish. *Bromatologia i Chemia Toksykologiczna* **1973**, *6*, 13–22.

23. García-Moreno, P.; Pérez-Gálvez, R.; Espejo-Carpio, F.; Muñío, M.; Guadix, A.; Guadix, E. Lipid characterization and properties of protein hydrolysates obtained from discarded Mediterranean fish species. *J. Sci. Food Agric.* **2013**, *93*, 3777–3784. [CrossRef]

24. Vidacek, S.; Medic, H.; Marusic, N.; Tonkovic, S.; Petrak, T. Influence of different freezing regimes on bioelectrical properties of Atlantic chub mackerel (*Scomber colias*). *J. Food Proc. Eng.* **2012**, *35*, 735–741. [CrossRef]

25. Alberio, G.; Barbagallo, R.; Todaro, A.; Bono, G.; Spagna, G. Effect of freezing/thawing process in different sizes of blue fish in the Mediterranean through lysosomal enzymatic tests. *Food Chem.* **2014**, *148*, 47–53. [CrossRef]

26. Stamatis, N.; Arkoudelos, J. Quality assessment of *Scomber colias japonicus* under modified atmosphere and vacuum packaging. *Food Cont.* **2007**, *18*, 292–300. [CrossRef]

27. Miranda, J.M.; Carrera, M.; Pastén, A.; Vega-Gálvez, A.; Barros-Velázquez, J.; Aubourg, S.P. The impact of quinoa (*Chenopodium quinoa* Willd.) ethanolic extracts in the icing medium on quality loss of Atlantic chub mackerel (*Scomber colias*) under chilling storage. *Eur. J. Lipid Sci. Technol.* **2018**, *120*, 1800280. [CrossRef]

28. Bligh, E.G.; Dyer, W.J. A rapid method of total extraction and purification. *Can. J. Biochem. Physiol.* **1959**, *37*, 911–917. [CrossRef]

29. Lowry, R.; Tinsley, I. Rapid colorimetric determination of free fatty acids. *J. Am. Oil Chem. Soc.* **1976**, *53*, 470–472. [CrossRef]

30. Chapman, R.; McKay, J. The estimation of peroxides in fats and oils by the ferric thiocyanate method. *J. Am. Oil Chem. Soc.* **1949**, *26*, 360–363. [CrossRef]

31. Vyncke, W. Direct determination of the thiobarbituric acid value in trichloracetic acid extracts of fish as a measure of oxidative rancidity. *Fette, Seifen, Anstrichm.* **1970**, *72*, 1084–1087. [CrossRef]

32. Aubourg, S.P.; Medina, I. Quality differences assessment in canned sardine (*Sardina pilchardus*) by fluorescence detection. *J. Agric. Food Chem.* **1997**, *45*, 3617–3621. [CrossRef]

33. Tozawa, H.; Erokibara, K.; Amano, K. Proposed modification of Dyer's method for trimethylamine determination in codfish. In *Fish Inspection and Quality Control*; Kreuzer, R., Ed.; Fishing News Books Ltd.: London, UK, 1971.

34. Ashie, I.; Smith, J.; Simpson, B. Spoilage and shelf-life extension of fresh fish and shellfish. *Crit. Rev. Food Sci. Nutr.* **1996**, *36*, 87–121. [CrossRef]

35. Miyashita, K.; Tagagi, T. Study on the oxidative rate and prooxidant activity of free fatty acids. *J. Am. Oil Chem. Soc.* **1986**, *63*, 1380–1384. [CrossRef]

36. Naseri, M.; Rezaei, M. Lipid changes during long-term storage of canned sprat. *J. Aquat. Food Prod. Technol.* **2012**, *21*, 48–58. [CrossRef]

37. Aubourg, S.P.; Gallardo, J.M.; Medina, I. Changes in lipids during different sterilising conditions in canning albacore (*Thunnus alalunga*) in oil. *Int. J. Food Sci. Technol.* **1997**, *32*, 427–431. [CrossRef]

38. Rodríguez, A.; Carriles, N.; Aubourg, S.P. Effect of chill storage under different icing conditions on sensory and physical properties of canned farmed salmon (*Oncorhynchus kisutch*). *Int. J. Food Sci. Technol.* **2010**, *45*, 295–304. [CrossRef]

39. Barbosa, R.G.; Trigo, M.; Fett, R.; Aubourg, S.P. Impact of a packing medium with alga *Bifurcaria bifurcata* extract on canned Atlantic mackerel (*Scomber scombrus*) quality. *J. Sci. Food Agric.* **2018**, *98*, 3462–3467. [CrossRef] [PubMed]

40. Tironi, V.A.; Tomás, M.C.; Añón, M.C. Structural and functional changes in myofibrillar proteins on sea salmon (*Pseudopercis semifasciata*) by interaction with malondialdehyde (RI). *J. Food Sci.* **2002**, *67*, 930–935. [CrossRef]

41. Selmi, S.; Monser, L.; Sadok, S. The influence of local canning process and storage on pelagic fish from Tunisia: Fatty acids profile and quality indicators. *J. Food Proc. Preserv.* **2008**, *32*, 443–457. [CrossRef]

42. Losada, V.; Rodríguez, A.; Ortiz, J.; Aubourg, S.P. Quality enhancement of canned sardine (*Sardina pilchardus*) by a preliminary chilling treatment. *Eur. J. Lipid Sci. Technol.* **2006**, *108*, 598–605. [CrossRef]

43. Farvin, K.H.S.; Jacobsen, C. Phenolic compounds and antioxidant activities of selected species of seaweeds from Danish coast. *Food Chem.* **2013**, *138*, 1670–1681. [CrossRef]

44. Kuda, T.; Ikemori, T. Minerals, polysaccharides and antioxidant properties of aqueous solutions obtained from macroalgal beach-casts in the Noto Peninsula, Ishikawa, Japan. *Food Chem.* **2009**, *112*, 575–581. [CrossRef]

45. Pereira, L.; Amado, A.; Critchley, A.; van de Velde, F.; Ribeiro-Claro, P. Identification of selected seaweed polysaccharides (phycocolloides) by vibrational spectroscopy (FTIR-ATR and FT-Raman). *Food Hydroc.* **2009**, *23*, 1903–1909. [CrossRef]

46. Little, A. Effect of pre- and post-mortem handling on reflectance characteristics of skipjack tuna. *J. Food Sci.* **1972**, *37*, 502. [CrossRef]

47. Lee, S.; Joo, S.T.; Alderton, A.L.; Hill, D.W.; Faustman, C. Oxymioglobin and lipid oxidation in yellowfin tuna (*Thunnus albacares*) loins. *J. Food Sci.* **2003**, *68*, 1664–1668. [CrossRef]

48. Wetterskog, D.; Undeland, I. Loss of redness (*a**) as a tool to follow hemoglobin-mediated lipid oxidation in washed cod mince. *J. Agric. Food Chem.* **2004**, *52*, 7214–7221. [CrossRef] [PubMed]

49. Gallardo, J.M.; Pérez-Martín, R.; Franco, J.M.; Aubourg, S.P.; Sotelo, C.G. Changes in volatile bases and trimethylamine oxide during the canning of albacore (*Thunnus alalunga*). *Int. J. Food Sci. Technol.* **1990**, *25*, 78–81. [CrossRef]

Article

Free Amino Acids and Biogenic Amines in Canned European Eels: Influence of Processing Step, Filling Medium and Storage Time

Lucía Gómez-Limia [1,2], Roxana Cutillas [1,2], Javier Carballo [1,2], Inmaculada Franco [1,2] and Sidonia Martínez [1,2,*]

[1] Food Technology, Faculty of Science, University Campus as Lagoas s/n, University of Vigo, 32004 Ourense, Spain; lugomez@uvigo.es (L.G.-L.); rcutillas@uvigo.es (R.C.); carbatec@uvigo.es (J.C.); inmatec@uvigo.es (I.F.)

[2] CITACA-Agri-Food Research and Transfer Cluster, Campus Auga, University of Vigo, 32004 Ourense, Spain

* Correspondence: sidonia@uvigo.es

Received: 16 September 2020; Accepted: 28 September 2020; Published: 29 September 2020

Abstract: This study evaluated the effects of the canning process and different filling media on the free amino acid and biogenic amine contents of eels. The main free amino acids were histidine, taurine and arginine, which constituted 72% of the free amino acids in raw eels. All steps in the canning process significantly altered the free amino acid content of eels, relative to raw samples. The changes were influenced by the step, the composition of the frying or filling medium and the storage time. The biogenic amine contents were very low in all samples. Histamine was not detected in either raw eels or canned eels. The highest values were obtained for 2-phenylethylamine. The step of the canning process, the composition of the frying or filling medium and storage time also determined the changes in the biogenic amine contents. The biogenic amines indices were low, indicating the good quality of canned eels.

Keywords: free amino acids; biogenic amines; canning; filling medium; European eels

---

1. Introduction

Fish is a good source of protein, rich in essential amino acids, micro- and macroelements, lipids rich in unsaturated fatty acids and fat-soluble vitamins. However, it is extremely perishable due to the high contents of water, protein and non-protein nitrogen compounds and to the activities of autolytic enzymes, which cause spoilage.

Free amino acids (FAAs) represent one of the most important fractions of non-protein nitrogen in fish. FFAs play a very important role in bacterial spoilage of fish. They undergo important changes during processing and storage, and therefore the different technologies used greatly influence the amino acid profiles. On the other hand, FAAs play an important role in the sweetness, sourness, bitterness and umami taste of foods. In addition, they can interact with reducing sugars (Maillard reaction or non-enzymatic browning), giving rise to different tastes, colors and aromas [1,2]. The FAA content of food and loss of FAA due to processing and storage are also of interest in relation to nutritional aspects. Some amino acids, such as taurine, which is present in high amounts in fish, can have beneficial effects on health [3].

Once the fish is dead, the quality of the flesh degrades rapidly due to chemical reactions (changes in protein and lipid fractions, formation of biogenic amines and hypoxanthine) and microbiological spoilage. Biogenic amines are nitrogenous, low molecular mass organic bases. They are found in a variety of foods due to microbial decarboxylation of the corresponding amino acids or to transamination of aldehydes and ketones by amino acid transaminases [4,5]. Biogenic amine formation depends on factors such as the FAA content, development of microbial activity, processing and preservation conditions during the pre- and post-mortem period [6,7]. However, storage temperature is the most important factor contributing to the formation of biogenic amines [7]. The most important biogenic amines in fish are histamine, tyramine, cadaverine and putrescine, which are formed by the enzymatic decarboxylation of histidine, tyrosine, lysine and ornithine, respectively [8]. Biogenic amines can have negative effects on health. Histamine is important in fish intoxication and is considered an indicator of fish spoilage. Consumption of fish containing high levels of histamine can produce a tingling, burning sensation in the mouth, urticaria, flushing, vomiting and diarrhea [9]. Histamine, tyramine, 2-phenylethylamine and tryptamine can also affect the nervous and vascular systems [10]. Putrescine and cadaverine potentiate histamine toxicity [11]. Some biogenic amines such as putrescine, cadaverine, spermidine and spermine are also considered precursors of carcinogens. Cadaverine has also been found to be a useful indicator of the initial stage of fish decomposition. In canned fish, biogenic amines are used to indicate the freshness of the fish prior to processing and the sanitary conditions during the canning process. Biogenic amines are closely correlated with sensory alterations in fish [12,13]. The profile and content of biogenic amines vary in different fish species [2].

Canning has long been used as a method of preserving fish. However, owing to the thermal sensitivity of some chemical constituents, different breakdown and hydrolytic reactions occur in canned fish. The associated changes during the different steps may be desirable or undesirable, and they can cause important modifications in the nutritional, sensory and safety quality of the final product [9,12,14]. Each step of the canning process contributes to the resulting quality of the final product. The major steps involved in canning fish are pre-cooking, packing with a filling medium in hermetically sealed cans and sterilizing to reach commercial sterility. After these steps, a maturation process begins on storage and continues until the cans reach the consumer.

Various types of filling medium are used for canned fish, including brine and oil. Although vegetable oils are widely used for canning fish, available information on the effects is very limited.

The European eel (*Anguilla anguilla*) is a catadromous species, i.e., it spends most of its life in rivers, during which it grows and matures before migrating to the sea to spawn. Eels are commercially very valuable in Europe (mainly Spain, Portugal, Italy and Netherlands) and Asia (mainly Japan, China, Korea and Taiwan). The female can reach a maximum weight of 9 kg, while the males are smaller, reaching a maximum weight of 2.8 kg, although the specimens in greatest demand are much smaller (<200 g). The eel population has fallen drastically due to various factors threatening its survival, including an increase in contamination of human origin and environmental impacts associated with the construction of diverse obstacles in rivers, such as hydroelectric dams [15]. Canning enables eels to be consumed throughout the year while respecting the state limitations aimed at protecting the species. The use of larger eels for canning (less valued as a fresh product) could also help to increase the reproductive success of the species.

The development of new canned products is also of interest due to the considerable economic importance of such products, their extensive acceptance by consumers, ease of transport and export and nutrient supply.

Although a large number of papers published in recent years have addressed the amino acid and biogenic amine contents of food, information about the loss of free amino acids and formation of biogenic amines during canning and the effects of different steps and different filling media on these components is scarce. In addition, existing studies are limited to a few species, such as tuna, mackerel and sardines, with no reports considering eels.

The present research was undertaken to study the effects of the different steps during canning (frying, sterilization and storage for 2 and 12 months) and of different filling media (sunflower oil, olive oil and spiced olive oil) on the free amino acid content and the formation of biogenic amines in European eels.

2. Materials and Methods

2.1. Selection and Preparation of Samples

The European eels included in the study were caught in the River Ulla (Galicia, NW Spain) and purchased at a local market ("Plaza de Abastos, Mariscos vivos del Grove") in Ourense (Galicia, Northwest Spain).

All of the eels used in this study weighed between 200 and 600 g. The eels were eviscerated and transferred to the laboratory, where they were frozen (−20 °C) until canning.

Some randomly selected samples of the frozen eels, hereafter referred to as "raw eels", were thawed in a refrigerator at +4 °C for 12 h and processed as control samples. The other samples were thawed in 12% brine at room temperature for 45 min before being cut into slices (1.5–2 cm). All slices were mixed to ensure a homogeneous product. The fish slices were fried for 2 min at 190 °C in a deep fryer, to eliminate the water present in the eel and to prevent formation of a water–oil mixture in the cans. Two different frying media were used: sunflower oil (refined sunflower oil) and olive oil (refined olive oil + virgin olive oil). The fried slices of eel were cooled (30 °C) and placed in cans (6 or 7 slices in each). The hot filling medium was then added: sunflower oil (eels previously fried in sunflower oil), olive oil or olive oil plus chili and pepper (eels previously fried in olive oil). The olive oil plus chili and pepper filling medium will be referred to hereafter as "spiced olive oil". The cans were then vacuum-sealed and sterilized. The time/temperature combination used for sterilization was 118 °C and 30 min (F0 = 11). Finally, the cans were cooled and stored at room temperature. The gross weight of each can was 185.68 ± 4.29 g and the drained weight 50.00 ± 3.91 g.

Eels were sampled raw (control), after each processing step (frying and sterilization treatment) and at two different times throughout the storage of each final product (2 and 12 months of room storage). For biogenic amine determinations, the eels were also sampled after thawing-salting, as biogenic amines can be formed during this step.

2.2. Moisture and pH Determination

The moisture content was determined after drying the eels to constant weight in an oven for 16 h at 105 ± 1 °C [16]. The pH was measured using a digital pH meter (Crison, model GLP21, Barcelona, Spain) according to the AOAC (Association of Official Analytical Chemists) methods [16].

2.3. Free Amino Acid (FAA) Determination

The FAAs were extracted and derivatized following the procedure described by Franco et al. [17], with some modifications. Fish samples (2 g) were thoroughly homogenized in 20 mL of 0.6 N $HClO_4$ in a lab blender (Ultra-Turrax® T25, IKA, Staufen, Germany) for 2 min. The samples were then centrifuged at 1800× g for 20 min. The supernatant was collected and then filtered. The pH of the filtrate was adjusted to 7.1 ± 0.2 with 30% KOH and then cooled for 10 to 20 min until reaching a temperature of 2 °C.

For derivatization, 0.2 mL of standard solution or hydrolyzed sample was dried in a vacuum centrifuge concentrator at 37 °C. A 20 μL aliquot of the derivatizing solution (ethanol + Milli-Q water + triethylamine + phenyl isothiocyanate) was then added to the samples, and the solution was mixed and left to stand at room temperature for 20 min. The resulting solution was evaporated in a vacuum centrifuge concentrator at 37 °C. The dry residue was resuspended in 500 μL of diluent solution disodium acid phosphate (675.4 mg) + Milli-Q water (950 mL) and filtered (through Waters 0.45 μm pore diameter filters).

The FAAs were identified and quantified by HPLC, under the conditions described by Franco et al. [17], with some minor modifications. The liquid chromatography equipment consisted of a chromatograph (ThermoFinnigan, Silicon Valley, CA, USA) with a UV/VISIBLE photodiode array detector (Spectrasystem UV6000LP). The samples were separated on a reversed phase column of diameter 4.6 mm and length 25 cm (C18 Ultrasphere 5–ODS, from Beckman, Fullerton, CA, USA).

The temperature of the column was maintained at 50 ± 1 °C with a column heater (Spectrasystem 3000). The wavelength of the detector was 254 nm. Standards of the 19 different amino acids were supplied by Sigma Chemical Co. (St Louis, MO, USA). Data regarding free amino acid composition were expressed in mg/100 g of muscle. All the samples and standards were injected into the column at least in duplicate.

Repeatability tests were performed by injecting a standard and a sample into the column consecutively six times in a day. Reproducibility tests were also carried out by injecting the standard and the sample into the column twice a day for three days under the same experimental conditions. The results obtained in these tests were not significantly different ($p < 0.05$).

2.4. Biogenic Amine Analysis

Biogenic amines were analyzed by the method described by Lorenzo et al. [18]. Fish muscle (5 g) was mixed with 10 mL of 0.6 N $HCLO_4$ and 1 mL of internal standard (1–7 diaminoheptane). The mixture was homogenized with a lab blender (Ultraturrax) for 2 min and then centrifuged at 3000 rpm for 10 min at 4 °C. The supernatant was collected, and the same process was repeated with the residue for complete extraction. Finally, the two supernatants were placed in a 25 mL volumetric flask and 0.6 N $HClO_4$ was added to make up the final volume.

For derivatization, an aliquot (0.5 mL) of each extracted sample or of the standard solution of any biogenic amine was immediately placed in a tube, and 100 µL of 2 N NaOH, 150 µL of a saturated solution of $NaHCO_3$ and 1 mL of dansyl chloride were added consecutively. The tube was shaken gently and placed in a water bath at 40 °C for 45 min. In order to remove dansyl chloride residue, 50 µL of ammonia was then added and the mixture was left to stand for 30 min. Finally, the volume was made up to 2.5 mL with acetonitrile and the mixture was filtered through 0.25 µm pore-size filters prior to HPLC analysis.

Separation, identification and quantification of the biogenic amines were carried out by HPLC, following the procedure described by Eerola et al. [19], with the aforementioned HPLC equipment. The different biogenic amines were separated on a reversed phase column of diameter 4.6 mm and length 25 cm (C18 Ultrasphere 5–ODS, from Beckman, Fullerton, CA, USA). The temperature of the column was 40 ± 1 °C and the wavelength of the detector 254 nm.

Separation was achieved at a flow rate of 1 mL/min, with a gradient between two solvents: a solution of 0.1 M ammonium acetate was used as eluent A and acetonitrile as eluent B.

A standard solution containing appropriate amounts of tryptamine, 2-phenylethylamine, putrescine, cadaverine, histamine, tyramine, spermidine and 1,7-diaminoheptane (as an internal standard) was used to quantify the biogenic amines present in the samples. Each biogenic amine was expressed in mg/kg.

Repeatability and reproducibility tests were also performed in the biogenic amine analysis. The results of the tests were not significantly different ($p < 0.05$).

The biogenic amine index (BAI) was estimated using the equation described by Veciana-Nogués et al. [13]:

$$\text{Biogenic amine index} = (\text{histamine} + \text{cadaverine} + \text{putrescine} + \text{tyramine}) \tag{1}$$

2.5. Statistical Analysis

All analyses were carried out at least in triplicate. The data were examined by one-way analysis of variance (ANOVA), and the least squares test (LSD) was used ($p < 0.05$) to compare the mean values. The tests were implemented using Statistica software version 8.0 (Statsoft © Inc., Tulsa, OK, USA). Canonical discriminant analysis (CDA) was used to classify the eel samples. The CDA variables were selected by principal components extraction and linear discriminant analysis, and the variables with the highest discriminatory capacity were selected to establish which FAAs and biogenic amines can be used to discriminate and classify eels according to the canning step and type of filling medium.

3. Results and Discussion

3.1. Moisture and pH Values

The moisture contents of raw eels and eels packed in sunflower oil, olive oil and spiced olive oil, at each step of the canning process and after room storage (for 2 and 12 months), are shown in Figure 1A. The moisture content of raw eel was 74.15 ± 0.64%. Frying in sunflower oil and in olive oil resulted in water content losses of 11.14 and 15.75%, respectively. Sterilization treatment did not cause changes in the moisture content of canned eels packed in sunflower oil or in olive oil. However, sterilization caused a decrease in the moisture content of canned eels packed in spiced olive oil.

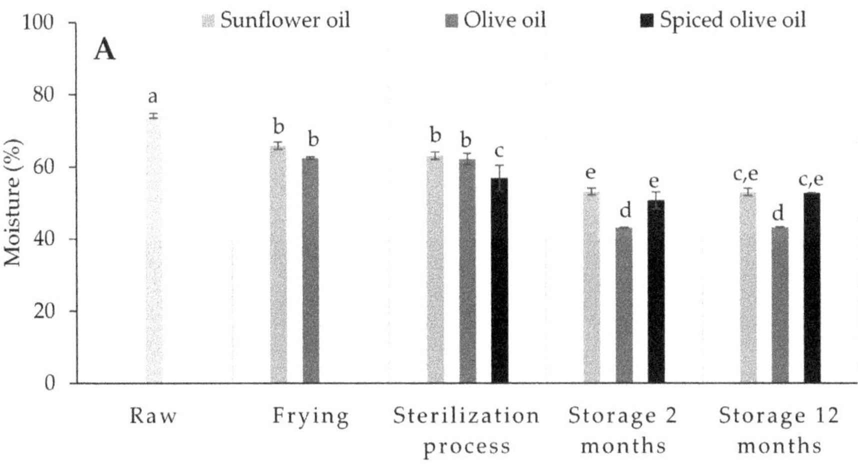

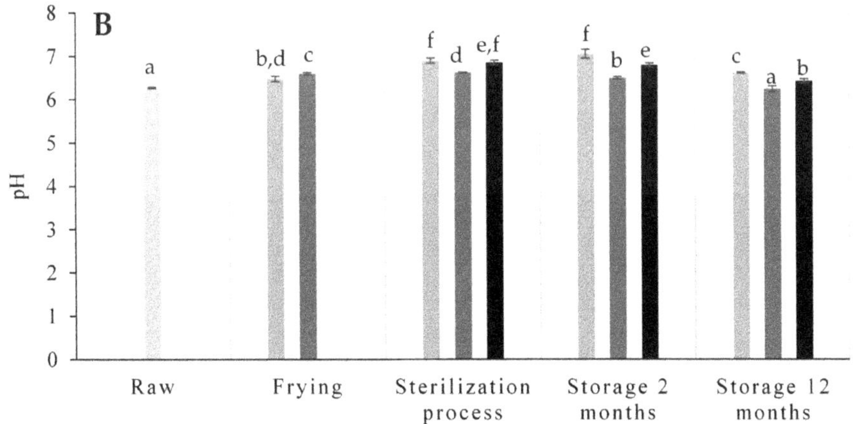

Figure 1. Moisture content (%) (**A**) and pH values (**B**) of raw and canned European eels packed in sunflower oil, olive oil and spiced olive oil throughout the different steps of the canning process and after 2 and 12 months of storage. Values with different superscripts were significantly different ($p < 0.05$).

After storage for two months, the moisture content of the eels decreased and then remained stable for up to twelve months of storage. The eels packed in olive oil had a lower moisture content (43.21%) than canned eels packed in sunflower oil and in spiced olive oil. Water can be lost due to the thermal treatment and to denaturation of proteins in muscle, which causes a decrease in the water holding capacity of the myofibrillar protein fraction [14]. In addition, frying, sterilization and storage cause an interchange between the water of the fish and filling medium, causing an increase in lipid content and a decrease in moisture content. In the canned eels packed in spiced olive oil, some active components could also pass into the muscle.

The pH can significantly affect protease activity and production of biogenic amines. The pH of raw eels was high, between 6.25 and 6.31 (Figure 1B). The pH increased significantly during frying (6.48 and 6.60 in eels fried in sunflower oil and olive oil, respectively). Sterilization caused a further increase in the pH of canned eels packed in sunflower oil and in spiced olive oil; however, it did not vary in canned eels in olive oil. The increases can be attributed to proteolysis, breakdown of protein and enzymatic activity. During storage, the pH remained stable at the beginning of storage and decreased after 12 months of storage in canned eels packed in sunflower oil and in spiced olive oil. In canned eels packed in olive oil, the pH decreased throughout storage. This decrease may be due to the degradation of lipids and other components during storage. The change in pH varied depending on the filling medium. The pH was lowest in canned eels packed in olive oil.

Several studies have reported the relationships between pH and biogenic amine accumulation. A pH between 4 and 5.5 favors decarboxylase activity [10]. In this study, significant ($p < 0.05$) positive correlations between pH and putrescine ($r = 0.75$), between pH and cadaverine ($r = 0.45$) and between pH and spermidine ($r = 0.48$) were observed.

3.2. Free Amino Acids

The FAA content is very important in order to evaluate protein-rich food. The FAA contents of the muscle tissue of the raw eel and of eels at each step of the canning process and after room storage (for 2 and 12 months) are presented in Tables 1 and 2.

The main FAAs in raw eels were histidine (222.76 ± 6.99 mg/100 g), taurine (91.63 ± 6.62 mg/100 g) and arginine (80.58 ± 6.07 mg/100 g), which constituted 72% of the total FAA content.

Histidine accounts for 41.11% of the total FAAs. This result is consistent with those of previous studies [1,20]. The histidine contents of fish muscles can vary significantly due to differences in fish species, sex, season, feeding, living environment, swimming activity and stage of maturity [21]. Histidine is important because of its physiological and nutritional roles in fish. Some migratory species such as tuna, skipjack and mackerel contain high amounts of histidine that maintain the muscle pH level during swimming and that act as an energy source during prolonged starvation [21]. The eel is also a migratory species. This may be the reason for the high histidine content in the eels analyzed, although it is lower than in other species such as tuna and mackerel [20]. Antoine et al. [20] reported important variations in histidine content of individual fish of the same species and between red and white muscle. On the other hand, histidine is the precursor of histamine, which mainly appears when extreme temperatures are used.

Taurine represented 16.92% of total FAAs in raw eels. The values obtained are similar to those reported for megrim (95 mg/100 g) by Gormley, Neumann, Fagan and Brunton [22]. These authors reported a wide range of taurine content in 14 fish species (between 7 and 176 mg/100 g). Taurine is a derivative of cysteine. It is not strictly an amino acid, and as it contains a sulfonic acid group, it is referred to as sulfonic acid. Some fish such as mackerel, seabream and tuna have a high proportion of taurine in their muscle tissue [21,23]. Although taurine does not have an important impact on taste, this free amino acid is known to benefit human health due to anti-inflammatory, anti-atherosclerotic effect and anti-obesity effects [3].

Arginine accounts for 14.87% of the FAA content of raw eels. Arginine has anti-aging and anti-fatigue effects, but it is a precursor of putrescine. Arginine and histidine are associated with a bitter taste [1].

Table 1. Free amino acid content of raw and canned European eels packed in sunflower oil throughout the different steps of the canning process and after 2 and 12 months of storage (expressed mg/100 g muscle).

	Raw	Frying	Sterilization	Storage 2 Months	Storage 12 Months
NON-ESSENTIAL AMINO ACIDS					
Aspartic acid	14.24 ± 0.54 [a]	11.23 ± 2.60 [a]	26.34 ± 1.39 [b,c]	28.22 ± 0.39 [c]	23.64 ± 1.93 [b]
Glutamic acid	21.60 ± 1.75 [a,b]	23.11 ± 1.07 [a]	15.90 ± 0.45 [c]	20.37 ± 0.84 [a,b]	19.92 ± 1.09 [b]
Hydroxyproline	0.86 ± 0.04 [a,b]	1.02 ± 0.01 [a]	0.79 ± 0.03 [b]	1.33 ± 0.14 [c]	2.13 ± 0.09 [d]
Serine	3.81 ± 0.03 [a]	4.67 ± 0.12 [b]	2.83 ± 0.45 [c]	1.62 ± 0.02 [d]	3.04 ± 0.44 [c]
Glycine	50.24 ± 2.29 [a]	57.10 ± 2.40 [a]	39.89 ± 2.22 [b]	56.12 ± 5.78 [a]	42.30 ± 4.88 [b]
Arginine	80.58 ± 6.07 [a]	76.44 ± 2.02 [a]	128.16 ± 0.22 [b]	140.34 ± 5.90 [c]	159.91 ± 8.90 [d]
Alanine	12.37 ± 1.13 [a]	12.20 ± 0.50 [a]	13.54 ± 1.98 [a]	15.72 ± 3.65 [a]	22.40 ± 1.77 [b]
Proline	2.68 ± 0.28 [a]	4.77 ± 0.26 [b]	5.20 ± 0.18 [b]	6.66 ± 1.23 [c]	8.09 ± 0.71 [d]
Tyrosine	5.98 ± 0.32 [a]	8.97 ± 0.33 [b]	9.37 ± 0.19 [b]	10.25 ± 1.62 [b]	12.21 ± 1.03 [c]
Taurine	91.63 ± 6.62 [a]	169.73 ± 0.57 [b]	65.08 ± 2.93 [c]	76.21 ± 7.04 [c]	75.57 ± 6.88 [c]
Ornithine	1.58 ± 0.08 [a]	0.27 ± 0.02 [b]	0.33 ± 0.13 [b,c]	0.48 ± 0.01 [c,d]	0.59 ± 0.15 [d]
Total non-essential amino acids (NEAA)	286.31 ± 19.25 [a]	370.75 ± 4.48 [b]	283.79 ± 20.01 [a]	357.31 ± 25.58 [b]	368.43 ± 20.47 [b]
ESSENTIAL AMINO ACIDS					
Histidine	222.76 ± 6.99 [a]	333.86 ± 3.11 [b]	221.74 ± 1.69 [a]	240.24 ± 5.10 [a]	280.36 ± 22.41 [c]
Threonine	9.07 ± 0.06 [a]	11.47 ± 0.96 [a,b]	11.92 ± 0.71 [a,b]	13.50 ± 0.26 [b,c]	16.14 ± 2.23 [c]
Valine	1.73 ± 0.15 [a]	4.49 ± 0.18 [b]	7.48 ± 0.35 [c]	6.05 ± 0.10 [d]	8.17 ± 1.09 [c]
Isoleucine	2.47 ± 0.32 [a]	3.53 ± 0.54 [a,b]	4.66 ± 0.14 [b,c]	4.48 ± 1.30 [b,c]	5.27 ± 0.73 [c]
Leucine	4.08 ± 0.32 [a]	5.26 ± 1.52 [a]	7.60 ± 0.25 [b]	5.03 ± 0.05 [a]	10.57 ± 0.74 [c]
Phenylalanine	4.36 ± 1.23 [a,b]	4.17 ± 0.18 [a]	5.42 ± 0.08 [a,b,c]	5.57 ± 0.53 [b,c]	6.10 ± 0.67 [c]
Lysine	4.48 ± 0.52 [a]	15.80 ± 0.50 [b]	16.39 ± 1.76 [b]	15.96 ± 0.46 [b]	29.67 ± 0.47 [c]
Tryptophan	5.24 ± 0.17 [a]	5.81 ± 0.14 [a]	7.06 ± 0.35 [b]	7.16 ± 1.00 [b]	8.33 ± 0.25 [c]
Total essential amino acids (EAA)	255.53 ± 6.59 [a]	379.98 ± 2.13 [b]	274.57 ± 13.05 [a]	299.49 ± 2.08 [a]	346.72 ± 34.79 [b]
EAA/NEAA ratio	0.90 ± 0.08 [a,b]	1.02 ± 0.01 [a]	0.95 ± 0.04 [a,b]	0.84 ± 0.07 [b]	0.94 ± 0.06 [a,b]
Total free amino acids	541.83 ± 12.66 [a]	750.73 ± 6.30 [b]	563.84 ± 37.95 [a]	656.80 ± 23.50 [c]	743.10 ± 26.87 [b,c]

[a–d] Mean values of at least three determinations ± standard deviation with different superscripts in the same row were significantly different ($p < 0.05$).

Table 2. Free amino acid content of raw and canned European eels packed in olive oil or spiced olive oil throughout the different steps of the canning process and after 2 and 12 months of storage (expressed as mg/100 g muscle).

	Raw	Frying (in Olive Oil)	Sterilization		Storage 2 Months		Storage 12 Months	
			Olive Oil	Spiced Olive Oil	Olive Oil	Spiced Olive Oil	Olive Oil	Spiced Olive Oil
NON-ESSENTIAL AMINO ACIDS								
Aspartic acid	14.24 ± 0.54 [a]	11.97 ± 2.54 [a]	37.49 ± 2.01 [b]	39.39 ± 5.66 [b]	29.80 ± 1.11 [c]	32.33 ± 1.38 [c]	32.26 ± 0.40 [c]	23.96 ± 0.43 [d]
Glutamic acid	21.60 ± 1.75 [a,b]	23.21 ± 0.95 [a,b]	21.35 ± 0.46 [a,c]	22.79 ± 4.38 [a,b]	20.62 ± 1.38 [a,c,d]	24.02 ± 0.36 [b]	18.79 ± 0.20 [c,d]	17.80 ± 1.26 [d]
Hydroxyproline	0.86 ± 0.04 [a]	0.82 ± 0.19 [a]	0.84 ± 0.07 [a]	0.85 ± 0.20 [a]	0.81 ± 0.03 [a]	0.90 ± 0.04 [a]	1.32 ± 0.14 [b]	1.77 ± 0.24 [c]
Serine	3.81 ± 0.03 [a]	3.33 ± 0.23 [a]	2.26 ± 0.27 [b,c]	2.69 ± 0.33 [b]	1.98 ± 0.01 [c]	3.38 ± 0.19 [a]	2.19 ± 0.15 [b,c]	2.31 ± 0.32 [b,c]
Glycine	50.24 ± 2.29 [a]	35.82 ± 3.40 [b]	48.26 ± 1.24 [a]	54.17 ± 6.91 [a]	37.33 ± 2.20 [b]	50.08 ± 5.24 [a]	29.85 ± 2.48 [c]	33.74 ± 2.34 [b]
Arginine	80.58 ± 6.07 [a]	100.67 ± 6.97 [b]	115.08 ± 7.07 [c]	130.44 ± 3.88 [d]	137.07 ± 16.04 [d]	143.58 ± 2.32 [d]	319.70 ± 5.93 [e]	247.61 ± 11.62 [f]
Alanine	12.37 ± 1.13 [a]	14.29 ± 0.92 [b]	13.90 ± 0.50 [a,b]	18.74 ± 0.26 [c]	13.58 ± 1.36 [a,b]	17.78 ± 1.02 [c]	13.57 ± 1.12 [a,b]	17.84 ± 0.07 [c]
Proline	2.68 ± 0.28 [a]	5.99 ± 0.52 [c,d]	5.31 ± 0.30 [b,c]	4.82 ± 0.19 [b]	5.52 ± 0.56 [b,c]	6.81 ± 0.67 [e]	5.66 ± 0.32 [b,c]	6.46 ± 0.57 [d,e]
Tyrosine	5.98 ± 0.32 [a]	10.29 ± 0.94 [c]	10.54 ± 0.30 [c]	9.73 ± 0.26 [c]	8.40 ± 0.11 [b]	9.98 ± 0.20 [c]	8.55 ± 0.38 [b]	10.78 ± 0.35 [c]
Taurine	91.63 ± 6.62 [a]	119.00 ± 14.70 [b]	103.84 ± 6.23 [a,b]	100.06 ± 23.40 [a,b]	76.71 ± 11.62 [c]	100.57 ± 14.61 [a,b]	66.77 ± 6.73 [c]	60.23 ± 1.73 [c]
Ornithine	1.58 ± 0.08 [a]	0.33 ± 0.03 [b]	0.73 ± 0.12 [c]	0.48 ± 0.15 [b,d]	0.65 ± 0.03 [c,d]	0.61 ± 0.15 [c,d]	0.54 ± 0.06 [d]	0.43 ± 0.05 [b,c]
Total non-essential amino acids (NEAA)	286.31 ± 19.25 [a]	325.14 ± 21.72 [a,b]	346.70 ± 35.87 [b]	373.26 ± 20.80 [b,d]	332.48 ± 4.31 [a,b]	357.76 ± 30.75 [b]	499.09 ± 14.42 [c]	408.42 ± 28.36 [d]
ESSENTIAL AMINO ACIDS								
Histidine	222.76 ± 6.99 [a]	225.23 ± 13.63 [a,c]	264.47 ± 3.01 [b]	270.09 ± 5.48 [b]	279.62 ± 4.74 [b]	259.97 ± 1.19 [b,c]	96.36 ± 8.98 [c]	139.64 ± 7.28 [e]
Threonine	9.07 ± 0.06 [a]	9.61 ± 1.42 [a]	13.85 ± 1.29 [b]	13.91 ± 0.15 [b]	14.12 ± 0.53 [b]	12.51 ± 0.39 [b]	12.71 ± 0.55 [b]	12.22 ± 1.08 [b]
Valine	1.73 ± 0.15 [a]	4.40 ± 0.71 [b]	10.81 ± 0.73 [d]	7.46 ± 0.48 [c]	6.64 ± 0.05 [c]	7.40 ± 0.35 [c]	26.73 ± 4.01 [e]	11.26 ± 1.47 [d]
Isoleucine	2.47 ± 0.32 [a]	4.36 ± 0.50 [b,c,e]	3.91 ± 0.47 [b,c,d]	3.77 ± 0.19 [b,d]	3.09 ± 0.02 [a,d]	4.65 ± 0.41 [c,e]	3.90 ± 0.36 [c,b,d]	4.68 ± 0.63 [e]
Leucine	4.08 ± 0.32 [a]	6.20 ± 0.59 [b]	6.83 ± 0.33 [b,c]	6.70 ± 0.55 [b]	4.27 ± 0.02 [a]	7.67 ± 0.01 [c]	4.77 ± 0.08 [a]	6.05 ± 0.10 [b]
Phenylalanine	4.83 ± 1.18 [a,b,c]	5.36 ± 0.94 [c]	3.84 ± 0.98 [a,b]	5.36 ± 1.06 [c,d]	3.69 ± 0.24 [a,b]	6.60 ± 0.28 [d]	3.59 ± 0.10 [a]	5.11 ± 0.52 [b,c]
Lysine	4.48 ± 0.52 [a]	23.33 ± 2.53 [b]	23.32 ± 2.70 [b]	24.89 ± 1.35 [b,d]	22.55 ± 2.15 [b]	28.35 ± 1.36 [d]	16.79 ± 0.85 [c]	17.17 ± 0.72 [c]
Tryptophan	5.24 ± 0.17 [a]	6.90 ± 0.38 [b]	7.06 ± 0.43 [b]	7.15 ± 1.06 [b]	6.91 ± 0.17 [b]	7.48 ± 0.66 [b]	7.77 ± 0.17 [b]	6.95 ± 1.00 [b]
Total essential amino acids (EAA)	255.53 ± 6.59 [a]	288.41 ± 17.03 [b]	335.50 ± 0.58 [c]	326.25 ± 13.43 [c]	340.90 ± 7.44 [c]	316.30 ± 12.03 [c]	174.83 ± 3.23 [d]	186.71 ± 8.21 [d]
EAA/NEAA ratio	0.90 ± 0.08 [a]	0.95 ± 0.17 [a]	0.94 ± 0.04 [a]	0.94 ± 0.13 [a]	1.03 ± 0.04 [a]	0.90 ± 0.20 [a]	0.35 ± 0.04 [b]	0.44 ± 0.03 [b]
Total free amino acids	541.83 ± 12.66 [a]	601.49 ± 25.99 [b]	702.91 ± 0.71 [c]	699.52 ± 7.37 [c]	673.38 ± 3.13 [c,d]	713.84 ± 10.36 [c]	661.21 ± 36.39 [d]	611.12 ± 0.45 [b]

[a–f] Mean values of at least three determinations ± standard deviation with different superscripts in the same row were significantly different ($p < 0.05$).

Glycine was detected in relatively high amounts in the raw eels (50.24 ± 2.29 mg/100 g). This amino acid is an indicator, together with hydroxyproline, of the presence of connective tissue, principally collagen, and plays a key role in its stability [24]. Glycine is associated with a sweet taste [1].

Glutamic acid (21.60 ± 1.75 mg/100 g) and aspartic acid (14.24 ± 0.54 mg/100 g) were also quantitatively important. They play an important role in the umami taste of food.

Other amino acids such as alanine, serine and threonine, which are associated with a sweet taste, were also found in appreciable quantities (12.37, 3.81 and 9.07 mg/100 g, respectively). Valine, isoleucine, leucine, phenylalanine and tryptophan contents of raw eels were 1.73, 2.47, 4.08, 4.83 and 5.24 mg/100 g, respectively. These amino acids contribute a bitter taste.

The other amino acids were found in lower quantities in raw eels.

The non-essential amino acid (NEAA) content was 286.31 ± 19.25 mg/100 g in raw samples.

The essential amino acid (EAA) content is a major factor that affects the nutritional value of protein. In raw eels, the EAA content was 255.53 ± 6.59 mg/100 g. The value of the essential/non-essential (EAA/NEAA) ratio for FAAs was 0.90 ± 0.08.

The FAA composition can vary significantly according to the production conditions during the preservation and processing of the food. Significantly different ($p < 0.05$) quantities of some FAAs were detected at all steps of the canning process.

Frying greatly altered the FAA content of eels, relative to raw samples. The changes were influenced by the composition of the frying oil and were not homogeneous for the different FAAs, because the concentrations of some amino acids decreased, while those of others increased. In the eels fried in sunflower oil (Table 1), significant increases ($p < 0.05$) in serine, proline, tyrosine, taurine, histidine, valine and lysine contents and a decrease in the ornithine content were observed. Total essential and non-essential free amino acids increased during frying in sunflower oil.

In eels fried in olive oil (Table 2), the EAA and NEAA contents both increased significantly. Arginine, alanine, proline, tyrosine, taurine, valine, isoleucine, leucine, phenylalanine, lysine and tryptophan contents increased during frying in olive oil, while glycine and ornithine contents decreased.

The increase in FAAs during frying may be derived from the moisture lost during process. However, some FAAs increased and other decreased, suggesting that moisture was not the only factor affecting the FAA content during the frying process. Some changes can be attributed to the heat-induced denaturalization of protein, to the decomposition of FAAs and to the formation of different volatile compounds such as pyrazines and sulfides, which have important impacts on the aroma and flavor of fried fish [25]. On the other hand, amino acids can react with sugar, giving rise to the Maillard reaction. The changes during frying can cause variations in fish flavor, especially those associated with sweet and bitter tastes, as the amino acids associated with these parameters underwent the most notable changes.

In general, the sterilization process also had a significant effect on the FAA composition of the European eel samples. The changes were also influenced by the composition of the filling medium and were not homogenous for the different free amino acids. In canned eel packed in sunflower oil (Table 1), aspartic acid, arginine, proline, valine, leucine and tryptophan contents increased significantly after the sterilization process ($p < 0.05$). However, glutamic acid, serine, glycine, taurine and histidine contents decreased. Loss of glutamic acid is of great interest because, although glutamic acid is a NEAA, it is an important source of nitrogen and it is involved in taste perception, contributing to the umami taste [26]. Both total EAA and NEAA contents were lower in canned eels packed in sunflower oil than in those packed in the other media.

In canned eels packed in olive oil (Table 2), significant increases ($p < 0.05$) in aspartic acid, glycine, arginine, ornithine, histidine, threonine and valine, and decreases in serine and phenylalanine, were observed after the sterilization process. In the canned eels packed in spiced olive oil, aspartic acid, glycine, arginine, alanine, histidine, threonine and valine contents increased, while serine and proline contents decreased after the sterilization process.

Therefore, important variations between canned eels packed in different filling media were observed. This may be due to different interactions between FAAs and filling oil. There were also differences between canned eels packed in olive oil and canned eels packed in spiced olive oil, and therefore the spices influenced the changes in amino acids during sterilization.

Aubourg [14] reported that FAAs could be lost as a result of extraction by the filling medium and/or interaction reactions with oxidized lipids. The different fatty acid compositions of the filling media also condition the heat penetration, which also conditions the effect of the treatments on the different components [27]. On the other hand, the different biogenic amines levels in canned eels could be due to the different degrees of heat-induced degradation of FAA in different filling media.

The EAA/NEAA ratio is considered a good way of estimating the production of free essential amino acids during processing. However, in canned eels, this ratio did not vary during the sterilization process, as the contents of both total FAAs and total NEAAs increased.

Some studies have reported changes in individual amino acids caused by heating [14]. Heat processing causes denaturalization of protein. The denatured proteins are more reactive and can interact with other constituents. Other FAAs form amines, volatile acids and other nitrogenous substances.

Canned storage is normally necessary to produce satisfactory textural and optimal palatability of canned fish [14]. During storage, many compounds migrate from the fish to the filling medium and vice versa, in a dynamic equilibrium influenced by the characteristics of the fish muscle and the filling medium as well as by the type of processing and storage. The values reported in Tables 1 and 2 show that the FAAs underwent greater changes during storage. The changes were also influenced by the composition of the filling medium and storage time. In canned eels packed in sunflower oil, storage for 2 months caused increases in glutamic acid, hydroxyproline, glycine, arginine and proline and decreases in serine, valine and leucine.

In canned eels packed in olive oil, storage for 2 months led to loss of aspartic acid, glycine, tyrosine, taurine, valine and leucine and an increase in the arginine content. In canned eels packed in spiced olive oil, the losses were lower during 2 months of storage. Only the arginine content decreased, and the serine, proline, isoleucine and leucine contents increased.

The greatest changes occurred after 12 months of storage. In canned eels packed in sunflower oil, the hydroxyproline, serine, arginine, alanine, proline, tyrosine, histidine, valine, leucine, lysine and tryptophan contents increased, and the aspartic acid and glycine contents decreased after storage for 12 months.

In canned eels packed in olive oil and spiced olive oil, hydroxyproline, arginine and valine contents increased, and glycine, histidine and lysine contents decreased. In the canned eels packed in spiced olive oil, aspartic acid, glutamic acid, serine, taurine, leucine and phenylalanine contents also decreased after 12 months of storage. In canned eels packed in olive oil and in spiced olive oil, there was a significant decrease in EAA and an increase in NEAA and a consequent significant decrease in the EAA/NEAA ratio.

These results suggest that the hydrolysis reactions and interactions continued during room storage. On the other hand, the exchange of FAAs and the interactions appear to differ depending on the filling medium.

3.3. Biogenic Amine Composition

Biogenic amines are found in very low levels in fresh fish, and their formation is associated with bacterial spoilage. Exposure of fish muscle to high temperatures causes an increase in biogenic amine formation in fish [7]. The levels of the biogenic amines found in the samples of fish are shown in Table 3. In all samples studied, the biogenic amine contents were generally very low.

Table 3. Biogenic amines (mg/kg muscle) and biogenic amine index of raw and canned European eels in each step during the canning process and at 2 and 12 months of storage, in sunflower oil, in olive oil and spice olive oil.

		Biogenic Amine Content						Total Biogenic Amines	Biogenic Amine Index
		Histamine	β-Phenylethylamine	Tyramine	Putrecine	Cadaverine	Spermidine		
Raw		ND	3.10 ± 0.73 [a]	1.61 ± 0.07 [a]	0.46 ± 0.06 [a]	1.40 ± 0.06 [a]	2.36 ± 0.17 [a,f]	8.91 ± 0.46 [a]	3.50 ± 0.36 [a]
Salting		ND	3.05 ± 0.72 [a]	5.69 ± 0.95 [b]	1.89 ± 0.29 [b]	4.73 ± 0.29 [b]	5.33 ± 0.22 [b]	20.67 ± 1.69 [b]	12.30 ±1.18 [b]
Sunflower oil	F	ND	4.54 ± 0.25 [b,c,f]	4.98 ± 0.50 [b]	1.28 ± 0.27 [c]	4.59 ± 0.33 [b]	2.90 ± 0.85 [a,d]	18.59 ± 1.49 [c]	10.99 ± 0.72 [c]
	S	ND	4.01 ± 0.21 [c,g]	1.90 ± 0.58 [a]	0.86 ± 0.19 [d,f]	4.10 ± 0.19 [b]	2.68 ± 0.32 [a,f]	13.87 ± 0.52 [e]	7.49 ±0.45 [d]
	ST2	ND	6.13 ± 0.11 [d]	1.15 ± 0.04 [c]	1.00 ± 0.22 [d]	4.60 ± 0.22 [b]	2.71 ± 0.12 [a]	15.66 ± 0.08 [f]	6.86 ± 0.25 [d]
	ST12	ND	8.23 ± 0.55 [e]	ND	0.62 ± 0.07 [e]	1.36 ± 0.10 [a]	0.34 ± 0.04 [c]	10.52 ± 0.79 [g,h]	2.12 ± 0.06 [e]
Olive oil	F	ND	4.97 ± 0.30 [f]	1.41 ± 0.21 [a,d]	0.88 ± 0.13 [d,f]	4.32 ± 0.16 [b]	3.60 ± 0.67 [a,d]	14.51 ± 0.57 [e,f]	6.64 ± 0.73 [d]
	S	ND	3.64 ±0.44 [a,c]	1.60 ± 0.17 [a]	0.69 ± 0.15 [e,f]	4.30 ± 0.15 [b]	2.76 ± 0.42 [a,d]	12.55 ± 0.34 [d]	6.63 ± 0.68 [d]
	ST2	ND	3.56 ± 0.08 [a,g]	1.81 ± 0.03 [a]	0.63 ± 0.04 [e]	5.51 ± 0.06 [c]	3.14 ± 0.03 [d]	15.04 ± 1.19 [e,f]	7.22 ± 0.31 [d]
	ST12	ND	4.35 ± 0.17 [b,c]	0.83 ± 0.03 [e]	0.42 ± 0.04 [a]	6.28 ± 0.31 [c]	1.79 ± 0.18 [e,f]	14.13 ± 0.28 [e]	7.51 ± 0.35 [d]
Spiced olive oil	S	ND	3.02 ± 0.12 [a]	1.33 ± 0.07 [a,d]	0.63 ± 0.01 [e]	4.64 ± 0.02 [b]	2.16 ± 0.22 [f]	11.69 ± 0.57 [d,g]	6.60 ± 0.37 [d]
	ST2	ND	3.21 ± 0.55 [a]	1.80 ± 0.08 [a]	1.09 ± 0.03 [d]	5.76 ± 0.04 [c]	3.37 ± 0.17 [d]	15.04 ± 0.73 [e,f]	8.66 ± 0.02 [f]
	ST12	ND	7.19 ± 0.54 [h]	ND	0.30 ± 0.01 [g]	1.93 ± 0.01 [e]	0.31 ± 0.03 [c]	9.99 ± 0.13 [a,h]	2.28 ± 0.67 [e]

[a–h] Means in the same column with different letters differ significantly ($p < 0.05$). F: frying, S: sterilization process; ST2: storage 2 months; ST12: storage 12 months. ND: not detected.

Histamine is produced in raw fish due to the action of bacterial histidine decarboxylase at high temperatures and/or prolonged exposure. Thawing of frozen fish and long storage time at room temperature before canning can also lead to accumulation of histamine. Histamine is very heat resistant and it can remain intact during the sterilization process or in other processed fish products [27]. The legal limit for histamine established for fish products by the US Food and Drug Administration [28] is 50 mg/kg, and that established by the European Commission [29] is 100 mg/kg. Although eels have a high histidine content, histamine was not detected in either raw eels or canned eels (<LOD), which showed that the fish used in these products were fresh, and the products were produced using good manufacturing practices. Adequate storage temperatures and duration of refrigeration prevent the formation of histamine [7,28]. Özogul, Özogul and Gökbulut [30] reported 0.5 mg histamine/100 g of muscle of eel stored on ice for 1 day, and the content increased with temperature and time of storage. Veciana-Nogués et al. [13] observed very low histamine content in both fresh and canned tuna. In a study of various different fish products, Zhai et al. [5] reported the maximum histamine levels in canned anchovies (26.95 mg/kg) and canned sardines (22.38 mg/kg) and a histamine content of less than 10 mg/kg in all other canned samples tested. Mohan et al. [27] and Barbosa et al. [12] observed that cooking tuna before canning increased the histamine content.

The 2-phenylethylamine and tyramine have an aromatic structure [31], which indicate a vasoconstrictor activity.

The 2-phenylethylamine content of raw eels was 3.10 ± 0.73 mg/kg. Similar results were reported for eels by Özogul et al. [30]. Salting did not cause changes in 2-phenylethylamine content (3.05 ± 0.72 mg/kg); however, frying increased the content. The increase was similar in eels fried in sunflower oil (4.54 ± 0.25 mg/kg) and in eels fried in olive oil (4.97± 0.30 mg/kg). No significant differences ($p > 0.05$) were observed before and after the sterilization step in eels fried in sunflower oil (4.01 ± 0.21 mg/kg). However, the 2-phenylethylamine content decreased in canned eels packed in olive oil (3.64 ±0.44 mg/kg) and in spiced olive oil (3.02 ± 0.12 mg/kg) after the sterilization process.

The different 2-phenylethylamine levels in canned eels could be due to the different degrees of heat-induced degradation of FAA in different filling media and the different protective effects of the filling media.

Veciana-Nogués et al. [13] observed a slight increase in 2-phenylethylamine content of canned tuna after cooking and after packing, and they did not observe this biogenic amine after the sterilization step. Storage caused a significant increase ($p < 0.05$) in the 2-phenylethylamine content of canned eels, mainly after 12 months of storage in sunflower oil and in spiced olive oil. Generally, 2-phenylethylamine is formed when tyramine concentration is high, and its presence can be related to a non-specific activity of tyrosine decarboxylase [32].

The tyramine content of raw eels was 1.61 ± 0.07 mg/kg. During salting, the tyramine concentrations increased significantly (by 5.69 ± 0.95 mg/kg) ($p < 0.05$). Subsequent frying did not cause significant changes in the tyramine content when sunflower oil was used but caused a decrease when olive oil was used. In canned eels packed in sunflower oil, the sterilization process caused a significant decrease ($p < 0.05$) in the tyramine content. However, the tyramine content did not vary during the sterilization step in canned eels packed in olive oil or in spiced olive oil. During storage, significant decreases were observed in tyramine content, mainly after 12 months of storage. Tyramine was not detected even after 12 months of storage in canned eels in sunflower oil and in spiced olive oil, and it decreased to 0.83 mg/kg in canned eels packed in olive oil. Bilgin and Gençcelep [10] observed that tyramine was detected in canned tuna, chunk canned tuna, marinated anchovies, canned mackerel and canned sardines at levels ranging between ND and 48.63 mg/kg.

The concentration of putrescine ranged between 0.41 and 0.52 mg/kg in raw eels. Özogul et al. [30] reported higher values in eels stored in ice for 1 day (8.6 mg/kg), and the content increased with an increase in the temperature and time of storage. Salting increased the putrescine content (1.89 ± 0.29 mg/kg). However, the subsequent frying and sterilization process decreased putrescine content in both olive oil and sunflower oil. Storage for 12 months caused a significant decrease in putrescine content in all samples. The putrescine levels can decrease because this biogenic amine is an intermediate product in the synthesis of spermidine and spermine [33]. Zarei et al. [4] found that putrescine was detected in different canned tuna samples in the range 0.29–52.83 mg/kg. Zhai et al. [5] reported that putrescine was detected within the range ND-25.01 mg/kg. Bilgin and Gençcelep [10] observed that putrescine was detected in canned sardines, canned mackerel and marinated anchovies in the range ND-57.30 mg/kg.

Cadaverine content is a good indicator of spoilage. The results in Table 3 show that the mean value of cadaverine was 1.40 ± 0.15 mg/kg in raw eels. Cadaverine was formed, especially after the salting process. The increase in cadaverine content after salting can be explained by microbial contamination during handling and salting. Frying and sterilization did not cause changes in cadaverine content. The cadaverine content of canned eels packed in sunflower oil was stable after sterilization up to 2 months of storage and then decreased. However, in the canned eels packed in olive oil, it increased throughout storage. When spices were used in the filling medium, the cadaverine content increased during the first two months of storage but then decreased for up to 12 months. The different behavior of the cadaverine, as with other biogenic amines, may be due to the reactions with the filling medium and the different extraction processes. Prester [9] pointed out that biogenic amine content may decrease due to the loss to the filling medium and elimination prior to analysis of canned muscle. Barbosa et al. [12] observed an increase during cooking and a marked decrease during canning of skipjack tuna.

The spermidine content of raw eels was 2.36 ± 0.17 mg/kg. The content increased during the thawing-salting process. As for other biogenic amines, the increase may be due to handling and mechanical preparation during thawing-salting. Barbosa et al. [12] also observed an increase in spermidine content during thawing of skipjack tuna.

Frying caused a decrease in the spermidine content. The sterilization step did not affect the spermidine content of canned eels packed in sunflower oil or olive oil; however, this step decreased the spermidine content in canned eels packed in spiced olive oil. Veciana-Nogués et al. [13] observed a decrease in spermidine content of canned tuna after cooking and sterilization. Barbosa et al. [12] reported very variable spermidine contents in cooked and canned samples, which depended on the degree of handling. Storage for 2 months did not cause changes in spermidine content in canned eels packed in sunflower oil or olive oil; however, it increased the spermidine content of canned eels in spiced olive oil. Storage for 12 months caused a decrease in the content, mainly in canned eels packed in sunflower oil or spiced olive oil.

Different studies have pointed out that the presence of spermidine and other biogenic amines such as putrescine and cadaverine in canned samples can be considered a health hazard, as biogenic amine in combination with nitrite can lead to the formation of nitrosamines, which are known to be carcinogenic [34].

The presence of biogenic amines is not only important from a health point of view but because these substances can be used as indicators of the degree of freshness of food, as well as spoilage and sensory quality [12,13].

Zhai et al. [5] observed that the total biogenic amine content in different canned fish products ranged from 1.94 to 112.54 mg/kg, with a mean value of 46.43 mg/kg.

As noted above, salting caused an increase in total biogenic amines, probably due to handling during the process. However, during frying and sterilization, a decrease in the biogenic amine content was observed. As biogenic amines are heat resistant, the decrease may be due to a loss to the oil medium. The total biogenic amine content increased during the first two months of storage, before decreasing to 12 months. This may be due to the breakdown of protein, reactions with the filling medium and the different extraction processes. Differences between filling media were observed. The final product with the highest amine content was canned eels in olive oil. Nevertheless, the values obtained for both total biogenic amine content and biogenic amine index are very low, well below recommended levels in all samples. The FDA [28] recommended maximum values of 1000 mg/kg of total biogenic amines in fish. The European Commission [35] recommended that the maximum level of total biogenic amines should be less than 200 mg/kg in fish and fish products.

The biogenic amine index varied in a similar way. All values of biogenic amine index were below 50 mg/kg, indicative of good quality food [13], so that good quality can be concluded in all cases.

However, the quality indices depend on many factors, mainly concerning the nature of the product (species, fresh, salted, heat treatment, storage, canned). Therefore, more studies on biogenic amines are required in order to establish the limit for fish acceptability.

3.4. Statistical Analysis

Multivariate statistical techniques were used with the aim of discriminating between eels at different stages of processing and canning with different filling media. The data on all free amino acids and biogenic amines studied were included in the factorial analysis to obtain the variables that contributed most to the classification.

The comparison for free amino acids is shown in Figure 2. Two discriminating functions were statistically significant ($p < 0.05$) in each case. In all cases, the most significant function ($p < 0.05$) was F1. Five groups were clearly separated in canned eels packed in sunflower oil, each corresponding to a specific step during canning (Figure 2A). In the canned eels packed in olive oil (Figure 2B) or in spiced olive oil (Figure 2C), some overlap between fried eels, sterilized eels and canned eels stored for 2 months was observed. These results indicate that the greatest changes occurred during frying and these were lower during sterilization and the first two months of storage. The linear discriminant analysis clearly classified the canned eels stored for 12 months (Figure 2D), indicating that the greatest changes take place at this last stage. In the eels packed in sunflower oil (Figure 2A), the FAAs with the highest discriminatory power were proline, tyrosine, taurine, histidine, threonine, lysine and tryptophan. In the eels packed in olive oil (Figure 2B), hydroxyproline, arginine, tyrosine and lysine were the fatty acids with the highest discriminatory power. In the eels packed in spiced olive oil (Figure 2C), these corresponded to glutamic acid, tyrosine, histidine, isoleucine and tryptophan. In addition, discriminant analysis selected 10 free amino acids (serine, histidine, threonine, lysine, glycine, hydroxyproline, arginine, tyrosine, leucine and proline) able to discriminate the different types of canned eels packed with different filling media after storage for 12 months (Figure 2D).

Discriminant analysis was also used to differentiate raw and canned eels according to biogenic amine content (Figure 3). Two discriminatory functions were statistically significant ($p < 0.05$), indicating the ability of biogenic amines to discriminate the eel samples. Figure 3A shows the classification of eels packed in sunflower oil. Two biogenic amines, phenylalanine and putrescine, showed high discriminatory power. The percentage of correctly classified eels with these variables is 99%. Figure 3B shows the classification of eels packed in olive oil. Putrescine, tyramine and spermidine displayed the highest discriminatory power, with correct classification of 98% of the different eel products. Figure 3C shows the classification of eels packed in spiced olive oil. In this case, putrescine and spermidine also displayed a high level of discriminatory power, with correct classification of 98% of the products.

In the canned eels stored for 12 months (Figure 3D), some overlap between canned eels packed in sunflower oil and canned eels packed in spiced olive oil was observed. Phenylalanine, tyrosine and spermidine displayed a high level of discriminatory power.

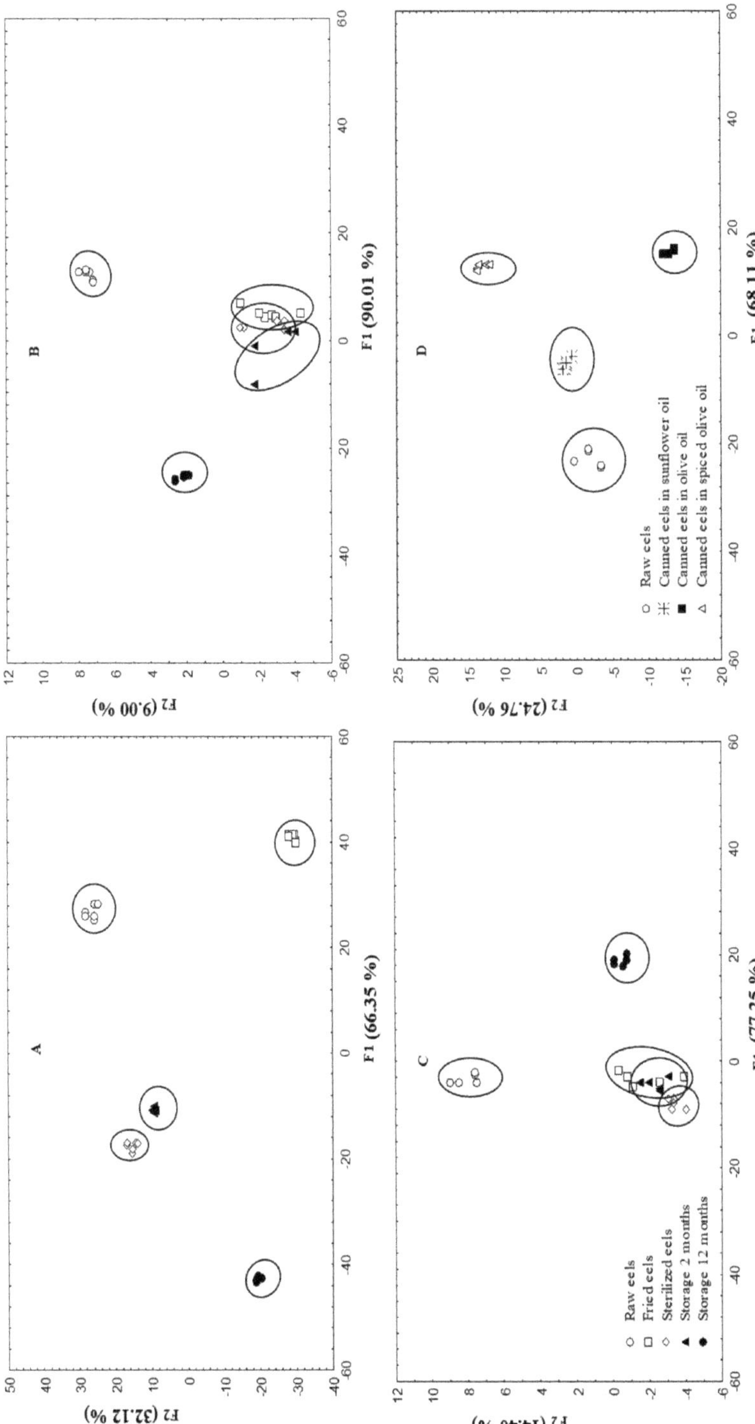

Figure 2. Scatter plot of data obtained in linear discriminant analysis (LDA) for raw and canned European eels packed in sunflower oil (**A**), olive oil (**B**) and spiced olive oil (**C**), at each step of the canning process and after storage for 2 and 12 months. (**D**) represents the plot of the three types of canned eels after storage for 12 months.

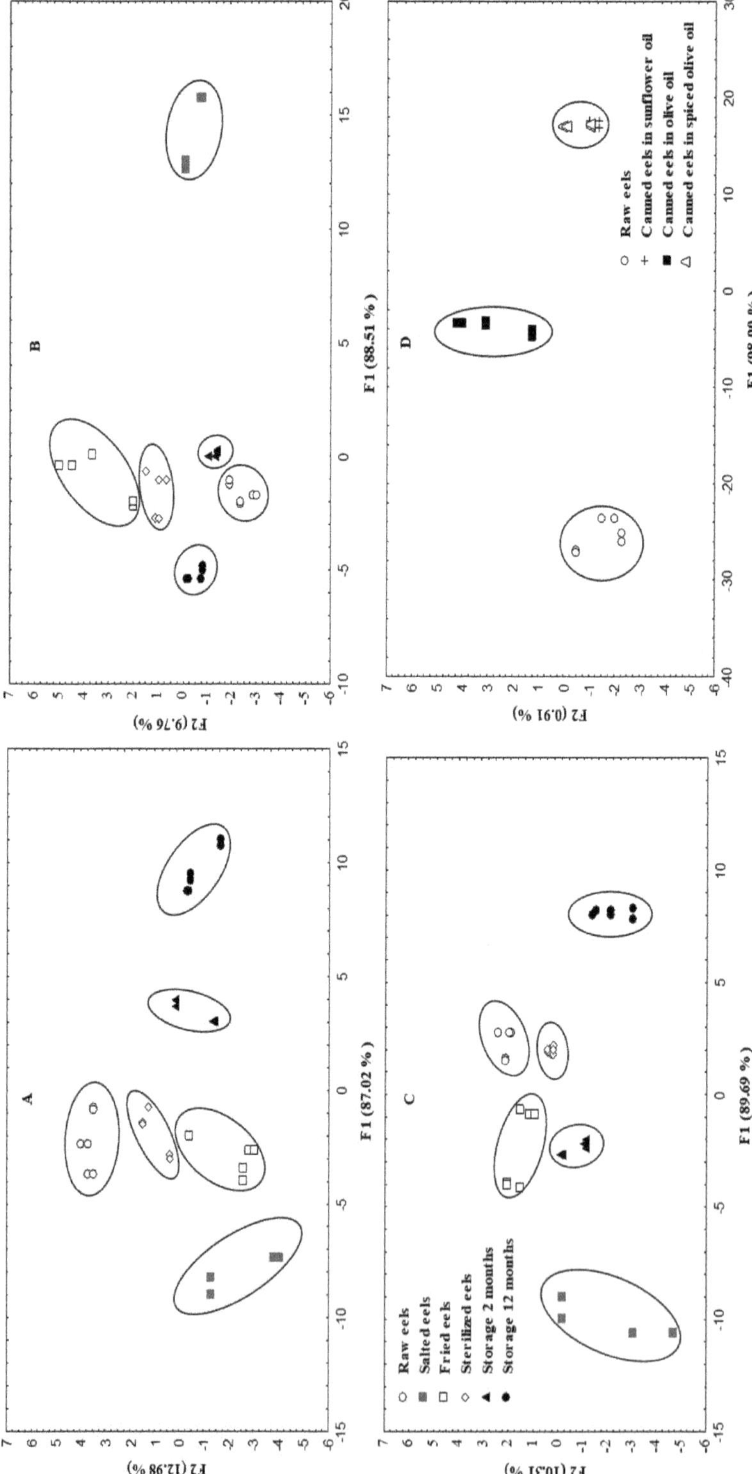

Figure 3. Scatter plot of amine content data obtained in linear discriminant analysis (LDA) for raw and canned European eels packed in sunflower oil (**A**), olive oil (**B**) and spiced olive oil (**C**), at each step of the canning process and after storage for 2 and 12 months. (**D**) represents the plot of the three types of canned eels after storage for 12 months.

4. Conclusions

The study findings showed that canned European eels are a good source of free amino acids, with a good balance of essential amino acids, and that they provide very small amounts of biogenic amines.

In general, canning, filling medium and duration of storage of European eels had significant effects on the free amino acid and biogenic amine contents of the fish. The changes varied depending on the free amino acid or biogenic amine considered.

The main free amino acids were histidine, taurine and arginine. Significant differences ($p < 0.05$) in the quantities of some free amino acids were detected at all steps of the canning process. The changes were influenced by the composition of the frying oil and filling medium and continued during room storage.

The biogenic amine contents of all samples were very low. Histamine was not detected in raw eels or in canned eels. The highest values were obtained for 2-phenylethylamine. The addition of spices to canned eels packed in olive oil appeared to reduce the biogenic amine content of the final product, except that of 2-phenylethylamine. Canned eels packed in spiced olive oil had a similar biogenic amine content as the canned eels packed in sunflower oil after 12 months of storage.

The findings provide a greater understanding of the influence of canning and filling medium on free amino acids and biogenic amine composition and can be used to improve the quality of this type of product.

Author Contributions: Conceptualization, S.M. and I.F.; methodology, S.M. and I.F.; validation, S.M. and I.F.; investigation, L.G.-L. and R.C.; data curation, L.G.-L. and S.M.; writing—original draft preparation, L.G.-L. and S.M.; writing—review and editing, S.M., I.F. and J.C.; visualization, L.G.-L.; supervision, S.M. and I.F. All authors have read and agreed to the published version of the manuscript.

Funding: This research received no external funding.

Acknowledgments: Thanks to the Xunta de Galicia for its financial support under the Consolidation and Restructuring Program of Competitive Research Units: Strategic Research Partnerships (2009/060) (CITACA Strategic Partnership ED431E 2018/07). The first author acknowledges financial support from the University of Vigo through a pre-doctoral fellowship.

Conflicts of Interest: The authors declare no conflict of interest.

References

1. Duan, W.; Huang, Y.; Xiao, J.; Zhang, Y.; Tang, Y. Determination of free amino acids, organic acids, and nucleotides in 29 elegant spices. *Int. J. Food Sci. Nutr.* **2020**, *8*, 3777–3792. [CrossRef]
2. Ruiz-Capillas, C.; Moral, A. Free amino acids and biogenic amines in red and white muscle of tuna stored in controlled atmospheres. *Amino Acids* **2004**, *26*, 125–132. [CrossRef] [PubMed]
3. Murakami, S. Taurine and atherosclerosis. *Amino Acids* **2014**, *46*, 73–80. [CrossRef] [PubMed]
4. Zarei, M.; Najafzadeh, H.; Enayati, A.; Pashmforoush, M. Biogenic amines content of canned tuna fish marketed in Iran. *Am. Eurasian J. Toxicol. Sci.* **2011**, *3*, 190–193.
5. Zhai, H.; Yang, X.; Li, L.; Xia, G.; Cen, J.; Huang, H.; Hao, S. Biogenic amines in commercial fish and fish products sold in southern China. *Food Control* **2012**, *25*, 303–308. [CrossRef]
6. Mohammed, G.I.; Bashammakh, A.S.; Alsibaai, A.A.; Alwael, H.; El-Shahawi, M.S. A critical overview on the chemistry, clean-up and recent advances in analysis of biogenic amines in foodstuffs. *Trends Anal. Chem.* **2016**, *78*, 84–94. [CrossRef]
7. Visciano, P.; Schirone, M.; Tofalo, R.; Suzzi, G. Biogenic amines in raw and processed seafood. *Front. Microbiol.* **2012**, *3*, 188. [CrossRef]
8. Silva, T.M.; Sabaini, P.S.; Evangelista, W.P.; Gloria, M.B.A. Occurrence of histamine in Brazilian fresh and canned tuna. *Food Control* **2011**, *22*, 323–327. [CrossRef]
9. Prester, L. Biogenic amines in fish, fish products and shellfish: A review. *Food Addit. Contam. Part A* **2011**, *28*, 1547–1560. [CrossRef]
10. Bilgin, B.; Gençcelep, H. Determination of biogenic amines in fish products. *Food Sci. Biotechnol.* **2015**, *24*, 1907–1913. [CrossRef]

11. Tsai, Y.; Kung, H.; Lee, T.; Chen, H.; Chou, S.; Wei, C.; Hwang, D. Determination of histamine in canned mackerel implicated in a food borne poisoning. *Food Control* **2005**, *16*, 579–585. [CrossRef]

12. Barbosa, R.G.; Gonzaga, L.V.; Lodetti, E.; Olivo, G.; Costa, A.C.O.; Aubourg, S.P.; Fett, R. Biogenic amines assessment during different stages of the canning process of skipjack tuna (*Katsuwonus pelamis*). *Int. J. Food Sci. Technol.* **2018**, *53*, 1236–1245. [CrossRef]

13. Veciana-Nogués, M.T.; Mariné-Font, A.; Vidal-Carou, M.C. Biogenic amines as hygienic quality indicators of tuna. Relationships with microbial counts, ATP-related compounds, volatile amines, and organoleptic changes. *J. Agric. Food Chem.* **1997**, *45*, 2036–2041. [CrossRef]

14. Aubourg, S.P. Loss of quality during the manufacture of canned fish products. *Food Sci. Technol. Int.* **2001**, *7*, 199–215. [CrossRef]

15. Cobo, F.; Hernández, J.S.; Vieira, R.; Servia, M.J. Seasonal downstream movements of the European eel in a Southwestern Europe river (River Ulla, NW Spain). *Nova Acta Científica Compostel.* **2014**, *21*, 77–84.

16. AOAC. Official methods of analysis of AOAC international. In *Anonymous*; AOAC International: Gaithersburg, MD, USA, 2006.

17. Franco, I.; Prieto, B.; Bernardo, A.; González Prieto, J.; Carballo, J. Biochemical changes throughout the ripening of a traditional Spanish goat cheese variety (Babia-Laciana). *Int. Dairy J.* **2003**, *13*, 221–230. [CrossRef]

18. Lorenzo, J.M.; Cachaldora, A.; Fonseca, S.; Gómez, M.; Franco, I.; Carballo, J. Production of biogenic amines "in vitro" in relation to the growth phase by *Enterobacteriaceae* species isolated from traditional sausages. *Meat Sci.* **2010**, *86*, 684–691. [CrossRef]

19. Eerola, S.; Hinkkanen, R.; Lindfors, E.; Hirvi, T. Liquid chromatographic determination of biogenic amines in dry sausages. *J. AOAC Int.* **1993**, *76*, 575–577. [CrossRef]

20. Antoine, F.R.; Wei, C.I.; Littell, R.C.; Quinn, B.P.; Hogle, A.D.; Marshall, M.R. Free amino acids in dark- and white-muscle fish as determined by O-phthaldialdehyde precolumn derivatization. *J. Food Sci.* **2001**, *66*, 72–77. [CrossRef]

21. Osako, K.; Kurokawa, T.; Kuwahara, K.; Nozaki, Y. Seasonal variations in taurine and histidine levels of horse mackerel caught in the East China Sea. *Fish. Sci.* **2004**, *70*, 1180–1182. [CrossRef]

22. Gormley, T.R.; Neumann, T.; Fagan, J.D.; Brunton, N.P. Taurine content of raw and processed fish fillets/portions. *Eur. Food Res. Technol.* **2007**, *225*, 837–842. [CrossRef]

23. Calanche, J.; Tomas, A.; Martinez, S.; Jover, M.; Alonso, V.; Roncalés, P.; Beltrán, J.A. Relation of quality and sensory perception with changes in free amino acids of thawed seabream (*Sparus aurata*). *Food Res. Int.* **2019**, *119*, 126–134. [CrossRef]

24. Chen, Y.; Ye, R.; Wang, Y. Acid-soluble and pepsin-solublecollagens from grass carp (*Ctenopharyngodon idella*) skin: A com-parative study on physicochemical properties. *Int. J. Food Sci. Technol.* **2015**, *50*, 186–193. [CrossRef]

25. Özyurt, G.; Kafkas, E.; Etyemez, M. Effect of the type of frying oil on volatile compounds of goatfish (*Upeneus pori*) during cold storage. *Int. J. Food Sci. Technol.* **2011**, *46*, 2598–2602. [CrossRef]

26. Sarower, M.G.; Hasanuzzaman, A.F.M.; Biswas, B.; Abe, H. Taste producing components in fish and fisheries products: A review. *Int. J. Food Ferment. Technol.* **2012**, *2*, 113–121.

27. Mohan, C.O.; Remya, S.; Murthy, L.N.; Ravishankar, C.N.; Kumar, K.A. Effect of filling medium on cooking time and quality of canned yellowfin tuna (*Thunnus albacares*). *Food Control* **2015**, *50*, 320–327. [CrossRef]

28. FDA (United States Food and Drug Administration). Ch.7: Scombrotoxin (histamine) formation. In *Fish and Fishery Products Hazards and Controls Guidance*, 4th ed.; Department of Health and Human Service, Public Health Service, Food and Drug Administration, Center for Food Safety and Applied Nutrition, Office of Seafood: Washington, DC, USA, 2011; pp. 113–152.

29. EEC. Council Directive (EEC) no 5556/2003/EEC 10 January 2003. Concerning A Coordinated Programme for the Official Control of Foodstuffs for 2003. 2003. Available online: http://extwprlegs1.fao.org/docs/pdf/eur39620.pdf (accessed on 9 September 2020).

30. Özogul, Y.; Özogul, F.; Gökbulut, C. Quality assessment of wild European eel (*Anguilla anguilla*) stored in ice. *Food Chem.* **2006**, *95*, 458–465. [CrossRef]

31. Rabie, M.A.; Toliba, A.O.; Sulieman, A.R.; Malcata, F.X. Changes in biogenic amine contents throughout storage of canned fish products. *Pak. J. Food Sci.* **2014**, *24*, 137–150.

32. Lorenzo, J.M.; Martínez, S.; Franco, I.; Carballo, J. Biogenic amine content during the manufacture of dry-cured lacón, a Spanish traditional meat product: Effect of some additives. *Meat Sci.* **2007**, *77*, 287–293. [CrossRef]

33. Gloria, M.B.A.; Engeseth, N.J. Bioactive amines. In *Handbook of Food Science, Technology and Engineering*; Hui, Y.H., Ed.; CRC Press–Taylor and Francis: New York, NY, USA, 2005; Volume 4, pp. 1–38.

34. Drabik-Markiewicz, G.; Dejaegher, B.; De Mey, E.; Kowalska, T.; Paelinck, H.; Vander Heyden, Y. Influence of putrescine, cadaverine, spermidine or spermine on the formation of N-nitrosamine in heated cured pork meat. *Food Chem.* **2011**, *126*, 1539–1545. [CrossRef]

35. European Commission. European Commission (EC) No 2073/2005 of 15 November 2005 on microbiological criteria for foodstuffs. *Off. J. Eur. Union* **2005**, *338*, 1–26.

Article

Oxidative Stability and Antioxidant Activity in Canned Eels: Effect of Processing and Filling Medium

Lucía Gómez-Limia [1], Nicolás Moya Sanmartín [1], Javier Carballo [1], Rubén Domínguez [2], José M. Lorenzo [1,2] and Sidonia Martínez [1,*]

[1] Área de Tecnología de los Alimentos, Facultad de Ciencias de Ourense, Universidad de Vigo, 32004 Ourense, Spain; lugomez@uvigo.es (L.G.-L.); Nmoya@alumnos.uvigo.es (N.M.S.); carbatec@uvigo.es (J.C.); jmlorenzo@ceteca.net (J.M.L.)

[2] Centro Tecnológico de la Carne de Galicia, Rúa Galicia N_4, Parque Tecnológico de Galicia, San Cibrao das Viñas, 32900 Ourense, Spain; rubendominguez@ceteca.net

* Correspondence: sidonia@uvigo.es

Citation: Gómez-Limia, L.; Sanmartín, N.M.; Carballo, J.; Domínguez, R.; Lorenzo, J.M.; Martínez, S. Oxidative Stability and Antioxidant Activity in Canned Eels: Effect of Processing and Filling Medium. *Foods* **2021**, *10*, 790. https://doi.org/10.3390/foods10040790

Academic Editor: Trine Kastrup Dalsgaard

Received: 9 March 2021
Accepted: 5 April 2021
Published: 7 April 2021

Abstract: The effect of canning and the use of different filling media (sunflower oil, olive oil, and spiced olive oil) on oxidation parameters (acidity, peroxide value (PV), and thiobarbituric acid reactive substances (TBARS) index), antioxidant capacity, and total phenol and vitamin E contents in eels was studied. A preliminary frying treatment caused a decrease in titratable acidity and an increase in TBARS, antioxidant capacity, and vitamin E in the eel muscle. During sterilization, TBARS also increased significantly. The magnitude of the changes depended on the filling medium. Storage also had a significant effect on oxidation parameters in eel muscle and in filling media. After one year of storage, the sunflower oil and canned eels packed in this oil presented higher antioxidant capacity and vitamin E content than olive oil, spiced olive oil, or canned eels packed in these oils. However, the total phenol contents were higher when olive oil or spiced olive oil were used as filling media. Despite the losses, the results show that the canning process and subsequent storage preserved a great part of the antioxidant capacity and vitamin E content of the filling medium, which is of interest to the consumer. Both sunflower oil and olive oil as filling media are of great nutritional interest.

Keywords: canned eels; filling medium; olive oil; sunflower oil; oxidation; antioxidants; total phenols; vitamin E

1. Introduction

Canned fish are products obtained from various marine species, packed in airtight containers with different filling media, and then sterilized by heat treatment. Canned fish is one of the most popular forms of fish consumption due to its properties of healthy and high nutritional value, storability, availability, and ready-to-eat nature. However, the quality of canned fish can vary with different factors such as type of fish, fresh fish quality, treatment conditions (time, temperature, and container type), filling medium, etc. [1,2]. The type of filling medium and its state strongly affect determine the quality of canned fish. The filling medium influences the nutrient content of the final product and can produce dilution and extraction of some components of the fish muscle [3]. Moreover, during and after heat treatment, complex interactions between components of fish and the filling medium take place, which also affect the characteristics of the canned fish. Different filling media such as brine, olive oil, sunflower oil, soybean oil, and other vegetable oils can be used in the packing of canned fish, with the olive and sunflower oils being the most frequently used.

During processing and storage, fish and filling mediums undergo modifications because of mechanical, thermal, hydrolytic, and oxidative degradations. Particularly, oxidative reactions during heat treatment produce a deterioration of the oil, decreasing its quality and affecting canned fish. Lipid oxidation is very important to the quality of foods, especially of those having highly unsaturated fatty acid contents, and it can cause nutrient

43

losses, unpalatable flavors and odors, shortening of shelf life, and possible production of unhealthy molecules [4,5]. In this regard, the rancidity of the oils has been linked to harmful health effects such as cancer and neurological disorders because of the production of potentially toxic compounds [6].

In addition, fish lipids are rich in long-chain, highly unsaturated fatty acids, which are highly susceptible to oxidation [7,8]. However, oxidation can be minimized in the canned fish by using filling media containing natural antioxidants such as tocopherols or polyphenols [9].

On the other hand, in order to improve the flavor, some spices and aromatic herbs can also be added to the filling medium. These spices and herbs also contribute with their antioxidant and antimicrobial capacities to the nutritional, organoleptic, and healthy characteristics of the final products [10–12].

The European eel (*Anguilla anguilla*) is an important fish species in some European and Asian countries despite the fact that the excessive capture of juveniles and the impossibility of captive breeding make its availability limited. The canning process would enable eels to be consumed throughout the year while respecting the regulations aimed at protecting the species.

The European eel is a fish with a high fat content. It has a high content of monoun-saturated (51.77%) and polyunsaturated (15.22%) fatty acids. The canning process and filling medium relate to its fatty acid profile [1]. Canning can also cause oxidation during fish processing.

Due to the considerations previously made about the peculiarities of the fish lipids and the effects of the canning process and filling media on the quality of the final product, the aim of this work was the study of the effect of canning and subsequent storage, and of different filling media (sunflower oil, olive oil, and spiced olive oil), on the oxidation processes, antioxidant capacity, and content in some antioxidant compounds such as phenols and vitamin E, both in canned eels and in the filling media.

2. Materials and Methods

2.1. Sample Preparation

European eels (*Anguilla anguilla*) were obtained from a local market (Mariscos Vivos del Grove, Plaza de Abastos) in Ourense (Galicia, Spain). All of the individual eels used in the study weighed between 200 and 600 g, and they were purchased immediately after capture. They were transferred to the laboratory, where they were eviscerated and frozen stored for a maximum of one month at −20 °C. Some randomly selected samples of the frozen eels, hereafter referred to as "raw eels," were thawed in a refrigerator at 4 °C for 12 h and processed as control samples.

In order to make canned eels, the fish were thawed in 12% brine at room temperature for 45 min. The eels were cut into slices and, after mixing to ensure product homogeneity, the slices were fried at 190 °C for 2 min in a conventional frying pan using refined sunflower or olive oil as frying media. The fried slices were then cooled (until 30 °C) and placed into 125 mL glass cans (6 or 7 slices in each). The hot filling medium (sunflower oil for eels previously fried in sunflower oil and olive oil or olive oil plus chili and pepper for eels previously fried in olive oil) was then added. The filling medium with olive oil plus chili and pepper will be referred to throughout this paper as "spiced olive oil." The drained weight of eel and the amount of oil added in each can were 50.00 ± 3.91 g and 34.9 ± 3.4 g, respectively. Next, the cans were vacuum-sealed, sterilized at 118 °C for 30 min ($F_0 = 11$), and finally cooled and stored at room temperature.

Eels were sampled raw (control), after each processing step (frying and sterilization treatment), and throughout the storage of the canned product (after 2 and 12 months of storage). The same sampling protocol was followed for the oil samples. Before analysis, the cans were opened and the filling oil was carefully drained off gravimetrically through a 3 mm pore sieve and then filtered for 3 min, thus obtaining the fish and the oil separately.

2.2. Measurement of the Acidity, Peroxide Value, and Thiobarbituric Acid Reactive Substances Index in Eels and Oils

The acidity of the eel muscle was determined according to the official method of the Association of Official Agricultural Chemists (AOAC) [13]. The sample (2.5 g) was homogenized with distilled water (25 mL) and left standing for 1 min. Subsequently, the homogenate was filtered, and the filtrate (25 mL) was titrated with NaOH (0.1 N) until pH 8.12. Titratable acidity was expressed as % of lactic acid.

The acidity degree of the oils was assessed according to the procedure described by the European Regulation [14]. Results were expressed as a percentage of oleic acid.

Fat from eel was extracted according to Bligh and Dyer [15] using a mixture of chloroform–methanol (2:1 v/v) as solvent. The peroxide value (PV) of eel fat and filling oils was determined according to standard methods for the oils analysis proposed by the American Oil Chemists' Society (AOCS) [16] and was expressed as meq. O_2/kg of the sample (fat).

The thiobarbituric acid reactive substances (TBARS) index was determined according to the Kirk and Sawyer [17] method. Malondialdehyde (MDA) is the main product resulting from the degradation of hydroperoxides generated by lipid oxidation, and the reaction between MDA and the thiobarbituric acid (TBA) generates a pink-red compound measurable by spectrophotometry using wavelengths between 532–535 nm. The results were expressed as mg malondialdehyde/kg of sample.

2.3. Determination of Antioxidant Activity

For the determination of antioxidant activity, eel or oil extracts were prepared in ethyl acetate (eel muscle: 0.5 g/mL; oils: 0.1 g/mL). The mixtures were homogenized for 1 min (IKA T25 digital Ultra-Turrax, Wilmington, EEUU), sonicated for 15 min (Branson 3510 Ultrasonic Cleaner; Branson Ultrasonics Corporation, Danbury, CT, USA), and subsequently centrifuged at 14,000$\times$ g for 10 min in an Eppendorf centrifuge 5804R (Eppendorf AG, Hamburg, Germany). The supernatants were collected, filtered, and finally stored at $-30\,^{\circ}$C until analysis.

The antioxidant activity was determined by assessing the 2,2-diphenyl-1-picryl-hydrazyl-hydrate (DPPH) free radical scavenging activity of the eel or oil extracts according to the method described by Brand-Williams et al. [18], with some modifications. When the DPPH radical reacts with an antioxidant compound, it is reduced and changes its color. Two mL of different extract concentrations (muscle of eel: 100, 50 and 12 mg/mL; oils: 25, 10 and 5 mg/mL) were mixed with 0.5 mL of a DPPH solution in ethyl acetate (0.2 mmol/L, v/v). The reaction mixture was maintained in the dark for 60 min. The color changes were read as absorbance at 517 nm in a Shimadzu UV-1800 spectrophotometer (Shimadzu Corp., Kyoto, Japan) against a control solution prepared by mixing pure ethyl acetate (2 mL) and the DPPH radical solution (0.5 mL).

The antioxidant activity was expressed in terms of IC_{50} DPPH. The IC_{50} DPPH is defined as the extract concentration (mg/mL) required to decrease the initial DPPH concentration by 50%. This value was calculated by linear regression analysis of the dose–response curve obtained by plotting the radical scavenging activity against extract concentration [19]. To calculate the IC_{50} DPPH different percentages of inhibition were calculated as follows:

$$\% \text{ Inhibition} = [(A_{\text{control}} - A_{\text{sample}})/A_{\text{control}}] \times 100,$$

where A_{control} is the absorbance of the blank at 0 time, and A_{sample} is the absorbance of the sample.

2.4. Determination of Total Phenolic Content

Samples (3 g muscle or oil) were mixed with 10 mL aqueous methanol (80/20 v/v). After extraction following the previously described procedure, total phenolic content was determined by the Folin–Ciocalteu colorimetric method assay described by Singleton and Rossi with some modifications [20]. Aliquots of 0.5 mL of extract were mixed with 2.5 mL of

Folin–Ciocalteu reagent and 2 mL of $NaCO_3$ (7.5%; p/v). The mixture was heated at 45 °C for 15 min in a water bath in darkness and left standing for 30 min before measurement of the absorbance at 765 nm. The total phenolic content was expressed as gallic acid equivalents (GAE)/100 g of sample, after interpolating the absorbance values in a standard curve obtained using different concentrations of gallic acid (5–500 mg/L).

2.5. Determination of Vitamin E Content

The vitamin E content was determined by HPLC techniques, using a ThermoFinnigan (Silicon Valley, CA, USA) chromatograph equipped with a UV/VISIBLE photodiode array detector Spectrasystem UV6000LP.

Two and a half g of eel muscle were saponified by mixing with 4 mL of 50% KOH and 6 mL of ethanol. The mix was then left for 30 min in a water bath at 80 °C under dark conditions. To avoid deterioration of vitamin E during saponification, 0.25 g of ascorbic acid was added to each sample. Then, the mix was cooled and after 10 mL of hexane and 5 mL of distilled water were added, it was homogenized in a vortex and centrifuged at $4200\times g$ for 5 min. Next, the supernatant (4 mL) was collected and dried in a stream of N_2. The dried residue was then reconstituted with 1.5 mL of HPLC quality methanol and filtered into a vial. The vitamin E was identified and quantified using a C18 reversed-phase column of 5 μm particle size, diameter 4.6 mm, and length 25 cm (Ultrasphere 5–ODS, Beckman, Fullerton, CA, USA). The detector wavelength was 294 nm. The chromatographic conditions were 1 mL/min flow for 25 min with a mixture of 96:4 (Methanol:MilliQ water). The standards were prepared by successive dissolution of α-tocopherol in methanol and saponification of the mixture as for the samples.

For the determination of vitamin E in oils, 1:10 dilutions of oil in isopropanol were used. The same HPLC system with a fluorescence (FLD-3100) detector was used. The detector was set at an excitation wavelength of 290 nm and an emission wavelength of 330 nm. A normal phase silica column (SunFire™ Prep Silica, 4.6 mm ID × 250 mm, 5 μm particle size, Waters, Milford, MA, USA) was used. The column oven was thermostated at 30 °C. A total of 10 μL of each sample and standard were injected. The chromatographic conditions used were 1 mL/min flow of hexane:isopropanol (98:2) during 15 min. Standards were prepared by successively dissolving α-tocopherol in isopropanol.

2.6. Statistical Analysis

All analyses were carried out at least in triplicate. The data were examined by analysis of variance (ANOVA), and the least-squares test (LSD) was used ($p < 0.05$) to compare the mean values. The tests were implemented using the Statistica software, version 7.1 (Statsoft© Inc., Tulsa, OK, USA). Correlations between the different parameters analyzed were determined by multiple regressions, with confidence intervals of 95% ($p < 0.05$), 99% ($p < 0.01$) and 99.9% ($p < 0.001$).

3. Results and Discussion

3.1. Effect of Processing and Filling Medium on Acidity

Acidity has important effects on muscle quality. Its changes can be used as an indicator of postmortem transformation of glycogen into lactic acid and of the degradation of muscles during storage.

The acidity in the eel muscle was calculated as titratable acidity (% lactic acid). Values of titratable acidity values of raw eels and eels packed in sunflower oil, olive oil, or spiced olive oil, at each step of the canning process and after room storage (for 2 and 12 months) are shown in Figure 1A. Titratable acidity (0.36% lactic acid in raw eels) decreased after frying and sterilization processes. This can be due to the loss of organic acids to the frying oils and the destruction of some heat-labile acids. There were no significant differences associated with the type of oil (sunflower or olive) used in the processes. The lower acidity values in canned fish compared to the raw samples could be due to the formation and accumulation of some dibasic amino acid and volatile basic nitrogenous compounds due to breakdown and proteolysis during heat treatment [21].

A)

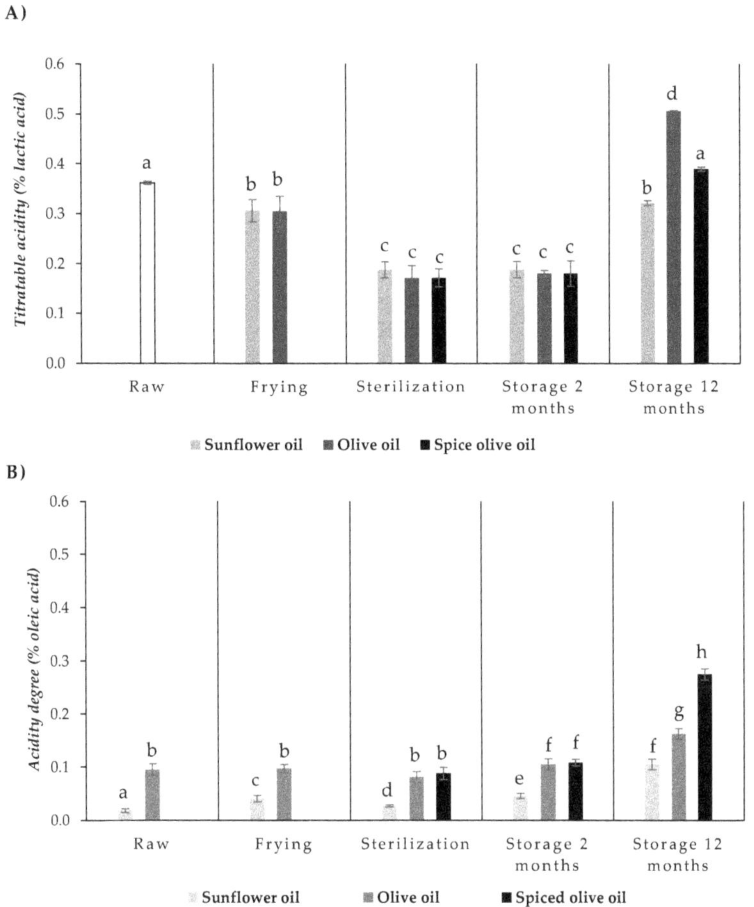

Figure 1. Titratable acidity (% lactic acid) of eels (**A**) and acidity degree (% oleic acid) of filling oils (**B**) throughout the different steps of the canning process (raw, after frying, and after sterilization process) and after 2 and 12 months of storage. Plotted values are mean of at least three determinations ± standard deviation. [a–h] Mean values with different letters are significantly different ($p < 0.05$).

During the first two months of storage, titratable acidity remained stable in all canned eels. However, the values increased after 12 months of storage. This increase may be due to the release of organic acids from the food matrix or to the formation of new acid compounds during storage. Titratable acidity values at the end of the storage were significantly ($p < 0.05$) affected by the filling medium, and the highest values were observed in canned eels packed in olive oil (0.51% lactic acid).

In order to check the variations in the acidity of the oils, this parameter was measured in raw sunflower and olive oils, and also after frying, sterilization treatment, and storage during 2 and 12 months (Figure 1B). The acidity degree of the oils depends on their content of free fatty acids. It is expressed as a percentage of oleic acid, the main fatty acid present in oils, and allows the evaluation of the behavior of the oil during heat treatment and storage. The acidity of the raw oils was 0.018 % oleic acid and 0.101% oleic acid for the sunflower and olive oils, respectively, and the differences were significant ($p < 0.05$). Usually, the susceptibility to oxidation is higher in the oils having higher initial acidity values. The acidity degree of oils is a very important parameter and depends on the intensity of the refining process carried out since during this process the free fatty acids were partially removed using different chemical or physical processes. In the case of olive oil, the acidity degree is used as a classification criterion that allows distinguishing among refined (<0.3%), extra virgin (not more than 0.8%), virgin (<2%), and olive oil (<3.3%); the oil is considered unsuitable for consumption when acidity degree is >2% [22].

Frying resulted in a significant ($p < 0.05$) increase in the acidity degree value in sunflower oil (to 0.04% oleic acid); however, no significant changes were observed in the olive oil (0.09% oleic acid) after this process. During frying, the oils undergo three different reactions—hydrolysis, oxidation, and thermal degradation. These reactions do not occur to the same extent in all vegetable oils [23]. The hydrolysis also leads to an increase in acidity, with further formation of methyl ketones and lactones that cause unpleasant smells and tastes. The high temperatures can significantly deteriorate the oils, especially if they are highly unsaturated since oxidation products are formed.

The sterilization process caused a decrease in the acidity degree of sunflower oil (0.026% oleic acid) with respect to the fried oil; however, no significant changes were observed in olive oil and spiced olive oil (0.08% and 0.09% oleic acid, respectively), compared to fresh olive oil. During the frying and sterilization processes, the composition of the fried or packed food and that of the oil/fat used as frying or filling medium change continuously, mainly due to the alteration of the oil, but also due to changes in the food, and reactions of interaction between oil and food. A higher degree of acidity indicating a higher release of fatty acids from triglycerides means a greater degree of deterioration due to the heat damage. Additionally, the decrease in the acidity values seems to be the result of free fatty acid degradation via oxidation processes.

Usually, sunflower oil has a higher content of α-tocopherols, but in our case, these did not seem to protect the oil against oxidation, probably due to the higher content of easily oxidizable linoleic acid [24]. In the case of olive oil, heat treatments caused fewer changes in the acidity degree than in sunflower oil. This trend has also been pointed out in other studies since olive oil is rich in antioxidant compounds such as phenolic compounds and tocopherols [25,26]. On the other hand, olive oil has a higher concentration of monounsaturated fatty acids, such as oleic acid, which are more resistant to degradation than polyunsaturated fatty acids, such as linoleic acid. Sunflower oil had greater amounts of polyunsaturated fatty acids (PUFAs); thus, taking into account that oxidation susceptibility is correlated exponentially with the number of unsaturation of fatty acids [5], this could partially explain our results.

Storage significantly ($p < 0.005$) increased the acidity degree in the three oils studied. This increase was higher in the spicy olive oil after 12 months of storage (to 0.27% oleic acid). During storage, many compounds migrate from fish to filling medium and vice versa. The balance established depends on the type of fish and previous treatments of it, the type of filling medium, the canning technology, and storage conditions. Spices and condiments can sometimes act as pro-oxidants in canned food [10].

In the present study, the spices used in the preparation of spiced olive oil, pepper, and chili, can also undergo an oxidation process during processing and storage, which may contribute to the increase in the degree of acidity of the spiced olive oil. Additionally, during storage, the spices could release some components and these can migrate from spices to the oil and increase acidity to the filling medium [27].

3.2. Effect of Processing and Filling Medium on Peroxide and TBARS Values

The peroxide value is often used to measure the deterioration of oils. This value is related to treatments and storage conditions (oxygen, light, temperature, metals, enzymes, presence of antioxidants or pro-oxidants, fatty acid composition, the use of oxygen-permeable packages, etc.) [28]. It allows estimating the degree of oxidation of fatty acids.

Thermal treatment and high storage temperatures can increase the susceptibility of unsaturated fatty acids toward oxidation resulting in losses of these compounds. Dimerization and polymerization are important reactions in the thermal oxidation in oil [29]. The heat temperature and time, filling oil or antioxidant contents, affect the hydrolysis, oxidation, and polymerization of the fat. The degree of fat oxidation in fish can also be determined by assessing the thiobarbituric acid reactive substances (TBARS) values that evaluate the formation of secondary oxidation products, especially malonaldehyde (MDA) and its further transformation and/or reaction with the processed products. Lipid oxidation consists of three stages—initiation, propagation, and termination. The peroxides that are normally formed as primary products can subsequently undergo scission to form secondary oxidation products such as carbonyls, alcohols, hydrocarbons, and furans through different reactions such as cyclization, hydrogen abstraction, addition reaction, recombination, scission, and polymerization [30]. The aldehydes, such as hexanal malondialdehyde (MDA), are one of the most abundant products found [31].

PV (expressed as meq. O_2/kg of the sample) and TBARS (mg malonaldehyde/kg of the sample) of raw and canned eels sampled at each stage of the canning process (after frying, after sterilization process, and after room storage for 2 and 12 months) are shown in Figure 2.

Peroxide values of raw eels ranged from 1.71 to 2.30 meq. O_2/kg. These data were consistent with those reported by Selmi et al. [32] in fresh tuna (2.50 meq. O_2/kg) but lower than the values observed by these same authors in sardine (8.14 meq. O_2/kg). In that study, fresh sardine showed higher PUFAs content and a lower monounsaturated fatty acids (MUFAs) fraction than fresh tuna. The European eel has a high content of monounsaturated fatty acids (51.77%), followed by saturated fatty acids (SFAs) (32%) and polyunsaturated fatty acids (PUFAs) (15.22%) [1].

It has been reported that the maximum level of TBARS values indicating good quality of the fresh fish during storage is 1–2 mg MDA/kg [33]. In the present work, the TBARS values observed in raw eels were very lower (0.17 mg MDA/kg) than those indicated figures. Ehsani and Jasour [34] found even lower values in rainbow trout fillets (0.06 mg MDA/kg) such as Naseri et al. [35] in silver carp muscle (0.01 mg MDA/kg). However, Kong et al. [36] reported higher TBARS values in pink salmon muscle (4.59 mg MDA/kg). These differences could be due to different concentrations of fat in the fish muscle [1] and different storage times and conditions before analysis.

Frying did not have a significant effect on PV when sunflower oil or olive oils were used as frying media. Peroxides are unstable under frying conditions, and the use of oil for frying does not lead to substantial increases in peroxide values because peroxides decompose spontaneously above 150 °C to form secondary oxidation products [37]. However, frying caused a significant increase in TBARS values. No significant differences were observed between TBARS values of the samples fried in sunflower oil (1.19 mg MDA/kg) and the samples fried in olive oil (1.06 mg MDA/kg).

The sterilization process did not increase PVs either. The PV levels were lower than those observed by El-Shehawy and Farag [33] in canned tuna, canned sardine, and canned mackerel in different filling media.

During the treatment of sterilization, TBARS values increased significantly. In this case, the increase was higher in canned eels packed in sunflower oil (2.10 mg MDA/kg) or in spiced olive oil (2.57 mg MDA/kg) than in canned eels packed in olive oil (1.34 mg MDA/kg). Selmi et al. [32] reported that cooking increased PV in sardine, but canning did not cause significant changes. In this same line, Uriarte-Montoya et al. [38] indicated that the canning did not have a significant effect on the oxidation in Pacific sardine (*Sardinops sagax caerulea*).

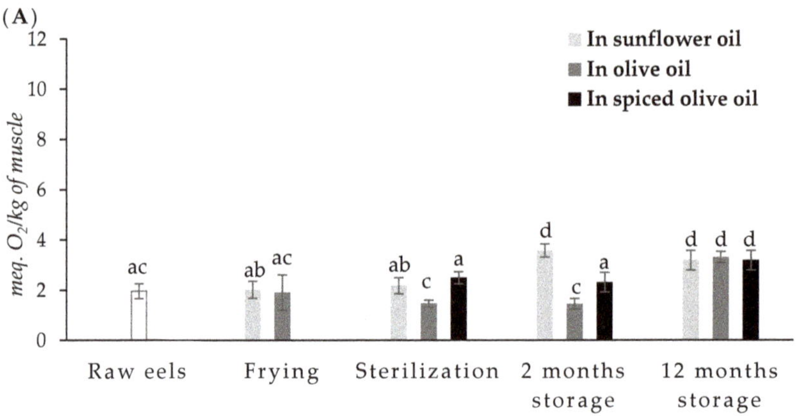

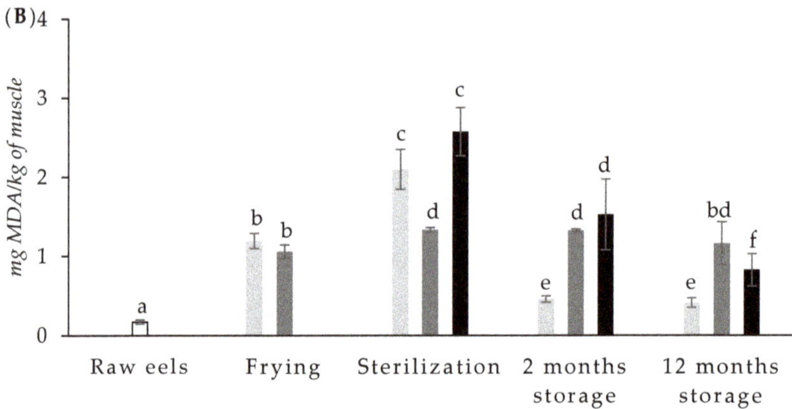

Figure 2. Peroxide values (PVs) (meq. O_2/kg of the sample) (**A**) and thiobarbituric acid reactive substances (TBARS) values (mg malonaldehyde/kg of the sample) (**B**) of raw and canned European eel in each step during the canning process and at 2 and 12 months of stored. [a–f] Mean values with different letters are significantly different ($p < 0.05$).

The high treatment temperatures or long treatment times cause an increase in secondary oxidation products, such as malonaldehyde [39]. Naseri et al. [35] did not observe significant differences between the TBARS values of raw silver carp and precooked samples before the canning process. However, they observed a significant increase in TBARS values during sterilization (canning) in silver carp packed in sunflower oil and soybean oil, but not in canned silver carp packed in olive oil. Kong et al. [36] observed a slight increase during the first 10 min of heating, followed by a significant decrease in TBARS values as heating progressed.

Medina et al. [40] reported significant differences between TBARS values in canned tuna when different filling media were used. TBARS values were higher in canned tuna muscle using brine as the filling medium than in canned tuna packed in extra virgin olive oil. These authors suggested that the antioxidant activity of phenolic compounds of olive oil could be responsible for these differences. Bilgin and Gençcelep [41] also found significant differences among TBARS values of fish canned in different filling media.

Canned storage is normally necessary to produce satisfactory textural and optimal palatability of canned fish. Although the shelf life of a canned fish is between one and five years, storage time can be variable. In this study, the effect of short storage (two months) and longer storage (12 months) was evaluated. Storage had a significant effect on PV in canned eels. The PV increased significantly after two months of storage in canned eels packed in sunflower oil, and then it was stable until 12 months of storage. In the case of canned eels packed in olive oil or in spiced olive oil, PV remained stable during the first two months of storage but significantly ($p < 0.05$) increased after 12 months of storage. No significant differences were observed between the three types of canned eels at 12 months storage. Selmi et al. [32] reported an increase in PV levels after six months of storage in canned tuna. The high temperatures during processing can accelerate oxidation, but filling medium can dissolve oxidation products, or these can be transformed into secondary oxidation products, reducing their concentration in the fish [29]. It has been pointed out that at high temperatures, the initial hydroperoxides formed exist only briefly and will be quickly decomposed into various volatile and nonvolatile products [42]. Moreover, lipid oxidation products can interact with proteins, which could explain in some cases a decrease in primary and secondary oxidation products [43]. Heat and storage can also prompt lipid hydrolysis. The free fatty acids, which appear as a result of hydrolysis, can further undergo oxidation reactions. A direct correlation between lipolysis and lipid oxidation of seafood products has been reported [44]. These free fatty are more susceptible to undergo oxidative reactions in order to form primary and secondary oxidative products [45].

In addition, the filling medium used can determine the peroxide value of the canned fish. Leung et al. [46] found in salmon subjected to different treatments that frying, baking, or using old oil led to an increase in peroxide values, while cooking did not produce significant changes. Al-Saghir et al. [47], in salmon and trout, reported that during frying, salmon fried in olive oil or corn oil increased its peroxide value, but the increase was much greater in baked trout and fried trout in sunflower oil. However, Talab [43] observed that different treatments produced a decrease in peroxide values in carps. The variation in the hydroperoxides formed in the fried and canned samples can be attributed to the influence of the frying or filling medium. In order to evaluate the effect of the filling medium, the PV in the oils was also determined.

Storage during two months caused a decrease in TBARS values in canned eels packed in sunflower oil and canned eels packed in spiced olive oil; this decrease was more important in canned eels packed in sunflower oil. TBARS values continued to decrease in canned eels packed in spiced olive oil until 12 months of storage. In canned eels packed in olive oil or sunflower oil, TBARS values did not undergo significant changes throughout the remaining 10 months of storage. The decrease of the TBARS values can be due to the disappearance of the malondialdehyde through reactions with amines, nucleosides, nucleic acids, amino-containing phospholipids, proteins, or other aldehydes that are also by-products of lipid oxidation [41]. The further oxidation of primary products and the formation of carboxylic acids and other compounds that are not reactive to the 2-thiobarbituric acid can be also the cause of the decrease of the TBARS values during storage [5,48]. The changes in TBARS values may also be explained by the different interactions between fish muscle and filling medium, by the diluting effect of the filling medium on secondary oxidation products, and/or by the effect of the thermal treatment.

PV (meq. O_2/kg of the sample) and TBARS (mg MDA/kg of the sample) of oils sampled at each stage of the canning process (raw, after frying, after sterilization process, and after room storage for 2 and 12 months) are shown in Figure 3.

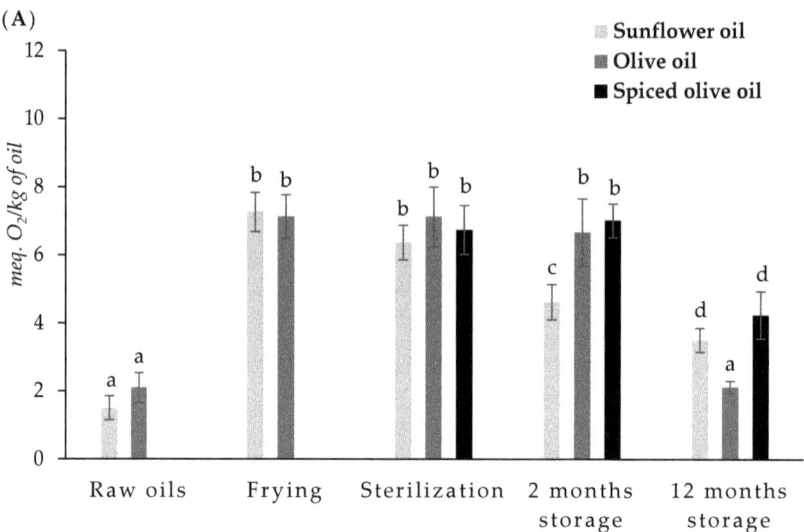

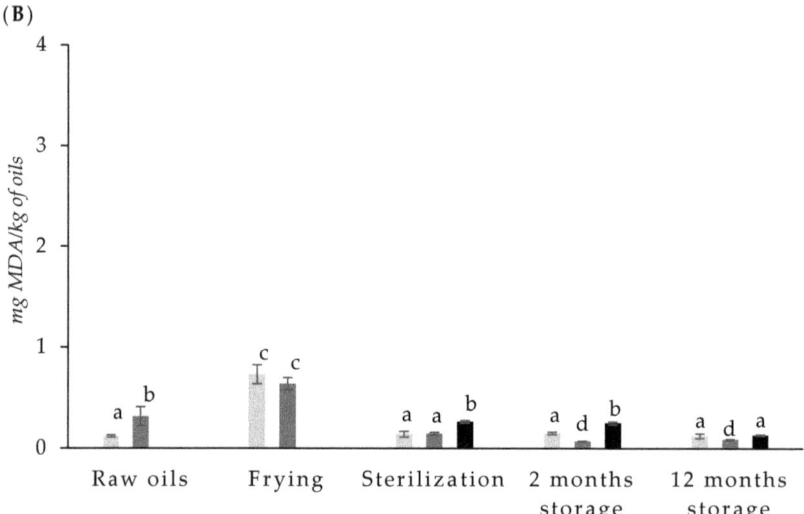

Figure 3. Peroxide values (PVs) (meq. O_2/kg of the sample) (**A**) and thiobarbituric acid reactive substances (TBARS) values (mg malonaldehyde/kg of the sample) (**B**) of oils in each step during the canning process and at 2 and 12 months of stored. [a–d]- Mean values with different letters are significantly different ($p < 0.05$).

Peroxide values in raw oils were 1.50 and 2.49 meq. O_2/kg in sunflower oil and olive oil, respectively. The raw oils presented TBARS values of 0.12 mg MDA/kg and 0.31 mg MDA/kg for sunflower and olive oil, respectively.

Frying caused a significant increase in PV with respect to crude oils (7.26 and 7.13 meq. O_2/kg of oil in sunflower and olive oils, respectively). However, the sterilization process did not cause significant changes in PVs of the oils.

TBARS values also increased after frying, with the increase being higher in sunflower oil (0.73 mg MDA/kg) than in olive oil (0.64 mg MDA/kg). However, sterilization treatment resulted in a significant decrease in the TBARS index in all oils. In this case, this decrease was higher in sunflower oil and in olive oil than in spiced olive oil.

Storage produced a decrease in PV in sunflower oil. In the case of olive oil or spiced olive oil, no significant differences were observed after two months of storage. However, after 12 months of storage, a decrease in PV was also observed in olive and spiced olive oil. The lowest PVs were observed in olive oil after 12 months of storage.

No significant changes in TBARS values of sunflower oil were observed during storage. In the case of olive oil, TBARS values decreased after two months of storage. On the other hand, in the spicy olive oil, no changes were observed after two months of storage, but there was a decrease after one year of storage in relation to the values obtained after sterilization.

Alhibshi et al. [49] pointed out that when the peroxide values are between 30 and 40 meq. O_2/kg the rancid flavor is clearly noticeable. In the samples under study, these indicated values were not observed. Different studies are carried out on the evolution of the peroxide value in oils after heat treatment. Naz et al. [50] found that the peroxide value increased in olive oil during frying, and this increase was proportional to the treatment time. Alajtal et al. [51] also observed an increase in the peroxide value in sunflower and olive oils after frying. The increase in the peroxide value during frying and sterilization is due to oxidation processes. A possible explanation for the peroxide values measured during the frying and sterilization process is that the rate of degradation is similar to the rate of formation, increasing TBARS values.

The subsequent decrease during storage occurs when the peroxides are transformed into other chemical compounds, such as aliphatic carbonyls [5,28]. Since the peroxides formed are not stable compounds, the peroxide value is not always correlated with the degree of oxidation of oils [5].

The results seem to be related to the fatty acid composition of oils. When oils are exposed to high temperatures, the polyunsaturated fatty acids are affected, and the oils with a high n-3 PUFA level experience a greater effect on TBARS values than oils with a low n-3 PUFA level. Heat treatments cause the oil to experience a series of chemical reactions such as oxidation, hydrolysis, and polymerization [52].

Caponio et al. [53] evaluated the effect of the replacement of refined oils with extra virgin olive oil in bakery products rich in fat. They observed that the evolution of the oxidation levels in the analyzed samples during storage was related to the type of oil used. The use of extra virgin olive oil led to significantly lower values of hydroperoxides, ultraviolet absorption constants, triacylglycerol, oligopolymers, and oxidized triacylglycerols.

3.3. Effect of Processing and Filling Medium on Antioxidant Capacity

Fish tissues can contain different antioxidant components such as enzymes, carotenoids, vitamin E, vitamin C, peptides, amino acids, and phenolic compounds that contribute to their antioxidant defense mechanism. These antioxidants can inhibit the initiation and propagation steps of lipid oxidation. They act acting as free radical scavengers and metal ion chelators. They can have multiple effects, and their mechanisms of action are therefore frequently difficult to interpret. These components are part of the cell plasma, mitochondria, and cell membranes [54]. However, these endogenous antioxidants are consumed sequentially after the death of the fish. Moreover, in the case of canned fish, some of these antioxidant components present in the raw fish can be reduced during processing. In canned fish, the oxidative stability of fish lipids is variable depending on the antioxidant content and stability of the filling medium. In the case of oils, their stability is particularly related to the content of different components, such as polyphenols and tocopherols, which can act as radical scavengers. However, in the case of refined oils, some antioxidant components are lost during the refining operations. Additionally, spices and herbs have been used as a natural source of antioxidant and antimicrobial compounds [11,12].

The DPPH radical is commonly used as substrate in estimating antioxidant activity because of the ability of the antioxidant compounds to reducing the DPPH free radical.

The DPPH radical scavenging activity (expressed in % inhibition of the DPPH radical) of different extracts of muscle eels (100, 50, and 12 mg/mL), both raw and after each step of processing and storage, and canned in different filling media, is shown in Supplementary Materials (Figure S1).

The inhibition at 50% (IC_{50}) values are shown in Table 1.

Table 1. The inhibition at 50% (IC_{50}; mg/mL) values of European eel and filling oils after each step during the canning process and after 2 and 12 months of subsequent storage.

Steps	Eel Muscle		Oils	
Raw	—	107.05 ± 9.63 [a]	Sunflower oil	7.88 ± 0.17 [a]
			Olive oil	13.09 ± 1.08 [b]
Frying	In sunflower oil	25.24 ± 5.39 [b]	Sunflower oil	7.29 ± 0.08 [a]
	In olive oil	28.39 ± 4.94 [b,d]	Olive oil	13.96 ± 1.07 [b]
Sterilization	In sunflower oil	36.43 ± 4.32 [c,e]	Sunflower oil	7.51 ± 0.48 [a]
	In olive oil	34.71 ± 1.42 [c,d]	Olive oil	12.88 ± 0.18 [b]
	In spiced olive oil	30.28 ± 1.77 [c,d]	Spiced olive oil	16.85 ± 0.62 [c]
2 months storage	In sunflower oil	31.77 ± 2.83 [c,d]	Sunflower oil	8.57 ± 0.48 [d]
	In olive oil	38.31 ± 3.32 [c,e]	Olive oil	12.58 ± 0.44 [b]
	In spiced olive oil	35.03 ± 6.02 [c,e]	Spiced olive oil	16.11 ± 0.28 [c,f]
12 months storage	In sunflower oil	36.77 ± 5.44 [c,e]	Sunflower oil	10.34 ± 0.24 [e]
	In olive oil	40.67 ± 0.53 [e,f]	Olive oil	15.75 ± 0.67 [f]
	In spiced olive oil	43.54 ± 1.59 [f]	Spiced olive oil	15.92 ± 0.26 [f]

Mean values of at least three determinations $\pm$ standard deviation. [a–f] Mean with different superscripts in the same column were significantly different ($p < 0.05$); IC_{50}: extract concentration (mg/mL) required to decrease the initial 2,2-diphenyl-1-picryl-hydrazyl-hydrate (DPPH) concentration by 50%.

The antioxidant activity of the muscle eel and oil extracts is concentration dependent, and lower IC_{50} values indicate better protective action. The antioxidant capacity of raw eels was low (IC_{50} = 107.05 mg/mL). At the concentration of 100 mg/mL, the percentage of inhibition was 40.48%. The antioxidant capacity of plants has been extensively studied. However, research on the antioxidants from animal sources is limited so far.

In the case of fish, the few existing studies that have been carried out mainly focused on protein hydrolysates of fish. On the other hand, the antioxidant capacity can vary significantly between different fish particularly due to factors such as diet and environmental conditions [55]. A high percentage of fat in some fish, such as eel, can imply the presence of high content of antioxidant compounds. In addition, endogenous antioxidants are sequentially degraded after the death of the fish [56], and the rate of degradation depends on the packaging method and on the storage temperature and time. Therefore, the storage conditions prior to canning have a significant influence on the antioxidant compounds in raw fish and on the final content of oxidation products.

Frying caused a significant increase in antioxidant capacity probably due to the incorporation of some components of the oil into the muscle. In addition, it has been pointed out that the Maillard reaction products and the peptides generated during frying have antioxidant activities [57]. The increase was similar in eels fried in sunflower oil (IC_{50} = 25.24 mg/mL) and eels fried in olive oil (IC_{50} = 28.39 mg/mL). At the highest concentration of the extract (100 mg/mL), the percent of inhibition was 87.9 and 93.6% in eels fried in sunflower oil and olive oil, respectively.

The sterilization process caused a loss of antioxidant capacity in the eel muscle, although it was only significant in eels packed in sunflower oil. These losses can be due to the leaching of soluble components, such as vitamins and proteins, into filling medium and thermal damage during heat treatments [58]. Losses of antioxidant capacity during storage were low. A significant decrease was only observed after twelve months of storage in eels packed in olive oil and spiced olive oil.

The filling media influenced the nutritional quality and antioxidant capacity of canned fish. Medina et al. [40], in canned tuna, found protective effects against lipid oxidation of the extra virgin olive oil rich in phenols and also of the soybean oil rich in tocopherols. Medina et al. [59] investigated the ability of polyphenols extracted from extra virgin olive oil to inhibit lipid oxidation in canned tuna. They observed that a high concentration (400 ppm) of these polyphenols was an effective antioxidant. However, a low concentration (100 ppm) promoted hydroperoxide formation and decomposition. After one year of storage, canned eels packed in sunflower oil presented higher antioxidant capacity than canned eels packed in olive oil or spiced olive oil. Mohan et al. [2] reported that canned tuna packed in sunflower oil as the filling medium offered higher protection compared to other oils such as groundnut oil and coconut oil, possibly due to its protective antioxidant activity. These results can be explained by a high content of natural antioxidants such as tocopherols in sunflower oil.

The DPPH radical scavenging activity of different extracts of oils (25, 10, and 5 mg/mL) is shown in Supplementary Materials (Figure S2).

Concentration-dependent scavenging activity was also found for the studied oils. The resulting DPPH inhibition percentages at the highest concentration (25 mg/mL) were 87.8 and 78.2 % in the raw sunflower and olive oils, respectively. The IC_{50} values were 7.88 and 13.09 mg/mL in raw sunflower and olive oils, respectively. The higher antioxidant activity of the raw sunflower oil can be explained by its richness in tocopherols.

Antioxidant capacity remained stable during frying. No significant differences were detected between the antioxidant capacity of the raw and sterilized oils, except in the case of spiced olive oil. In this case, antioxidant capacity decreased during the sterilization process. As noted above, spices and condiments can sometimes act as pro-oxidants in canned food [10].

In the case of sunflower oil and olive oil, the antioxidant capacity decreased during storage, while in spiced olive oil it remained constant.

3.4. Effect of Processing and Filling Medium on Vitamin E Content

Vitamin E is a common term designing tocopherols and tocotrienols. Vitamin E can be found in eight chemical forms (α, β, γ, δ-tocopherol and α, β, γ, δ-tocotrienol) having different levels of biological activity. α-Tocopherol is the predominant form of vitamin E in fish muscle, mainly in marine fish. It has the highest bioavailability and is the most important lipid-soluble antioxidant.

Due to this reason, the α-tocopherol content was determined in the eel muscle after each treatment, and results are shown in Figure 4A.

The α-tocopherol content ranged from 0.69 to 0.88 mg/100 g in raw eels. The vitamin E content in fish can differ widely between species and it is also affected by the season, genetic differences, tocopherols in feed, maturation stage, or storage conditions [4,60]. Additionally, vitamin E content in fish seems to be directly related to lipid content and n-3-PUFA content [4].

The vitamin E content of fried samples increased significantly ($p < 0.05$) when compared to the raw eels. This is probably due to the enrichment from the oil in which eels were fried. The increase was higher in eels fried in sunflower oil than in eels fried in olive oil.

Sterilization process did not have a significant effect on the vitamin E content of fish, even though canned eels packed in spiced olive oil showed higher average values than the fried eels. Ersoy and Özeren [61] reported a significant increase in tocopherol content in African catfish after different treatments such as baking, grilling, microwaving, and frying. Merdzhanova et al. [60] also found a significant increase in tocopherol content after frying in Black Sea horse mackerel.

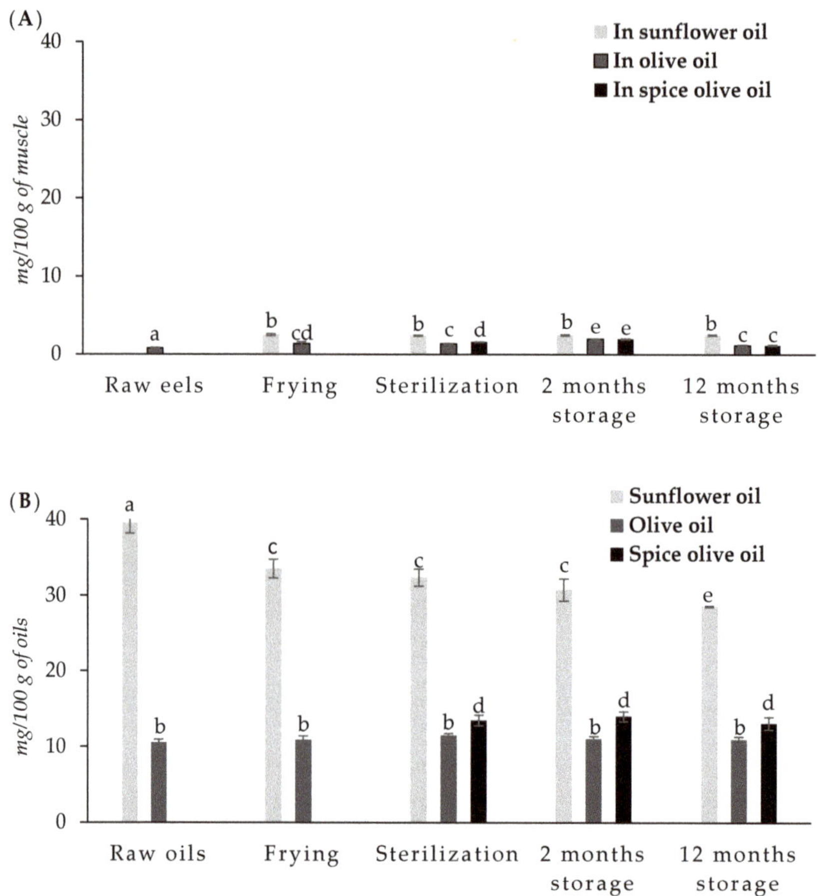

Figure 4. Vitamin E (mg/100 g) of European eel (**A**) and filling oils (**B**) after each step during the canning process and after 2 and 12 months of subsequent storage. [a–e] Mean values with different letters are significantly different ($p < 0.05$).

Vitamin E is sensitive to oxygen, light, and temperature. However, fried and sterilized eels can be enriched in vitamin E due to uptake from the oil and the absorbed quantity depends on the type of oil used in these operations. Sunflower oil has a higher content of vitamin E than olive oil; hence, the highest values in eels fried and packed in sunflower oil. Chilli and pepper can also have a high content of vitamin E [62], which could explain the higher vitamin E content of the samples packed in spiced olive oil.

Vitamin E content remained higher and stable during storage in canned eels packed in sunflower oil. This higher vitamin E content may be related to the higher antioxidant capacity and lower TBARS values of eels packed in sunflower oil at 12 months of storage. However, vitamin E increased during the first two months of storage and then decreased in canned eels packed in olive oil and spiced olive oil. Losses in vitamin E content during storage appear to be due to degradation processes enhanced by factors that influence this degradation such as heat temperatures, time, and exposure to light and oxidative conditions. However, vitamin E is quite stable if foods are adequately protected from conditions favoring lipid oxidation.

Medina et al. [40] reported that there is an inhibition of peroxide decomposition in canned tuna due to polyphenols and tocopherols of filling oil (extra virgin olive oil and refined olive oil). However, they also reported that the protective effect of the polyphenols was higher than that of tocopherols.

The amounts of vitamin E absorbed by fish during frying and/or canning processes depends on the type of oil used [63], and on the number of antioxidants present in the oil. With the purpose of verifying this, the α-tocopherol content was determined in the oils during the frying and canning. Results are summarized in Figure 4B.

Sunflower oil showed a high content of vitamin E (39.51 mg/100 mL) in the present study. These values agree with the concentration found in sunflower oil by Ortiz et al. [64] (35.9 mg/100 mg). Olive oil presented lower content (10.56 mg/100 mL). High variability in the vitamin E content was observed in vegetable oils, depending on different factors such as botanic species, edaphoclimatic environmental conditions, agronomic practices, and extraction procedures [64].

Vitamin E levels significantly decreased in sunflower oil during frying, coinciding with the increase of vitamin E content in the fried eel muscle. Thermal degradation during heating, however, should also contribute to this decrease. No significant changes were observed in the olive oil during the heating process. Quiles et al. [64] reported that the decrease in tocopherol with the frying time depends on the type of oil and time of frying. They reported that olive oils are more stable than sunflower oil during frying processes.

The sterilization process did not cause significant changes in the vitamin E content of the sunflower oil or olive oil. However, this treatment caused an increase in spiced olive oil. Spices such as pepper and chili can contain significant quantities of fat-soluble vitamins, such as tocopherols (mainly vitamin E) [65], which can be dissolved in the oil during the sterilization treatment, thus increasing their concentrations in the oil.

Storage did not modify vitamin E content in olive oil or spiced olive oil; however, it slightly but significantly decreased the vitamin E content in sunflower oil.

3.5. Effect of Processing and Filling Medium on Total Phenolic Content

Vegetable oils also contain other natural antioxidants such as phenolic compounds that strongly contribute to their antioxidant capacity. The Folin–Ciocalteau assay is often used to determine the total phenolic content of food [66–68].

The total phenol contents were also quantified in fresh fish and raw oils, and in both fish and oils after each stage of processing and storage. Phenols were not detected in either the eels or the sunflower oil. Phenolic compounds can be completely destroyed during the refining process of the oils [69]. Matthaus and Spener [70] reported that carotenoids and phenolic compounds are removed almost totally, while vitamin E-active compounds and phytosterols are reduced by about 10 to 40%.

On the other hand, the phenols present in animal tissues come mainly from food.

Total phenolic contents in olive oil and spiced olive oil are shown in Figure 5.

The number of total phenol compounds in olive oil was 13.33 mg GAE/100 g.

According to the literature, the total phenol content in vegetable oils can vary within a wide range depending on the cultivar, environmental edaphoclimatic conditions, cultural practices, and ripening stage of the olives [67,71]. The refining process of virgin olive oil can eliminate practically all the phenolic compounds. In the present study, the olive oil used was a mix of refined olive oils and virgin olive oils, and therefore, phenolic compounds are found.

The frying and sterilization processes caused high losses of phenolic compounds. During sterilization, the highest losses occurred in spiced olive oil. These losses could be due to thermal oxidation reactions, hydrolysis, and polymerization, or to covalent linking between oxidized phenols and proteins or amino acids [25,72]. On the other hand, it has been pointed out that when vegetables are present during heating treatment, they also contribute to the leaching of substances that are released from disrupted cell walls and subcellular compartments [68].

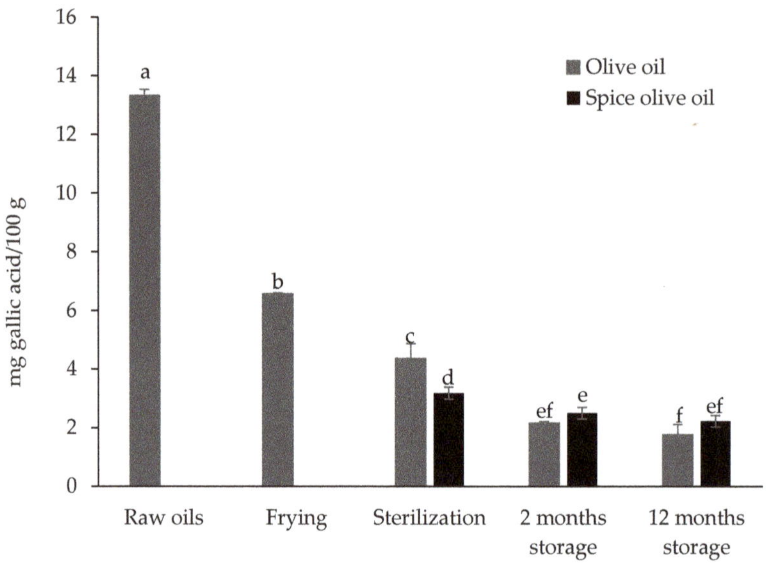

Figure 5. Total phenols (mg gallic acid/100 g) of olive oil and spiced olive oil after each step during the canning process and after 2 and 12 months of subsequent storage. [a–f] Mean values with different letters are significantly different ($p < 0.05$).

Baiano et al. [73] reported that the phenolic concentration was higher in the extract of the unflavored olive oil than in olive oil flavored with lemons, hot pepper, oregano, rosemary, and garlic. They pointed out that this could be explained based on interactions between the olive oil and the flavoring agents during the extraction phase, which are responsible for the formation of bonds between phenolics and components of fruits, spices, and herbs. This could explain the lower content of phenols detected in the spiced oil in the present work.

A decrease in the total phenolic content was also observed after the first two months of storage in olive oil or spiced olive oil, but from this point, the contents remained stable until 12 months of storage.

4. Conclusions

The different processes during eel canning led to changes in the parameters reflecting the oxidation and in the antioxidant properties, both in the eel muscle and in the filling oils. However, canned eels packed in sunflower, olive, or spiced olive oils presented acceptable values for the parameters indicating fat oxidation after 12 months of storage.

Peroxide and TBARS values were determined by the filling medium and the storage time. At the end of storage, the peroxide indices were similar in the three types of preserves, while the TBARS values were higher in the preserves in olive oil.

Canning involves high heat treatments during precooking and sterilization steps, which promote changes in the fish. The fish quality after processing and storage is a result of a combination of the phytonutrient extractability, loss of compounds by degradation and leaching, and interaction between the fish and the filling medium. Canning caused the exchange of different components between the eel muscle and the filling oil used, increasing antioxidant capacity and vitamin E content in eel muscle after canning. The magnitude of these exchanges was dependent on the filling medium.

The antioxidant capacity of fresh eel is very low, but it increased significantly in canned food due to the filling oil. After one year of storage, canned eels packed in sunflower oil and the sunflower oil used as filling medium presented higher antioxidant capacity and vitamin E content than canned eels packed in olive oil or spiced olive oil and their respective filling oils. However, canned eels packed in olive oil or spiced olive oil and their filling oils presented higher total phenol contents.

The canning process and subsequent storage preserved a great part of the antioxidant capacity and vitamin E content of the filling medium, which is of interest to the consumer. All oils used as filling media (sunflower oil, olive oil, and spiced olive oil) are of great nutritional interest, and they could be used in canned eels. Therefore, the decision on the oil to be used must be based mainly on its availability and price, and the market to which the product may be destined.

Supplementary Materials: The following are available online at https://www.mdpi.com/article/10.3390/foods10040790/s1, Figure S1: DPPH radical scavenging activity of different extracts (100, 50 and 12 mg/mL) of muscle eels packed in sunflower (**A**), olive (**B**) and spiced olive (**C**) oils after each stage of processing and storage., Figure S2: DPPH radical scavenging activity of different extracts (25, 10 and 5 mg/mL) of the filling media after each stage of canning process and subsequent storage. (**A**) sunflower oil; (**B**) olive oil; (**C**) spiced olive oil.

Author Contributions: Conceptualization, S.M.; formal analysis, L.G.-L. and N.M.S.; data curation, L.G.-L., J.C., J.M.L., and S.M.; writing—original draft preparation, L.G.-L., J.C., J.M.L., and S.M.; writing—review and editing, J.C., R.D., J.M.L., and S.M.; supervision, S.M. All authors have read and agreed to the published version of the manuscript.

Funding: This research was funded by Xunta de Galicia, Spain (CITACA Strategic Partnership), grant number ED431E 2018/07.

Data Availability Statement: Not applicable.

Acknowledgments: Lucía Gómez-Limia acknowledges receiving financial support from the University of Vigo through a predoctoral scholarship. The authors would also like to thank GAIN (Axencia Galega de Innovación) for supporting this study (grant number IN607A2019/01).

Conflicts of Interest: The authors declare no conflict of interest.

References

1. Gómez-Limia, L.; Cobas, N.; Franco, I.; Martínez-Suárez, S. Fatty acid profiles and lipid quality indices in canned European eels: Effects of processing steps, filling medium and storage. *Food Res. Int.* **2020**, *136*, 109601. [CrossRef]
2. Mohan, C.O.; Remya, S.; Murthy, L.N.; Ravishankar, C.N.; Asok Kumar, K. Effect of filling medium on cooking time and quality of canned yellowfin tuna (*Thunnus albacares*). *Food Control* **2015**, *50*, 320–327. [CrossRef]
3. Gómez-Limia, L.; Franco, I.; Martínez-Suárez, S. Effects of processing step, filling medium and storage on amino acid profiles and protein quality in canned European eels. *J. Food Compos. Anal.* **2021**, *96*, 103710. [CrossRef]
4. Secci, G.; Parisi, G. From farm to fork: Lipid oxidation in fish products. A review. *Ital. J. Anim. Sci.* **2016**, *15*, 124–136. [CrossRef]
5. Domínguez, R.; Pateiro, M.; Gagaoua, M.; Barba, F.J.; Zhang, W.; Lorenzo, J.M. A comprehensive review on lipid oxidation in meat and meat products. *Antioxidants* **2019**, *8*, 429. [CrossRef]
6. Kaleem, A.; Aziz, S.; Iqtedar, M.; Abdullah, R.; Aftab, M.; Rashid, F.; Shakoori, F.R.; Naz, S. Investigating changes and effect of peroxide values in cooking oils subject to light and heat. *FUUAST J. Biol.* **2015**, *5*, 191–196.
7. Munekata, P.E.S.; Pateiro, M.; Domínguez, R.; Zhou, J.; Barba, F.J.; Lorenzo, J.M. Nutritional characterization of sea bass processing by-products. *Biomolecules* **2020**, *10*, 232. [CrossRef] [PubMed]
8. Pateiro, M.; Munekata, P.E.S.; Domínguez, R.; Wang, M.; Barba, F.J.; Bermúdez, R.; Lorenzo, J.M. Nutritional profiling and the value of processing by-products from gilthead sea bream (*Sparus aurata*). *Mar. Drugs* **2020**, *18*, 101. [CrossRef]
9. Choe, E.; Min, D.B. Mechanisms and factors for edible oil oxidation. *Compr. Rev. Food Sci. Food Saf.* **2006**, *5*, 169–186. [CrossRef]
10. Bravi, E.; Mangione, A.; Marconi, O.; Perretti, G. Comparative study of oxidation in canned foods with a combination of vegetables and covering oils. *Ital. J. Food Sci.* **2015**, *27*, 132–142.
11. Pateiro, M.; Barba, F.J.; Domínguez, R.; Sant'Ana, A.S.; Mousavi Khaneghah, A.; Gavahian, M.; Gómez, B.; Lorenzo, J.M. Essential oils as natural additives to prevent oxidation reactions in meat and meat products: A review. *Food Res. Int.* **2018**, *113*, 156–166. [CrossRef]
12. Pateiro, M.; Munekata, P.E.S.; Sant'Ana, A.S.; Domínguez, R.; Rodríguez-Lázaro, D.; Lorenzo, J.M. Application of essential oils as antimicrobial agents against spoilage and pathogenic microorganisms in meat products. *Int. J. Food Microbiol.* **2021**, *337*, 108966. [CrossRef]
13. AOAC Official Methods of Analysis of AOAC International. AOAC. 2005. Available online: https://www.aoac.org/ (accessed on 5 March 2021).

14. European Commission Regulation. European Regulation EU 2016/1227 of 27 July 2016 amending Regulation (EEC) No 2568/91 on the characteristics of olive oil and olive-residue oil and on the relevant methods of analysis. *Off. J. Eur. Union* **2016**, *202*.

15. Bligh, E.G.; Dyer, W.J. A rapid method of total lipid extraction and purification. *Can. J. Biochem. Physiol.* **1959**, *37*, 911–917. [CrossRef] [PubMed]

16. AOCS Official methods and recommended practices of the American Oil Chemists' Society 2009. Available online: https://www.aocs.org/attain-lab-services/methods (accessed on 5 March 2021).

17. Kirk, S.; Sawyer, R. *Pearson's Composition and Analysis of Foods*, 9th ed.; Longman: New York, NY, USA, 1991.

18. Brand-Williams, W.; Cuvelier, M.E.; Berset, C. Use of a free radical method to evaluate antioxidant activity. *Lebensm. Wiss. Technol.* **1995**, *28*, 25–30. [CrossRef]

19. Armesto, J.; Gómez-Limia, L.; Carballo, J.; Martínez, S. Impact of vacuum cooking and boiling and refrigerated storage on the quality of galega kale (*Brassica oleracea* var. *acephala* cv. Galega). *LWT Food Sc.Tech.* **2017**, *79*, 267–277. [CrossRef]

20. Singleton, V.L.; Rossi, J.A. Colorimetry of total phenolics with phosphomolybdic-phosphotungstic acid reagents. *Am. J. Enol. Vitic.* **1965**, *16*, 144–158.

21. El Lahamy, A.A. Mohamed changes in fish quality during canning process and storage period of canned fish products: Review article. *Glob. J. Nutr. Food Sci.* **2020**, *3*, 1–7.

22. European Commission Regulation. Commission Implementing Regulation (EU) No 1348/2013 of 16 December 2013 amending Regulation (EEC) No 2568/91 on the characteristics of olive oil and olive-residue oil and on the relevant methods of analysis 2013. *Off. J. Eur. Union* **2013**, *338*, 31–67.

23. Santos, C.S.P.; Cruz, R.; Cunha, S.C.; Casal, S. Effect of cooking on olive oil quality attributes. *Food Res. Int.* **2013**, *54*, 2016–2024. [CrossRef]

24. Silva, L.; Pinto, J.; Carrola, J.; Paiva-Martins, F. Oxidative stability of olive oil after food processing and comparison with other vegetable oils. *Food Chem.* **2010**, *121*, 1177–1187. [CrossRef]

25. Cerretani, L.; Bendini, A.; Rodriguez-Estrada, M.T.; Vittadini, E.; Chiavaro, E. Microwave heating of different commercial categories of olive oil: Part I. Effect on chemical oxidative stability indices and phenolic compounds. *Food Chem.* **2009**, *115*, 1381–1388. [CrossRef]

26. Malheiro, R.; Oliveira, I.; Vilas-Boas, M.; Falcão, S.; Bento, A.; Pereira, J.A. Effect of microwave heating with different exposure times on physical and chemical parameters of olive oil. *Food Chem. Toxicol.* **2009**, *47*, 92–97. [CrossRef] [PubMed]

27. Ayadi, M.A.; Grati-Kamoun, N.; Attia, H. Physico-chemical change and heat stability of extra virgin olive oils flavoured by selected Tunisian aromatic plants. *Food Chem. Toxicol.* **2009**, *47*, 2613–2619. [CrossRef]

28. Ahmed, M.; Pickova, J.; Ahmad, T.; Liaquat, M.; Farid, A.; Jahangir, M. Oxidation of lipids in foods. *Sarhad J. Agric.* **2016**, *32*, 230–238. [CrossRef]

29. Weber, J.; Bochi, V.C.; Ribeiro, C.P.; Victório, A.d.M.; Emanuelli, T. Effect of different cooking methods on the oxidation, proximate and fatty acid composition of silver catfish (*Rhamdia quelen*) fillets. *Food Chem.* **2008**, *106*, 140–146. [CrossRef]

30. Schaich, K.M. Lipid oxidation: Theoretical aspects. In *Bailey's Industrial Oil and Fat Products*; Wiley, A., Ed.; Interscience Publication, John Wiley and Sons: New York, NY, USA, 2005; pp. 269–299.

31. Estevez, M. Oxidative damage to poultry: From farm to fork. *Poult. Sci.* **2015**, *94*, 1368–1378. [CrossRef]

32. Selmi, S.; Monser, L.; Sadok, S. The influence of local canning process and storage on pelagic fish from Tunisia: Fatty acid profiles and quality indicators. *J. Food Process. Preserv.* **2008**, *32*, 443–457. [CrossRef]

33. El Shehawy, S.M.; Farag, Z.S. Safety assessment of some imported canned fish using chemical, microbiological and sensory methods. *Egypt. J. Aquat. Res.* **2019**, *45*, 389–394. [CrossRef]

34. Ehsani, A.; Jasour, M.S. Improvement of lipid stability of refrigerated rainbow trout (Oncorhynchus mykiss) fillets by pre-storage α-tocopherol acetate dipping treatment. *Vet. Res. Forum* **2012**, *3*, 269–73.

35. Naseri, M.; Rezaei, M.; Moieni, S.; Hosseini, H.; Eskandari, S. Effects of different filling media on the oxidation and lipid quality of canned silver carp (Hypophthalmichthys molitrix). *Int. J. Food Sci. Technol.* **2011**, *46*, 1149–1156. [CrossRef]

36. Kong, F.; Oliveira, A.; Tang, J.; Rasco, B.; Crapo, C. Salt effect on heat-induced physical and chemical changes of salmon fillet (O. gorbuscha). *Food Chem.* **2008**, *106*, 957–966. [CrossRef]

37. Marmesat, S.; Morales, A.; Velasco, J.; Dobarganes, M.C. Action and fate of natural and synthetic antioxidants during frying. *Grasas Aceites* **2010**, *61*, 333–340.

38. Uriarte-Montoya, M.H.; Villalba-Villalba, A.G.; Pacheco-Aguilar, R.; Ramirez-Suarez, J.C.; Lugo-Sánchez, M.E.; García-Sánchez, G.; Carvallo-Ruíz, M.G. Changes in quality parameters of Monterey sardine (Sardinops sagax caerulea) muscle during the canning process. *Food Chem.* **2010**, *122*, 482–487. [CrossRef]

39. Shanmugasundaram, B.; Singh, V.K.; Shrestha, S.; Sarkar, S.K.; Jeevaratnam, K.; Koner, B.C. Comparative evaluation of peroxidation of sesame and cottonseed oil induced by different methods of heating. *J. Sci. Ind. Res.* **2019**, *78*, 162–165.

40. Medina, I.; Sacchi, R.; Biondi, L.; Aubourg, S.P.; Livio, P. Effect of packing media on the oxidation of canned tuna lipids. antioxidant effectiveness of extra virgin olive oil. *J. Agric. Food Chem.* **1998**, *46*, 1150–1157. [CrossRef]

41. Bilgin, B.; Gençelep, H. Determination of biogenic amines in fish products. *Food Sci. Biotechnol.* **2015**, *24*, 1907–1913. [CrossRef]

42. Saguy, I.S.; Dana, D. Integrated approach to deep fat frying: Engineering, nutrition, health and consumer aspects. *J. Food Eng.* **2003**, *56*, 143–152. [CrossRef]

43. Talab, A.S. Effect of cooking methods and freezing storage on the quality characteristics of fish cutlets. *Adv. J. Food Sci. Technol.* **2014**, *6*, 468–479. [CrossRef]

44. Bakar, J.; Zakipour Rahimabadi, E.; Che Man, Y.B. Lipid characteristics in cooked, chill reheated fillets of Indo-Pacificking mackerel (*Scomberomorous guttatus*). *LWT Food Sci. Tech.* **2008**, *41*, 2144–2150. [CrossRef]

45. Kaneniwa, M.; Miao, S.; Yuan, C.; Iida, H.; Fukuda, Y. Lipid components and enzymatic hydrolysis of lipids in muscle of Chinese fresh-water fish. *J. Am. Oil Chem. Soc.* **2000**, *77*, 825–830. [CrossRef]

46. Leung, K.S.; Galano, J.M.; Durand, T.; Lee, J.C.Y. Profiling of omega-polyunsaturated fatty acids and their oxidized products in salmon after different cooking methods. *Antioxidants* **2018**, *7*, 96. [CrossRef]

47. Al-Saghir, S.; Thurner, K.; Wagner, K.H.; Frisch, G.; Luf, W.; Razzazi-Fazeli, E.; Elmadfa, I. Effects of different cooking procedures on lipid quality and cholesterol oxidation of farmed salmon fish (Salmo salar). *J. Agric. Food Chem.* **2004**, *52*, 5290–5296. [CrossRef]

48. Taghvaei, M.; Jafari, S.M.; Mahoonak, A.S.; Nikoo, A.M.; Rahmanian, N.; Hajitabar, J.; Meshginfar, N. The effect of natural antioxidants extracted from plant and animal resources on the oxidative stability of soybean oil. *LWT Food Sci. Technol.* **2014**, *56*, 124–130. [CrossRef]

49. Alhibshi, E.A.; Ibraheim, J.A.; Hadad, A.S. Effect of heat processing and storage on characteristic and stability of some edible oils. In Proceedings of the 6th International Conference on Agriculture, Environment and Biological Sciences, Kuala Lumpur, Malaysia, 21–22 December 2016.

50. Naz, S.; Siddiqi, R.; Sheikh, H.; Sayeed, S.A. Deterioration of olive, corn and soybean oils due to air, light, heat and deep-frying. *Food Res. Int.* **2005**, *38*, 127–134. [CrossRef]

51. Alajtal, A.I.; Sherami, F.E.; Elbagermi, M.A. Acid, Peroxide, ester and saponification values for some vegetable oils before and after frying. *AASCIT J. Mater.* **2018**, *4*, 43–47.

52. Santos, J.C.O.; Santos, I.M.G.; Souza, A.G. Effect of heating and cooling on rheological parameters of edible vegetable oils. *J. Food Eng.* **2005**, *67*, 401–405. [CrossRef]

53. Caponio, F.; Giarnetti, M.; Summo, C.; Paradiso, V.M.; Cosmai, L.; Gomes, T. A comparative study on oxidative and hydrolytic stability of monovarietal extra virgin olive oil in bakery products. *Food Res. Int.* **2013**, *54*, 1995–2000. [CrossRef]

54. Grim, J.M.; Miles, D.R.B.; Crockett, E.L. Temperature acclimation alters oxidative capacities and composition of membrane lipids without influencing activities of enzymatic antioxidants or susceptibility to lipid peroxidation in fish muscle. *J. Exp. Biol.* **2010**, *213*, 445–452. [CrossRef]

55. Bhouri, A.M.; Fekih, S.; Hanene, J.H.; Madiha, D.; Imen, B.; ElCafsi, M.; Hammami, M.; Chaouch, A. Antioxidant property of wild and farmed sea bream (Sparus aurata)cooked in different ways. *Afr. J. Biotechnol.* **2011**, *10*, 19623–19630.

56. Pazos, M.; González, M.J.; Gallardo, J.M.; Torres, J.L.; Medina, I. Preservation of the endogenous antioxidant system of fish muscle by grape polyphenols during frozen storage. *Eur. Food Res. Technol.* **2005**, *220*, 514–519. [CrossRef]

57. Devi, W.S.; Sarojnalini, C. Effect of cooking on the polyunsaturated fatty acid and antioxidant properties of small indigenous fish species of the Eastern Himalayas. *Int. J. Eng. Res. Appl.* **2014**, *4*, 146–151.

58. Roe, M.; Church, S.; Pinchen, H.; Finglas, P. *Nutrient Analysis of Fish and Fish Products—Analytical Report*; Institute of Food Research: Surrey, UK, 2013.

59. Medina, I.; Satué-Gracia, M.T.; German, J.B.; Frankel, E.N. Comparison of natural polyphenol antioxidants from extra virgin olive oil with synthetic antioxidants in tuna lipids during thermal oxidation. *J. Agric. Food Chem.* **1999**, *47*, 4873–4879. [CrossRef] [PubMed]

60. Merdzhanova, A.; Stancheva, M.; Dobreva, D.A.; Makedonski, L. Fatty acid and fat soluble vitamins composition of raw and cooked Black Sea horse mackerel. *Ovidius Univ. Ann. Chem.* **2013**, *24*, 27–34. [CrossRef]

61. Ersoy, B.; Özeren, A. The effect of cooking methods on mineral and vitamin contents of African catfish. *Food Chem.* **2009**, *115*, 419–422. [CrossRef]

62. Emmanuel-Ikpeme, C.; Henry, P.; Augustine Okiri, O. Comparative evaluation of the nutritional, phytochemical and microbiological quality of three pepper varieties. *J. Food Nutr. Sci.* **2014**, *2*, 80. [CrossRef]

63. Ortiz, J.; Romero, N.; Robert, P.; Araya, J.; Lopez-Hernández, J.; Bozzo, C.; Navarrete, E.; Osorio, A.; Rios, A. Dietary fiber, amino acid, fatty acid and tocopherol contents of the edible seaweeds Ulva lactuca and Durvillaea antarctica. *Food Chem.* **2006**, *99*, 98–104. [CrossRef]

64. Quiles, J.L.; Ramírez-Tortosa, M.C.; Gómez, J.A.; Huertas, J.R.; Mataix, J. Role of vitamin E and phenolic compounds in the antioxidant capacity, measured by ESR, of virgin olive, olive and sunflower oils after frying. *Food Chem.* **2002**, *76*, 461–468. [CrossRef]

65. Meckelmann, S.W.; Riegel, D.W.; Van Zonneveld, M.J.; Ríos, L.; Peña, K.; Ugas, R.; Quinonez, L.; Mueller-Seitz, E.; Petz, M. Compositional characterization of native Peruvian chili peppers (*Capsicum* spp.). *J. Agric. Food Chem.* **2013**, *61*, 2530–2537. [CrossRef]

66. Bayram, B.; Esatbeyoglu, T.; Schulze, N.; Ozcelik, B.; Frank, J.; Rimbach, G. Comprehensive analysis of polyphenols in 55 extra virgin olive oils by HPLC-ECD and their correlation with antioxidant activities. *Plant Foods Hum. Nutr.* **2012**, *67*, 326–336. [CrossRef]

67. Borges, T.H.; Serna, A.; López, L.C.; Lara, L.; Nieto, R.; Seiquer, I. Composition and antioxidant properties of spanish extra virgin olive oil regarding cultivar, harvest year and crop stage. *Antioxidants* **2019**, *8*, 217. [CrossRef]

68. Yeo, J.D.; Shahidi, F. Effect of hydrothermal processing on changes of insoluble-bound phenolics of lentils. *J. Funct. Foods* **2017**, *38*, 716–722. [CrossRef]
69. Nergiz, C.; Unal, K. Determination of phenolic acids in virgin olive oil. *Food Chem.* **1991**, *39*, 237–240. [CrossRef]
70. Matthaus, B.; Spener, F. Introduction: What we know and what we should know about virgin oils—A general introduction. *Eur. J. Lipid Sci. Technol.* **2008**, *110*, 597–601. [CrossRef]
71. Franco, M.N.; Galeano-Díaz, T.; López, Ó.; Fernández-Bolaños, J.G.; Sánchez, J.; De Miguel, C.; Gil, M.V.; Martín-Vertedor, D. Phenolic compounds and antioxidant capacity of virgin olive oil. *Food Chem.* **2014**, *163*, 289–298. [CrossRef]
72. Kishimoto, N. Microwave heating induces oxidative degradation of extra virgin olive oil. *Food Sci. Technol. Res.* **2019**, *25*, 75–79. [CrossRef]
73. Baiano, A.; Terracone, C.; Gambacorta, G.; Notte, E. La Changes in quality indices, phenolic content and antioxidant activity of flavored olive oils during storage. *JAOCS J. Am. Oil Chem. Soc.* **2009**, *86*, 1083–1092. [CrossRef]

Article

The Effects of Select Hydrocolloids on the Processing of Pâté-Style Canned Pet Food

Amanda N. Dainton, Hulya Dogan and Charles Gregory Aldrich *

Department of Grain Science & Industry, Kansas State University, 201 Shellenberger Hall, 1301 Mid Campus Drive, Manhattan, KS 66506, USA; adainton@ksu.edu (A.N.D.); dogan@ksu.edu (H.D.)

* Correspondence: aldrich4@ksu.edu

Abstract: Hydrocolloids are commonly used in canned pet food. However, their functional effects have not been quantified in this food format. The objective was to determine the effects of select hydrocolloids on batter consistency, heat penetration, and texture of canned pet food. Treatments were added to the formula as 1% dextrose (D) and 0.5% guar gum with 0.5% of either dextrose (DG), kappa carrageenan (KCG), locust bean gum (LBG), or xanthan gum (XGG). Data were analyzed as a 1-way ANOVA with batch as a random effect and separated by Fisher's LSD at $p < 0.05$. Batter consistency (distance traveled in 30 s) thickened with increasing levels of hydrocolloids (thinnest to thickest: 23.63 to 2.75 cm). The D treatment (12.08 min) accumulated greater lethality during the heating cycle compared to all others (average 9.09 min). The KCG treatment (27.00 N) was the firmest and D and DG (average 8.75 N) the softest with LBG and XGG (average 15.59 N) intermediate. Toughness was similar except D (67 N·mm) was less tough than DG (117 N·mm). The D treatment showed the greatest expressible moisture (49.91%), LBG and XGG the lowest (average 16.54%), and DG and KCG intermediate (average 25.26%). Hydrocolloids influenced heat penetration, likely due to differences in batter consistency, and affected finished product texture.

Keywords: expressible moisture; gel; gum; heat penetration; thermally processed; texture; wet pet food

Citation: Dainton, A.N.; Dogan, H.; Aldrich, C.G. The Effects of Select Hydrocolloids on the Processing of Pâté-Style Canned Pet Food. *Foods* **2021**, *10*, 2506. https://doi.org/10.3390/foods10102506

Academic Editors: Sidonia Martinez and Javier Carballo

Received: 16 September 2021
Accepted: 15 October 2021
Published: 19 October 2021

Publisher's Note: MDPI stays neutral with regard to jurisdictional claims in published maps and institutional affiliations.

1. Introduction

Canned pet foods are commercially sterilized, low-acid products and come in formats similar to stews and pâtés in appearance. These products primarily consist of meats and water and contain binding/structural ingredient systems similar to restructured meat products for human consumption [1,2]. Hydrocolloids, such as carrageenan and guar, locust bean, and xanthan gums, are common choices. They are able to increase the viscosity of unprocessed meat batters and emulsions [3,4] and may increase or decrease the firmness of a finished product, depending on the formulation [5,6]. These differences in functionality are driven by their chemical structures. Briefly, kappa carrageenan is a linear molecule with repeating galactose units and 3,6-anhydrogalactose units connected with alternating α-1,3 and β-1,4 bonds [2]. Xanthan gum consists of a 1,4 linked β-D-glucose backbone with a side chain of a glucuronic acid and two mannose units every other glucose unit. Locust bean gum and guar gum are the most similar as they have the same linear 1,4-β-D-mannose backbone with galactose side chains connected with 1,6-α-glycosidic bonds. However, they differ in their galactose content; locust bean gum contains less galactose by weight compared to guar gum (17–26% vs. 33–40%) [2]. Though it is likely that these ingredients are included in commercial pet foods for similar functional benefits, this has not been a primary area of investigation. Instead, research has focused on the nutritional value of carbohydrate hydrocolloids as soluble fibers [7–9].

The incorporation of hydrocolloids in commercial pet foods has waned as companies wish to differentiate their products from their competitors' offerings [10]. There are no reports of why pet food companies use the inclusion of hydrocolloids or lack thereof

Foods **2021**, *10*, 2506. https://doi.org/10.3390/foods10102506   https://www.mdpi.com/journal/foods

to distinguish their formulas from others. Mixed findings suggest that the addition of similar ingredients to canned pet foods causes softer stools in dogs [11] while others reported firmer stools compared to a control diet [12]. A small segment of the population believes some hydrocolloids to be toxic or carcinogenic to dogs and cats, though this is not confirmed in the literature [13,14]. Regardless, pet food companies are looking for new ingredients to replace the commonly used hydrocolloids. However, there are no reports quantifying and differentiating the functional effects of these ingredients in canned pet foods. This limits the ability to identify alternative, label friendly ingredients with similar functionality. Research with sausage, meatballs, and other restructured meat products can provide some insight, but many of these products utilize hydrocolloids for their ability to mimic the mouthfeel of fat [2]. As such, conclusions from restructured meat products for human consumption may not be directly applicable due to differences in formulation and processing methods.

The objective of this experiment was to characterize the physicochemical and processing effects of guar gum and blends of guar gum with another hydrocolloid on the processing of canned pet foods. The hypothesis was that the addition of hydrocolloids would decrease heat penetration and alter the color and texture of processed foods. Additionally, systems containing guar gum with an additional hydrocolloid would have measurable differences in texture driven by the mechanism of the additional hydrocolloid.

2. Materials and Methods

2.1. Formulation of Canned Pet Foods

Five treatments were designed to show the effects of guar gum, kappa carrageenan, locust bean gum, and xanthan gum on functional properties of canned pet food. Dextrose was chosen as a space-holding control ingredient for its aqueous solubility [15] and its similar moisture content compared to the carbohydrate hydrocolloids of interest. Controlling the moisture content of the samples was a concern to minimize confounding effects on heat penetration [16]. Guar gum was specifically chosen because of its high thickening power [2] and its prevalence in commercial canned pet foods. As such, guar gum was also included in treatments containing either kappa carrageenan, locust bean gum, and xanthan gum in an attempt to mimic commercial pet foods.

There are limited recommendations for the inclusion level of these ingredients in canned pet food. Locust bean gum is used in pet foods at 0.2–0.5% [2] for its binding effects, or the ability to increase interactions between macromolecules themselves and with the solvent [17]. Additionally, a model lunch meat formula contained 0.6% kappa carrageenan for its gelling effects [2]. The inclusion level of 0.5% for individual hydrocolloids was chosen based on these recommendations as well as preliminary data wherein higher inclusion levels were firmer than commercial canned pet foods (data not presented).

This led to the creation of five experimental treatments (Table 1): 1% dextrose (D), 0.5% dextrose and 0.5% guar gum (DG), 0.5% guar gum and 0.5% kappa carrageenan (KCG), 0.5% guar gum and 0.5% locust bean gum (LBG), and 0.5% guar gum and 0.5% xanthan gum (XGG).

Prior to diet production, frozen blocks of mechanically deboned chicken (CJ Foods, Bern, KS, USA) were ground with a lab-scale meat grinder (Weston Pro Series #32, Southern Pine, NC, USA) fitted with a die plate with 7 mm diameter holes. Treatments were replicated three times over three days of production, with each treatment made on each day. Water was heated in a stock pot until it reached 40 °C, at which point the tempered ground chicken was added and the mixture was brought back up to 40 °C. Brewer's rice (Lortscher Animal Nutrition, Bern, KS, USA), spray-dried egg white (Rembrandt Foods, Okoboji, IA), sunflower oil (Kroger, Manhattan, KS, USA), potassium chloride (Lortscher Animal Nutrition, Bern, KS, USA), vitamin premix (Lortscher Animal Nutrition, Bern, KS, USA), and trace mineral premix (Lortscher Animal Nutrition, Bern, KS, USA) were added to the stock pot and heated to 60 °C with continuous stirring. Once the batter reached the target temperature, dextrose (Fairview Mills, Seneca, KS, USA) and/or the respective hydrocolloid ingredient(s) (Danisco, New Century, KS, USA) were added and the batter was stirred continuously for 5 min while maintaining temperature. Treatment order was randomized each day to maintain similar initial internal can temperatures, which can influence heat penetration and the required length of processing [18,19]. After mixing, 21 cans (size 300 × 407; House of Cans, Lincolnwood, IL, USA) were filled with 405 ± 5 g of batter for each treatment.

Table 1. Ingredient composition of thermally processed canned pet foods [1] containing select hydrocolloids.

Ingredient, % *w/w*	D	DG	KCG	LBG	XGG
Mechanically separated chicken	56.00	56.00	56.00	56.00	56.00
Water	38.35	38.35	38.35	38.35	38.35
Brewer's rice	3.00	3.00	3.00	3.00	3.00
Potassium chloride	0.50	0.50	0.50	0.50	0.50
Spray-dried egg white	0.50	0.50	0.50	0.50	0.50
Sunflower oil	0.50	0.50	0.50	0.50	0.50
Vitamin premix [2]	0.10	0.10	0.10	0.10	0.10
Trace mineral premix [3]	0.05	0.05	0.05	0.05	0.05
Dextrose	1.00	0.50	-	-	-
Guar gum	-	0.50	0.50	0.50	0.50
Kappa carrageenan	-	-	0.50	-	-
Locust bean gum	-	-	-	0.50	-
Xanthan gum	-	-	-	-	0.50

[1] D = 1% dextrose; DG = 0.5% guar gum and 0.5% dextrose; KCG = 0.5% guar gum and 0.5% kappa carrageenan; LBG = 0.5% guar gum and 0.5% locust bean gum; XGG = 0.5% guar gum and 0.5% xanthan gum. [2] One kg of vitamin premix supplies 17,163,000 IU vitamin A, 920,000 IU vitamin D3, 79,887 IU vitamin E, 22.0 mg vitamin B12 (cobalamin), 4719 mg vitamin B2 (riboflavin), 12,186 mg vitamin B5 (d-pantothenic acid), 14,252 mg vitamin B1 (thiamin), 64,730 mg vitamin B3 (niacin), 5537 mg vitamin B6 (pyridoxine), 720 mg vitamin B9 (folic acid), and 70.0 mg vitamin B7 (biotin). [3] One kg of trace mineral premix supplies 88,000 mg zinc sulfate, 38,910 mg ferrous sulfate, 11,234 mg copper sulfate, 5842 mg manganous oxide, 310 mg sodium selenite, and 1584 mg calcium iodate.

2.2. Analysis of Pre-Thermal Processing Batters of Canned Pet Food

Three consistency measurements were taken per treatment replication using a Bostwick consistometer (CSC Scientific Company, Fairfax, VA, USA). Briefly, this analysis utilized a sloped trough and slide gate to determine how thick or thin a sample of a set volume was. Measurements of distance in centimeters traveled in 30 s were recorded. A sample that traveled farther was considered to have thinner consistency and a sample that did not travel as far was considered to have thicker consistency. The Bostwick consistometer methodology was chosen because it does not require room temperature samples, which is a concern for viscosity analysis. Generally, viscosity of food samples is greater (i.e., thicker) at cooler temperatures [3] and those values would not be relevant for samples collected directly from production. In the present experiment, all samples were analyzed at the same temperature immediately after the complete batter was mixed for 5 min at 60 °C. Additionally, the Bostwick consistometer is widely used by the pet food industry because of its low cost and limited required training [20,21].

Three pH measurements were taken with a pH meter (P/N 54X002608; Oakton Instruments, Vernon Hills, IL, USA) fitted with a pointed pH probe (model #FC240B; Hanna Instruments, Smithfield, RI, USA). Finally, three samples for water activity were collected and stored in covered containers to return to room temperature for measurement with a water activity meter (Decagon CX-2; Meter Group, Pullman, WA, USA).

2.3. Analysis of Processing Control Measures and Thermal Processing Calculations

Treatments prepared on the same day were processed in the retort at the same time. Four thermocouples (Ecklund-Harrison Technologies, Fort Meyers, FL) per treatment were placed in cans prior to filling and connected to a data capture system (CALSoft v. 5; TechniCAL LLC, Metairie, LA, USA) to record temperature in the center of the cans during processing. Fill weight and gross headspace were recorded for these cans as well. Specifically, gross headspace was measured as the distance from the top of the can body to the top of the batter inside the container. Once measurements were taken, lids (size 300 × 407 sanitary lids; House of Cans, Lincolnwood, IL, USA) were sealed onto the cans with a seamer (Dixie Seamer, 91118; Athens, GA, USA). Cans were randomly loaded into a still retort (Dixie, 00-43; Athens, GA, USA) and processed at 144.79 kPa and 121 °C. Thermocouple-containing cans were randomly distributed among all other cans. Temperature inside the retort was also recorded by the data capture system. The intent was for the data capture system to record temperature inside the retort and inside each can every 15 s. However, the data capture system could not consistently record temperature at this rate during production days 1 and 2. The longest length of time between temperature measurements was 7.75 min and the average ± standard deviation excluding the normal time intervals was 1.23 ± 1.30 min. The cooling cycle was started once the coldest can among all treatments in the retort containing a thermocouple achieved a minimum lethality, or the relative amount of time at a constant reference temperature of 121.11 °C [22], of 8 min. This value has been reported as a minimum for commercial canned pet food [23] and was chosen to remain consistent with pet food industry practices. Cans were cooled in the retort with municipal water (20 °C) until the last can containing a thermocouple dropped below 50 °C before removal from the retort.

Calculations for lethality (Equation (1)) [22] and cook value (C_{100}; Equation (2)) were made using the thermocouple temperature data (Figure S1). $T_C(t)$ is the internal can temperature at any given time t and Δt represents the length of time between temperature measurements. The reference temperature and the z-value representing the change in temperature required to see a 1 log reduction in the D value, or the amount of time required to see a 1 log reduction, were specific to the item of interest [22,24]. These values were 121.11 and 10 °C, respectively, for the calculation of lethality. The z value came from experiments measuring the heat resistance of *Clostridium botulinum* 213-B in a pH7 phosphate buffer [25] and had been used in a preliminary experiment with thermally processed pet food [26]. The C_{100} calculation utilized the reference temperature (100 °C) and z value (33 °C) for thiamin, the weakest nutrient. Both equation integrals were solved using the trapezoid rule (Equations (3) and (4), respectively). These calculations were used to discuss the effect of the treatments on heat penetration and dissipation. For example, higher lethality and C_{100} during the heating retort cycle were indirect indicators of faster heat penetration rates. This methodology was used in a preliminary study of the effects of container size and type on lethality values of wet pet food processed for the same amount of time [26].

$$Lethality = \int_0^t 10^{\frac{T_C(t)-121.11\ °C}{10\ °C}} \Delta t, \tag{1}$$

$$C_{100} = \int_0^t 10^{\frac{T_C(t)-100\ °C}{33\ °C}} \Delta t, \tag{2}$$

$$Lethality = \sum_0^t 10^{\frac{T_C(t)-121.11\ °C}{10}} \Delta t, \tag{3}$$

$$C_{100} = \sum_0^t 10^{\frac{T_C(t)-100\ °C}{33}} \Delta t, \tag{4}$$

2.4. Analysis of Processed Canned Pet Foods

Four cans per combination of treatment and production day were blended (14-speed Osterizer; Sunbeam Products, Boca Raton, FL, USA) and freeze dried (model #HR7000-L; Harvest Right, LLC, Salt Lake City, UT, USA) for analysis of moisture content (AOAC 934.01) in duplicate. Can vacuum was measured on 4 cans per treatment replicate with a glycerin-filled vacuum gauge (#25.300/30; Fisher Scientific, Hampton, NH USA) fitted with a rubber collar (#10816-11; Wilkens-Anderson Co., Chicago, IL, USA) and metal tip.

Three cans per treatment replicate were analyzed for pH (meter: P/N 54X002608, Oakton Instruments, Vernon Hills, IL; probe: model #FC240B, Hanna Instruments, Smithfield, RI, USA), free liquid, and expressible moisture by centrifugation as an indication of water holding capacity [27]. Briefly, free liquid was quantified as the mass of the liquid phase, if present, upon opening the can. Expressible moisture was determined by weighing approximately 1 g of sample into two Whatman Grade 3 filter papers (GE Healthcare Life Sciences, Piscataway, NJ, USA) and one Whatman Grade 50 filter paper (GE Healthcare Life Sciences, Piscataway, NJ, USA) folded into a thimble shape and centrifuging in a 50 mL polypropylene centrifuge tube (Globe Scientific Inc., Mahwah, NJ, USA) at $2000 \times g$ (Sorvall® RC 6 Plus; Thermo Electron Corporation, Waltham, MA, USA). As-is and dried filter paper weights were recorded before and after centrifuging to account for residue transfer to the filter papers [Equations (5)–(7)]. Analysis of expressible moisture was conducted in quadruplicate for each can per treatment replicate.

$$Expressed\ moisture + residue,\ g = As - is\ filter\ paper\ after\ centrifugation,\ g - As - is\ initial\ filter\ paper,\ g, \quad (5)$$

$$Expressed\ residue, g = Dried\ filter\ paper\ after\ centrifugation,\ g - Dried\ initial\ filter\ paper,\ g, \quad (6)$$

$$Expressible\ moisture,\ \% = \frac{Equation\ (5) - Equation\ (6)}{As - is\ weight\ of\ sample} * 100, \quad (7)$$

Texture was characterized with a modified back extrusion test using a texture analyzer (TA-XT2; Texture Technologies Corp., Hamilton, MA, USA) fitted with a 5.08 cm diameter and 2 cm tall cylindrical probe and a 30 kg load cell. The trigger force was set to 5 g and the test speeds (pre-, test-, and post-) were set to 1 mm/s. The probe was pressed into the center of the products in cans to a depth of 2 cm. Firmness was recorded as the largest force measurement observed during the 2 cm compression. Often this value was similar to or not different from the force measurement recorded at the end of the compression. Toughness was calculated as the area under the curve of the compression peak using the trapezoid rule (Figure S2). Five cans from each treatment replicate were analyzed and values were averaged together to generate composite values for each replicate. Cans were selected from the beginning, middle, and end of the filling sequence for each treatment to accurately capture texture for the entire production. This methodology was selected instead of a texture profile analysis procedure because some of the treatments did not form structures that would remain stable if removed from the can. A similar methodology was applied to canned cat foods processed to comparable lethality values with different processing conditions [28].

Three cans per replicate of treatment were analyzed for color with a CIELAB colorspace colorimeter (CR-410 Chroma Meter, Konica Minolta, Chiyoda, Tokyo, Japan) with five measurements taken from four evenly spaced regions of pâté from each can. Sections were created by removing the product from the can and slicing 3 times with a knife. In cases where slices could not be created, color of the top and bottom of the product were measured while inside the can and internal color by separating product into 3 sections of roughly the same size. Color was described in terms of L* (brightness), a* (red to green scale), and b* (yellow to blue scale). Values of L* closer to 100 indicated lighter products, whereas values of L* closer to 0 indicated darker products. The red to green and yellow to blue scales contained negative and positive values. More negative values of a* and b* indicated greener and bluer color, respectively. On the other hand, more positive values of a* and b* indicated redder and yellower color, respectively. All three scales form a three-dimensional space with the intersection point in the middle of each scale.

2.5. Statistical Analysis

Data were analyzed as a randomized complete block design with treatment as the fixed effect and day as the random block using statistical analysis software (SAS 9.4; SAS Institute, Cary, NC, USA). Values were presented as least square means ± the standard error of the mean and differences were calculated using Fisher's LSD in the GLIMMIX procedure. The CORR procedure was used to calculate r and p-values for Pearson correlations. All tests were considered significant if $p < 0.05$.

3. Results

3.1. Pre-Thermal Processing Batter Analyses

Batter pH and batter water activity were not different ($p > 0.05$) among treatments and averaged 5.94 and 0.990, respectively (Table 2). Consistency was affected ($p < 0.05$) by treatment. The D treatment was the thinnest (23.64 cm) and often traveled the full 24 cm in less than 30 s. Guar gum alone (DG = 6.60 cm) decreased ($p < 0.05$) consistency, resulting in a thicker batter, in comparison to D. The KCG, LBG, and XGG batters exhibited the lowest consistency with no differences ($p > 0.05$) between them (average = 2.75 cm).

Table 2. Batter characteristics of thermally processed canned pet foods [1] containing select hydrocolloids.

Measurement	D	DG	KCG	LBG	XGG	SEM	*p*-Value
Consistency, cm/30 s	23.64 [a]	6.60 [b]	1.69 [c]	3.63 [c]	2.94 [c]	0.719	<0.0001
pH	5.90	5.93	5.95	5.94	5.97	0.081	0.7411
Water activity	0.984	0.996	0.992	0.987	0.992	0.0067	0.2528

[abc] Treatment means with unlike superscripts are different ($p < 0.05$). [1] D = 1% dextrose; DG = 0.5% guar gum and 0.5% dextrose; KCG = 0.5% guar gum and 0.5% kappa carrageenan; LBG = 0.5% guar gum and 0.5% locust bean gum; XGG = 0.5% guar gum and 0.5% xanthan gum.

3.2. Processing Control Analysis and Thermal Processing Values

Data from three thermocouples were removed from statistical analysis due to failure (2) and low fill weight (1). No differences ($p > 0.05$) were noted in can fill weight (average = 406.6 g), gross headspace (average = 13.87 mm), initial internal can temperature (average = 55.67 °C), or post-processing can vacuum (average = -12.9 kPa) across treatments (Table 3).

Table 3. Processing controls of thermally processed canned pet foods [1] containing select hydrocolloids.

Measurement	D	DG	KCG	LBG	XGG	SEM	*p*-Value
Number of thermocouples	12	12	11	11	11	-	-
Initial internal can temperature, °C	55.32	56.13	58.51	55.01	53.40	1.768	0.2038
Can fill weight, g	404.6	404.9	405.1	404.3	404.0	0.54	0.3900
Gross headspace, mm	14.83	14.53	13.29	13.17	13.53	0.350	0.4248
Post-processing can vacuum, kPa	-12.4	-11.9	-16.0	-12.7	-11.4	1.76	0.4605

[1] D = 1% dextrose; DG = 0.5% guar gum and 0.5% dextrose; KCG = 0.5% guar gum and 0.5% kappa carrageenan; LBG = 0.5% guar gum and 0.5% locust bean gum; XGG = 0.5% guar gum and 0.5% xanthan gum.

Heating cycle length and cooling cycle length across the three days averaged 81.25 ± 1.392 and 79.58 ± 22.735 min, respectively. The total, heating, and cooling lethalities were all affected ($p < 0.05$) by the treatments (Table 4). Heating lethality was greater ($p < 0.05$) for D (12.08 min) compared to the four other treatments (average = 9.09 min). The same relationship was observed for total lethality (D = 20.24 min; average of all others = 18.46 min). On the other hand, D (8.17 min) accumulated lower ($p < 0.05$) cooling lethality than LBG and XGG (average = 9.60 min) with DG and KCG not different ($p > 0.05$; average = 8.97 min) from any of the treatments. The total C_{100} was not affected ($p > 0.05$; average = 197.10 min) by the treatments. However, D accumulated more ($p < 0.05$) C_{100} during the heating cycle (137.01 min) and less ($p < 0.05$) C_{100} during the cooling cycle (64.90 min) compared to all other treatments (averages = 117.24 and 78.66 min, respectively). Total lethality and C_{100} were very strongly correlated ($r = 0.98$; $p < 0.0001$). The same relationship was observed between heating lethality and C_{100} ($r = 1.00$; $p < 0.0001$) and between cooling lethality and C_{100} ($r = 0.95$; $p < 0.0001$).

Table 4. Lethality and cook values (C_{100}) for thermally processed canned pet foods [1] containing select hydrocolloids.

Measurement, min	D	DG	KCG	LBG	XGG	SEM	*p*-Value
Total lethality	20.24 [a]	18.63 [b]	18.27 [b]	18.33 [b]	18.26 [b]	0.470	0.0121
Heating lethality	12.08 [a]	9.71 [b]	9.25 [b]	8.74 [b]	8.66 [b]	0.566	0.0177
Cooling lethality	8.17 [b]	8.92 [ab]	9.01 [ab]	9.59 [a]	9.61 [a]	1.714	0.0428
Total C_{100}	201.90	196.41	195.19	196.42	195.56	2.923	0.2307
Heating C_{100}	137.01 [a]	120.67 [b]	118.84 [b]	115.38 [b]	114.06 [b]	3.870	0.0196
Cooling C_{100}	64.90 [b]	75.74 [a]	76.36 [a]	81.04 [a]	81.50 [a]	9.233	0.0099

[ab] Treatment means with unlike superscripts are different ($p < 0.05$). [1] D = 1% dextrose; DG = 0.5% guar gum and 0.5% dextrose; KCG = 0.5% guar gum and 0.5% kappa carrageenan; LBG = 0.5% guar gum and 0.5% locust bean gum; XGG = 0.5% guar gum and 0.5% xanthan gum.

3.3. Physicochemical Quality of Processed Canned Cat Food

Many processed treatment characteristics were affected by the differences in carbohydrate hydrocolloid content (Table 5). The only experimental treatment to exhibit two phases was D, with 16.91% ± 1.629% of the product mass as a free liquid phase. The finished product pH was greatest ($p < 0.05$) for KCG (6.38) and lowest for dextrose-containing treatments (D and DG; average = 5.96) with LBG and XGG (average = 6.26) intermediate. Total moisture was greater ($p < 0.05$) for D and DG (average = 79.21%) than for LBG and XGG (average = 77.29%) with KCG (77.83%) intermediate and not different ($p > 0.05$) from any other treatment. Expressible moisture as a percentage of the total sample mass was greatest ($p < 0.05$) for D (49.91%) and lowest for LBG and XGG (average = 16.54%) with DG and KCG intermediate (average = 25.26%). The KCG treatment exhibited the greatest ($p < 0.05$) firmness and toughness (27.00 N and 370 N·mm, respectively) of all experimental treatments, followed by LBG and XGG (average = 15.59 N and 235 N·mm, respectively). The replacement of 0.5% dextrose with guar gum nearly doubled ($p < 0.05$) toughness (D = 67 N·mm; DG = 117 N·mm) but did not affect ($p > 0.05$) firmness (average = 8.75 N). Increasing the level of dextrose darkened ($p < 0.05$) the product. Additionally, LBG was lighter ($p < 0.05$) than DG with KCG and XGG intermediate and not different ($p > 0.05$). No differences ($p > 0.05$) in a* and b* were noted between KCG, LBG, and XGG (averages = 4.41 and 15.39, respectively). However, the inclusion of dextrose in D and DG resulted in redder ($p < 0.05$; average = 8.37) and yellower ($p < 0.05$; average = 22.05) color.

Table 5. Finished product characteristics of thermally processed canned pet foods [1] containing select hydrocolloids.

Measurement	D	DG	KCG	LBG	XGG	SEM	*p*-Value
pH	5.95 [c]	5.97 [c]	6.38 [a]	6.27 [b]	6.24 [b]	0.080	<0.0001
Total moisture, %	79.37 [a]	79.04 [a]	77.83 [ab]	77.30 [b]	77.28 [b]	0.798	0.0418
EM [2], % of sample	49.91 [a]	26.93 [b]	23.59 [b]	15.92 [c]	17.16 [c]	1.905	<0.0001
Firmness, N	9.03 [c]	8.47 [c]	27.00 [a]	16.30 [b]	14.87 [b]	2.673	<0.0001
Toughness, N·mm	67 [d]	117 [c]	370 [a]	245 [b]	225 [b]	32.5	<0.0001
L* [3]	53.61 [c]	56.88 [b]	57.59 [ab]	59.09 [a]	58.65 [ab]	1.044	0.0023
a* [4]	8.18 [a]	8.56 [a]	4.03 [b]	4.68 [b]	4.51 [b]	1.244	0.0108
b* [5]	21.40 [a]	22.69 [a]	14.64 [b]	15.93 [b]	15.59 [b]	1.511	<0.0001

[abcd] Treatment means with unlike superscripts are different ($p < 0.05$). [1] D = 1% dextrose; DG = 0.5% guar gum and 0.5% dextrose; KCG = 0.5% guar gum and 0.5% kappa carrageenan; LBG = 0.5% guar gum and 0.5% locust bean gum; XGG = 0.5% guar gum and 0.5% xanthan gum. [2] EM = expressible moisture. [3] L* represents the lightness/darkness scale of color; values closer to 100 indicate lighter products and values closer to zero indicate darker products. [4] a* represents the red/green scale of color; more positive values indicate redder colors and more negative values indicate greener colors. [5] b* represents the yellow/blue scale of color; more positive values indicate yellower colors and more negative values indicate bluer colors.

4. Discussion

The aim of this experiment was to quantify the functional characteristics present in canned pet food containing select hydrocolloids, specifically guar gum, kappa carrageenan, locust bean gum, and xanthan gum. Treatments were designed to show the effects of common hydrocolloid systems and guar gum alone at inclusion levels mimicking commercial canned pet food.

4.1. Characteristics of Pre-Thermal Processing Batters of Canned Pet Food

Consistency was affected by the treatments in the present experiment and generally thickened when the total hydrocolloid content of the treatment increased. Thickness of a hydrocolloid solution is dependent in the interactions between the hydrocolloid molecules and the solvent or liquid component of the system [17]. As such, increasing the amount of carbohydrate hydrocolloids increased the number of reactions possible with the solvent. The DG was approximately 3.5 times thicker than D, while KCG, LBG, and XGG were only an average of 2.4 times thicker than DG. Guar gum has a high thickening power compared to many carbohydrate hydrocolloids [2] because it contains many hydroxyl groups that form hydrogen bonds with water. There are no published reference values for consistency of pre-thermal processing batters of canned pet food. A batter with thinner consistency may be easier to mix and pump to the container filling station in a commercial pet food facility. However, a thinner consistency batter may splash more when containers are filled. This would contaminate the seam area, prevent a proper hermetic seal from forming, and expose the pet food to potential external contamination during and after thermal processing [29]. As such, consistency of pre-thermal processing batters should be considered when formulating new commercial pet foods.

Consistency was chosen as the metric to describe viscosity, which is known to affect the rate of heat penetration and the time required to reach lethality in food products [30]. The Bostwick consistometer does not measure viscosity directly and is influenced by other factors including gravitational forces, though it does allow for analysis of samples during can filling. Consistency is also listed in U.S. federal regulations as a potential critical factor for scheduled processes for thermally processed low-acid foods [31]. Studies with other food products have found conflicting results regarding the correlation between direct viscosity and Bostwick consistency measurements [32,33]. Similar research should be conducted with pet foods to validate the Bostwick consistometer as a method for apparent viscosity analysis.

Batter pH and batter water activity were not different among the treatments. The lack of difference in batter pH suggested that minimal reactions occurred during the 5 min mixing after the addition of hydrocolloids. It is also possible that differences may have been detected if the data were analyzed as the concentration of hydrogen ions instead of as pH values (i.e., the negative logarithm of the concentration of pH values). This would have yielded a range of average concentration of hydrogen ions from $1.07 \cdot 10^{-6}$–$1.25 \cdot 10^{-6}$ hydrogen ions in the batters. The lack of difference observed in batter water activity was mainly influenced by the low concentration of hydrocolloids. Generally, carbohydrate hydrocolloids do not affect water activity when their inclusion level is less than 2% [1].

4.2. Thermal Processing Controls and Characteristics of Canned Pet Food

The intention of this experiment was to begin the cooling cycle after the last can containing a thermocouple reached a lethality value of 8 min. However, treatments appear slightly over-processed as the lowest average heating lethality was 8.69 min (Table 3). Similarly, another experiment struggled to achieve their targets when processing canned foods to different F_0 values [23]. Three thermocouples failed during the present experiment, but more than the minimum 10 thermocouples recommended by the Institute for Thermal Processing Specialists [34] were successful across the three replicates for each treatment. Nevertheless, thermal processing parameters of initial internal can temperature, fill weight, gross headspace, and post-processing vacuum were constant. This indicated that differences in lethality were due to the treatments and not influenced by confounding factors. There are no published reference values for these parameters for commercial canned pet food. However, a preliminary study of canned pet food with initial internal can temperatures around 30 °C and can volumes of 88.7 and 162.7 mL observed post-processing can vacuums of -0.8 kPa [26]. Initial internal can temperatures in that experiment were roughly 50% colder than the present experiment and likely influenced the differences in post-processing can vacuum.

Differences were observed in the heating and cooling of the experimental treatments. Specifically, D obtained greater lethality and C_{100} during the heating phase and lower values during the cooling phase of retort processing. This could indicate a faster rate of heat penetration and heat dissipation compared to all other treatments. The thickening of pre-thermal processing batters due to the increase in hydrocolloid content likely increased the resistance to heat, leading to lower lethality and C_{100} when the foods were processed under the same time and temperature conditions. Previous research of the effect of viscosity on heat penetration found that increased food thickness decreased the average heating slope and increased the amount of time required to thermally process food [35]. This suggests that thinner food consistencies may benefit production facilities by decreasing the amount of time to process a food product, which could allow for more products to be made in the same amount of time. It is likely that the heating and cooking lag factors (j_h and j_c, respectively) and the heating and cooling penetration factors (f_h and f_c, respectively) were influenced by the treatments. The lag factors describe how long a food product initially takes to begin heating or cooling while the penetration factors describe the rate of heating or cooling [25]. The present experiment did not investigate these parameters, however, future experiments should do so to provide more understanding of how hydrocolloids affect thermal processing.

The C_{100} calculation has never been applied in literature to canned pet foods. This metric can describe the detrimental effect of increased thermal processing on quality changes such as texture and nutrients. Thiamin degradation is an important concern for pet foods, especially those for cats. Consumption of a thiamin deficient diet can be deadly within a few weeks [36,37]. Deficient pet foods should be recalled to prevent illness and death but recalls are costly to pet food companies. As such, this is a great concern for the pet food industry. However, there are no published reference values for acceptable and unacceptable cook values as it relates to thiamin content or other quality factors of canned pet food. The data presented in this study suggest that canned pet foods processed under commercial conditions have cook values of at least 195.91–201.90 min. Future experimentation needs to determine a maximum C_{100} before texture and thiamin are degraded to unacceptable levels.

4.3. Physicochemical Quality of Processed Canned Pet Food

Color was largely similar between KCG, LBG, and XGG, which was expected. Hydrocolloids are rarely involved in browning reactions. For example, an experiment with chicken sausages found that the level of carbohydrate hydrocolloids only explained 26.5% of the variation in lightness, 6.6% of the variation in redness, and none of the variation in yellowness [5]. Instead, other factors, such as fat inclusion level, were more influential. Differences in color were identified in pâté-style canned pet foods containing different soluble proteins at a 2.5% inclusion level [38]. Companies wishing to alter the color of their products with ingredients at low inclusion levels may have more success changing soluble proteins than carbohydrate hydrocolloids. Regardless, values for the lightness, redness, and yellowness of canned pet foods containing different carbohydrate hydrocolloids have never been published. A pilot study with canned pet foods presented CIELAB color-space values for commercial products, but the ingredient compositions were not disclosed [39]. As such, the values presented for KCG, LBG, and XGG could serve as reference values for chicken-based canned pet foods containing the respective hydrocolloids.

The D and DG treatments appear to have confounding factors influencing their color. First, D was processed to a higher total lethality, which increased the redness and yellowness of thermally processed shrimp in curry [40]. The DG treatment was similar to D in redness and yellowness, which suggested that degree of processing is not the only confounding factor. It is highly likely that the dextrose in both treatments participated in Maillard reactions during thermal processing. This reaction occurs between α-amino groups in proteins and reducing sugars [41] and is associated with increased redness and yellowness and more acidic pH levels in infant formula [42]. This suggested that D and DG could not serve as controls for redness and yellowness in the present experiment. There are no published values for the redness and yellowness of canned pet foods containing dextrose. The data for these two treatments are useful benchmarks for pet food companies who wish to use dextrose to increase the redness and yellowness of their products.

The pH levels of the processed foods were affected by the treatments. The D and DG treatments had more acidic pH, which could be tied to the production of Maillard reaction products mentioned in the previous paragraph. It also appeared that pH became more basic after thermal processing with the degree of change dependent on the carbohydrate hydrocolloids present. This would suggest that thermal processing caused a degree of hydrogen bonding, thus decreasing the amount of free hydrogen ions and explaining the shift to a more neutral pH for treatments LBG and XGG. The KCG treatment shifted even more because the sulfate half-ester groups in kappa carrageenan are negatively charged [1] and shift the pH even more. It may also be that the differences in pH are related to the differences in color. Chicken breasts classified by visual color assessment were further differentiated by pH and CIELAB color values [43]. Specifically, pH was slightly more acidic for lighter chicken breasts and slightly more neutral for darker chicken breasts. This may be related to denaturation of myoglobin due to processing [44], however, this effect would be small in the present experiment due to the low amounts of myoglobin present in chicken meat [45]. Unfortunately, the water activity of the processed foods was not measured in the present experiment. Even though it was not anticipated that water activity would be different due to the low inclusion levels of carbohydrate hydrocolloids [1], this information would have enhanced the discussion.

Firmness, toughness, and expressible moisture were affected by the experimental treatments. Specifically, firmness was higher when guar gum was included with another hydrocolloid and toughness increased with higher total carbohydrate hydrocolloid inclusions. Increasing the level of hydrocolloids in a product would increase the gel strength [2]; this is observed in the toughness parameter. Experiments with 0.5–1.5% carbohydrate hydrocolloids in meatballs [6] and restructured hams [4] observed this phenomenon as increased firmness. It may be that 0.5% guar gum, as in DG, in canned pet foods is not enough to influence firmness compared to a sample without a hydrocolloid. It is also possible that the D treatment exhibited enough variability in firmness that the difference compared to DG was not detectable. This is supported by the visually wider spread in the force deformation curves for D vs. all other treatments (Figure S2) and visual inspection of the D treatment cans prior to compression. It was observed that cans filled later in the sequence for that treatment progressively contained more of the solid phase and less of the liquid phase. This likely contributed to the variation in force deformation curves for the D treatment and further illustrated the importance of guar gum keeping ingredients in suspension and evenly distributed. The effect of increased carbohydrate hydrocolloid content was also observed in expressible moisture. As was mentioned in the discussion of consistency, hydrocolloids with hydroxyl groups can form hydrogen bonds with water [2]. Increasing the level of those hydrocolloids introduces more hydroxyl groups, resulting in more bonding with water. This would lower the amount of water that could be expressed and has been observed in restructured hams [4]. The difference between D and DG highlights this well and illustrates the strong power of guar gum to interact with water. It is possible that processing D to a higher total lethality decreased the overall protein functionality [23] and confounded the observed lower toughness and higher level of expressible moisture. In future experiments, treatments with significantly different consistencies could be thermally processed separately to ensure that all treatments receive the same level of processing.

The KCG treatment was firmer and tougher with lower levels of expressible moisture compared to LBG and XGG. This is caused by the different gel structures formed with these hydrocolloids. Gels created by the combination of kappa carrageenan and potassium ions (i.e., from potassium chloride) can withstand substantial application of force before fracturing [46,47]. These gel systems are typically described as "firm" and "brittle" [2]. On the other hand, guar gum, locust bean gum, and xanthan gum form bonds with the hydrogen atoms in water to form a gel structure [2]. Specifically, gels containing xanthan gum and a galactomannan are described as "firm" and "rubbery" [46]. These different gelation mechanisms are defining features in this analysis. For example, a difference in expressible moisture between DG and KCG was not observed or expected because kappa carrageenan does not participate in many hydrogen bonds with water. This concept was echoed in the force deformation curves produced by the texture analysis procedure (Figure S2). The KCG treatment exhibited larger peaks during the deformation vs. the LBG and XGG treatments. As kappa carrageenan gels are "brittle," the KCG treatment may have fractured multiple times during the compression as increasing levels of force were applied. The LBG and XGG treatments contained gels described as "rubbery" and were able to deform more elastically with less resistance compared to the KCG treatment. As such, the LBG and XGG treatments exhibited smoother force deformation curves compared to the KCG treatment.

Quantitative texture analysis of canned pet food is rarely reported, and expressible moisture has never been documented. As such, the values presented in this manuscript can serve as references for commercial product development and improvement. These metrics may be important to pet owner acceptability and pet palatability and food preference. Reports suggest that cats prefer a softer food requiring less work to chew in the first 7 days of consuming a canned food [28]. The softer textures for LBG and XGG vs. KCG may be preferred by cats, but this was not a focus of the present study. Future work should expand upon this study and utilize palatability testing with dogs and cats as well as consumer testing with pet owners to determine which textures are preferred and why they are preferred.

4.4. Proposed Future Research

This work highlighted multiple areas for future research. First, the effect the hydrocolloid concentration has on pre-thermal processing batter consistency, heat penetration, and finished product texture and expressible moisture should be investigated. This would aid in determining the optimal levels of the hydrocolloids evaluated in the present experiment. As learned from this research, dextrose is not a good control ingredient and another should be used to avoid the changes in pH and color that were observed in the present experiment. The use of the primary meat as the control ingredient is standard practice in evaluating the effects of hydrocolloids in restructured meat products for human consumption. Another alternative control ingredient could be cellulose, which is a carbohydrate ingredient but has no effect on viscosity [48]. Second, the Bostwick consistometer should be validated against direct apparent viscosity methods. This could be done simultaneously to other work in an experiment evaluating hydrocolloids. Findings from such an experiment may confirm that Bostwick consistency is an appropriate methodology or suggest that a different method should be the standard. Finally, the changes due to dextrose inclusion at low levels were unexpected. The effect of inclusion level on pH, color, and Maillard reaction products should be explored in the event that dextrose is essential for future experiments or for commercial products.

5. Conclusions

Hydrocolloid inclusions affected canned pet foods before, during, and after thermal processing. Thickening batter consistency to 6.60 cm traveled in 30 s or thicker likely decreased the rate of heat penetration and lowered the accumulation of lethality and C_{100}. The addition of at least 0.5% guar gum toughened wet pet foods and decreased expressible moisture, but at least 1% hydrocolloid content was needed to observe differences in firmness. Dextrose inclusion at either 0.5% or 1% resulted in lower product pH and increased red and yellow color hues. Replacement of guar gum alone may need to focus on increased consistency prior to thermal processing. On the other hand, researchers should address the greater firmness and toughness and lower expressible moisture observed when kappa carrageenan and guar gum were used in combination compared to guar gum with either xanthan gum or locust bean gum. The differences observed in the present experiment illustrated the importance of hydrocolloids to canned pet foods. These distinctions may influence pet palatability and pet owner preference. Additionally, methodologies for quantifying differences in firmness, toughness, and expressible moisture of canned pet food were described. These methods should be utilized when evaluating new functional and structural ingredient systems for canned pet food.

Supplementary Materials: The following are available online at https://www.mdpi.com/article/10.3390/foods10102506/s1, Figure S1: Internal can temperatures for thermally processed wet pet foods containing different carbohydrate hydrocolloid ingredients [1]; Figure S2: Force deformation curves from modified back extrusion procedure applied to thermally processed wet pet foods containing different carbohydrate hydrocolloid ingredients [1].

Author Contributions: Conceptualization, A.N.D. and C.G.A.; methodology, A.N.D. and H.D.; investigation, A.N.D.; data curation, A.N.D.; writing—original draft preparation, A.N.D.; writing—review and editing, A.N.D., H.D. and C.G.A.; visualization, A.N.D.; supervision, C.G.A.; project administration, C.G.A. All authors have read and agreed to the published version of the manuscript.

Funding: This research received no external funding.

Data Availability Statement: The data presented in this study are available upon request from the corresponding author.

Conflicts of Interest: The authors declare no conflict of interest.

References

1. BeMiller, J.N. *Carbohydrate Chemistry for Food Scientists*, 3rd ed.; Woodhead Publishing: Swaston, Cambridge, UK, 2019.
2. Phillips, G.O.; Williams, P.A. (Eds.) *Handbook of Hydrocolloids*, 2nd ed.; CRC Press LLC: Boca Raton, FL, USA, 2009.
3. Casas, J.A.; Mohedano, A.F.; García-Ochoa, F. Viscosity of guar gum and xanthan/guar gum mixture solutions. *J. Sci. Food Agric.* **2000**, *80*, 1722–1727. [CrossRef]
4. Kim, T.-K.; Shim, J.-Y.; Hwang, K.-E.; Kim, Y.-B.; Sung, J.-M.; Paik, J.-D.; Choi, Y.-S. Effect of hydrocolloids on the quality of restructured hams with duck skim. *Poult. Sci.* **2018**, *97*, 4442–4449. [CrossRef]
5. Andrès, S.; Zaritzky, N.; Califano, A. The effect of whey protein concentrates and hydrocolloids on the texture and colour characteristics of chicken sausages. *Int. J. Food Sci. Technol.* **2006**, *41*, 954–961. [CrossRef]
6. Demirci, Z.O.; Yılmaz, I.; Demirci, A.Ş. Effects of xanthan, guar, carrageenan and locust bean gum addition on physical, chemical and sensory properties of meatballs. *J. Food Sci. Technol.* **2014**, *51*, 936–942. [CrossRef]
7. Kienzle, E.; Schrag, I.; Butterwick, R.; Opitz, B. Calculation of gross energy in pet foods: New data on heat combustion and fibre analysis in a selection of foods for dogs and cats. *J. Anim. Physiol. Anim. Nutr.* **2001**, *85*, 148–157. [CrossRef]
8. Sunvold, G.D.; Fahey, G.C.; Merchen, N.R.; Bourquin, L.D.; Titgemeyer, E.C.; Bauer, L.L.; Reinhart, G.A. Dietary fiber for cats: In vitro fermentation of selected fiber sources by cat fecal inoculum and in vivo utilization of diets containing selected fiber sources and their blends. *J. Anim. Sci.* **1995**, *73*, 2329–2339. [CrossRef]
9. Sunvold, G.D.; Fahey, G.C.; Merchen, N.R.; Titgemeyer, E.C.; Bourquin, L.D.; Bauer, L.L.; Reinhart, G.A. Dietary fiber for dogs: IV. In vitro fermentation of selected fiber sources by dog fecal inoculum and in vivo digestion and metabolism of fiber-supplemented diets. *J. Anim. Sci.* **1995**, *73*, 1099–1109. [CrossRef]
10. Phillips-Donaldson, D. Pet Food Round-Up: Ingredients in the News. *Petfood Industry Magazine.* 15 December 2016. Available online: https://www.petfoodindustry.com/blogs/7-adventures-in-pet-food/post/6185-pet-food-round-up-ingredients-in-the-news (accessed on 20 September 2020).
11. Zentek, J.; Kaufmann, D.; Pietrzak, T. Digestibility and effects on fecal quality of mixed diets with various hydrocolloid and water contents in three breeds of dogs. *J. Nutr.* **2002**, *132*, 1679S–1681S. [CrossRef]
12. Karr-Lilienthal, L.; Merchen, N.R.; Grieshop, C.; Smeets-Peeters, M.; Fahey, G.C. Selected gelling agents in canned dog food affect nutrient digestibilities and fecal characteristics of ileal cannulated dogs. *Arch. Tierernaehrung* **2002**, *56*, 141–153. [CrossRef]
13. Cohen, S.M.; Ito, N. A critical review of the toxicological effects of carrageenan and processed Eucheuma seaweed on the gastrointestinal tract. *Crit. Rev. Toxicol.* **2002**, *32*, 413–444. [CrossRef]
14. Woodard, G.; Woodard, M.W.; McNeely, W.H.; Kovacs, P.; Cronin, M.T.I. Xanthan gum: Safety evaluation by two-year feeding studies in rats and dogs and a three-generation reproduction study in rats. *Toxicol. Appl. Pharmacol.* **1973**, *24*, 30–36. [CrossRef]
15. Jackson, R.F.; Silsbee, C.G. The solubility of dextrose in water. *Sci. Pap. Bur. Stand.* **1992**, *17*, 715–724. [CrossRef]

16. Lee, K.-Y.; Park, S.-M.; An, H.-W.; Cho, H.-D.; Han, B.-H. Prediction of thermal diffusivities of meat products containing fish meat. *Bull. Korean Fish. Soc.* **1993**, *26*, 26.
17. Eliasson, A.-C. (Ed.) *Carbohydrates in Food*; Marcel Dekker, Inc.: New York, NY, USA, 1996.
18. Ahmad, M.N.; Kelly, B.P.; Magee, T.R.A. Measurement of heat transfer coefficients using stationary and moving particles in tube flow. *Food Bioprod. Process.* **1999**, *77*, 213–222. [CrossRef]
19. Berry, M.R.; Bush, R.C. Establishing thermal processes for products with straight-line heating curves from data taken at other retort and initial temperatures. *J. Food Sci.* **1989**, *54*, 1040–1042. [CrossRef]
20. Côté, C.; Germain, I.; Dufresne, T.; Gagnon, C. Comparison of two methods to categorize thickened liquids for dysphagia management in a clinical care setting context: The Bostwick consistometer and the IDDSI Flow Test. Are we talking about the same concept? *J. Texture Stud.* **2019**, *50*, 95–103. [CrossRef]
21. Mouquet, C.; Greffeuille, V.; Treche, S. Characterization of the consistency of gruels consumed by infants in developing countries: Assessment of the Bostwick consistometer and comparison with viscosity measurements and sensory perception. *Int. J. Food Sci. Nutr.* **2006**, *57*, 459–469. [CrossRef]
22. Singh, R.P.; Heldman, D.R. *Introduction to Food Engineering*, 5th ed.; Elsevier: London, UK, 2014; ISBN 9780123985309.
23. Hendriks, W.H.; Emmens, M.M.A.; Trass, B.; Pluske, J.R. Heat processing changes the protein quality of canned cat foods as measured with a rat bioassay. *J. Anim. Sci.* **1999**, *77*, 669–676. [CrossRef]
24. Morris, S.A. *Food and Package Engineering*, 1st ed.; Wiley-Blackwell: Oxford, UK, 2011; ISBN 9781119949794.
25. Toledo, R.T.; Singh, R.K.; Kong, F. *Fundamentals of Food Process Engineering*, 4th ed.; (Food Science Text Series); Springer International Publishing: Cham, Switzerland, 2018; ISBN 978-3-319-90097-1.
26. Molnar, L.M.; Donadelli, R.A.; Aldrich, C.G. 238 The effect of container size and type on lethality values during production of thermally processed wet pet foods. *J. Anim. Sci.* **2017**, *95*, 117. [CrossRef]
27. Jauregui, C.A.; Regenstein, J.M.; Maker, R.C. A simple centrifugal method for measuring expressible moisture, a water-binding property of muscle foods. *J. Food Sci.* **1981**, *46*, 1271–1273. [CrossRef]
28. Hagen-Plantinga, E.A.; Orlanes, D.F.; Bosch, G.; Hendriks, W.H.; van der Poel, A.F.B. Retorting conditions affect palatability and physical characteristics of canned cat food. *J. Nutr. Sci.* **2017**, *6*, 1–5. [CrossRef]
29. Black, D.G.; Barach, J.T. (Eds.) *Canned Foods: Principles of Thermal Process Control, Acidification and Container Closure Evaluation*, 8th ed.; GMA Science and Education Foundation: Washington, DC, USA, 2015; ISBN 978-0-937774-23-6.
30. Singh, A.; Pratap Singh, A.; Ramaswamy, H.S. A controlled agitation process for improving quality of canned green beans during agitation thermal processing. *J. Food Sci.* **2016**, *81*, E1399–E1411. [CrossRef]
31. U.S. Food and Drug Administration (FDA). 21 Code of Federal Regulations. Part 113—Thermally Processed Low-Acid Foods Packaged in Hermetically Sealed Containers. Available online: https://www.ecfr.gov/cgi-bin/text-idx?SID=d961c4db3fc4f8bba567269da74fcf90&mc=true&node=pt21.2.113&rgn=div5 (accessed on 11 December 2020).
32. Ordiz, M.I.; Ryan, K.N.; Cimo, E.D.; Stoner, M.E.; Loehnig, M.E.; Manary, M.J. Effect of emulsifier and viscosity on oil separation in ready-to-use therapeutic food. *Int. J. Food Sci. Nutr.* **2015**, *66*, 642–648. [CrossRef]
33. Vercruysse, M.C.M.; Steffe, J.F. On-line viscometry for pureed baby food: Correlation of Bostwick consistometer readings and apparent viscosity data. *J. Food Process Eng.* **1989**, *11*, 193–202. [CrossRef]
34. Institute for Thermal Processing Specialists (IFTPS). Guidelines for Conducting Thermal Processing Studies. Available online: http://iftps.org/wp-content/uploads/2017/12/Retort-Processing-Guidelines-02-13-14.pdf (accessed on 20 September 2020).
35. MacNaughton, M.S.; Whiteside, W.S.; Rieck, J.R.; Thomas, R.L. The effects of residual air and viscosity on the rate of heat penetration of retort food simulant in pouch when using static and oscillating motions. *J. Food Sci.* **2018**, *83*, 922–928. [CrossRef]
36. Davidson, M. Thiamin deficiency in a colony of cats. *Vet. Rec.* **1992**, *130*, 94–97. [CrossRef]
37. Loew, F.M.; Martin, C.L.; Dunlop, R.H.; Mapletoft, R.J.; Smith, S.I. Naturally-occurring and experimental thiamin deficiency in cats receiving commercial cat food. *Can. Vet. J.* **1970**, *11*, 109–113.
38. Polo, J.; Rodríguez, C.; Ródenas, J.; Morera, S.; Saborido, N. The use of spray-dried animal plasma in comparison with other binders in canned pet food recipes. *Anim. Feed Sci. Technol.* **2009**, *154*, 241–247. [CrossRef]
39. Hsu, C.; He, F.; Mangian, H.; White, B.; Lambrakis, L.; de Godoy, M.R.C. 77 Chemical composition, texture and color analyses, and true metabolizable energy of green banana flour in retorted diets for domestic cats. *J. Anim. Sci.* **2020**, *98*, 57. [CrossRef]
40. Mallick, A.K.; Srunuvasa Gopal, T.K.; Ravishankar, C.N.; Vijayan, P.K.; Geethalakshmi, V. Changes in instrumental and sensory properties of Indian white shrimp in curry medium during retort pouch processing at different F_0 values. *J. Texture Stud.* **2010**, *41*, 611–632. [CrossRef]
41. Benjakul, S.; Chantakun, K.; Karnjanapratum, S. Impact of retort process on characteristics and bioactivities of herbal soup based on hydrolyzed collagen from seabass skin. *J. Food Sci. Technol.* **2018**, *55*, 3779–3791. [CrossRef]
42. Takeda, Y.; Shimada, M.; Ushida, Y.; Saito, H.; Iwamoto, H.; Okawa, T. Effects of sterilization process on the physicochemical and nutritional properties of liquid enteral formula. *Food Sci. Technol. Res.* **2015**, *21*, 573–581. [CrossRef]
43. Fletcher, D. Broiler breast meat color variation, pH, and texture. *Poult. Sci.* **1999**, *78*, 1323–1327. [CrossRef]
44. Andrés-Bello, A.; Barreto-Palacios, V.; García-Segovia, P.; Mir-Bel, J.; Martínez-Monzó, J. Effect of pH on color and texture of food products. *Food Eng. Rev.* **2013**, *5*, 158–170. [CrossRef]
45. Kranen, R.W.; Van Kuppevelt, T.H.; Goedhart, H.A.; Veerkamp, C.H.; Lambooy, E.; Veerkamp, J.H. Hemoglobin and myoglobin content of muscles of broiler chickens. *Poult. Sci.* **1999**, *78*, 467–476. [CrossRef]

46. Phillips, G.O.; Wedlock, D.J.; Williams, P.A. (Eds.) *Gums and Stabilisers for the Food Industry 5*; Oxford University Press: New York, NY, USA, 1990.
47. Wang, Y.; Yuan, C.; Cui, B.; Liu, Y. Influence of cations on texture, compressive elastic modulus, sol-gel transition and freeze-thaw properties of kappa-carrageenan gel. *Carbohydr. Polym.* **2018**, *202*, 530–535. [CrossRef]
48. National Research Council (NRC). *Nutrient Requirements of Dogs and Cats*; The National Academies Press: Washington, DC, USA, 2006.

 *foods*

Article

Effect of *Crocus sativus* L. Stigmas Microwave Dehydration on Picrocrocin, Safranal and Crocetin Esters

Aarón García-Blázquez [†], Natalia Moratalla-López [†], Cándida Lorenzo, M. Rosario Salinas and Gonzalo L. Alonso *

Cátedra de Química Agrícola, E.T.S.I. Agrónomos y Montes, Universidad de Castilla-La Mancha, Campus Universitario, 02071 Albacete, Spain; Aaron.Garcia@uclm.es (A.G.-B.); Natalia.Moratalla@uclm.es (N.M.-L.); Candida.Lorenzo@uclm.es (C.L.); Rosario.Salinas@uclm.es (M.R.S.)
* Correspondence: Gonzalo.Alonso@uclm.es; Tel.: +34-967-59-92-10; Fax: +34-967-59-92-38
† Both authors contributed equally to this work.

Abstract: The dehydration process is the basis to obtain high quality saffron and to preserve it for a long time. This process modifies saffron's main metabolites that define its quality, and are responsible for the characteristic color, taste, and aroma of the spice. In this work, the effect of microwave dehydration on saffron main metabolites (picrocrocin, safranal and crocetin esters) from *Crocus sativus* L. stigmas at three determinate powers and different time lapses was evaluated. The results showed that this dehydration process obtained similar or lower crocetin esters content, and after three months of storage, higher concentration was shown in treatments at 440 W for 36 s, 55 s, and 73 s; at 616 W for 90 s; and at 800 W for 20 s. Picrocrocin content was lower and safranal content was higher in all treatments compared to the control both before and after storage. Regarding to commercial quality, microwave dehydration obtained Category I of saffron according to International Standard Organization (ISO) 3632. After three months of storage, treatments at 616 W for 83 s and 800 W for 60 s obtained lower categories. The results obtained suggest that microwave dehydration is a suitable process for obtaining high quality saffron, 800 W with 6 lapses of 20 s being the best conditions studied.

Keywords: saffron quality; secondary metabolites; drying; high performance liquid chromatography-diode array detection (HPLC-DAD); spectrophotometry

Citation: García-Blázquez, A.; Moratalla-López, N.; Lorenzo, C.; Salinas, M.R.; Alonso, G.L. Effect of *Crocus sativus* L. Stigmas Microwave Dehydration on Picrocrocin, Safranal and Crocetin Esters. *Foods* **2021**, *10*, 404. https://doi.org/10.3390/foods10020404

Academic Editor: Sidonia Martinez and Javier Carballo
Received: 25 January 2021
Accepted: 8 February 2021
Published: 12 February 2021

Publisher's Note: MDPI stays neutral with regard to jurisdictional claims in published maps and institutional affiliations.

1. Introduction

Saffron is the dried stigmas of *Crocus sativus* L. and its high value is due to its colour, flavour, and aroma [1]. There is confusion about the names of the spice and the plant. *C. sativus* L. is the plant, while saffron refers to the spice obtained from the dehydrated stigmas of the plant itself [2,3]. Therefore, the dehydration process is necessary to convert *C. sativus* L. stigmas into saffron.

Traditionally, saffron in Spain is obtained by *C. sativus* L. stigmas dried by a process called "toasting", in which stigmas are put on a sieve with a silk bottom placed over a heating source, whereas in other countries, in terms of temperature, stigmas are dehydrated at room temperature under sunlight or in the shade. Stigmas' dehydration is the key process to obtaining the spice and is responsible for saffron composition [4–6]. During this process, stigmas lose around 80% of their weight and, according to the International Standard ISO 3632 (2011) [7], moisture must be lower than 12% in order to preserve the spice for a long time [8]. Its quality is based on the capacity of the spice to give colour, taste, and aroma to foods and beverages, and these organoleptic properties are influenced by the dehydration process [8–12]. Saffron colour comes from crocetin esters also known as crocins, its bitter taste from picrocrocin, and its distinctive aroma from safranal [13,14]. Depending on the physical and chemical characteristics of saffron, it is classified into one of the three categories established by the ISO 3632 (2011) standard. This international standard is principally used to determine the saffron quality in international commercial agreements, colouring strength being the main parameter from which its price is set. Although the quality of this spice is measured via Ultraviolet-visible (UV-vis) spectrometry, according to ISO 3632 (2011), this method does not provide a precise determination of picrocrocin and safranal concentration [15].

Crocins are a group of water-soluble carotenoids responsible for saffron's colour strength [16–19]. For each crocetin ester, there could be various geometric isomers, *trans* isomers being the most abundant and more stable than *cis* [20]. During the dehydration process, safranal is formed from picrocrocin due to high temperatures. It can also be obtained by extreme pH or a two-step enzymatic process [8,21]. Storage is also a determinant in saffron quality, as safranal concentration is higher in saffron stored longer than one month due to its formation from crocetin esters and picrocrocin during storage time [22,23].

Alternatives to traditional dehydration methods are being researched, as this process is known for being unproductive and slow [24,25]. Some techniques that are studied today are microwave dehydration, oven dehydration, freeze dehydration (lyophilization), vacuum dehydration, and infrared dehydration, among others to be applied to saffron dehydration [26]. These alternative dehydration methods show greater crocin content in freeze dried saffron and in processes where high temperature is applied [5,14]. Microwave dehydration provides great efficiency, controllability over the process, and a higher rate at a lower temperature due to water molecules absorbing the energy and evaporating quickly [13,27,28]. Compared to heat dehydration, microwave provides higher speed and volumetric heating, instead of superficial heating on account of traditional dehydration [29,30].

Previous studies have proposed microwave dehydration as a good alternative to traditional dehydration methods, as it provides higher concentrations of safranal and crocins in a shorter time under determinate conditions [5,6,14]. However, the effect of microwave dehydration on the main metabolites of saffron has not been studied in depth. Thus, the aim of this work was to study the effect of microwave dehydration on saffron's main metabolites and its commercial quality. To achieve this, the content of the most important compounds in saffron—picrocrocin, safranal, and crocetin esters—were measured after carrying out the microwave dehydration of fresh stigmas at different powers and times, using high performance liquid chromatography-diode array detection (HPLC-DAD) and UV-vis spectrometry.

2. Materials and Methods

2.1. Samples and Reagents

Fresh stigmas of *Crocus sativus* L. were purchased from the company "Molineta de Minaya" (Minaya, Spain). The control sample consists of a part of fresh stigmas that were dehydrated according to the own company's internal procedures, considering therefore that this saffron belongs to Protected Designation of Origin (PDO) "Azafrán de La Mancha" [4,31]. Acetonitrile used in HPLC gradient was purchased from Scharlau (Barcelona, Spain). Ultrahigh-purity water was produced using a Milli-Q system (Millipore, Bedford, MA, USA).

2.2. Dehydration Process

Fresh stigmas were distributed into 15 portions of 1.5 g each and placed into a filter paper box. The dehydration process was carried out at low, medium, and high power (440 W, 616 W, and 800 W, respectively) and different time lapses for each (36 s, 55 s, 73 s, and 130 s for 440 W; 26 s, 39 s, 52 s, 79 s, 83 s, and 90 s for 616 W; 20 s, 30 s, 40 s, and 60 s for 800 W) in order to control the decrease of humidity and prevent vegetal material from burning out. Between each lapse, a 10 s rest period was maintained, during which time the mass was measured to monitor weight loss. Time lapses were repeated until weight was reduced by 80% ± 2, resulting in the total energy applied at every treatment (shown in Table 1).

Table 1. Dehydration treatments applied to fresh stigmas of *Crocus sativus* L.

Treatment	Power (W)	Seconds (s)/Lapse	N° Lapses	Total Time (s)	Joules (J)
440-36	440	36	6	216	95,040
440-55	440	55	4	220	96,800
440-73	440	73	3	219	96,360
440-130	440	130	2	260	114,400
616-26	616	26	6	156	96,096
616-39	616	36	4	144	88,704
616-52	616	52	3	156	96,096
616-79	616	78	2	156	96,096
616-83	616	83	2	166	102,256
616-90	616	90	2	180	110,880
800-20	800	20	6	120	96,000
800-30	800	30	5	150	120,000
800-40	800	40	3	120	96,000
800-60	800	60	2	120	96,000

2.3. Saffron Extract Preparation

The saffron extracts were prepared according to ISO 3632 (2011) [7] slightly modified. They were then ground to a powder and passed through a sieve 0.5 mm in pore diameter; 50 mg was then placed in a 100 mL flask, adding 90 mL of Milli-Q water. The solution was stirred for 1 h at 1000 rpm using a magnetic stir bar in the dark. The flask was filled to 100 mL and homogenised through agitation. The solution was filtered through a 0.45 μm pore sized hydrophilic polytetrafluoroethylene (PTFE) filter (Millipore, Bedford, MA, USA). Two extracts were obtained from each dehydration treatment, including the control.

2.4. Nomenclature for Crocetin Esters

Abbreviations for crocetin esters were adopted from Carmona et al. [20]: *trans*-5-tG, *trans*-crocetin (tri-β-D-glucosyl)-(β-D-gentibiosyl) ester; *trans*-5-nG, *trans*-crocetin (β-D-neapolitanosyl)-(β-D-gentibiosyl) ester; *trans*-4-GG, *trans*-crocetin di-(β-D-gentibiosyl) ester; *trans*-4-ng, *trans*-crocetin (β-D-neapolitanosyl)-(β-D-glucosyl) ester; *trans*-3-Gg, *trans*-crocetin (β-D-glucosyl)-(β-D-gentibiosyl) ester; *trans*-2-gg, *trans*-crocetin di-(β-D-glucosyl) ester; *cis*-4-GG, *cis*-crocetin di-(β-D-gentibiosyl) ester; *trans*-2-G, *trans*-crocetin (β-D-gentibiosyl) ester; *cis*-3-Gg, *cis*-crocetin (β-D-glucosyl)-(β-D-gentibiosyl) ester; *trans*-1-g, *trans*-crocetin (β-D-glucosyl) ester.

2.5. HPLC-DAD Analysis

This analysis was performed according to the method by García-Rodríguez et al. [11]. 20 μL of each sample was injected into the Agilent 1200 HPLC chromatograph (Palo Alto, CA, USA) equipped with a 250 mm × 4.6 mm diameter, 5 μm Develosil Octadecyl System-Trifunctional (ODS-HG-5) chromatographic column (Teknokroma, Sant Cugat del Vallès, Barcelona, Spain) equilibrated at 40 °C. The eluents were water (A) and acetonitrile (B) with the following gradients: 20% B, 0–5 min; 20–80% B, 5–15 min; and 80% B, 15–20 min at 0.8 mL/min of flow rate. The DAD detector (Hewlett Packard, Waldbronn, Germany) was set at 250, 330, and 440 nm to detect picrocrocin, safranal, and crocetin esters, respectively. All analyses were performed in duplicate for each replicate (n = 4).

Identification of crocetin esters, picrocrocin, and safranal was carried out as previously reported [11,16]. Quantification was based on the following calibration curves [11]: $C_i = (0.00746 \pm 0.00004)A_i - (0.00571 \pm 0.12863)$, correlation coefficient (R^2) = 0.9999 for *trans*-5-tG, *trans*-5-nG, *trans*-4-GG and *trans*-4-ng; $C_i = (0.00713 \pm 0.00003)A_i - (0.00472 \pm 0.05608)$, R^2 = 0.9999 for *trans*-3-Gg, *trans*-2-gg, *trans*-2-G and *trans*-1-g; $C_i = (0.00531 \pm 0.0004)A_i - (0.00571 \pm 0.12863)$, R^2 = 0.9999 for *cis*-4-GG; $C_i = (0.00500 \pm 0.00003)A_i - (0.00331 \pm 0.05608)$, R^2 = 0.9999 for *cis*-3-Gg; $C_i = (0.02900 \pm 0.00002)A_i + (0.51940 \pm 0.02631)$, R^2 = 0.9999 for picrocrocin, and $C_i = (0.03227 \pm 0.00063)A_i + (0.05101 \pm 0.03103)$,

$R^2 = 0.9989$ for safranal. Limits of detection (LOD) and quantification (LOQ) were taken into consideration [11].

2.6. UV-Vis Spectrometry

In order to perform UV-vis spectrometry analysis according to ISO 3632 (2011) [7], the same extract used for HPLC-DAD analysis was diluted to 1:10 (v/v) and then scanned by duplicate at a wavelength of 440, 330, and 257 nm by a Perkin Elmer Lambda 20 UV-Vis spectrophotometer (Perkin Elmer, Norwalk, CT, USA) in a 1 cm pathway quartz cell.

2.7. Statistical Analysis

The statistical analysis was performed using SPSS Statistics 25 for Windows (IBM, Armond, NY, USA). Data were analysed by performing one-way analysis of variance (ANOVA), with Duncan's test for multiple comparisons and Dependent *t*-test for paired samples with degree of freedom of 3, considering $p < 0.05$ as statistically significant (95% confidence interval).

3. Results and Discussion
3.1. Content of the Main Metabolites of Microwave-Dehydrated Saffron

To determine the effect of microwave dehydration on saffron's main metabolites, the stigmas were dehydrated at different powers and time lapses, and subsequently analysed through HPLC-DAD, which is the only way to determine these compounds [11].

Picrocrocin, safranal, and crocetin esters concentrations are shown in Table 2. With regards to picrocrocin, the first metabolite to be detected in the analysis, control values were significantly higher in traditional dehydration than the microwave treatments. Safranal was not quantified in traditional "toasting" because its content was below the limit of quantification. However, all microwave dehydration treatments showed safranal. Tong et al. [6] observed that an increase in the time of microwave dehydration of fresh stigmas (from 3 to 6 min) at 600 W registered a decrease of safranal concentration, while at 450 W a longer time (from 6 to 10 min) obtained a higher concentration. In our study, it is essential to supply an energy of the same order by combining powers and times (providing different time lapses), so that within each power studied there is no great difference in total time to be able to compare our results with those showed by these authors. Another study described the effect on different dehydration methods, including microwave dehydration [14], in which safranal reported the highest concentration in compared to electric oven dehydration and vacuum oven dehydration; however, the results in the mentioned study cannot be compared to ours as the dehydration time employed was excessive (1.9 h).

The total content of crocetin esters showed significant variances between the control and the other microwave treatments, except for 440-36, 440-130, 616-39, 616-52, and 800-20. This also happened in the sum of *trans* crocetin esters with the exception for the treatment 440-130, which also showed significant differences to the control. The *cis* isomers showed significant differences in microwave-treated samples against the control, and those samples which showed the highest concentration of *cis* isomers were also the ones that had the highest safranal concentration. In this sense, Carmona et al. [23] reported that a high temperature for the dehydration process promotes the isomerization of *trans* crocetin esters to *cis*, as well as safranal synthesis from these carotenoids and picrocrocin. Speranza and Dadá [32] observed the formation of 13-*cis*-crocin, after 1 h of exposition to light. On the other hand, this is the first time that the concentration of the main metabolites of stigmas dehydrated by microwave is evaluated in detail. There is no previous study showing the concentration of *trans* and *cis* isomers, which could be used to compare our results, but it seems that the energy supplied by microwave dehydration of stigmas may be involved in the formation of *cis* crocetin esters. Thus, all these results could indicate that different energy sources could influence the isomerization process and the formation of safranal from picrocrocin and the cycling of *cis* isomers.

Table 2. Content of the main compounds of saffron obtained under different microwave dehydration treatments.

Treatment	Compounds (g/kg Saffron ± SD)					
	Picrocrocin	**Safranal**	**Total CE**	**∑ *Trans*-CE**	**∑ *Cis*-CE**	***Trans*/*Cis***
Control	244.2 ± 8.9 i	<LOQ	262.7 ± 8.7 f	260.4 ± 8.9 f	2.35 ± 0.17 a	113 ± 12 g
440-36	215.4 ± 0.1 e–g	0.41 ± 0.03 ab	252.0 ± 1.7 ef	245.3 ± 1.6 d–f	6.70 ± 0.14 de	37 ± 1 de
440-55	192.6 ± 3.2 bc	0.71 ± 0.07 bc	239.1 ± 6.4 b–e	231.8 ± 6.1 b–e	7.30 ± 0.37 e	32 ± 1 cd
440-73	174.6 ± 2.9 a	1.97 ± 0.15 f	226.6 ± 2.8 a–c	216.3 ± 2.6 ab	10.35 ± 0.36 fg	21 ± 1 ab
440-130	201.3 ± 0.4 cd	0.40 ± 0.01 ab	248.9 ± 1.5 d–f	242.4 ± 1.5 c–e	6.43 ± 0.05 de	38 ± 1 d–f
616-26	208.4 ± 3.8 d–f	0.90 ± 0.11 cd	243.4 ± 7.6 c–e	235.6 ± 7.6 c–e	7.72 ± 0.10 e	31 ± 1 b–d
616-39	217.9 ± 0.4 fg	1.39 ± 0.20 e	250.9 ± 3.9 d–f	244.2 ± 3.9 d–f	6.71 ± 0.28 de	37 ± 2 de
616-52	221.7 ± 2.2 g	1.34 ± 0.15 e	255.2 ± 4.0 ef	248.7 ± 4.2 ef	6.53 ± 0.26 de	38 ± 2 d–f
616-79	202.6 ± 2.0 cd	0.23 ± 0.02 a	231.0 ± 6.0 a–c	225.1 ± 5.8 a–c	5.94 ± 0.17 cd	38 ± 1 d–f
616-83	177.8 ± 2.7 a	0.21 ± 0.03 a	216.9 ± 6.0 a	212.5 ± 5.9 a	4.47 ± 0.30 b	48 ± 3 f
616-90	174.3 ± 2.0 a	0.41 ± 0.04 ab	221.6 ± 3.8 ab	214.7 ± 3.6 ab	6.93 ± 0.19 de	31 ± 2 b–d
800-20	233.7 ± 2.7 h	1.15 ± 0.14 de	252.7 ± 8.5 ef	245.7 ± 8.2 d–f	7.00 ± 0.37 de	35 ± 1 e
800-30	207.3 ± 0.7 de	2.25 ± 0.22 f	238.3 ± 3.6 b–e	228.4 ± 3.6 a–d	9.97 ± 0.78 f	23 ± 2 a–c
800-40	202.2 ± 0.2 cd	2.69 ± 0.35 g	227.5 ± 4.1 a–c	216.2 ± 4.5 ab	11.36 ± 0.99 g	20 ± 2 a
800-60	188.7 ± 4.4 b	0.38 ± 0.05 ab	233.4 ± 6.6 a–d	228.4 ± 6.4 a–d	4.97 ± 0.19 bc	46 ± 1 ef

LOQ = limit of quantification. Total CE = Total crocetin esters. ∑ *trans*-CE = sum of *trans* isomers of crocetin esters. ∑ *cis*-CE = sum of *cis* isomers of crocetin esters. Values are the mean of two extracts conducted in duplicate (2 × 2 n), SD = standard deviation. One-way analysis of variance (ANOVA) for each column is included. Different letters within each column represent statistically significant variances, according to Duncan test ($p < 0.05$).

The proportion between *trans* and *cis* crocetin esters was analysed, showing that *trans* crocetin esters are the predominant form of crocins. All the microwave-treated samples showed significant variance with the control, the values of which ranged from 20 in 800-40 to 48 in 616-83. Therefore, the saffron obtained by microwave dehydration contains less content of *trans* crocetin esters than those obtained from traditional "toasting", resulting in saffron with less bioactive capacity, as *trans* crocetin esters compounds are more bioactive than *cis* isomers [3,12,33].

Crocins are a wide group of glycosyl esters, of which the predominant ones are *trans*-4-GG and *trans*-3-Gg. Concentration values of the crocetin esters are shown in Table 3. One of the predominant crocetin esters is *trans*-4-GG, which showed differences to the control in all microwave dehydration treatments, except for 440-73, 616-83, 616-90, and 800-40. The other main crocetin ester is *trans*-3-Gg, which showed the highest concentration values of all the reported glycosides and obtained significant differences between the control and all the microwave dehydration treatments studied. Previous studies showed that *trans*-4-GG concentration is higher than *trans*-3-Gg in traditionally dehydrated stigmas [11,16,34,35]. Other works are in accordance with our results as a greater value of *trans*-3-Gg concentration than *trans*-4-GG is also registered when stigmas are dried in the shade or freeze-dried [35,36]. However, under these different methods of dehydration, contrary results have also been obtained [36,37]. Therefore, it can be mentioned that these two compounds are the main crocetin esters of saffron.

In the other crocetin esters studied, significant differences to the control were shown for *trans*-5-nG, *trans*-2-gg, *trans*-2-G and *trans*-1-g. All these compounds showed higher content in control except for *trans*-1-g, whose content was higher in all the microwave treatments [3].

Table 3. : Crocetin esters content of saffron obtained by different dehydration treatments.

Treatment	Compounds (g/kg Saffron ± SD)									
	Trans-5-tG	*Trans*-5-nG	*Trans*-4-GG	*Trans*-4-ng	*Trans*-3-Gg	*Trans*-2-gg	*Cis*-4-GG	*Trans*-2-G	*Cis*-3-Gg	*Trans*-1-g
Control	0.33 ± 0.09 ab	1.18 ± 0.12 e	90.2 ± 4.3 a	1.37 ± 0.15 a	125.8 ± 4.5 d	39.11 ± 4.62 d	1.39 ± 0.11 a	2.26 ± 0.57 b	0.96 ± 0.08 a	0.16 ± 0.01 a
440-36	0.83 ± 0.02 f	0.37 ± 0.02 a–d	107.9 ± 0.6 e	2.08 ± 0.16 bc	113.1 ± 1.9 c	20.62 ± 0.32 a–c	4.20 ± 0.06 de	<LOD	2.50 ± 0.09 d-f	0.42 ± 0.01 g
440-55	0.45 ± 0.05 a–c	0.24 ± 0.03 a–c	97.5 ± 1.3 bc	1.63 ± 0.16 ab	110.8 ± 3.9 c	20.74 ± 1.36 a–c	4.50 ± 0.18 de	<LOD	2.80 ± 0.19 fg	0.46 ± 0.01 h
440-73	0.38 ± 0.04 ab	0.14 ± 0.04 a	94.6 ± 0.4 ab	1.85 ± 0.20 a–c	97.5 ± 2.0 a	21.38 ± 0.38 bc	6.49 ± 0.18 f	<LOD	3.86 ± 0.18 h	0.46 ± 0.00 h
440-130	0.73 ± 0.01 ef	0.33 ± 0.02 a–d	104.9 ± 0.1 de	2.32 ± 0.13 c	111.6 ± 1.3 c	21.48 ± 0.20 bc	3.98 ± 0.03 cd	0.53 ± 0.02 a	2.45 ± 0.01 d–f	0.54 ± 0.01 i
616-26	0.67 ± 0.07 ef	0.43 ± 0.09 cd	101.4 ± 2.2 cd	1.54 ± 0.13 ab	108.3 ± 4.1 bc	22.17 ± 1.02 c	5.13 ± 0.14 e	0.73 ± 0.17 a	2.59 ± 0.05 ef	0.37 ± 0.01 f
616-39	0.71 ± 0.06 ef	0.43 ± 0.07 cd	105.5 ± 0.6 de	1.75 ± 0.19 a–c	112.0 ± 2.8 c	22.82 ± 0.57 c	4.47 ± 0.21 de	0.72 ± 0.20 a	2.24 ± 0.12 c–e	0.29 ± 0.01 d
616-52	0.71 ± 0.04 ef	0.46 ± 0.08 cd	107.6 ± 1.6 e	2.07 ± 0.30 bc	113.9 ± 2.8 c	23.00 ± 0.47 c	4.31 ± 0.22 de	0.60 ± 0.12 a	2.23 ± 0.10 c–e	0.33 ± 0.01 e
616-79	0.67 ± 0.04 ef	0.46 ± 0.03 cd	101.5 ± 1.7 cd	1.54 ± 0.04 ab	100.5 ± 3.2 ab	19.31 ± 0.97 a–c	3.88 ± 0.10 cd	0.72 ± 0.02 a	2.06 ± 0.07 cd	0.34 ± 0.01 e
616-83	0.44 ± 0.04 a–c	0.20 ± 0.05 ab	95.7 ± 1.8 a–c	1.37 ± 0.14 a	97.4 ± 3.3 a	16.62 ± 0.90 a	2.86 ± 0.19 b	0.45 ± 0.04 a	1.61 ± 0.12 b	0.32 ± 0.00 e
616-90	0.30 ± 0.02 a	0.17 ± 0.02 a	94.3 ± 1.0 ab	1.37 ± 0.07 a	100.5 ± 2.2 ab	17.09 ± 0.49 ab	4.33 ± 0.08 de	0.44 ± 0.02 a	2.60 ± 0.11 ef	0.55 ± 0.05 i
800-20	0.64 ± 0.06 de	0.56 ± 0.10 d	105.2 ± 2.9 de	1.84 ± 0.27 a–c	114.9 ± 4.2 c	21.89 ± 1.10 c	4.66 ± 0.25 de	0.43 ± 0.06 a	2.33 ± 0.14 de	0.22 ± 0.01 b
800-30	0.50 ± 0.04 b–d	0.42 ± 0.10 b–d	98.4 ± 1.0 bc	1.75 ± 0.20 a–c	105.1 ± 2.2 a–c	21.42 ± 0.40 bc	6.93 ± 0.76 fg	0.52 ± 0.13 a	3.04 ± 0.02 g	0.25 ± 0.01 c
800-40	0.40 ± 0.08 ab	0.35 ± 0.08 a–d	95.2 ± 1.7 ab	1.70 ± 0.22 ab	96.4 ± 2.6 a	21.11 ± 0.33 a–c	7.66 ± 0.70 g	0.80 ± 0.27 a	3.70 ± 0.37 h	0.25 ± 0.01 c
800-60	0.60 ± 0.06 c–e	0.33 ± 0.08 a–d	100.3 ± 1.7 b–d	1.49 ± 0.14 ab	104.8 ± 3.9 a–c	20.26 ± 0.72 a–c	3.14 ± 0.14 bc	0.39 ± 0.02 a	1.83 ± 0.05 bc	0.26 ± 0.00 c

LOD: limit of detection. *trans*-5-tG: *trans*-crocetin (tri-β-D-glucosyl)-(β-D-gentibiosyl) ester; *trans*-5-nG: *trans*-crocetin (β-D-neapolitanosyl)-(β-D-gentibiosyl) ester; *trans*-4-GG: *trans*-crocetin di-(β-D-gentibiosyl) ester; *trans*-4-ng: *trans*-crocetin (β-D-neapolitanosyl)-(β-D-glucosyl) ester; *trans*-3-Gg: *trans*-crocetin (β-D-glucosyl)-(β-D-gentibiosyl) ester; *trans*-2-gg: *trans*-crocetin di-(β-D-glucosyl) ester; *cis*-4-GG: *cis*-crocetin di-(β-D-gentibiosyl) ester; *trans*-2-G: *trans*-crocetin (β-D-gentibiosyl) ester; *cis*-3-Gg: *cis*-crocetin (β-D-glucosyl)-(β-D-gentibiosyl) ester; *trans*-1-g: *trans*-crocetin (β-D-glucosyl) ester; Values are the mean of two extracts conducted in duplicate (2 × 2 n), SD = standard deviation. One-way analysis of variance (ANOVA) for each column is included. Different letters within each column represent statistically significant variances, according to Duncan test ($p < 0.05$).

Cis-4-GG and *cis*-3-Gg were the *cis*-crocetin esters identified and quantified in this work. Both showed significant differences to the control with higher content in the treatments studied. These crocetin esters are known for being less bioactive than the *trans* esters, which results in a saffron with lower bioactive capacity.

Saffron aroma is enhanced after at least a month of its storage in stigmas traditionally dehydrated, the reason why saffron is not sold immediately after dehydration. In addition, previous studies observed that saffron's main compounds' content evolves over time [4,35]. Thus, microwave dehydrated saffron was stored for three months in order to study the evolution of the main metabolites, after which these compounds were analysed by HPLC-DAD again.

Picrocrocin, safranal, and total crocetin esters concentration after three months of storage are shown in Table 4. Both picrocrocin and safranal maintained the significant differences shown between the control and the treatments studied before the storage. In practically all dehydration parameters studied, safranal values were higher after storage, which matches with the previous results of storage effect on safranal according to Maggi et al. [38]. Moreover, Sereshti et al. [39] reported that safranal content is lower in freshly dried stigmas, and at least a month of storage is known to be necessary for the development of saffron aroma [22,23].

Table 4. Content of the main compounds of saffron obtained under different dehydration treatments after being stored for three months.

Treatment	Compounds (g/kg Saffron ± SD)					
	Picrocrocin	Safranal	Total CE	$\sum$ *Trans*-CE	$\sum$ *Cis*-CE	*Trans/Cis*
Control	247.3 ± 3.9 g *	0.33 ± 0.11 a	232.2 ± 8.5 de	231.2 ± 8.5 ef	1.04 ± 0.03 a	223 ± 1 h
440-36	212.5 ± 2.1 e	0.88 ± 0.01 bc	268.2 ± 1.5 g	264.3 ± 1.6 hi	3.90 ± 0.14 c	68 ± 3 f
440-55	209.4 ± 1.0 e	0.81 ± 0.00 b	266.6 ± 7.2 g	262.8 ± 6.9 hi	3.76 ± 0.29 c	71 ± 4 f
440-73	195.0 ± 0.2 e	1.59 ± 0.00 f	254.9 ± 0.4 g	249.9 ± 0.3 i	5.03 ± 0.02 e	50 ± 1 e
440-130	186.6 ± 2.5 b	1.24 ± 0.02 d	238.8 ± 2.7 e	235.9 ± 2.5 fg	2.90 ± 0.16 b	82 ± 4 g
616-26	208.4 ± 0.1 de *	0.98 ± 0.01 c *	226.3 ± 1.0 c–e	219.0 ± 0.8 c–e	7.28 ± 0.19 f	30 ± 1 bc *
616-39	203.1 ± 1.3 c–e	1.76 ± 0.02 g	215.2 ± 2.6 c	207.3 ± 2.4 c	7.87 ± 0.12 f	26 ± 0 b
616-52	197.8 ± 0.5 b–d	1.65 ± 0.01 fg	220.6 ± 0.5 cd	212.8 ± 0.4 cd	7.79 ± 0.16 f	27 ± 1 b
616-79	189.0 ± 0.7 b	0.78 ± 0.00 b	174.1 ± 4.9 b	169.7 ± 4.8 b	4.40 ± 0.04 cd	39 ± 1 d *
616-83	158.1 ± 2.5 a	1.45 ± 0.02 e	130.8 ± 1.7 a	126.7 ± 1.5 a	4.07 ± 022 cd *	31 ± 1 bc
616-90	197.4 ± 0.1 bc	1.18 ± 0.00 d	275.1 ± 0.5 fg	267.1 ± 0.5 gh	7.77 ± 0.12 f	34 ± 1 bc *
800-20	235.6 ± 7.0 f *	1.59 ± 0.29 f	262.0 ± 3.3 g *	252.5 ± 3.3 hi *	9.54 ± 0.07 g	26 ± 0 b
800-30	205.7 ± 2.9 de *	3.59 ± 0.04 h	241.6 ± 3.6 ef *	228.2 ± 3.1 d–f *	13.37 ± 0.51 h	17 ± 0 a
800-40	193.2 ± 1.7 bc	3.82 ± 0.09 i	233.8 ± 3.5 de *	218.9 ± 3.0 c–e *	14.85 ± 0.51 i	15 ± 0 a
800-60	161.5 ± 1.8 a	1.42 ± 0.04 e	164.4 ± 0.7 b	159.6 ± 0.7 b	4.81 ± 0.04 d *	33 ± 0 c

Total CE: Total crocetin esters. $\sum$ *trans*-CE: sum of *trans* isomers of crocetin esters. $\sum$ *cis*-CE: sum of *cis* isomers of crocetin esters. Values are the mean of two extracts conducted in duplicate (2 × 2 n), SD = standard deviation. One-way analysis of variance (ANOVA) for each column is included. Different letters within each column represent statistically significant variances, according to Duncan test ($p < 0.05$). Data marked with * does not show significant variances to same treatment and compound before storage ($p < 0.05$).

The content of total crocetin esters was less in the control and in some microwave dehydration treatments compared to the results obtained before its storage. The treatment 616-83 showed significant variances with the rest of the microwave dehydration treatments and with the control for total crocetin esters and for the sum of *trans* crocetin esters. All dehydration treatments showed significantly higher *cis* content than the control, resulting in a lower *trans/cis* proportion. The treatments that showed the highest concentration of total *cis* isomers also registered the highest safranal content, which was also observed before the storage, reinforcing the previously mentioned relationship between safranal and *cis* crocetin esters.

Crocetin esters were analysed after three months of storage to analyse whether deterioration of these compounds had taken place. Main crocetin esters' concentration values are shown in Table 5.

Table 5. Crocetin esters content of saffron obtained under different dehydration treatments after being stored for three months.

Treatment	Compounds (g/kg Saffron ± SD)									
	Trans-5-tG	Trans-5-nG	Trans-4-GG	Trans-4-ng	Trans-3-Gg	Trans-2-gg	Cis-4-GG	Trans-2-G	Cis-3-Gg	Trans-1-g
Control	0.43 ± 0.03 bc *	1.44 ± 0.02 i	90.0 ± 1.1 c *	1.20 ± 0.07 c *	110.4 ± 6.2 ef	4.75 ± 0.73 d	1.04 ± 0.10 a	21.32 ± 0.26 g	<LOD	1.67 ± 0.18 d
440-36	0.85 ± 0.02 f *	0.74 ± 0.04 gh	115.0 ± 1.1 g	1.80 ± 0.12 de *	121.7 ± 0.8 g	2.88 ± 0.33 c	3.90 ± 0.14 c	20.16 ± 0.11 g	<LOD	1.24 ± 0.05 bc
440-55	0.77 ± 0.02 ef	0.69 ± 0.02 f–h	114.6 ± 1.2 g	2.02 ± 0.10 de	122.9 ± 4.5 g	2.53 ± 0.20 bc	3.76 ± 0.29 c	18.25 ± 0.88 f	<LOD	1.11 ± 0.09 b
440-73	0.72 ± 0.00 f	0.57 ± 0.00 gh	110.2 ± 2.9 g	2.22 ± 0.01 f	111.5 ± 0.2 g	2.36 ± 0.01 c	5.03 ± 0.02 f	21.05 ± 0.04 h	<LOD	1.28 ± 0.00 c
440-130	0.65 ± 0.01 df	0.58 ± 0.02 ef	105.3 ± 0.4 f *	1.86 ± 0.02 de	107.4 ± 1.7 de	2.52 ± 0.10 bc	2.90 ± 0.16 b	16.53 ± 0.32 e	<LOD	1.05 ± 0.05 b
616-26	0.45 ± 0.04 bc	0.36 ± 0.02 bc *	99.3 ± 1.2 de *	2.36 ± 0.23 f	103.9 ± 1.1 c–e *	2.36 ± 0.05 a–c	4.37 ± 0.11 cd	9.10 ± 0.43 c	2.91 ± 0.09 c	1.17 ± 0.08 bc
616-39	0.37 ± 0.01 b	0.36 ± 0.02 bc *	95.2 ± 0.4 d	1.93 ± 0.27 de *	97.5 ± 1.5 c	2.14 ± 0.06 a–c	5.04 ± 0.05 e	8.70 ± 0.32 c	2.83 ± 0.06 c	1.07 ± 0.04 b
616-52	0.48 ± 0.06 bc	0.44 ± 0.01 cd*	97.0 ± 0.6 de	2.11 ± 0.06 ef *	100.3 ± 1.5 cd	2.31 ± 0.05 a–c	4.89 ± 0.17 de	9.05 ± 0.09 c	2.90 ± 0.02 c	1.11 ± 0.02 b
616-79	0.09 ± 0.04 a	0.24 ± 0.02 ab	78.1 ± 2.5 b	1.85 ± 0.01 de	82.5 ± 2.1 b	1.85 ± 0.01 ab	2.45 ± 0.02 b	4.32 ± 0.26 ab	1.95 ± 0.05 b *	0.73 ± 0.01 a
616-83	<LOQ	<LOQ	60.2 ± 0.3 a	0.47 ± 0.04 a	60.9 ± 1.0 a	1.73 ± 0.12 a	2.43 ± 0.15 b	2.81 ± 0.33 a	1.64 ± 0.07 a *	0.58 ± 0.09 a
616-90	0.82 ± 0.02 de	0.82 ± 0.02 e–g	118.3 ± 0.2 g	1.72 ± 0.01 d	128.2 ± 0.3 g	2.48 ± 0.02 a–c	4.16 ± 0.03 cd	13.48 ± 0.01 d	3.81 ± 0.00 d	1.28 ± 0.01 bc
800-20	0.83 ± 0.07 f	0.83 ± 0.07 h	112.8 ± 1.5 g	1.98 ± 0.05 de *	118.8 ± 1.8 fg *	2.73 ± 0.06 c	5.78 ± 0.04 f	13.25 ± 0.08 d	3.76 ± 0.02 d	1.30 ± 0.05 bc
800-30	0.53 ± 0.03 cd *	0.53 ± 0.03 de *	101.4 ± 0.4 ef	1.88 ± 0.14 de *	104.6 ± 1.8 c–e *	2.47 ± 0.11 a–c	9.22 ± 0.38 g	15.14 ± 0.60 e	4.14 ± 0.12 e	1.70 ± 0.08 d
800-40	0.43 ± 0.01 bc *	0.43 ± 0.01 cd *	96.5 ± 0.4 de *	1.38 ± 0.08 c	99.9 ± 1.7 cd *	2.48 ± 0.11 a–c	10.53 ± 0.38 h	16.02 ± 0.66 e	4.32 ± 0.13 e	1.82 ± 0.11 d
800-60	0.15 ± 0.05 a	0.15 ± 0.05 a	76.4 ± 0.3 b	0.90 ± 0.04 b	74.6 ± 0.3 b	1.82 ± 0.06 ab	2.76 ± 0.01 b	4.85 ± 0.17 b	2.05 ± 0.03 b	0.66 ± 0.02 a

LOQ: limit of quantification; LOD: limit of detection. *trans*-5-tG: *trans*-crocetin (tri-β-D-glucosyl)-(β-D-gentibiosyl) ester; *trans*-5-nG: *trans*-crocetin (β-D-neapolitanosyl)-(β-D-gentibiosyl) ester; *trans*-4-GG: *trans*-crocetin di-(β-D-gentibiosyl) ester; *trans*-4-ng: *trans*-crocetin (β-D-neapolitanosyl)-(β-D-glucosyl) ester; *trans*-3-Gg: *trans*-crocetin (β-D-glucosyl)-(β-D-gentibiosyl) ester; *trans*-2-gg: *trans*-crocetin di-(β-D-glucosyl) ester; *cis*-4-GG: *cis*-crocetin di-(β-D-gentibiosyl) ester; *trans*-2-G: *trans*-crocetin (β-D-gentibiosyl) ester; *cis*-3-Gg: *cis*-crocetin (β-D-glucosyl)-(β-D-gentibiosyl) ester; *trans*-1-g: *trans*-crocetin (β-D-glucosyl) ester; Values are the mean of two extracts conducted in duplicate (2×2 n), SD = standard deviation. One-way analysis of variance (ANOVA) for each column is included. Different letters within each column represent statistically significant variances, according to Duncan test ($p < 0.05$). Data marked with * did not show significant variance with same treatment and compound before storage ($p < 0.05$).

The two main crocetin esters, *trans*-4-GG and *trans*-3-Gg kept the previous proportion, *trans*-3-Gg concentration being higher than *trans*-4-GG in all microwave dehydration treatments studied and in the control. The concentration value of *trans*-4-GG compared to the content obtained before storage decreased slightly in the control, while it increased in some of the microwave dehydration treatments. For *trans*-3-Gg, the control also obtained lower value compared to its content before storage, and the treatments increased their concentration only in some of them. Regarding *trans*-4-GG, at 440 W for 36 s, 55 s, and 73 s, along with 616-90 and 800-20, there was significant variance compared to the rest of the treatments. *Trans*-3-Gg showed similar significant differences to those described for *trans*-4-GG in relation to the microwave dehydration treatments. As before the storage of the saffron, *cis*-4-GG and *cis*-3-Gg were identified and quantified. Regarding *cis*-4-GG, all microwave treated samples showed significant variances with the control, which obtained the lowest concentration (1.04 g/kg). *Cis*-3-Gg, however, could not be detected in the control and in the treatments performed at 440 W.

Therefore, after storage, new significant variances were observed. Some of the microwave treated samples presented different trends compared to the control. It is noteworthy that microwave dehydrated saffron showed higher total crocetin esters content in some treatments after storage (at 440 W for 36 s, 55 s and 73 s, 616-90 and at 800 W for 20 s, 40 s and 60 s) and safranal content decreased in 440-73, although the storage is known for improving safranal content but also diminishing carotenoids due to its oxidation [23,24,40]. Considering the main metabolites' content of saffron obtained from the different microwave dehydration treatments studied, and compared to the control (traditionally obtained saffron), the treatments that obtained an increase in total crocetin esters after storage also showed high contents of picrocrocin and safranal. These treatments were able to dehydrate stigmas of *C. sativus* L. and obtain saffron with a content of main metabolites equal to or superior to those obtained by traditionally dehydrated saffron. Among them, treatment 800-20 stands out for obtaining saffron with the highest content of picrocrocin, safranal, and total crocetin esters. The 440-130 treatment would stand out for obtaining saffron with a great bioactive capacity. In addition, it would have high content of the main metabolites. On the other hand, 616-83 would not be recommended for use due to the lowest content of all evaluated compounds in the saffron obtained from this treatment.

Discriminant function analysis was performed on results from HPLC-DAD grouped together according to the power used in the dehydration process, in order to identify a relationship between the compounds during the dehydration process and determine canonical functions that separate samples within two functions (Figure 1).

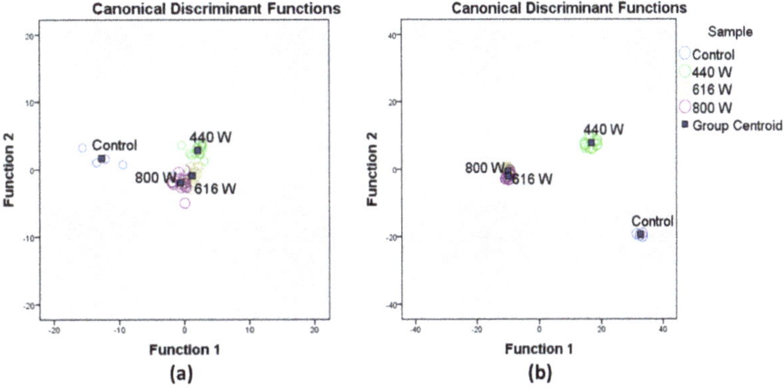

Figure 1. Graphical plot of the results from the stepwise canonical analysis of saffron samples obtained by microwave dehydration at different powers and control samples, before (**a**) and after three months of storage (**b**).

After the dehydration process, the control sample was separated from microwave dehydrated stigmas by function 1 (71.4% of variance) and function 2 (92.8% of cumulative variance) (Figure 1a). Function 1 depended on *trans*-4-GG, *trans*-3-Gg, and *trans*-5-nG mainly, and function 2 depended on *trans*-4-GG, *trans*-5-tG, and *trans*-2-G primarily. Similar analysis was performed after three months of storage, and in this case, control sample and stigmas dehydrated at 440 W were separated from the rest of the treatments by function 1 (82.9% of variance) and function 2 (99.5% of cumulative variance) (Figure 1b). In this case, function 1 depended on *cis*-3-Gg, *trans*-4-GG, and *trans*-2-G mainly, and function 2 depended on *trans*-3-Gg, *trans*-5-tG, and *trans*-5-nG, principally. After three months of storage, a higher separation of the treatments at 440 W due to *cis*-3-Gg, mainly, was observed. Considering Tables 3 and 5, *cis*-3-Gg content decreased at 440 W after storage, while the isomer *trans*-3-Gg increased. It seems that the dehydration process at this power produces a less stable isomerization than the rest of the powers studied.

Therefore, the microwave dehydration process separated saffron obtained from the stigmas dehydrated by "toasting", and this separation was more pronounced after three months of storage, due mainly to the content of crocetin esters and more specifically to the most abundant crocins.

3.2. Commercial Quality of Microwave-Dehydrated Saffron

The commercial quality of saffron is classified into three categories established by ISO 3632 [7]. This standard indicates that the highest quality category must be composed of saffron with a minimum value of 200 of $A^{1\%}_{1\,cm}$ 440 nm, 70 of $A^{1\%}_{1\,cm}$ 257 nm, and a $A^{1\%}_{1\,cm}$ 330 nm value between 20 and 50. Results of UV-vis spectrophotometric analyses of the control and saffron dehydrated at different powers and time lapses are shown in Table 6.

Table 6. UV-vis spectrophotometric parameter values of saffron samples obtained under different microwave dehydration treatments.

Treatment	($A^{1\%}_{1\,cm}$ ± SD) Initial				($A^{1\%}_{1\,cm}$ ± SD) after Storage			
	440 nm	330 nm	257 nm	Cat.	440 nm	330 nm	257 nm	Cat.
Control	238 ± 8 ab	15 ± 1 a	146 ± 24 b	-	235 ± 12 c	22 ± 1 ab	97 ± 3 e–h	I
440-36	260 ± 3 de	24 ± 1 b–d	96 ± 1 a	I	262 ± 3 ef	26 ± 1 b–e	97 ± 1 e–h	I
440-55	256 ± 4 cd	25 ± 1 b–e	96 ± 2 a	I	268 ± 3 fg	26 ± 1 b–e	100 ± 1 gh	I
440-73	239 ± 4 ab	25 ± 2 de	88 ± 2 a	I	259 ± 1 g	27 ± 1 c–e	95 ± 1 e–g	I
440-130	265 ± 1 d–f	28 ± 1 f	99 ± 1 a	I	246 ± 2 c–e	24 ± 1 bc	87 ± 1 c	I
616-26	256 ± 5 cd	25 ± 1 c–e	96 ± 2 a	I	248 ± 1 c–e	19 ± 1 a	90 ± 1 cd	I
616-39	269 ± 2 ef	25 ± 2 de	100 ± 1 a	I	262 ± 2 ef	30 ± 2 ef	99 ± 3 f–h	I
616-52	274 ± 3 f	26 ± 1 ef	102 ± 1 a	I	258 ± 2 d–f	25 ± 1 b–d	94 ± 2 d–f	I
616-79	259 ± 1 de	28 ± 1 f	95 ± 1 a	I	246 ± 1 c–e	29 ± 1 de	93 ± 2 de	I
616-83	237 ± 2 ab	23 ± 1 bc	87 ± 1 a	I	156 ± 1 a	35 ± 1 g	82 ± 1 b	III
616-90	235 ± 1 a	26 ± 1 d–f	89 ± 1 a	I	271 ± 3 fg	27 ± 1 c–e	97 ± 1 e–h	I
800-20	261 ± 2 de	26 ± 2 ef	101 ± 1 a	I	282 ± 8 g	30 ± 1 c–e	110 ± 4 i	I
800-30	248 ± 6 bc	25 ± 1 c–e	95 ± 3 a	I	262 ± 2 ef	34 ± 1 fg	101 ± 1 h	I
800-40	240 ± 2 ab	28 ± 2 f	98 ± 6 a	I	243 ± 3 cd	25 ± 1 b–d	90 ± 1 cd	I
800-60	247 ± 5 a–c	23 ± 1 b	92 ± 2 a	I	185 ± 1 b	25 ± 1 b–d	76 ± 1 a	II

Mean values of two extracts conducted in duplicate (2 × 2 n), SD = standard deviation. Cat. = Category. A dash in category means it could not be considered as saffron according to ISO 3632 (2011). One-way analysis of variance (ANOVA) for each column is included. Different letters within each column represent statistically significant variances, according to the Duncan test ($p < 0.05$).

The initial results showed that the control could not be classified as saffron according to ISO 3632 [7] due to its low value of $A^{1\%}_{1\,cm}$ 330 nm, while all the microwave treatments produced saffron belonging to Category I. Color strength ($A^{1\%}_{1\,cm}$ 440 nm) results showed that there were several differences across different treatments, although saffron obtained in 440-73, 616-83, 616-90, 800-30, 800-40, and 800-60 did not show significant differences between traditional and microwave dehydration. Values of ISO parameters obtained in microwave dehydration treatments are in concordance with those reported by Maghsoodi [5] in a previous essay, taking into account their results obtained at 200 W for 720 s (in total), since the same order of energy (144,000 J) as our work was applied. The values of $A_{1\,cm}^{1\%}$ 257 nm were significantly lower for all the microwave dehydration treatments compared to the control.

After storage, the treatments 616-83 and 800-60 showed a $A_{1\,cm}^{1\%}$ 440 nm value below 200, which relegated them to Category III ($\leq$170) and II ($\leq$200), respectively. These treatments did not maintain the quality previously observed before storage, which shows that they are not suitable for producing high quality saffron.

Saffron value is mainly determined by its $A_{1\,cm}^{1\%}$ 440 nm value, therefore, producers want the saffron with the highest 440 nm value as possible. Between all the treatments analyzed, those that showed the highest $A_{1\,cm}^{1\%}$ 440 nm value were 440-55, 440-73, 616-90, and 800-20. The last mentioned showed high crocetin esters content in HPLC-DAD analysis, proving to be a good alternative to traditional "toasting" for obtaining high quality saffron.

4. Conclusions

We demonstrated that microwave dehydration produces saffron with similar content of crocetin esters and more safranal than saffron obtained from "toasting" under conditions used in 800-20. This treatment maintained high metabolite content after three months of storage; therefore, microwave dehydration provides saffron with high metabolite content and better preservation.

Saffron's commercial quality was measured by UV-vis spectrophotometric analysis in order to classify the saffron obtained into ISO 3632 categories. Between all the treatments that produced saffron belonging to ISO Category I, 800-20 was the one that obtained the highest $A_{1\,cm}^{1\%}$ 440 nm value. The results indicate that saffron obtained from microwave dehydration is an adequate alternative to "toasting" because of the high metabolite content produced and the higher commercial quality obtained.

Author Contributions: Conceptualization, G.L.A. and N.M.-L.; methodology, G.L.A. and N.M.-L.; software, A.G.-B. and N.M.-L.; validation, A.G.-B., N.M.-L., C.L., M.R.S., and G.L.A.; formal analysis, G.L.A. and N.M.-L.; investigation, G.L.A., A.G.-B., and N.M.-L.; resources, G.L.A. and M.R.S.; data curation, G.L.A. and N.M.-L.; writing—original draft preparation, A.G.-B., N.M.-L., C.L., M.R.S., and G.L.A.; writing—review and editing, A.G.-B., N.M.-L., C.L., M.R.S., and G.L.A.; visualization, G.L.A.; supervision, G.L.A.; project administration, G.L.A. and M.R.S. All authors have read and agreed to the published version of the manuscript.

Funding: This research received no external funding.

Institutional Review Board Statement: Not applicable.

Informed Consent Statement: Not applicable.

Acknowledgments: The authors thank the Government of Castilla-La Mancha (Spain) in collaboration with FEDER for the project SBPLY/17/180501/000191.

Conflicts of Interest: The authors declare no conflict of interest.

References

1. Alonso, G.L.; Zalacain, A.; Carmona, M. Saffron. In *Handbook of Herbs and Spices*; Elsevier: Amsterdam, The Netherlands, 2012; Volume 1, pp. 469–498. ISBN 9780857095671.
2. Basker, D.; Negbi, M. Uses of saffron. *Econ. Bot.* **1983**, *37*, 228–236. [CrossRef]
3. Bagur, M.J.; Alonso-Salinas, G.L.; Jiménez-Monreal, A.M.; Chaouqi, S.; Llorens, S.; Martínez-Tomé, M.; Alonso, G.L. Saffron: An Old Medicinal Plant and a Potential Novel Functional Food. *Molecules* **2017**, *23*, 30. [CrossRef]
4. Carmona, M.; Zalacain, A.; Pardo, J.E.; López, E.; Alvarruiz, A.; Alonso, G.L. Influence of Different Drying and Aging Conditions on Saffron Constituents. *J. Agric. Food Chem.* **2005**, *53*, 3974–3979. [CrossRef]
5. Maghsoodi, V. Effect of Different Drying Methods on Saffron (*Crocus sativus* L) Quality. *IJCCE* **2012**, *31*, 85–89. [CrossRef]
6. Tong, Y.; Zhu, X.; Yan, Y.; Liu, R.; Gong, F.; Zhang, L.; Hu, J.; Fang, L.; Wang, R.; Wang, P. The influence of different drying methods on constituents and antioxidant activity of saffron from China. *Int. J. Anal. Chem.* **2015**, *2015*. [CrossRef] [PubMed]

7. ISO 3632-1:2011—Spices—Saffron (*Crocus sativus* L.)—Part 1: Specification. Available online: https://www.iso.org/standard/44523.html (accessed on 2 February 2021).
8. Carmona, M.; Zalacain, A.; Salinas, M.R.; Alonso, G.L. A New Approach to Saffron Aroma. *Crit. Rev. Food Sci. Nutr.* **2007**, *47*, 145–159. [CrossRef] [PubMed]
9. Alonso, G.L.; Salinas, M.R.; Garijo, J.; Sánchez-Fernández, M.A. Composition of crocins and picrocrocin from spanish saffron (*Crocus sativus* L.). *J. Food Qual.* **2001**, *24*, 219–233. [CrossRef]
10. Caballero-Ortega, H.; Pereda-Miranda, R.; Abdullaev, F.I. HPLC quantification of major active components from 11 different saffron (*Crocus sativus* L.) sources. *Food Chem.* **2007**, *100*, 1126–1131. [CrossRef]
11. García-Rodríguez, M.V.; Serrano-Díaz, J.; Tarantilis, P.A.; López-Córcoles, H.; Carmona, M.; Alonso, G.L. Determination of saffron quality by high-performance liquid chromatography. *J. Agric. Food Chem.* **2014**, *62*, 8068–8074. [CrossRef]
12. Moratalla-López, N.; Bagur, M.J.; Lorenzo, C.; Martínez-Navarro, M.E.; Salinas, M.R.; Alonso, G.L. Bioactivity and Bioavailability of the Major Metabolites of *Crocus sativus* L. Flower. *Molecules* **2019**, *24*, 2827. [CrossRef]
13. Sarfarazi, M.; Jafari, S.M.; Rajabzadeh, G.; Galanakis, C.M. Evaluation of microwave-assisted extraction technology for separation of bioactive components of saffron (*Crocus sativus* L.). *Ind. Crops Prod.* **2020**, *145*, 111978. [CrossRef]
14. Chen, D.; Xing, B.; Yi, H.; Li, Y.; Zheng, B.; Wang, Y.; Shao, Q. Effects of different drying methods on appearance, microstructure, bioactive compounds and aroma compounds of saffron (*Crocus sativus* L.). *LWT* **2020**, *120*, 108913. [CrossRef]
15. García-Rodríguez, M.V.; López-Córcoles, H.; Alonso, G.L.; Pappas, C.S.; Polissiou, M.G.; Tarantilis, P.A. Comparative evaluation of an ISO 3632 method and an HPLC-DAD method for safranal quantity determination in saffron. *Food Chem.* **2017**, *221*, 838–843. [CrossRef]
16. Carmona, M.; Zalacain, A.; Sánchez, A.M.; Novella, J.L.; Alonso, G.L. Crocetin esters, picrocrocin and its related compounds present in *Crocus sativus* stigmas and *Gardenia jasminoides* fruits. Tentative identification of seven new compounds by LC-ESI-MS. *J. Agric. Food Chem.* **2006**, *54*, 973–979. [CrossRef] [PubMed]
17. Sánchez, A.M.; Carmona, M.; Ordoudi, S.A.; Tsimidou, M.Z.; Alonso, G.L. Kinetics of individual crocetin ester degradation in aqueous extracts of saffron (*Crocus sativus* L.) upon thermal treatment in the dark. *J. Agric. Food Chem.* **2008**, *56*, 1627–1637. [CrossRef] [PubMed]
18. Moratalla-López, N.; Bouhadida, N.; Bagur, M.J.; García-Rodríguez, M.V.; Oueslati, S.; Alonso, G.L. Comparing Tunisian and Spanish saffron regarding their bioactive metabolites using HPLC and GC methods. *Acta Hortic.* **2017**, *1184*, 279–285. [CrossRef]
19. Jafari, S.-M.; Tsimidou, M.Z.; Rajabi, H.; Kyriakoudi, A. Bioactive ingredients of saffron: Extraction, analysis, applications. In *Saffron*; Elsevier: Amsterdam, The Netherlands, 2020; pp. 261–290.
20. Carmona, M.; Zalacain, A.; Alonso, G.L. *The Chemical Composition of Saffron: Color, Taste and Aroma*; Bomarzo SL: Albacete, Spain, 2006; ISBN 9788486977917.
21. Del Campo, C.P.; Carmona, M.; Maggi, L.; Kanakis, C.D.; Anastasaki, E.G.; Tarantilis, P.A.; Polissiou, M.G.; Alonso, G.L. Effects of mild temperature conditions during dehydration procedures on saffron quality parameters. *J. Sci. Food Agric.* **2010**, *90*, 719–725. [CrossRef] [PubMed]
22. Bolandi, M.; Shahidi, F.; Sedaghat, N.; Farhoush, R.; Mousavi-Nik, H. Shelf-life Determination of Saffron Stigma: Water Activity and Temperature Studies. *World Appl. Sci. J.* **2008**, *5*, 132–136.
23. Carmona, M.; Zalacain, A.; Salinas, M.R.; Alonso, G.L. Generation of Saffron Volatiles by Thermal Carotenoid Degradation. *J. Agric. Food Chem.* **2006**, *54*, 6825–6834. [CrossRef]
24. Funebo, T.; Ohlsson, T. Microwave-assisted air dehydration of apple and mushroom. *J. Food Eng.* **1998**, *38*, 353–367. [CrossRef]
25. Koocheki, A. Dehydration of saffron stigmas. In *Saffron*; Elsevier: Amsterdam, The Netherlands, 2020; pp. 291–299. ISBN 9780128186381.
26. Acar, B.; Sadikoglu, H.; Ozkaymak, M. Freeze Drying of Saffron (*Crocus sativus* L.). *Dry. Technol.* **2011**, *29*, 1622–1627. [CrossRef]
27. Crespo, R.J.; Castaño, J.A.; Capurro, J.A. Secado de forraje con el horno microondas: Efecto sobre el analisis de calidad. *Agric. Tec.* **2007**, *67*, 210–218. [CrossRef]
28. Soysal, Y. Microwave drying characteristics of parsley. *Biosyst. Eng.* **2004**, *89*, 167–173. [CrossRef]
29. De la Hoz, A.; Díaz-Ortiz, Á.; Moreno, A. Microwaves in organic synthesis. Thermal and non-thermal microwave effects. *Chem. Soc. Rev.* **2005**, *34*, 164–178. [CrossRef]
30. Galán, A.M.; Calinescu, I.; Trifan, A.; Winkworth-Smith, C.; Calvo-Carrascal, M.; Dodds, C.; Binner, E. New insights into the role of selective and volumetric heating during microwave extraction: Investigation of the extraction of polyphenolic compounds from sea buckthorn leaves using microwave-assisted extraction and conventional solvent extraction. *Chem. Eng. Process. Process Intensif.* **2017**, *116*, 29–39. [CrossRef]
31. OJEC Publication of an application for registration pursuant to Article 6(2) of regulation (EEC) No 2081/92 on the protection of geographical indications and designations of origin. *Off. J. Eur. Communities.* **2000**, *6*, 4–8.
32. Speranza, G.; Dada, G. 13-cis-crocin: A new crocinoid of saffron. *Gazz. Chim. Ital.* **1984**, *114*, 189–192.
33. Llorens, S.; Mancini, A.; Serrano-Díaz, J.; D'Alessandro, A.M.; Nava, E.; Alonso, G.L.; Carmona, M. Effects of Crocetin Esters and Crocetin from *Crocus sativus* L. on Aortic Contractility in Rat Genetic Hypertension. *Molecules* **2015**, *20*, 17570–17584. [CrossRef]
34. Del Campo, C.P.; Carmona, M.; Maggi, L.; Kanakis, C.D.; Anastasaki, E.G.; Tarantilis, P.A.; Polissiou, M.G.; Alonso, G.L. Picrocrocin content and quality categories in different (345) worldwide samples of saffron (*Crocus sativus* L.). *J. Agric. Food Chem.* **2010**, *58*, 1305–1312. [CrossRef]

35. Chaouqi, S.; Moratalla-López, N.; Lage, M.; Lorenzo, C.; Alonso, G.L.; Guedira, T. Effect of drying and storage process on Moroccan saffron quality. *Food Biosci.* **2018**, *22*, 146–153. [CrossRef]
36. Moratalla-López, N.; Parizad, S.; Habibi, M.K.; Winter, S.; Kalantari, S.; Bera, S.; Lorenzo, C.; García-Rodríguez, M.V.; Dizadji, A.; Alonso, G.L. Impact of two different dehydration methods on saffron quality, concerning the prevalence of Saffron latent virus (SaLV) in Iran. *Food Chem.* **2021**, *337*. [CrossRef] [PubMed]
37. Parizad, S.; Dizadji, A.; Habibi, M.K.; Winter, S.; Kalantari, S.; Movi, S.; Lorenzo, C.; Alonso, G.L.; Moratalla-López, N. The effects of geographical origin and virus infection on the saffron (*Crocus sativus* L.) quality. *Food Chem.* **2019**, *295*, 387–394. [CrossRef] [PubMed]
38. Maggi, L.; Carmona, M.; Zalacain, A.; Kanakis, C.D.; Anastasaki, E.G.; Tarantilis, P.A.; Polissiou, M.G.; Alonso, G.L. Changes in saffron volatile profile according to its storage time. *Food Res. Int.* **2010**, *43*, 1329–1334. [CrossRef]
39. Sereshti, H.; Ataolahi, S.; Aliakbarzadeh, G.; Zarre, S.; Poursorkh, Z. Evaluation of storage time effect on saffron chemical profile using gas chromatography and spectrophotometry techniques coupled with chemometrics. *J. Food Sci. Technol.* **2018**, *55*, 1350–1359. [CrossRef]
40. Raina, B.L.; Agarwal, S.G.; Bhatia, A.K.; Gaur, G.S. Changes in pigments and volatiles of saffron (*Crocus sativus* L) during processing and storage. *J. Sci. Food Agric.* **1996**, *71*, 27–32. [CrossRef]

 *foods*

 MDPI

Article

Effect of Humidity-Controlled Dehydration on Microbial Growth and Quality Characteristics of Fresh Wet Noodles

Jun-Jie Xing, Dong-Hui Jiang, Zhen Yang, Xiao-Na Guo and Ke-Xue Zhu *

State Key Laboratory of Food Science and Technology, School of Food Science and Technology, Jiangnan University, Wuxi 214122, China; jjxing@jiangnan.edu.cn (J.-J.X.); jiangdonghui@163.com (D.-H.J.); zhen.yang@jiangnan.edu.cn (Z.Y.); xiaonaguo@jiangnan.edu.cn (X.-N.G.)
* Correspondence: kxzhu@jiangnan.edu.cn; Tel./Fax: +86-510-8532-9037

Abstract: Humidity-controlled dehydration (HCD) was innovatively applied in this paper to control the growth of microorganisms in fresh wet noodles (FWN). Effects of HCD treatment with different temperatures (40, 60 or 80 °C), relative humidity (RH, 50%, 70% or 90%) and treatment time (5–32 min) on the total plate count (TPC), the shelf-life, and qualities of FWN were investigated. The results showed that HCD reduced the initial microbial load on the fresh noodles and extended the shelf-life up to 14–15 days under refrigeration temperature (10 °C). A 1.39 $\log_{10}$ CFU/g reduction for the initial TPC was achieved after HCD treatment at the temperature of 60 °C and RH of 90%. HCD with higher RH had a more positive influence on quality improvement. The L^* values, the apparent stickiness, and the cooking properties of the noodle body were improved by HCD while good sensory and texture quality of noodles were still maintained after the dehydration process.

Keywords: fresh wet noodles; humidity-controlled dehydration; microorganisms; shelf-life; noodle quality

Citation: Xing, J.-J.; Jiang, D.-H.; Yang, Z.; Guo, X.-N.; Zhu, K.-X. Effect of Humidity-Controlled Dehydration on Microbial Growth and Quality Characteristics of Fresh Wet Noodles. *Foods* **2021**, *10*, 844. https://doi.org/10.3390/foods10040844

Academic Editors: Javier Carballo and Sidonia Martinez

Received: 15 March 2021
Accepted: 11 April 2021
Published: 13 April 2021

Publisher's Note: MDPI stays neutral with regard to jurisdictional claims in published maps and institutional affiliations.

1. Introduction

Fresh wet noodles (FWN) are prone to be contaminated by microorganisms in production and distribution due to the high-water activity and rich nutrient content [1,2]. Therefore, the pasteurization of fresh noodles is always a key step to be assured in the industrial production of this traditional staple food, and it is also of great importance to effectively extend the shelf-life of noodle products without decreasing their edible quality [3]. Many studies have been done to adopt comprehensive measures to reduce microbial contamination by using raw materials with low bacteria load, controlling and improving sanitation during processing, packaging, and distribution of fresh noodle products [4–6].

A lot of research and attempt has been done for validation and improvement of the fresh-keeping techniques of FWN under room temperature. Decreasing pH by the addition of organic acid into fresh noodles had been proven to be an effective preservation method [7], while directly adding into or spraying edible alcohol on the surface of the noodles is currently the most widely used method for pasteurization during the fresh noodle production [8]. By adding a variety of natural and/or chemical preservatives, like sodium/calcium propionate, capryl monoglyceride, tea polyphenols, or chitosan into the noodle bodies, better antimicrobial and fresh-keeping results have been achieved in practice [5,9].

In addition to the above chemical methods, physical treatments, like heat and radiation could also kill microorganisms [6]. Most previous studies have been focused on non-thermal inactivation technologies including high hydrostatic pressure (HHP), pulsed electric field (PEF), intense pulsed light (IPL), low-temperature plasmas (LTP), ultrasound or ultraviolet light, etc. [1]. Other technologies like ozone water treatment, modified atmosphere packaging, control of water activity (a_w), and hurdle technology had been tried for preservation purposes in terms of process improvement in the FWN

91

field [10,11]. Moreover, heat treatment can destroy microorganisms by breaking the internal bonds of thermally activated molecules inside the organisms [12]. Although pasteurization by high-temperature heat treatment was reported not feasible for heat-sensitive food [1], moderate dehydration technology with high-temperature at 105–135 °C has been validated and applied in the production of Chinese semi-dried noodle to reduce the microbial load and the activity of the oxidase [2].

In general, the industrial drying process of noodle products was often at low temperature with divided steps, and the humidity of the hot air was usually taken into consideration for fine dried noodles, Udon noodles, and pasta products [13,14]. Empirically then, the relative humidity, as well as the temperature of the hot air, should be controlled during the drying process for both semi-dried noodles and fresh wet noodles. Nevertheless, heat treatment with excessive temperature and humidity may result in the over gelatinization or melting of starch, the generation of cracks or chaps in fresh noodles, thus deteriorating the noodle quality [15]. In terms of microbial control and quality improvements, it is hypothesized that the fresh noodles would benefit a lot from heat treatment with moderate or mild conditions. We, therefore, attempted a trial of thermal treatments for fresh noodles with a gentler method by employing humidity-controlled hot air to test that hypothesis [16]. And to the best of our knowledge, the method has not been done yet. Therefore, this study aims to investigate the effects of thermal dehydration treatment with medium temperature and controlled RH on the shelf-life and qualities of the fresh wet noodles.

2. Materials and Methods

2.1. Materials

Wheat flour was obtained from Yihai Kerry Grain and Oil Industry, and the protein, ash, and moisture contents were $11.28 \pm 0.15\%$, $0.44 \pm 0.09\%$, and $13.18 \pm 0.07\%$, respectively. The table salt used in the noodles was from the local supermarket. Other reagents used were of analytical grade.

2.2. Preparation of the Fresh Wet Noodles (FWN)

The formula of the fresh wet noodle in this study consisted of 1000 g of flour, 340 mL distilled water, and 10 g NaCl. For noodle processing, the flour, distilled water, and NaCl were put into a vacuum mixer (Model HWJZ-5, Nanjing, China) and mixed for 7 min under a vacuum degree of −0.08 MPa to form the noodle dough. The dough was put into a plastic bag that was sterilized by ultraviolet radiation beforehand and then rested for 25 min at 25 °C. Then, the dough was passed through a small roller noodle machine (Model JMTD-168/140, Beijing, China) gradually. The thickness was reduced from 2 mm to 1.5 mm with the roller gap and finally reduced to 1 mm to obtain a dough sheet. The dough sheet was passed through the noodle machine to get the resultant noodle strands with the dimensions of 1.0 mm in both width and thickness. The water content of the noodle body was about 34% after noodle preparation.

2.3. Humidity-Controlled Dehydration (HCD)

The above fresh noodles were thermally treated by using automatic noodle drying equipment (SYT-030, Beijing, China). The dehydration temperature and relative humidity were adjusted to 40, 60, 80 °C, and 50%, 70%, 90% with different combinations before operations. In this paper, the noodle sample treated under the temperature of 40 °C and RH of 50% was abbreviated as sample "40 °C-50%" and so on. Since the dehydration speed varied according to the thermal treatments, the dehydration time was different depending on the conditions when the water content of all the noodle samples was reduced from 34% to the same level (about 30%). The moisture content variations and the corresponding dehydration time of the HCD treated noodles were listed in Table 1 and the water content of the control sample was 34.53%. The obtained HCD noodles were cooled at room temperature (about 25 °C) for 1 h in a sterilized plastic bag and then the sealed noodles were stored at 10 °C before the microbial analysis.

Table 1. Humidity-controlled dehydration (HCD) and its effect on the final moisture content, gelatinization degree, and the changes of yeasts and molds count of fresh wet noodles *.

Temperature (°C)	HCD RH (%)	Time (min)	Final Moisture Content (%)	Gelatinization Degree (%)	Yeasts and Molds Count at Different Storage Time (CFU/g) 0 day	3 days	6 days	9 days	12 days	15 days	18 days
Control			34.00 ± 0.15 [a]	23.14 ± 0.74 [a]	150	2000	5000	-	-	-	-
	50	6	30.18 ± 0.20 [a]	23.46 ± 0.65 [a]	125	1450	3800	-	-	-	-
40	70	12	30.36 ± 0.19 [a]	23.63 ± 0.71 [a]	80	350	1750	3200	-	-	-
	90	32	30.27 ± 0.20 [a]	23.81 ± 0.47 [a]	42	200	1280	2200	-	-	-
	50	5	30.21 ± 0.22 [a]	24.66 ± 0.62 [ab]	78	190	1120	2000	-	-	-
60	70	10	30.33 ± 0.17 [a]	25.71 ± 0.38 [b]	20	25	65	310	1550	-	-
	90	23	30.25 ± 0.21 [a]	27.33 ± 0.41 [c]	10	10	10	20	20	400	-
	50	3	30.38 ± 0.16 [a]	31.39 ± 0.42 [d]	10	10	10	15	350	1800	-
80	70	7	30.29 ± 0.24 [a]	32.04 ± 0.53 [de]	<10	<10	<10	<10	<10	<10	20
	90	18	30.25 ± 0.23 [a]	33.83 ± 0.55 [e]	<10	<10	<10	<10	<10	<10	20

*, Values of moisture content represent mean $\pm$ S.D., n = 3. RH, relative humidity; The same lowercase letter means there was no significant difference within the same column; -, test was terminated by human intervention; <10, not detected or below the detectable limit.

2.4. Determination of Bacterial Content

The HCD treated noodles (25 g) in different storage stages were sampled, pulverized, and mixed with 225 mL of 0.85% aseptic physiological saline. The mixture was homogenized by a stomacher machine (Lab-blender 400; Seward Laboratory) for 60 s before transferred to make different series of dilutions using 0.85% aseptic physiological saline. 1 mL of appropriate dilutions was pipetted onto sterile plate count agar plates and incubated at 36 °C for 48 ± 2 h before the examination of the total plate counts (TPC) according to GB/T 4789.2-2016 [17]. The yeasts and molds count (YMC) were also calculated using Bengal red medium after incubation for 5 days at 28 °C according to GB/T 4789.15-2016 [18].

2.5. Color Measurement of the HCD Noodles

The change of L^* value of the HCD treated noodles in different storage days were measured by a Chroma Meter (Konica Minolta CR-400, Osaka, Japan) equipped with a D65 illuminant. Ten thin strips of the noodles were closely arrayed as a line, and measurement was carried out at 4 testing points on the surface of each noodle sample. L^* is a measurement of noodle brightness (100 = white, 0 = black) according to CIE L^*, a^*, and b^* system, and each of the sample points was tested three times.

2.6. Determination of the Degree of Gelatinization

The degree of gelatinization in the noodles was measured based on the starch-iodine complexes reaction, according to the method reported by Wootton et al. [19]. The HCD noodles were lyophilized, ground, and passed through a 100 mesh sieve. Two treated samples (1 g) were placed in tube 1 and tube 2, and the negative control in tube 1, respectively. Distilled water (50 mL) was then added to each tube, followed by shaking. The sample in tube 1 was heated in a boiling water bath for 20 min and then cooled to room temperature. Glucoamylase (2 mL) was added to each tube and incubated in a water bath at 50 °C for 1 h with shaking every 15 min. After 1 h of incubation, 2 mL of HCl (1 M) was added into each tube to terminate the reaction. Each sample was placed in a 100 mL volumetric flask filled with distilled water and filtered. The filtrate (10 mL) of each diluted sample was placed in a 250 mL volumetric flask, and 10 mL of iodine solution (0.1 M) and 18 mL of NaOH were added. The flask was then sealed for 15 min before the addition of 2 mL of sulfuric acid (10%, w/w). Sodium thiosulfate (0.1 M) was used as an indicator to determine the endpoint of titration and the volume consumption was recorded. The degree of gelatinization was calculated by the equation:

$$\text{Degree of gelatinization (\%)} = (V_0 - V_2)/(V_0 - V_1) \times 100$$

where V_0 is the volume (mL) of sodium thiosulfate consumed by a blank sample, V_1 is the volume (mL) of sodium thiosulfate consumed by a fully gelatinized noodle sample, V_2 is the volume (mL) of sodium thiosulfate consumed by the ungelatinized sample.

2.7. Cooking and Textural Properties of the Fresh Wet Noodles

2.7.1. Cooking Loss and Water Absorption Index

The cooking loss of the HCD noodles was determined as described by AACC Method 66–50 [20]. The raw noodle sample (Ws, about 25 g) was placed into 400 mL of boiling distilled water until the optimal cooking time (OCT) was reached. The boiled noodle was removed from the cooking water and drained for 3 min and weighed. The cooking water was cooled to room temperature and transferred in a 500-mL volumetric flask filled with distilled water, and then 100 mL of it was transferred to a 250-mL beaker. After that, the beaker was placed into an air oven to dry at 105 °C until a constant weight was reached. The residue was weighed and calculated as a percentage of the starting material (dry weight basis). In a separate test, the same amount of fresh noodles was weighted, cooked to OCT, and removed from the cooking water. The boiled noodle was then reweighted after absorbing the water on the surface with 5 layers of filter paper. The water absorption index was calculated as the ratio of cooked noodle weight over the dry sample weight.

2.7.2. Textural Properties

The texture properties of the uncooked and cooked HCD noodles were determined using a TA-XT2i texture analyzer (Stable Micro Systems, London, UK). The P/35 probe was used to measure the apparent stickiness of uncooked HCD noodles. The hardness of cooked HCD noodles was measured according to the description of Wu et al. with some modifications [21]. Fresh wet noodles were cooked to the OCT before removed from the cooking water and the remaining moisture on the surface of the boiled noodles was quickly dried by using five layers of filter paper after draining with cold water. Testing was completed within 15 min. The tensile testing was performed with an A/SPR probe and one strand of noodle was located through slots and was wound round parallel friction roller two or three times. The distance between the two rollers was set at 50 mm and the maximum operation distance was 100 mm. The pretest and test speed were 2.0 mm/s, and the post-test speed was 10.0 mm/s. For the hardness test, three strands of noodles were arrayed on the platform with HDP/FPS probe, and the pretest, test, post-test speed were 2, 0.8, 0.8 mm/s, respectively. The trigger force was 5 g and the compressive strain was 75%. The pause between the first and second compression was 2 s. Each of the noodle samples was tested at least 7 times and an average was applied.

2.8. Sensory Evaluation

The uncooked and cooked HCD fresh noodles were prepared for sensory evaluation using quantitative descriptive analysis (QDA) according to the method of Costa et al. [22]. The sensory evaluation panel consisted of 12 trained tasters (8 females and 4 males). The uncooked noodle samples were submitted to the panelists for estimation of color and appearance. A ten-point hedonic rating scale with "10" indicating "like extremely" and "1" indicating "dislike extremely". The chewiness ("20"), elasticity ("25"), adhesiveness ("15"), smoothness ("5"), flavor ("5"), and overall acceptability ("10") were assigned different weights and evaluated for the cooked noodles. The sensory tests were conducted in triplicate.

2.9. Scanning Electron Microscopy

The morphology of fresh wet noodles after the HCD treatment under different conditions was investigated using a scanning electron microscope (Quanta-200; FEI Ltd., Eindhoven, The Netherlands). Noodle samples were rinsed, eluted, and freeze-dried. The samples were then mounted with double-sided carbon adhesive tabs on aluminum stubs, and coated with gold-palladium. The surface and cross-section of the noodles were observed and photographed at an accelerating voltage of 1.0 kV.

2.10. Statistical Analysis

All of the above-mentioned tests were carried out in triplicate and the results are expressed as the mean $\pm$ standard deviation. SPSS 17.0 software (SPSS Inc., Chicago, CA, USA) was used for the analysis of variance (ANOVA), and a significant difference between any two means was calculated by using Duncan's multiple range tests. A 0.95 confidence level ($p < 0.05$) was applied to determine if a significant difference existed between the two mean values.

3. Results and Discussion

3.1. Microbial Analysis of HCD Fresh Wet Noodles

3.1.1. Initial Bacterial Content

The effect of HCD on the initial bacterial content including TPC and YMC was shown in Figure 1A. Under the same condition of temperature or relative humidity, the initial TPC of HCD noodle samples was significantly decreased ($p < 0.05$) with the increase of relative humidity or temperature, respectively. In this study, the moisture content of fresh wet noodles was settled to the same level of 30% to evaluate the effect of HCD on storage quality of fresh wet noodles. Therefore, the RH and HCD time would depend on each other, since, at higher RH, the time needed to dry the product is longer due to the lower moisture gradient between product and air. This result indicated that HCD could inhibit the growth of microbes to a certain degree, and it had a more positive effect under the condition of either higher RH and temperatures, or longer HCD time. By comparison, the initial TPC of 40 °C-70%, 60 °C-70% and 80 °C-70% samples were lower than that of the control by 0.37, 0.85, 1.54 $\log_{10}$ CFU/g, respectively, and a maximum inactivation level of 1.94 $\log_{10}$ CFU/g for the initial TPC of the HCD noodles was achieved in this study under the conditions of 80 °C-90%. As mentioned by Chau et al. [12], heat treatment in a high-temperature gas or environment could destroy the microorganism by activating the molecules of the organism and breaking down their internal bonds. In our previous study, the shelf-life of semi-dried noodles was extended 5 days after high-temperature-short-time (HTST, 120 °C) dehydration treatment, and the TPC was decreased significantly [2]. Inspired by this work, we tried to explore the applicability of thermal dehydration technology in the fresh wet noodles field, with some modifications. The results agreed with the previous study and suggested some thermolabile bacteria were inactivated during the HCD process and the relative humidity could affect the efficiency of the inactivation by heat. For instance, 60 °C of HCD significantly decreased the initial TPC of fresh wet noodles by 20.7, 23.9, and 39.5 percent as RH increased from 50% to 70% and 90%. It should be noted that the HCD time was also increased accordingly from 5 to 23 min, therefore, the effect of RH should be discussed in conjunction with the effects of HCD time. The YMC of the fresh noodles was reduced from 150 CFU/g to 125, 78, and drastically to 10 CFU/g as the temperature increased from 40 to 60 and 80 °C under the same RH of 50% (Table 1). All yeasts and molds have their optimal growth temperature: that is, if the temperature was raised above a certain point or critical point, most of the yeasts and molds would be killed by thermal treatment. The prime example was the progress of high-temperature short-time (HTST) pasteurization technology in the milk field, and most of the potentially harmful bacteria could be killed by increasing the temperature to about 60–80 °C [23,24]. In this study, the yeasts and molds in fresh noodles were effectively inactivated at 80 °C and the removal efficiency was above 90% (which was 1-$\log_{10}$ reduction) after 3 min of thermal dehydration. All the microorganisms, no matter it is spoilage or pathogenic bacteria, prefer to grow in high-a_w conditions and some foodborne pathogens were able to survive in foods with low a_w as well as in dry environments [25,26]. Except for the temperature, the relative humidity also had a great effect on the microbial thermo-tolerance in thermal treatment [26]. The results showed almost 50% of yeasts and molds were induced to death by thermal treatment at 60 °C and 50% of RH, while a higher inactivation rate of 90% (1-$\log_{10}$ reduction) was achieved for yeasts and molds in FWN when the RH was increased to 90%. A high level of relative humidity in environments should have led to a large number of microbes growing, but the high-moisture hot air during HCD induced microbial cell killing instead. Therefore, HCD with higher RH (longer HCD time) had a more positive influence on reducing both TPC and YMC, and we speculated that some heat-resistant bacteria could be inactivated under high temperatures with higher humidity.

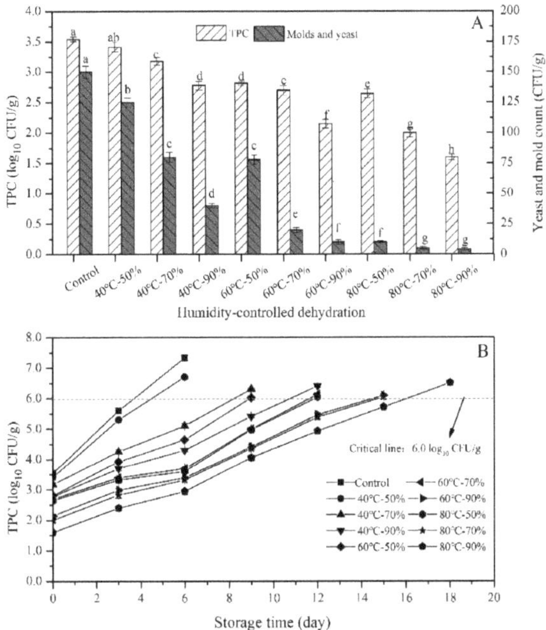

Figure 1. Application of humidity-controlled dehydration (HCD) in microbial control of fresh noodles. (**A**), effect of HCD on the initial total plate count (TPC) and yeast and mold count (YMC). (**B**), effect of HCD on the microorganism growth of fresh noodles during storage. Control, fresh wet noodles; 40 °C-50%, fresh wet noodle samples treated by humidity-controlled dehydration under the condition of 40 °C and relative humidity (RH) of 50%, etc. The different lowercase letter means there was a significant difference in TPC/YMC ($p < 0.05$).

3.1.2. Microbial Growth during Storage

According to the report of Li et al. [2], the fresh wet noodles would deteriorate and their quality lowered once the TPC was over 10^6 CFU/g and hence the quality analysis of noodle samples would be terminated in this study [11]. As shown in Figure 1B, the microbial load of the control sample with an initial TPC of 3.54 $\log_{10}$ CFU/g would quickly exceed the threshold level in 3–4 days during storage at 10 °C, and soon the fresh wet noodles would deteriorate. After HCD treatment, the shelf-life of FWN was extended up to 9–15 days with the increase of treating temperature and RH. The initial TPC is one of the most important factors that affect the shelf-life of fresh wet noodles, and in general, the shelf-life of FWN with a lower initial bacterial load could be prolonged. After HCD treatment, the initial TPC of sample 60 °C-90% was reduced by 1.39 $\log_{10}$ CFU/g compared with the control, and the shelf-life of which was extended up to more than 12 days. Since there was no significant difference ($p > 0.05$) in initial TPC between the sample 40 °C-90% and 60 °C-50%, and between the sample 60 °C-70% and 80 °C-50%, we could not distinguish the corresponding samples (with same water content of 30%) in terms of TPC variations and shelf-life.

The change of YMC of FWN during storage was similar to that of TPC and was listed in Table 1. HCD also had a great influence on the growth of yeasts and molds. Based on our experience, we assumed that the fresh noodles would deteriorate quickly after 3 days of storage if the YMC exceeded 2000 CFU/g, which was in step with the evolution of TPC. Therefore, it is important to take measures to reduce the initial microbial load to extend the shelf-life of FWN.

3.2. Quality Analysis of HCD Fresh Wet Noodles

3.2.1. Color Changes

Fresh wet noodles are prone to enzymatic and non-enzymatic browning during storage [2]; this would result in the noodle darkening and could be measured by the change of L^* value during storage (Figure 2A). Although excessive heat treatment has been reported to deteriorate the noodle quality including color degradation [26], the L^* value of HCD noodles in this study was much higher than the control (about 76, data not shown) and were increased to some extent with the increase of dehydration temperature which indicated a brighter color. This was mainly due to that moderate thermal treatment could inactivate enzymes especially the polyphenol oxidase [24]. The decrease in L^* value (ΔL) was 6.95 after 3 d of storage for sample 40 °C-50%, which was only 0.83 for sample 80 °C-90%. For HCD samples treated at 60 and 80 °C, the change of L^* value during storage was less with increasing relative humidity from 50% to 90% since the dehydration time was accordingly extended from 3–5 min to 18–23 min, respectively (Table 1). The experiment also showed that the color change was consistent with the results when applying HTST dehydration on semi-dried noodles [2]. The browning rate was slowed down by HCD treatment in this study, which was of great significance to reduce the color change during the storage of fresh noodles [27].

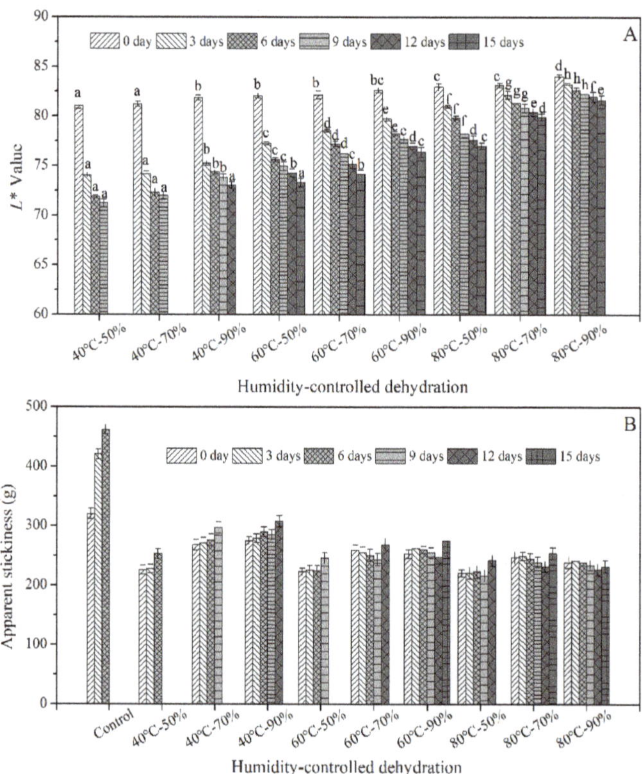

Figure 2. Effect of humidity-controlled dehydration on the L^* value (**A**) and apparent stickiness (**B**) of fresh wet noodles during storage. Control, fresh wet noodles; 40 °C-50%, fresh wet noodle samples treated by humidity-controlled dehydration under the condition of 40 °C and RH of 50%, etc. The different lowercase letter means there was a significant difference in the same storage time ($p < 0.05$).

3.2.2. Apparent Stickiness

As shown in Figure 2B, the apparent stickiness of fresh noodles was significantly decreased after the HCD treatment; this was mainly because the water evaporation on the surface of noodle bodies would result in the formation of a hard shell, thus the stickiness decreased. In general, the stickiness was decreased a little bit with the increase of dehydration temperature and increased with the increase of the relative humidity. Particularly, the apparent stickiness of the sample group (60 and 80 °C) was reduced and the increase of which was slowed down during storage. This may due to the aggregation of protein during the HCD treatment. The aggregated protein network would reduce the starch swelling, making the fresh wet noodles firmer and non-sticky [28]. This would be discussed in the next part based on the changes in the degree of gelatinization.

3.2.3. Degree of Gelatinization

The changes in the degree of gelatinization of the HCD fresh wet noodles were listed in Table 1. During the dehydration process, the starch in fresh noodles gelatinized when the temperature increased up to 60–80 °C. Besides temperature, the degree of gelatinization of the HCD noodles was also influenced by the relative humidity, since it needs more time to dehydrate under higher humidity, during which gelatinization may happen. However, a relatively high degree of gelatinization would make the characteristics of fresh noodles deviate from the characteristic of the raw noodles. And through the analysis of sensory evaluation of fresh noodles with different gelatinization degrees in preliminary experiments, the level of 35% could be considered as the acceptable threshold of starch gelatinization degree for fresh noodles. As listed in Table 1, there was no significant difference in gelatinization degree between and among the 40 °C sample groups and the control fresh noodles. With the increasing temperature at 60 and 80 °C, the gelatinization degreed of HCD noodles was increased up to 33.83% for the sample 80 °C-90%. This is mainly because the final water amount remaining in the HCD fresh wet noodles was about 30% and might not be sufficient to cause complete swelling and starch gelatinization [27]. On the other hand, the gluten matrix in the noodles was capable of entrapping starch granules and limiting their swelling [27]. Therefore, the HCD noodles in this study still possessed the characteristic of raw fresh noodles.

3.2.4. Microstructure

The microstructure of the fresh wet noodles treated with HCD under different conditions was investigated by scanning electron microscope (SEM). The micrographs of the surface (Figure 3A–F) and the cross-section (Figure 3a–f) part of the noodles in 600× and 300× magnification were shown in Figure 3. The obtained micrograph revealed that the 40 °C-50% HCD noodles presented a dispersed or incompact gluten network formation with sphere-shaped starch granules, which did not swell too much (similar to the control, figure not shown). The noodle surface gradually turned from initially rough into smooth as the temperature and RH rose to 60 °C and 90%, respectively, and the cross-sections turned from porous into compact. For the sample 80 °C-90%, the starch granules were observed to be encapsulated by protein and a dense and continuous network was formed, which is required for excellent cooking and edible quality.

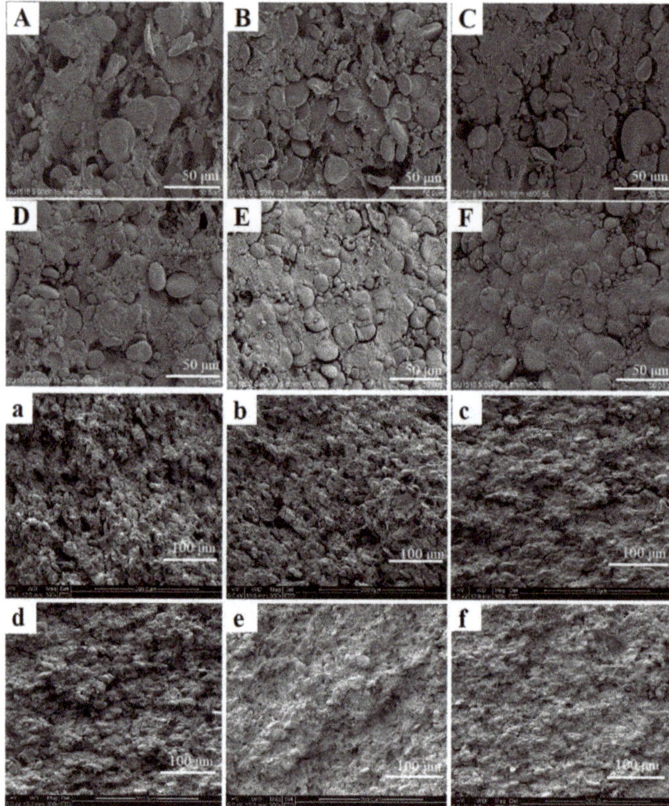

Figure 3. Microstructure changes of fresh noodles after humidity-controlled dehydration (HCD) treatment. (**A–F**), surface of HCD noodles with 40 °C-50%, 40 °C-90%, 60 °C-50%, 60 °C-90%, 80 °C-50%, 80 °C-90%; (**a–f**), cross sections of HCD noodles with 40 °C-50%, 40 °C-90%, 60 °C-50%, 60 °C-90%, 80 °C-50%, 80 °C-90%.

3.3. Cooking and Texture Properties of the HCD Fresh Wet Noodles

Maintaining integrity during cooking is crucially important for improving noodle quality [21]. After the HCD treatment, both cooking loss and water absorption of the HCD fresh wet noodles were decreased with increasing temperature and RH (Figure 4A,B). This indicated that the leaching of amylose and the dissolution of the water-soluble protein during boiling was restrained to some extent by HCD [2]. And the cooking loss and water absorption index of the 40 °C group samples were almost the same, and there was no statistically significant difference between and among samples. Only when the dehydration temperature increased to 80 °C, the cooking parameters of corresponding noodle samples would show a decrease with increasing relative humidity. The cooking quality could be highly affected by starch gelatinization and protein aggregation [27]. Baiano et al. [29] has reported that high temperature during the dehydration process could induce the protein aggregation in fresh noodles, and the dense gluten network would entrap the starch granules, and led to the reduced swelling of the starch granules and leaching of the starch molecules during cooking [2,28].

The tensile force and hardness of the cooked HCD noodles were primarily affected by the dehydration temperature, and then the relative humidity. Wagner et al. [28] have reported that disulfide-sulfhydryl (SH) would exchange for glutenin and gliadins when the temperature was above 55 and 70 °C, respectively. At the same time, starch gelatinization may happen during the HCD process at 60 or 80 °C. Therefore, the cooking and textural properties of the fresh noodles after humidity-controlled dehydration treatment in this study were improved and superior to those of the control [27,28].

3.4. Sensory Evaluation of HCD Fresh Wet Noodles

Changes in sensory properties for uncooked and cooked HCD noodles were listed in Table 2. Each indicator was assigned a corresponding weight based on its importance in meeting the demands of consumers. Overall, the fresh wet noodles without dehydration process scored the highest of 93.48 in almost all indicators except that the color and the total score of the sensory assessment for HCD noodles fell between 88.36 and 89.48. However, there was no significant difference in adhesiveness and smoothness among and between the HCD noodle samples and the control; little difference in other indicators among the 40, 60, or 80 °C sample groups. On the other hand, although there was no statistical difference between the 40 °C sample group and the control, both the color and flavor of FWN were deteriorated as the dehydration temperature increased to 60 and 80 °C. There is a strong correlation between the appearance and the overall acceptability and between chewiness and toughness of the fresh noodles; meanwhile, dehydration at middle temperature (60–80 °C) is more likely to give good results than at low temperature (40 °C). Heat treatment could induce changes in protein network formation and starch gelatinization in fresh noodle bodies, thus affects the cooking, textural, and sensory properties of cooked noodles [27]. Although various aspects of the sensory properties of FWN were not affected too much by HCD treatment, to some extent, this is just what we need. In conclusion, the fresh noodles could still maintain the characteristics of raw noodles in terms of sensory evaluation after humidity-controlled dehydration treatment, which was developed mainly for the control of microbial safety purposes.

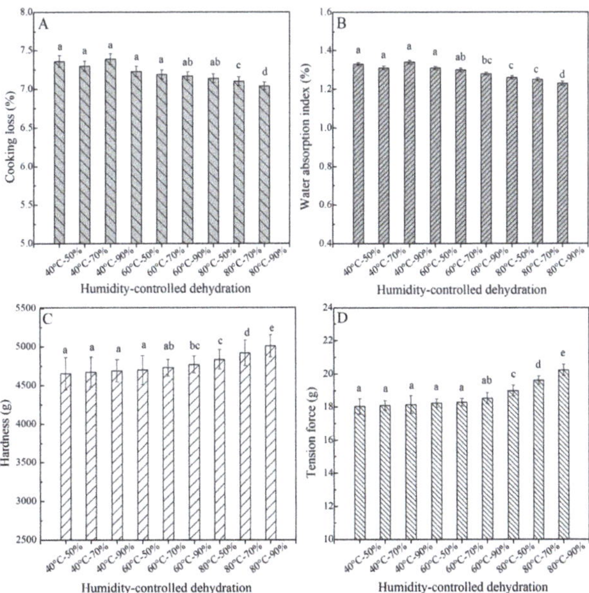

Figure 4. Effect of humidity-controlled dehydration on the textural and cooking quality of fresh wet noodles. (**A**), hardness; (**B**), tension force; (**C**), cooking loss; (**D**), water absorption index. 40 °-50%, fresh wet noodle samples treated by humidity-controlled dehydration under the condition of 40 °C and RH of 50%, etc.

Table 2. Effect of humidity-controlled dehydration (HCD) on the sensory evaluation of uncooked and cooked fresh wheat noodles *.

HCD	Value of Sensory Evaluation								
	Uncooked Noodles		Cooked Noodle Samples						
	Color	Appearance	Chewiness	Elasticity	Adhesiveness	Smoothness	Flavor	Overall Acceptability	Total Score
Control	9.70 ± 0.13 [a]	9.69 ± 0.19 [a]	18.79 ± 0.18 [a]	23.42 ± 0.23 [a]	13.72 ± 0.18 [a]	4.27 ± 0.18 [a]	4.71 ± 0.14 [a]	9.18 ± 0.15 [a]	93.48
40 °C-50%	9.83 ± 0.16 [a]	9.06 ± 0.19 [b]	17.20 ± 0.15 [c]	21.28 ± 0.20 [c]	13.70 ± 0.17 [a]	4.24 ± 0.21 [a]	4.46 ± 0.27 [a]	8.59 ± 0.18 [b]	88.36
40 °C-70%	9.79 ± 0.19 [a]	9.04 ± 0.17 [b]	17.18 ± 0.20 [c]	21.55 ± 0.18 [c]	13.71 ± 0.20 [a]	4.21 ± 0.23 [a]	4.44 ± 0.21 [a]	8.63 ± 0.22 [b]	88.55
40 °C-90%	9.78 ± 0.23 [a]	9.11 ± 0.22 [b]	17.22 ± 0.21 [c]	21.49 ± 0.23 [c]	13.71 ± 0.22 [a]	4.26 ± 0.24 [a]	4.17 ± 0.28 [b]	8.67 ± 0.24 [b]	88.41
60 °C-50%	9.49 ± 0.16 [b]	9.24 ± 0.24 [b]	17.35 ± 0.25 [b]	22.27 ± 0.24 [b]	13.71 ± 0.18 [a]	4.23 ± 0.26 [a]	4.05 ± 0.20 [b]	8.78 ± 0.15 [b]	89.12
60 °C-70%	9.42 ± 0.22 [b]	9.30 ± 0.29 [a]	17.47 ± 0.23 [b]	22.11 ± 0.19 [b]	13.69 ± 0.24 [a]	4.22 ± 0.18 [a]	3.94 ± 0.21 [b]	8.84 ± 0.27 [a]	88.99
60 °C-90%	9.43 ± 0.17 [b]	9.37 ± 0.26 [a]	17.45 ± 0.18 [b]	22.30 ± 0.21 [b]	13.70 ± 0.17 [a]	4.29 ± 0.25 [a]	3.90 ± 0.23 [b]	8.88 ± 0.27 [a]	89.32
80 °C-50%	9.48 ± 0.28 [ab]	9.41 ± 0.20 [a]	17.56 ± 0.17 [b]	22.05 ± 0.24 [b]	13.73 ± 0.16 [a]	4.25 ± 0.26 [a]	3.95 ± 0.17 [b]	8.81 ± 0.21 [a]	89.24
80 °C-70%	9.43 ± 0.17 [b]	9.39 ± 0.30 [a]	17.62 ± 0.19 [b]	22.25 ± 0.26 [b]	13.71 ± 0.25 [a]	4.25 ± 0.18 [a]	3.91 ± 0.18 [b]	8.92 ± 0.19 [a]	89.48
80 °C-90%	9.41 ± 0.16 [b]	9.42 ± 0.18 [a]	17.09 ± 0.20 [c]	22.18 ± 0.28 [b]	13.69 ± 0.15 [a]	4.29 ± 0.26 [a]	3.86 ± 0.16 [b]	8.62 ± 0.23 [b]	88.56

*, Control, fresh wheat noodles; 40 °C-50%, fresh wet noodles were treated by humidity-controlled dehydration under the conditions of 40 °C and RH of 50%, etc. Values represent mean ± S.D., n = 3. The same lowercase letter means there was no significant difference within the same column.

4. Conclusions

In this study, thermal dehydration with humidity-controlled hot air was innovatively applied as an effective way to control the initial microbial load in the fresh wet noodles before storage. HCD treatment could significantly reduce the microbial load in noodle bodies and prolong the shelf-life. The storage stability of color and apparent stickiness of the FWN was greatly improved after the humidity-controlled dehydration process, meanwhile, the textual and cooking properties also benefited a lot from the process. The humidity-controlled dehydration treatment seemed like a promising technique to pretreat the fresh wet noodles before packaging and distribution. However, further research is needed to optimize the conditions depending on the water content added in practical production when considering the economic benefits.

Author Contributions: Conceptualization, X.-N.G.; data curation, J.-J.X. and D.-H.J.; formal analysis, J.-J.X., D.-H.J. and Z.Y.; funding acquisition, J.-J.X. and K.-X.Z.; investigation, X.-N.G. and K.-X.Z.; methodology, X.-N.G. and K.-X.Z.; project administration, K.-X.Z.; resources, J.-J.X. and K.-X.Z.; software, Z.Y.; visualization, J.-J.X.; writing-original draft, D.-H.J.; writing-review and editing, J.-J.X., Z.Y. and X.-N.G. All authors have read and agreed to the published version of the manuscript.

Funding: This work was financed by the program of Natural Science Foundation of Jiangsu Province (BK20190590), the program of Fundamental Research Funds for the Central Universities (JUSRP11909), the National Key Research and Development Program of China (No.2018YFD0401003), and the program of Postdoctoral Research Foundation of China (2019M651706).

Data Availability Statement: The data that support the findings of this study are available on request from the corresponding author. The data are not publicly available due to privacy or ethical restrictions.

Conflicts of Interest: The authors declare no conflict of interest.

References

1. Chen, Y.; Chen, G.; Wei, R.; Zhang, Y.; Li, S. Quality characteristics of fresh wet noodles treated with nonthermal plasma sterilization. *Food Chem.* **2019**, *297*, 124900. [CrossRef]
2. Li, M.; Zhu, K.-X.; Sun, Q.-J.; Amza, T.; Guo, X.-N.; Zhou, H.-M. Quality characteristics, structural changes, and storage stability of semi-dried noodles induced by moderate dehydration: Understanding the quality changes in semi-dried noodles. *Food Chem.* **2016**, *194*, 797–804. [CrossRef]
3. Li, M.; Ma, M.; Zhu, K.X.; Guo, X.N.; Zhou, H.M. Delineating the physico-chemical, structural, and water characteristic changes during the deterioration of fresh noodles: Understanding the deterioration mechanisms of fresh noodles. *Food Chem.* **2017**, *216*, 374–381. [CrossRef]
4. Ma, M.; Han, C.-W.; Li, M.; Song, X.-Q.; Sun, Q.-J.; Zhu, K.-X. Inhibiting effect of low-molecular weight polyols on the physico-chemical and structural deteriorations of gluten protein during storage of fresh noodles. *Food Chem.* **2019**, *287*, 11–19. [CrossRef]

5. Tsiraki, M.I.; Karam, L.; Abiad, M.G.; Yehia, H.M.; Savvaidis, I.N. Use of natural antimicrobials to improve the quality charac-teristics of fresh "Phyllo"—A dough-based wheat product—Shelf life assessment. *Food Microbiol.* **2017**, *62*, 153–159. [CrossRef] [PubMed]

6. Chen, Y.; Zhang, Y.; Jiang, L.; Chen, G.; Yu, J.; Li, S.; Chen, Y. Moisture molecule migration and quality changes of fresh wet noodles dehydrated by cold plasma treatment. *Food Chem.* **2020**, *328*, 127053. [CrossRef] [PubMed]

7. Zhu, K.-X.; Dai, X.; Guo, X.; Peng, W.; Zhou, H.-M. Retarding effects of organic acids, hydrocolloids and microwave treatment on the discoloration of green tea fresh noodles. *LWT Food Sci. Technol.* **2014**, *55*, 176–182. [CrossRef]

8. Hou, G.G.; Otsubo, S.; Okusu, H.; Shen, L. Noodle Processing Technology. In *Asian Noodles: Science, Technology, and Processing*; John Wiley & Sons, Inc: Hoboken, NJ, USA, 2010; pp. 106–107.

9. Del Nobile, M.; Di Benedetto, N.; Suriano, N.; Conte, A.; Lamacchia, C.; Corbo, M.; Sinigaglia, M. Use of natural compounds to improve the microbial stability of Amaranth-based homemade fresh pasta. *Food Microbiol.* **2009**, *26*, 151–156. [CrossRef] [PubMed]

10. Bai, Y.-P.; Guo, X.-N.; Zhu, K.-X.; Zhou, H.-M. Shelf-life extension of semi-dried buckwheat noodles by the combination of aqueous ozone treatment and modified atmosphere packaging. *Food Chem.* **2017**, *237*, 553–560. [CrossRef] [PubMed]

11. Li, M.; Zhu, K.; Guo, X.; Peng, W.; Zhou, H. Effect of water activity (aw) and irradiation on the shelf-life of fresh noodles. *Innov. Food Sci. Emerg. Technol.* **2011**, *12*, 526–530. [CrossRef]

12. Chau, T.T.; Kao, K.C.; Blank, G.; Madrid, F. Microwave plasmas for low-temperature dry sterilization. *Biomaterials* **1996**, *17*, 1273–1277. [CrossRef]

13. Inazu, T.; Iwasaki, K.-I.; Furuta, T. Effect of Temperature and Relative Humidity on Drying Kinetics of Fresh Japanese Noodle (Udon). *LWT Food Sci. Technol.* **2002**, *35*, 649–655. [CrossRef]

14. Villeneuve, S.; Gélinas, P. Drying kinetics of whole durum wheat pasta according to temperature and relative humidity. *LWT Food Sci. Technol.* **2007**, *40*, 465–471. [CrossRef]

15. Piwińska, M.; Wyrwisz, J.; Kurek, M.A.; Wierzbicka, A. Effect of drying methods on the physical properties of durum wheat pasta. *CyTA J. Food* **2016**, *14*, 523–528. [CrossRef]

16. Xiang, Z.; Ye, F.; Zhou, Y.; Wang, L.; Zhao, G. Performance and mechanism of an innovative humidity-controlled hot-air drying method for concentrated starch gels: A case of sweet potato starch noodles. *Food Chem.* **2018**, *269*, 193–201. [CrossRef]

17. GB/T 4789.2, Code of National Food Safety Standard of China. *Food Microbiological Examination: Aerobic Plate Count*; The State Food and Drug Administration: Beijing, China, 2016.

18. GB/T 4789.15, Code of National Food Safety Standard of China. *Food Microbiological Examination: Mold and Yeast Count*; The State Food and Drug Administration: Beijing, China, 2016.

19. Wootton, M.; Weeden, D.; Munk, N. A rapid method for the estimation of starch gelatinization in processed foods. *Food Technol.* **1971**, *23*, 612–615.

20. American Association of Cereal Chemists. *AACC Method 66-50. AACC International Approved Methods*, 10th ed.; AACC Interna-tional: St. Paul, MN, USA, 2000.

21. Wu, J.; Corke, H. Quality of dried white salted noodles affected by microbial transglutaminase. *J. Sci. Food Agric.* **2005**, *85*, 2587–2594. [CrossRef]

22. Costa, C.; Lucera, A.; Mastromatteo, M.; Conte, A.; Del Nobile, M.A. Shelf life extension of durum semolina-based fresh pasta. *Int. J. Food Sci. Technol.* **2010**, *45*, 1545–1551. [CrossRef]

23. Fechner, K.; Dreymann, N.; Schimkowiak, S.; Czerny, C.-P.; Teitzel, J. Efficacy of dairy on-farm high-temperature, short-time pasteurization of milk on the viability of *Mycobacterium avium* ssp. paratuberculosis. *J. Dairy Sci.* **2019**, *102*, 11280–11290. [CrossRef] [PubMed]

24. Giribaldi, M.; Coscia, A.; Peila, C.; Antoniazzi, S.; Lamberti, C.; Ortoffi, M.; Moro, G.E.; Bertino, E.; Civera, T.; Cavallarin, L. Pasteurization of human milk by a benchtop High-Temperature Short-Time device. *Innov. Food Sci. Emerg. Technol.* **2016**, *36*, 228–233. [CrossRef]

25. Esbelin, J.; Santos, T.; Hébraud, M. Desiccation: An environmental and food industry stress that bacteria commonly face. *Food Microbiol.* **2018**, *69*, 82–88. [CrossRef]

26. Zhang, S.; Zhang, L.; Lan, R.; Zhou, X.; Kou, X.; Wang, S. Thermal inactivation of Aspergillus flavus in peanut kernels as influenced by temperature, water activity and heating rate. *Food Microbiol.* **2018**, *76*, 237–244. [CrossRef]

27. Güler, S.; Köksel, H.; Ng, P. Effects of industrial pasta drying temperatures on starch properties and pasta quality. *Food Res. Int.* **2002**, *35*, 421–427. [CrossRef]

28. Wagner, M.; Morel, M.-H.; Bonicel, J.; Cuq, B. Mechanisms of heat-mediated aggregation of wheat gluten protein upon pasta processing. *J. Agric. Food Chem.* **2011**, *59*, 3146–3154. [CrossRef]

29. Baiano, A.; Conte, A.; Del Nobile, M. Influence of drying temperature on the spaghetti cooking quality. *J. Food Eng.* **2006**, *76*, 341–347. [CrossRef]

 *foods*

Article

Microwave and Ultrasound Pre-Treatments for Drying of the "Rocha" Pear: Impact on Phytochemical Parameters, Color Changes and Drying Kinetics

Begüm Önal [1], Giuseppina Adiletta [1], Marisa Di Matteo [1], Paola Russo [2], Inês N. Ramos [3,*] and Cristina L. M. Silva [3,*]

[1] Department of Industrial Engineering, University of Salerno, via Giovanni Paolo II, 132, 84084 Fisciano, Italy; bonal@unisa.it (B.Ö.), gadiletta@unisa.it (G.A.); mdimatteo@unisa.it (M.D.M.)
[2] Department of Chemical Engineering Materials Environment, Sapienza University of Rome, via Eudossiana, 18, 00184 Rome, Italy; paola.russo@uniroma1.it
[3] Escola Superior de Biotecnologia, CBQF—Centro de Biotecnologia e Química Fina–Laboratório Associado, Universidade Católica Portuguesa, Rua Diogo Botelho 1327, 4169-005 Porto, Portugal
* Correspondence: iramos@porto.ucp.pt (I.N.R.); clsilva@porto.ucp.pt (C.L.M.S.)

Citation: Önal, B.; Adiletta, G.; Di Matteo, M.; Russo, P.; Ramos, I.N.; Silva, C.L.M. Microwave and Ultrasound Pre-Treatments for Drying of the "Rocha" Pear: Impact on Phytochemical Parameters, Color Changes and Drying Kinetics. *Foods* **2021**, *10*, 853. https://doi.org/10.3390/foods10040853

Academic Editors: Sidonia Martinez and Javier Carballo

Received: 18 February 2021
Accepted: 8 April 2021
Published: 14 April 2021

Publisher's Note: MDPI stays neutral with regard to jurisdictional claims in published maps and institutional affiliations.

Abstract: The objective of this research was to evaluate the effect of drying temperature and innovative pre-treatments (i.e., microwave and ultrasound) on "Rocha" pear drying behavior and quality characteristics, such as color, total phenolic content and antioxidant activity. Experiments were carried out with pear slabs subjected to microwaves (2450 MHz, 539 W, 4 min, microwave oven) and ultrasounds (35 kHz, 10 min, in an ultrasonic bath) as well as control samples. The drying process was conducted in a tray dryer at three different temperatures (50, 55 and 60 °C) and a fixed air velocity of 0.75 m/s. Microwave technology resulted in a higher quality deterioration in dried pear samples compared to those of controls and ultrasound pre-treated samples. The combined application of ultrasound pre-treatment and the higher drying temperature of 60 °C was characterized by the lowest color changes ($\Delta E = 3.86 \pm 0.23$) and higher preservation of nutritional parameters (total phenolic content, TPC = 345.60 ± 8.99; and antioxidant activity, EC_{50} = 8.80 ± 0.34). The drying characteristics of pear fruits were also analyzed by taking into account empirical models, with the Page model presenting the best prediction of the drying behavior. In conclusion, ultrasound application is a promising technology to obtain healthy/nutritious dried "Rocha" pear snacks as dietary sources for consumers.

Keywords: "Rocha" pear; ultrasound; microwave; drying; quality characteristics; empirical models

1. Introduction

The "Rocha" pear (*Pyrus communis* L.) is a traditional Portuguese cultivar classified with a Protected Designation of Origin (PDO) [1]. In addition to being a commercially important agricultural product throughout the world, the "Rocha" pear is characterized by its excellent organoleptic properties, nutritional quality and high resistance due to its initial rigidity [2–4]. Pear fruits are a rich source of vitamins and minerals, dietary fiber and phenolic substances. Therefore, pear consumption has received increasing interest due to its potential health-promoting benefits, in particular high antioxidant activity, as well as anti-inflammatory, anti-bacterial, anti-diabetic, uro-disinfectant, sedative and antipyretic properties [5–8]. The "Rocha" pear can be maintained up to 10 months under controlled atmosphere storage (2–3 kPa O_2 + 0.5–0.7 kPa CO_2 at −0.5 °C and 95% relative humidity) [9]. However, long-term storage under a controlled atmosphere (CA) may promote physiological deterioration such as internal browning disorders and superficial scald, influencing the overall aspects of the pear fruit and therefore consumer acceptability [3,10]. Nevertheless, post-harvest losses represent a major problem [11], not only from environmental but also from economic perspectives [12,13]. Hence, there is a need to develop alternative strategies to contribute to environmental and economic factors and sustain the nutritional and overall quality of agricultural products such as "Rocha" pears. In this context, the drying process is one of the most important preservation techniques to prolong shelf-life and availability, maintain the high-added-value components, valorize traditional food products, reduce transportation and storage costs and present new ways of consumption. As a consequence of the consumer's lifestyle/behavior changes, in recent years, the food dehydration industry, including dried pears, has become an important market segment due to increased demand for healthy/nutritious ready-to-eat products (i.e., snack preparations, integral breakfast cereals, bakery/confectionery products and rehydrated functional fruit recipes). Among the numerous techniques, conventional hot air drying is widely used in the industry as it enables uniform processing conditions, is easy-to-use, and produces better-dried products with optimal drying conditions [14–16]. However, this drying technique may promote the partial degradation of nutritional compounds and undesirable quality changes of the final product such as color, appearance and structural properties. These properties extensively depend upon drying conditions (e.g., air temperature and velocity). In this sense, recent advances in using novel food processing technologies have received much attention to reduce various adverse changes and prepare final dried products with fresh-like characteristics.

Among the so-called emerging technologies, ultrasound waves are applied as a pretreatment prior to convective drying of various fruit and vegetables to accelerate the drying process, reducing drying time [17,18], enhancing mass transfer phenomena [19,20], preserving functional components [21,22], inactivating enzymes [23] and improving rehydration characteristics [24,25]. There are two main phenomena induced by ultrasound that can affect solid-like materials: sponge effect and cavitation. The mechanical force and cavitation effects cause structural changes to the plant tissue such as the formation of micro-channels, which can be beneficial for water diffusion [20,23,26,27]. Much of the research on drying and quality characteristics by these authors have demonstrated that the application of ultrasound has different mechanisms and effects on various fruits.

Recently, utilization of microwave energy has drawn appreciable attention, as it is stated to enhance the drying process by lowering the processing time and operational costs [28,29]. Microwave technology penetrates the foodstuff quickly by heating it from inside to outside, resulting in an increase of the product temperature and consequently, rapid water evaporation [27,30]. Despite its benefits for drying efficiency, this technology has some major drawbacks regarding the dried product's quality, including color deterioration, the formation of hot/cold spots due to non-uniform heating, nutritional losses, reduction of volatile aroma compounds and possible textural damage [27,28,31,32]. Additionally, the high penetration power of microwave energy affects heat transfer behavior and may lead to overheating agricultural products, depending on the food's properties (nature, permittivity, water content), the geometry of the material, dielectric properties and oven design [27]. The application of microwave energy as a pre-treatment and its effects on the drying process and quality attributes have been addressed by a limited number of authors [31,33].

Obtaining an in-depth understanding of the drying process by means of models is a crucial tool for designing industrial drying systems and optimizing process conditions [34,35]. The most common empirical models are those of Newton, Henderson and Pabis, Page, Modified Page, two-term and Wang and Singh and have been utilized for various food materials including quince [36], cherry [37], grape [38], peach and strawberry [39] and persimmon [40] to describe several aspects of drying behavior These modeling approaches allow prediction and simulation of industrial drying systems, and help choose the most suitable operating conditions for the effective drying of fruits and vegetables.

However, no scientific work has been published concerning the effect of such kinds of pre-treatments (i.e., microwave and ultrasound applications) on drying behavior and quality characteristics of the "Rocha" pear. There is a need to study the application of emerging technologies to ensure process efficiency and product quality, satisfying the increased demand for healthy and nutritious snacks. Therefore, the focus of this work was to investigate the effects of microwave (2450 MHz, 539 W, 4 min, microwave oven) and ultrasound (35 kHz, 10 min, ultrasound bath) technologies as a pre-treatment on the convective air drying characteristics of "Rocha" pear slices, using drying temperatures of 50, 55 and 60 °C. Different empirical models were attempted to predict the drying process. Moreover, dried and raw "Rocha" pears quality attributes were assessed on water activity, color, total phenolic content and antioxidant activity.

2. Materials and Methods

2.1. Raw Material Preparation

"Rocha" pears (*Pyrus communis* L. 'Rocha') were acquired in a local market (Porto, Portugal) and stored in a refrigerator at 4 °C (BOSCH DUO SYSTEM, Bosch, Gerlingen, Germany) until drying or experimental analysis, up to a maximum of 2 days. Pears were selected taking into account sensory attributes such as freshness, uniform size, color and absence of mechanical damage and disease symptoms. Prior to drying, samples were washed with tap water, dried with absorbent paper and peeled. "Rocha" pears were sliced using a stainless steel slicer in order to maintain a uniform thickness of each sample. The average diameter and thickness of pear slices were 38 ± 0.12 mm and 6 ± 0.05 mm, respectively.

The average moisture content of raw pears, determined according to the AOAC official method [41] at 105 °C for 24 h, was found to be 6.42 kg water/kg d.m ($86.52 \pm 0.66\%$, w.b.).

2.2. Pre-Treatments Applied to the "Rocha" Pear

Three kinds of dried pear samples were compared in this study: (a) control (C) (without application of ultrasound or microwave pre-treatment), (b) ultrasound pre-treated (US) and (c) microwave pre-treated (MW) before drying.

Prior to the ultrasound application, portions of 40 g of pear slabs were hermetically packaged in eight polyethylene bags (VWR, $120 \times 170 \times 0.05$ mm) (6 slabs per bag). Bags were subjected to vacuum packaging at 1 mPa during 1.2 s of welding using a vacuum packaging machine (A300/41/42, Multivac, Wolfertschwenden, Germany). The vacuum-packaged samples were used to avoid adverse effects of rinsing out substances contained in the pear fruits. Ultrasound waves, when passing through a liquid medium, lead to mechanical vibration of the liquid. If the liquid medium contains dissolved gas, which is the case under normal conditions, microbubbles can form, grow and violently collapse due to the action of the sound wave. This phenomenon is called "acoustic cavitation" Therefore, the use of vacuum packaging might decrease cavitation effects in the pear samples by preventing direct contact with a liquid medium. This methodology is a key point before ultrasound application to food products, as recommended by Nowacka and Wedzik [24]. The pear samples were placed in a metal net at the bottom of an ultrasonic bath (Bandelin Sonorex RK 255H; 300 mm (L) × 150 mm (W) × 150 mm (H)) containing 3 L of distilled water and without any mechanical agitation. The ultrasound frequency was 35 kHz, and the nominal power of the ultrasounds was 160 W. The experiments were conducted for an ultrasonic processing time of 10 min at a temperature of 30 °C. The ultrasonic processing time and temperature conditions were chosen according to the literature [17,21,24,42]. During the ultrasound applications, the temperature increase was lower than 2 °C. The slabs of each treatment were removed from the ultrasound bath, drained with absorbent paper and then subjected to the drying process.

10 pear slabs (approximately 100 g) were placed in a microwave oven (Beko 20 L, P.C.R, dimensions: 454 mm (W) × 330 mm (D) × 262 mm (H)) with a rotating plate and then subjected to microwave energy at a frequency of 2450 MHz and power of 539 W for 4 min. [33]. The total fresh mass of pear slabs was approximately 300 g, and after microwave heating, the weight of the samples was 251.8 ± 0.6 g, therefore evaporating around 50 g of water. The energy per mass was around 1294 J/g. Each run was repeated

three times to prepare 30 pear slabs for each set. After microwave pre-treatment, all samples were cooled down and then subjected to drying.

2.3. Drying Experiments

Drying experiments of the control (C), microwave (MW) and ultrasound (US) pre-treated pear slabs were conducted in a pilot plant convective air tray drier with forced air (Armfield UOP8, Ringwood, England) and controlled air temperature and velocity, also including a metallic tube that allows air recirculation [43]. On-line acquisition of weight loss during the drying was recorded every 3 min by means of a digital balance (Sartorius, Goettingen, Germany) attached to the drying equipment and connected to a computer (Hewlett–Packard Vectra, Palo Alto, CA, USA). The tray dryer was previously heated to the set point temperature, and drying conditions were stabilized and maintained for at least 30 min before sample drying. An amount of approximately 300 g pear samples was divided into three similar portions and loaded onto the dryer's three top trays. The drying processes of the pears were conducted until a final constant moisture content of less than 0.1 ± 0.01 kg water/kg d.m., using air drying temperatures of 50, 55 and 60 °C at a fixed air velocity of 0.75 m/s. The air temperature was monitored with a squirrel data-logger (Grant Instruments 1023, Cambridge, England) and thermocouple wires connected to each tray. The air velocity was measured regularly by a vane anemometer (Airflow LCA 6000, Buckinghamshire, England). Outlet dry and wet bulb air temperatures were recorded to calculate its relative humidity values through a psychrometric chart calculator [44]. Average air relative humidity values were $33.97 \pm 0.81\%$, $31.67 \pm 0.46\%$ and $27.27 \pm 0.75\%$, respectively, for drying temperatures of 50, 55 and 60 °C.

Drying experiments were performed at each temperature in three independent sets (C, MW and US), each with three replicates. After drying, pear samples were cooled down to room temperature, further packaged using a vacuum packaging machine and allowed to stand in the dark until further analysis.

2.4. Modeling of Drying Kinetics

"Rocha" pears' drying kinetics was expressed as a moisture ratio (XR) as a function of time:

$$XR = \frac{X - X_e}{X_0 - X_e} \tag{1}$$

where X, X_0 and X_e were, respectively, the moisture contents at a given time, initial and equilibrium (kg water kg dry matter^{-1}). This equation can be simplified because X_e is small compared to X and X_0, and the error of considering it negligible is not significant [39,45]. Therefore, X_e was set equal to zero in all drying models. Moreover, this assumption was confirmed by calculating an average relative humidity value for all experiments and the corresponding X_e using sorption isotherms.

Several empirical models commonly applied to fruit materials were analyzed to fit "Rocha" pear drying data including the Newton model [46,47]:

$$\frac{X - X_e}{X_0 - X_e} = \exp(-kt) \tag{2}$$

the Henderson & Pabis model [48,49]:

$$\frac{X - X_e}{X_0 - X_e} = a \, \exp(-k \, t) \tag{3}$$

and the Page model [50]:

$$\frac{X - X_e}{X_0 - X_e} = \exp(-k \, t^N) \tag{4}$$

Drying data were normalized to each pear sample's initial moisture content to allow comparison between different pre-treatment effects. A non-linear regression analysis was used to estimate the coefficients (drying rate constant, k (min^{-1} or min^{-N}), and drying coefficients, a and N) of the tested models. For each model, the goodness of fit analysis was based on the following statistical parameter values: coefficient of determination (R^2) and the standard deviation of the experimental error (s). The model better describing each pear sample's drying characteristics was chosen according to a higher coefficient of determination (R^2) and a lower standard deviation of the experimental error (s).

2.5. Quality Parameters Evaluation

2.5.1. Water Activity

Each pear sample's water activity was determined using a water activity meter (Aqualab, Model Series 3 TE, Decagon Devices, Pullman, WA, USA) at 23 °C, previously calibrated with a standard solution of water activity of 0.760. Fresh and dried pears were inserted in a converter chamber, and a_w values were recorded after equilibration. Each measurement was carried out in triplicate and the mean values calculated.

2.5.2. Color Properties

The color parameters of fresh and dried pear slabs (control and microwave and ultrasound pre-treated) were evaluated using a Minolta CR-400 colorimeter (Konica-Minolta, Osaka, Japan) based on the CIELAB coordinate system. Before the measurements, the instrument was calibrated with a white ceramic plate. Three readings of three different replications were conducted for each sample.

Three color parameters (L*, a* and b*) were recorded. The L* value represented the brightness and varied from 0 (black) to 100 (white). The chromaticity coordinates indicated for a* the redness (+)/greenness (−) and for b* the yellowness (+)/blueness (−). The whiteness index (WI) and total color difference (ΔE) were derived from the values of L*, a* and b*. These values were crucial to obtaining an in-depth understanding of the impacts of pre-treatments and drying process conditions (i.e., temperature and time) on color changes.

The whiteness index (WI) represented the overall whiteness of fruits and vegetables that may indicate the extent of discoloration during the drying process [51]. WI was used to express the degree of whiteness of pear samples, and was calculated as follows [38]:

$$\text{WI} = 100 - \sqrt{(100 - \text{L}^*)^2 + (a^{*2}) + (b^{*2})} \tag{5}$$

Total color difference (ΔE) represented the degree of overall color changes of pear samples and was determined according to Equation (6) [52]. The reference value for total color difference (ΔE) was the fresh "Rocha" pear.

$$\Delta\text{E} = \sqrt[2]{(\text{L}^* - \text{L}_0^*)^2 + (a^* - a_0^*)^2 + (b^* - b_0^*)^2}, \tag{6}$$

where L_0^*, a_0^* and b_0^* were the color parameters for the fresh pear and L^*, a^* and b^* the corresponding values after drying.

2.5.3. Sample Extraction

The extraction methodology used was adapted from Salta et al. [53] and Önal et al. [54] with some modifications. The samples' extracts for the determination of total phenolics and total antioxidant capacity were prepared by homogenizing 5 g (±0.01) fresh sample or 2 g (±0.01) dried sample in 30 mL and 35 mL of methanol, respectively (100%, v/v) (CHROMASOLV®, for HPLC, ≥99.9%, Sigma-Aldrich, St. Louis, MO, USA), throughout an Ultra-turrax (Ika digital T25, IKA®-Werke GmbH & Co. KG, Staufen, Germany) at 11 rpm. The obtained mixture for each sample was stirred for 10 min by a vortex, and then extracts were filtered throughout a Whatman No:2 filter paper. Three extracts were prepared for each sample, and measurements were performed for each replicate.

2.5.4. Total Phenolic Content

The total phenolic content in pear samples was determined in accordance with the Folin-Ciocalteu colorimetric method previously reported by Önal et al. [54]. Describing briefly, the pear extracts were oxidized by the Folin-Ciocalteu reagent in the mixture of sodium carbonate solution and distilled water. The reaction mixture was allowed to stand at room temperature for 1 h, and then the absorbance was measured, in triplicate, at 760 nm, using a visible spectrophotometer (Novaspec II, Piscataway, NJ, USA). The calibration curve was constructed with standard Gallic acid, and results were reported as milligrams of gallic acid equivalents per 100 g of dry basis (mg of GAE/100 g d.m.).

2.5.5. DPPH Radical Scavenging Activity

The DPPH (1,1-diphenyl-2-picrylhydrazyl) radical scavenging assay was used to assess the antioxidant activity of fresh and dried pears according to the method proposed by Adiletta et al. [55] and Önal et al. [54]. Various volumes of pear extracts were prepared and then mixed with 3.5 mL of 6×10^{-5} m of DPPH methanol solution in cuvettes. The reaction mixtures were vortexed for 30 s. After 30 min incubating in the dark and at room temperature, absorbance values were recorded at 517 nm using a visible spectrophotometer (Novaspec II, Piscataway, NJ, USA). Observed color variation from purple to yellow demonstrated the progress of the reaction due to the reduction of the 1,1-diphenyl-2-picrylhydrazyl (DPPH) solution complex. Pure methanol (v/v) was used as the blank, and the preparation of the control sample was without adding any pear extract.

The total antioxidant activity of pear extracts was expressed as the percentage inhibition of the DPPH radical and calculated [56,57] as:

$$\% \text{ Antioxidant Activity} = \frac{\text{Abscontrol} - \text{Abssample}}{\text{Abscontrol}} \times 100 \tag{7}$$

where $\text{Abs}_{control}$ = the absorbance of control. Abs_{sample} = the absorbance of sample.

EC_{50} value, which was defined as the substrate concentration (mg mL^{-1}) needed to inhibit 50% of the DPPH radical scavenging activity, was employed to express the results. EC_{50} was determined from a graph of antioxidant activity (%) versus extract concentration of each pear (mg mL^{-1} sample).

2.6. Statistical Analysis

The significance of the pre-treatments (ultrasound and microwave) and drying process (i.e., temperature) effects were assessed employing a one-way analysis of variance (ANOVA). The significance level assumed in all situations was fixed at $p < 0.05$. Moreover, Pearson's correlation tests were performed to determine the relationship between the white index (WI) and total phenolic (TPC), the white index and antioxidant activity (EC_{50}) and total phenolic content and antioxidant activity. Concerning the modeling approaches, the 95% standard error of the parameter (SE) and statistical indicators of the quality of the regression (coefficient for determination (R^2) and standard deviation of the experimental error (s)) were also calculated [58]. All statistical analyses were carried out using IBM SPSS® Statistics® 24 for Windows (SPSS Inc., Chicago, IL, USA).

3. Results and Discussion

3.1. Color Evaluation

Color parameters of fresh pears and control (C), microwave (MW) and ultrasound (US) pre-treated dried samples at 50, 55 and 60 °C are presented in Table 1, and include the L*, the whiteness index (WI), and total color difference (ΔE). According to the results, color parameters were affected by both applied pre-treatments and drying temperatures. After the drying process, the L* value decreased in all microwave-treated samples at each drying temperature, and the lowest *L** value was in microwave-treated dried samples at 50 °C. It was visually observed that all microwave treated pears' color got browner after hot air drying compared to control and ultrasound treated dried ones. Similar observations were reported by Krokida et al. [59] wherein the lightness parameter (L*) decreased significantly in microwave treated apple, potato, banana and carrot samples conventionally dried at 70 °C. A similarity of L* values between the control and ultrasound pre-treated samples dried at 50 and 55 °C was observed. The highest L* value was observed in ultrasound pre-treated pears dried at 60 °C as 79.05 ± 0.25, while the L* values of fresh samples were 78.60 ± 0.25. The highest L* values of ultrasound pre-treated pears dried at 60 °C indicated that the ultrasound application with a higher drying temperature of 60 °C was sufficient to preserve the fresh fruit's lightness. Thus, the pear's original color could be better protected when the samples received the combination of ultrasound pre-treatment and higher drying temperatures of 60 °C.

Table 1. Color parameters of control (C), microwave (MW) and ultrasound (US) pre-treated "Rocha" pears dried at 50, 55 and 60 °C.

Sample	L*	WI	ΔE
Fresh	78.60 ± 0.90 [e]	72.11 ± 1.00 [f]	-
C50 °C	71.35 ± 1.98 [c]	61.21 ± 1.21 [c]	11.57 ± 0.71 [d]
MW50 °C	55.77 ± 2.99 [a]	48.35 ± 1.17 [a]	23.41 ± 2.18 [f]
US50 °C	73.77 ± 0.89 [cd]	63.23 ± 0.88 [cd]	9.80 ± 0.71 [cd]
C55 °C	72.46 ± 1.66 [c]	63.03 ± 0.60 [cd]	10.23 ± 0.78 [cd]
MW55 °C	58.35 ± 0.84 [ab]	51.38 ± 1.15 [b]	21.99 ± 0.92 [f]
US55 °C	77.15 ± 0.42 [de]	68.00 ± 0.48 [e]	6.11 ± 0.32 [ab]
C60 °C	75.06 ± 0.80 [cde]	64.96 ± 0.67 [d]	8.13 ± 0.71 [bc]
MW60 °C	60.53 ± 0.68 [b]	52.88 ± 0.95 [b]	14.91 ± 0.81 [e]
US60 °C	79.05 ± 0.25 [e]	70.96 ± 0.21 [f]	3.86 ± 0.23 [a]

Values in the same column with the same letter were not significantly different ($p > 0.05$).

Concerning the whiteness index, the drying process reduced its value for all dried samples. In particular, a significant decrease was observed in all microwave-treated dried pears. However, ultrasound-pre-treated dried pears at 60 °C demonstrated similar whiteness index values to those of fresh ones ($p > 0.05$).

The total color difference is a crucial parameter for dried fruits and vegetables, being an indicator of the human eye's ability to differentiate between products [60]. The drying temperature also influenced the values of total color changes (ΔE), revealing a decreasing value with the increase of temperature from 50 to 60 °C. These results showed the color deterioration during the drying process being more pronounced when the lowest drying temperature was employed. This tendency may be explained by the browning reactions or the formation of browning products occurring at lower-temperature and long-time exposure to the drying process at a low temperature (50 °C). The highest values of ΔE were observed for all microwave-pre-treated dried pears. These highest ΔEs were likely due to high differences in the L* (lightness) values. This trend may be attributed to non-uniform temperature distribution during microwave treatment such that few regions of pear samples could get heated very rapidly, while the remaining region could get heated to a lesser extent. Controlling heating uniformity is an important parameter for obtaining high quality microwave-dried fruits. During microwave applications, the constant microwave power can cause an increase in the average product temperature, and overheating of the food material can be prevented by controlling the microwave power [61]. A large number of factors may influence temperature distribution during microwave applications such as the thickness, geometry and dielectric properties of foods and microwave energy [62]. The highest color alterations of microwave-treated dried samples may be dependent on the thickness of pear, microwave time or loss of homogeneity during the microwave treatment.

The lowest value of ΔE was observed in ultrasound-pre-treated pears dried at 60 °C, indicating that the combined pre-treatment of the ultrasound application and higher drying temperature better helped to preserve the original color of pear in dried snacks (Figure 1). At this point, it could be concluded that ultrasound pre-treatment could be applied to control or inhibit browning reactions and attain the desirable color of final dried pears [63].

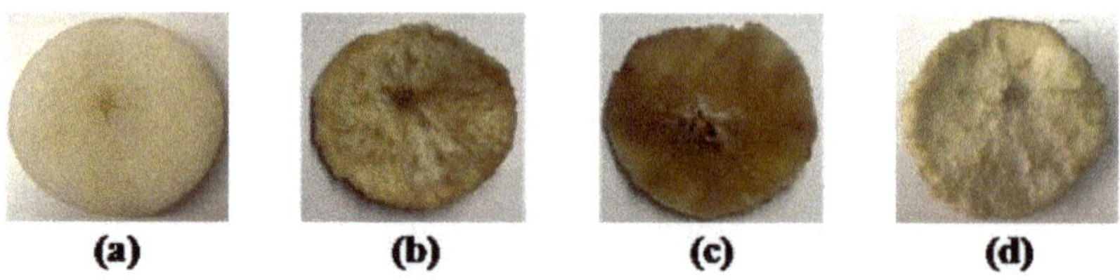

Figure 1. "Rocha" pear slabs: fresh (**a**), control (**b**), microwave treated (**c**) and ultrasound treated (**d**) dried at 60 °C.

3.2. Total Phenolic Content (TPC)

The total phenolic content of fresh pears and control (C), microwave (MW)- and ultrasound (US)-pre-treated pears dried at 50, 55 and 60 °C is summarized in Figure 2. The present experiments demonstrated that drying temperature and microwave and ultrasound pre-treatments had a remarkable effect on pear samples' total phenolic content. The total phenolic content of fresh pear slabs was found to be 336.82 mg GAE/100 g d.m. Total phenolic content levels generally decreased after the drying process. A significant reduction of total phenolic content was observed in all microwave-treated pears after drying, and there were no statistical differences ($p > 0.05$) between all microwave pre-treated dried samples. These results indicated that microwave pre-treatment had a negative impact on total phenolic content and damaged the nutritional composition of all dried pears. Moreover, microwave-treated dried pears' total phenolic content was slightly lower than control and ultrasound-treated dried pears ($p < 0.05$). The phenolic compounds' degradation by microwave treatment might be due to the heating effect caused by electromagnetic radiation during the microwave application.

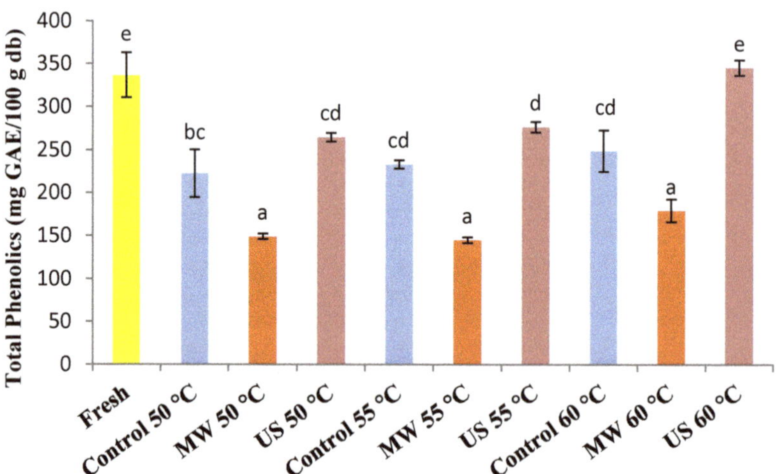

Figure 2. Total phenolic content of fresh and control (C), microwave (MW) and ultrasound (US) treated "Rocha" pears dried at 50, 55 and 60 °C. Values with the same letter were not significantly different ($p > 0.05$).

The most significant changes in microwave-treated dried pears' total phenolic composition might be attributed to the structural changes in fruits due to the combined drying method (MW application and hot air drying). During the microwave treatments, an observed shrinkage phenomenon could cause internal stress (non-uniform temperature) and surface tension, resulting in surface microcracks and leakage of exudate from the pears. Contrary to our findings, Hayat et al. [64] noticed that microwave energy could increase the bioavailability of some phenolic compounds (i.e., free phenolic compounds) by liberating them from the food matrix. Microwave heating is associated with an essential phenomenon of selective heating. The release of phenolic compounds is believed to be attributed to the selective heating of certain phenolic compounds in the microwave field or physical forces between the food matrix and phenolic compounds. The thermal instability of microwave heating could be one of the most important reasons for phenolic compound losses. Furthermore, the thermal instability of phenolic compounds in the food matrix can play an important role in the separation process [65]. However, not much knowledge about the impacts of microwave application on quality attributes of hot-air-dried fruits is currently available.

Concerning ultrasound pre-treatment, no significant differences ($p > 0.05$) were observed between all dried control pears and ultrasound-pre-treated ones up to 60 °C. Among all dried pears, the highest total phenolic content was 345.60 mg GEA/100 g d.m. in ultrasound pre-treated pears dried at 60 °C, and only this value was not statistically different from the fresh pears. Based on these results, the combined application of ultrasound pre-treatment and higher drying temperature of 60 °C affected the pears' total phenolic content positively. This behavior could be explained by the higher processing temperature and less exposure time, which contributed to a protective effect against oxidative and heat damage to pears' phenolic composition. Such situations may be attributed, in addition, to the better availability and extractability of antioxidant compounds, which can be enhanced by larger pores in the pear tissue, by an ultrasound application, thereby improving the extraction of polyphenols sample preparation [66,67].

The decrease of total phenolic compounds due to the drying process may be associated with the modifications in phenolic compounds' chemical structure or the binding of polyphenols with other compounds, such as proteins [68,69].

In this case, the ultrasound application with higher drying temperature (60 °C) exhibited better retention of pear samples' antioxidant activity.

3.3. DPPH Radical Scavenging Activity

The antioxidant activity of fresh and dried pears with different pre-treatments was evaluated by a DPPH radical scavenging activity assay (please see Figure 3). The lowest EC_{50} corresponds to the highest pears' antioxidant activity. The fresh samples provided a radical scavenging activity of 9.39 mg/mL. After the drying process, the samples had less antioxidant activity except for ultrasound-pre-treated samples dried at 60 °C. According to these results, dried pears' higher antioxidant activity value was positively correlated with the higher drying temperature. In this case, the lower drying temperature produced more degradation of pears' antioxidant activity. Previous studies stated that the lowest antioxidant activity of conventionally air-dried fruits and vegetables was linked to lower processing temperature and longer exposure time to heat [27,70,71]. Also, Nicoli et al. [72] stated that some fruits' antioxidant capacity could be maintained or enhanced by the formation of new antioxidant compounds. According to Kamiloglu and Capanoglu [73] and Vega-Gálvez et al. [71], generation and accumulation of Maillard-derived melanoidins, having a different degree of antioxidant activity, could also increase the antioxidant activity of fruits.

The microwave treatment resulted in significantly less antioxidant activity in dried pears than control and ultrasound-pre-treated dried pears, demonstrating that some phenolic acids were probably degraded by microwave application. The antioxidant activity values of control and ultrasound-pre-treated dried pears were significantly different ($p < 0.05$) than microwave-treated dried ones. The significantly highest values of EC50 (the lowest antioxidant activity) were obtained in "Rocha" pears subjected to microwave pre-treatment drying at 50 °C. This behavior could be due to the combined application of microwave pre-treatment and drying process conditions of exposure time and temperature: lower temperature and longer exposure time produced more degradation of polyphenols and consequently, the loss of antioxidant activity. Moreover, microwave pre-treatment probably caused significant structural changes in pear cells, which led to the loss of antioxidant activity on the fruit surface.

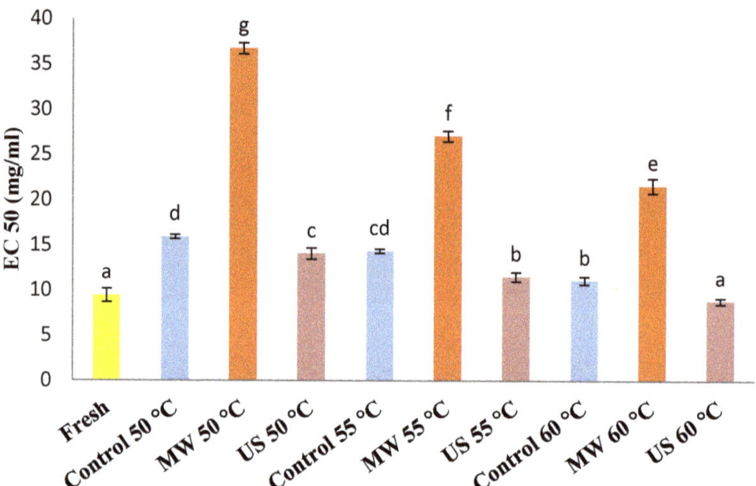

Figure 3. Antioxidant activity of fresh and control (C), microwave (MW) and ultrasound (US) treated "Rocha" pears dried at 50, 55 and 60 °C. Values with the same letter were not significantly different ($p > 0.05$).

All pear samples obtained from ultrasound pre-treatment exhibited better antioxidant properties. However, no statistical differences ($p > 0.05$) were observed between ultrasound-treated pears dried at 55 °C and control ones dried at 60 °C. The ultrasound-treated pears at a higher drying temperature of 60 °C showed the highest antioxidant activity retention. Such a situation could probably be attributed to the protection of pear cellular structure degradation by combined application (ultrasound treatment and higher drying temperature of 60 °C), which may have improved the samples' antioxidant activity.

Based on these results, microwave and ultrasound pre-treatments differentiated the phenolic contents and antioxidant potential in dried pears. Nevertheless, numerous factors may affect food products' antioxidant capacity, such as different drying methods, temperatures, applied pre-treatments, antioxidant assays, main antioxidant compounds' chemical structures and the interactions of several antioxidant reactions [55,70].

In our case, this proper combined drying method (ultrasound treatment and higher drying temperature of 60 °C) may be efficient, preserving the antioxidant activity and phenolic compounds in "Rocha" pear samples.

Pearson's correlation test was applied to determine the correlation between the white index (WI) and total phenolic content as well as the antioxidant activity (EC$_{50}$ value). The total phenolic content and antioxidant activity influenced the white index (WI) of "Rocha" pears. Total phenolic content showed a high positive correlation with the white index ($r = 0.928$). At the same time, the antioxidant activity (EC$_{50}$) had a high negative correlation with the white index ($r = -0.936$) because a lower EC$_{50}$ indicated higher antioxidant activity. If r is regarded as an absolute value, it is possible to observe a high correlation.

Moreover, it was confirmed herein by statistically significant negative Pearson's correlation among the total phenolic content and EC$_{50}$ value ($r = -0.845$). This result indicated that with a higher polyphenol content, a lower amount of "Rocha" pear tissue was required to scavenge 50% of initial free radical concentration (lower EC$_{50}$ value). The analogous relationship was reported by Rahman et al. [74] and Wiktor et al. [67].

3.4. Drying Kinetics: Experiments and Empirical Models

Figure 4a–c presents the drying curves of control, microwave and ultrasound pre-treated pear slices at the investigated temperatures of 50, 55 and 60 °C. The drying curves depended on the applied pre-treatments (i.e., microwave and ultrasound) and drying conditions (i.e., temperature and time). Fresh pears were characterized by moisture content of 6.42 ± 0.66 kg water/kg d.m. and water activity of 0.98 ± 0.004. The pear drying process was performed to a final moisture content of less than 0.1 ± 0.01 kg water/kg d.m. and average water activity of 0.41 ± 0.03 (<0.05).

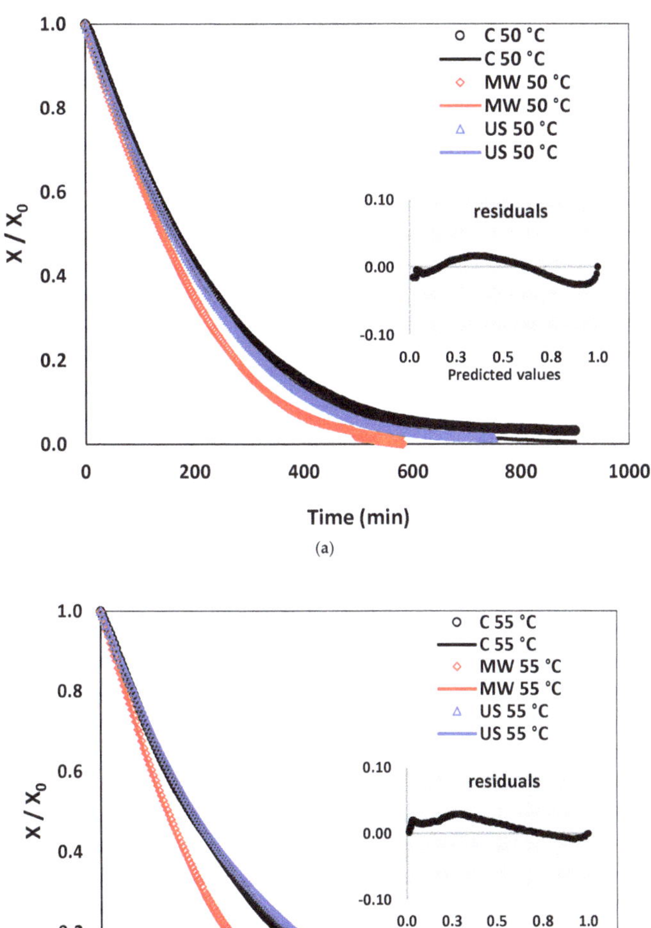

Figure 4. *Cont.*

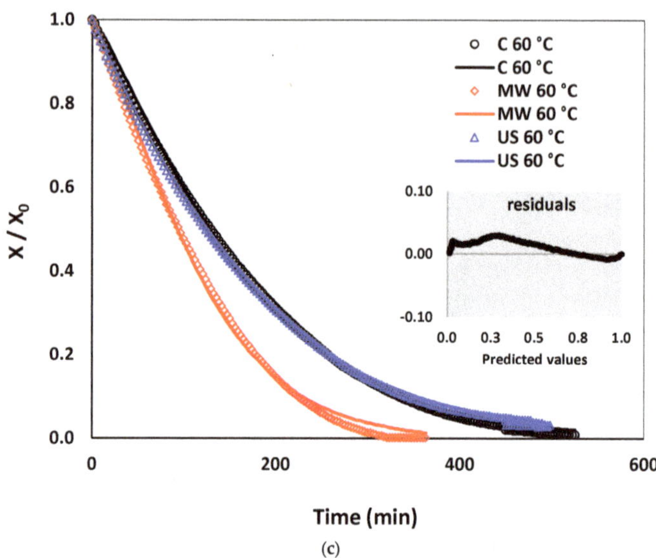

Figure 4. Experimental (symbols) and Page model-predicted (lines) drying curves of control and pre-treated (microwave—MW and ultrasound—US) samples at (**a**) 50 °C, (**b**) 55 °C and (**c**) 60 °C.

No constant rate period was detected in all samples within the studied drying experimental conditions. Only the falling rate period was observed in all samples at each drying temperature, indicating that the tray dryer's drying process was controlled by internal mass transfer diffusion (moisture migration within pear samples). Similar drying characteristics have been observed in several fruits' drying, such as the persimmon [75], apple [52], grape [38] and pineapple [76].

All pear samples' drying curves presented similar behavior: moisture content decreased and the rate of moisture loss decreased as time progressed. Moreover, the drying process was accelerated when the temperature increased from 50 to 60 °C. Drying behavior with increasing drying temperature was explained by Nascimento et al. [77] and Tao et al. [78]. The temperature rising reduces the relative humidity of the air, increases the moisture gradient between food materials and air and promotes the moisture movement within foods, thus enhancing drying time, but may cause also case-hardening. If case-hardening occurs, the diffusion path of moisture from the internal layers toward the product surface is impeded. As a consequence, diffusion will become slower, and the drying time will be prolonged. A similar phenomenon has been observed for yam slices' convective drying [79].

As shown in Figure 4, in microwave-pre-treated (539 W for 4 min) pear slices, the moisture content decreased faster than in control or ultrasound-pre-treated (US) samples, thus enhancing the drying process. In this way, the microwave pre-treated samples had shorter drying times at all investigated temperatures. Pear samples were dried to a final moisture content less than 0.1 kg water/kg d.m. Considering the practical application, the smaller time required to achieve the same final moisture content was 315 min for microwave pre-treated and drying at 60 °C. Corresponding values for control and ultrasound samples were 453 min and 492 min, respectively. Similar drying behavior was observed at drying temperatures of 50 and 55 °C. This behavior may be explained by the internal (volumetric) heating caused by moisture migration to the pear surface during microwave pre-treatment. Less drying time in the microwave and its combined applications may be associated with the rapid mass transfer within food materials during microwave heating. Heat was generated within the food due to the absorption of microwave energy, and it creates high internal pressure, temperature and concentration gradients. Therefore, the flow rate of the liquid through the food to the boundary is increased.

Concerning ultrasound-prior drying applications, at a temperature of 50 °C, samples had a shorter drying time than control ones. The lower drying temperature and longer exposure time may have produced higher damage to control samples. However, there was no drying time reduction in ultrasound-treated pears dried at 55 and 60 °C in comparison with control samples at the same temperatures. Under these conditions, the drying time of control and ultrasound pre-treated samples were quite similar. This could probably be caused by either the fact that ultrasound pre-treatments were applied through the vacuum packaging and not by the samples' immersion in the water medium or by structural changes of pear tissue that occurred during ultrasound pre-treatments. The structural properties of the food material are decisive for the ultrasound waves' action [77]. From this point of view, the pears' composition and structure may be modified by ultrasound application and can affect pears' ability to be dried.

The ultrasound application also causes a rapid alternative compression and expansion of the food matrix called a sponge effect. This mechanism is a direct effect of ultrasound, leading to the creation of microchannels and facilitating intracellular liquid removal to the surroundings. Another phenomenon associated with ultrasonic waves is the formation of thousands of cavitation bubbles in liquids. In our case, considering the effect of ultrasound application on "Rocha" pears 'drying process, these obtained findings could be associated with the pear samples' isolation by vacuum-packaging from the liquid medium (water) during the ultrasound application. This type of pre-treatment procedure probably caused a limited effect of cavitation, and only sponge effect occurred. Therefore, it can be noted that the occurrence of this sponge effect was insufficient to accelerate the drying process and moisture movement. These observations are in good agreement with results reported for ultrasound-pre-treated dried carrot slices [24]. On the contrary, in the cases where pre-treatment with ultrasound was applied before drying and food materials were directly immersed in a liquid medium, a significant reduction of processing time was observed for pineapple [42], melon [21] and mulberry [20].

"Rocha" pears' drying characteristics were affected by many factors such as the fruit's structure, applied pre-treatments, and drying temperature and time. From this viewpoint, obtaining an in-depth understanding and description of the microwave and ultrasound effects, and the involved mechanisms may also improve the drying process. Microwave application to the drying of pears had a remarkable impact on reducing process time, which may be attributed to the two microwave mechanisms of ionic polarization and dipole rotation. In ionic polarization, an electric field is applied to make ions move and collide with each other. As a consequence, their kinetic energy is converted into heat inside the food materials [80]. Both mechanisms are related to the "volumetric heating" of microwave energy. The ultrasound pre-treatment did not accelerate the drying process of pear fruits. Therefore, detailed information on ultrasound mechanisms (i.e., sponge effect and cavitation) and their effects on the drying process are needed.

In order to predict the moisture content (X) as a function of drying time (min), empirical models (presented in Section 2.4) were fitted to the experimental data. The drying kinetic parameters for each attempted model are reported in Table 2, where k is the drying rate constant (min^{-1} or min^{-N}), and a and N are drying parameters. The corresponding statistical parameters are summarized in Table 3. The coefficient of determination (R^2) and the standard deviation of the experimental error (s) were the primary criteria to select the best model to account for variation in the drying curves of pear samples. A good fitting corresponded to the highest R^2 value and the lowest value of s [47].

The statistical parameter estimations showed that R^2 and s values ranged, respectively, from 0.974 to 1.000 and from 0.0255 to 0.2038. According to model parameters in Table 2, for the Newton and Henderson & Pabis models, the drying rate k had values of 0.446×10^{-2} to 0.925×10^{-2} min^{-1}. The Page model gave the best results and demonstrated good agreement with experimental data (Figure 4), presenting the highest R^2 and the lowest s values among all tested empirical models. Hence, the Page model could predict with sufficient accuracy the evolution of moisture content for "Rocha" pear slabs at each drying temperature. Similar observations were reported by Doymaz and İsmail [37] and Senadeera et al. [40].

Table 2. Drying kinetics model parameters of "Rocha" pears dried at 50, 55 and 60 °C.

Model Name	Temperature	Parameters	Control	Microwaved	Ultrasound
Newton	50 °C	$k \times 10^2$ (1/min)	0.446 ± 0.005	0.558 ± 0.024	0.478 ± 0.009
	55 °C	$k \times 10^2$ (1/min)	0.578 ± 0.015	0.808 ± 0.043	0.546 ± 0.023
	60 °C	$k \times 10^2$ (1/min)	0.604 ± 0.013	0.851 ± 0.071	0.608 ± 0.014
Henderson & Pabis	50 °C	a	1.061 ± 0.007	1.074 ± 0.019	1.068 ± 0.010
		$k \times 10^2$ (1/min)	0.481 ± 0.006	0.598 ± 0.026	0.508 ± 0.010
	55 °C	a	1.068 ± 0.012	1.098 ± 0.025	1.058 ± 0.010
		$k \times 10^2$ (1/min)	0.615 ± 0.01	0.881 ± 0.046	0.577 ± 0.012
	60 °C	a	1.074 ± 0.018	1.090 ± 0.077	0.632 ± 0.017
		$k \times 10^2$ (1/min)	0.646 ± 0.026	0.925 ± 0.031	1.041 ± 0.011
Page	50 °C	$k \times 10^2$ (1/min^N)	0.169 ± 0.004	0.154 ± 0.015	0.165 ± 0.008
		N	1.184 ± 0.005	1.240 ± 0.021	1.191 ± 0.009
	55 °C	$k \times 10^2$ (1/min^N)	0.213 ± 0.014	0.170 ± 0.018	0.282 ± 0.029
		N	1.186 ± 0.013	1.310 ± 0.023	1.124 ± 0.022
	60 °C	$k \times 10^2$ (1/min^N)	0.184 ± 0.062	0.177 ± 0.029	0.312 ± 0.019
		N	1.224 ± 0.019	1.319 ± 0.031	1.126 ± 0.01

Table 3. Correlation coefficients (R^2 and s) of empirical drying models.

Model Name	Temperature	Correlation Coefficients	Control	Microwaved	Ultrasound
Newton	50 °C	R^2	0.995	0.984	0.991
		s	0.1168	0.1734	0.1622
	55 °C	R^2	0.991	0.978	0.993
		s	0.1477	0.1925	0.1321
	60 °C	R^2	0.986	0.974	0.994
		s	0.2038	0.1999	0.1282
Henderson & Pabis	50 °C	R^2	0.997	0.989	0.995
		s	0.0835	0.1432	0.1263
	55 °C	R^2	0.994	0.986	0.996
		s	0.1202	0.153	0.1001
	60 °C	R^2	0.991	0.983	0.996
		s	0.1612	0.1733	0.1071
Page	50 °C	R^2	1	0.997	0.999
		s	0.0255	0.0595	0.0481
	55 °C	R^2	0.999	0.998	0.998
		s	0.0541	0.0534	0.0776
	60 °C	R^2	0.999	0.997	0.999
		s	0.065	0.0737	0.0499

4. Conclusions

The influences of air drying temperature (50, 55 and 60 °C) and innovative pre-treatments (i.e., microwave and ultrasound applications) on quality attributes and drying behaviour of "Rocha" pear slices were investigated. Drying process conditions of pears significantly influenced all quality descriptors: lower drying temperature and longer exposure time resulted in more quality deterioration. Moreover, ultrasound pre-treatment combined with higher drying temperature (60 °C) improved the quality attributes of pear samples: better color preservation and retention of nutritional compounds (i.e., total phenolic content and antioxidant activity). On the other hand, microwave pre-treatment had a negative impact on the overall quality of dried pears.

Regarding the characteristic drying curve, only the falling rate was observed at all investigated conditions. The experimental results evidenced that the drying time decreased with increasing drying temperatures. Microwave pre-treatment was more effective in intensifying the drying process due probably to the volumetric heating of vapors generated inside the pear material. Therefore, microwave pre-treated pears had shorter drying times in comparison with control and ultrasound-pre-treated slices. On the other hand, ultrasound pre-treatment did not accelerate the drying process, which could be explained by the different ultrasound application mechanisms resulting probably from the following effects: the limited cavitation phenomenon and only the sponge effect occurring in vacuum-packaged pear samples. Among the applied empirical models, the Page model provided the best fit results (with the highest correlation factor, R^2, and the lowest standard deviation of experimental error, s) to describe the drying behavior of "Rocha" pears at each drying temperature accurately.

It can be concluded that each pre-treatment exhibits a different effect mechanism on the drying characteristics and the preservation of nutritional and quality parameters. Despite the ultrasound not accelerating the drying process of pear fruits, quality attributes of dried pears were better retained. Current approaches to understanding and improving drying process efficiency state that more research is needed regarding ultrasound application conditions (i.e., time, power, etc.) and the ultrasound mechanism. In this context, ultrasound pre-treatment may be a promising technique to obtain high-quality dried fruits, being an alternative to traditional pre-treatments and suitable to an industrial context. Additional studies must be carried out to determine the influence of pre-treatments' mechanisms on the dried product's final quality.

Future research will be devoted to investigating the description and characterization of "Rocha" pear rehydration behavior. The selection of appropriate pre-treatments seems crucial for optimizing rehydration process conditions and new product design.

Author Contributions: C.L.M.S., I.N.R., B.Ö., M.D.M., P.R. and G.A. conceived and designed the experiments; I.N.R. and B.Ö. performed the experiments; C.L.M.S., I.N.R., B.Ö. and G.A. analyzed the data resources; I.N.R. and B.Ö. wrote the original draft of the manuscript; C.L.M.S., I.N.R. and B.Ö. wrote, reviewed and edited the manuscript. All authors have read and agreed to the published version of the manuscript.

Funding: This work was supported by National Funds from FCT—Fundação para a Ciência e a Tecnologia through project UIDB/50016/2020.

Data Availability Statement: The datasets generated for this study are available on request to the corresponding author.

Conflicts of Interest: The authors declare no conflict of interest.

References

1. European Commission. *Commission Regulation (EC) No 492/2003 of 18 March 2003*; Publications Office of the EU: Luxembourg, 2003.
2. Coelho, I.; Matos, A.S.; Teixeira, R.; Nascimento, A.; Bordado, J.; Donard, O.; Castanheira, I. Combining multielement analysis and chemometrics to trace the geographical origin of Rocha pear. *J. Food. Compost. Anal.* **2019**, *77*, 1–8. [CrossRef]
3. Deuchande, T.; Carvalho, S.M.P.; Giné-Bordonaba, J.; Vasconcelos, M.W.; Larrigaudière, C. Transcriptional and biochemical regulation of internal browning disorder in 'Rocha' pear as affected by O_2 and CO_2 concentrations. *Postharvest Biol. Technol.* **2017**, *132*, 15–22. [CrossRef]
4. Fundo, J.F.; Galvis-Sanchez, A.; Madureira, A.R.; Carvalho, A.; Feio, G.; Silva, C.LM.; Quintas, M.A.C. NMR water transwerse relaxation time approach to understand storage stability of fresh-cut 'Rocha' pear. *LWT-Food Sci. Technol.* **2017**, *74*, 280–285. [CrossRef]
5. Arzoo, P.M. Why is pear so dear. *Int. J. Res. Ayurveda Pharm.* **2016**, *7*, 108–113.
6. Proietti, N.; Adiletta, G.; Russo, P.; Buonocore, R.; Mannina, L.; Crescitelli, A.; Capitani, D. Evolution of physicochemical properties of pear during drying by conventional techniques, portable-NMR, and modelling. *J. Food Eng.* **2018**, *230*, 82–98. [CrossRef]

7. Saquet, A.A. Storage of pears. *Sci. Hortic.* **2019**, *246*, 1009–1016. [CrossRef]
8. Singh, V.; Jawendha, S.K.; Gill, P.P.S.; Gill, M.S. Suppression of fruits softening and extension of shelf life of pear by putresche application. *Sci. Hortic.* **2019**, *256*, 108623. [CrossRef]
9. Deuchande, T.; Carvalho, S.M.P.; Guterres, U.; Fidalgo, F.; Isidoro, N.; Larrigaudiere, C.; Vasconcelos, M.W. Dynamic controlled atmosphere for prevention of internal browning disorders in 'Rocha' pear. *LWT-Food Sci. Technol.* **2016**, *65*, 725–730. [CrossRef]
10. Silva, F.J.P.; Gomes, M.H.; Rodrigues, J.A.; Almeida, D.P. Antioxidant properties and fruit quality during long-term storage of 'Rocha' pear: Effects of maturity and storage conditons. *J. Food Qual.* **2010**, *33*, 1–20. [CrossRef]
11. Prusky, D. Reduction of incidence of postharvest quality losses, and future prospects. *Food Secur.* **2011**, *3*, 463–474. [CrossRef]
12. Marques, L.C.; Prado, M.M.; Freire, J.T. Rehydration characteristics of freeze-dried tropical fruits. *LWT-Food Sci. Technol.* **2009**, *42*, 1232–1237. [CrossRef]
13. Yao, L.; Fan, L.; Duan, Z. Effect of different pretreatments followed by hot-air and far-infrared drying on the bioactive compounds, phsicochemical property and microstructure of mango slices. *Food Chem.* **2020**, *305*, 125477. [CrossRef] [PubMed]
14. Doymaz, I. Effect of citric acid and blanching pre-treatments on drying and rehydration of Amasya red apples. *Food Bioprod. Process.* **2010**, *88*, 124–132. [CrossRef]
15. Kingsly, R.P.; Goyal, R.K.; ManiKaantan, M.R.; Ilyas, S.M. Effects of pre-treatments and drying air temperature on drying behaviour of peach slice. *Int. J. Food Sci. Tecnol.* **2007**, *42*, 65–69. [CrossRef]
16. Onwude, D.I.; Hashim, N.; Chen, G. Recent advances of novel thermal combined hot air drying of agricultural crops. *Trends Food Sci. Technol.* **2016**, *57*, 132–145. [CrossRef]
17. Bozkır, H.; Ergün, A.R.; Serdar, E.; Metin, G.; Baysal, T. Influence of ultrasound and osmotic dehydration pretreatments on drying and quality properties of persimmon fruit. *Ultrason. Sonochem.* **2019**, *54*, 135–141. [CrossRef]
18. Fijalkowska, A.; Nowacka, M.; Sledz, M.; Witrowa-Rajchert, D. Ultrasound as a pretreatment method to improve drying kinetics and sensory properties of dried apple. *J. Food Process Eng.* **2016**, *39*, 256–265. [CrossRef]
19. Miano, A.C.; Ibarz, A.; Augusto, P.E.D. Mechanisms for imroving mass transfer in food with ultrasound technology: Describing the phenomena in two model cases. *Ultrason. Sonochem.* **2016**, *29*, 413–419. [CrossRef]
20. Tao, Y.; Wang, P.; Wang, Y.; Kadam, S.U.; Han, Y.; Wang, J.; Zhou, J. Power ultrasound as a pre-treatment to convective drying of mulberry (*Morus alba* L.) leaves: Impact on drying kinetics and selected quality properties. *Ultrason. Sonochem.* **2016**, *31*, 310–318. [CrossRef]
21. Silva, G.D.; Barros, Z.M.P.; Medeiros, R.A.B.; Carvalho, C.B.O.; Brandão, S.C.R.; Azoubel, P.M. Pre-treatments for melon drying implementing ultrasound and vacuuum. *LWT-Food Sci. Technol.* **2016**, *74*, 14–119.
22. Romero, C.A.J.; Yépez, B.D.V. Ultrasound as pretreatment to convective drying of Andean blackberry (*Rubus glaucus* Benth). *Ultrason. Sonochem.* **2015**, *22*, 205–210. [CrossRef]
23. Zhang, Z.; Niu, L.; Li, D.; Liu, C.; Ma, R.; Song, J.; Zhao, J. Low intensity ultrasound as a pre-treatment to drying of daylilies: Impact on enzyme inactivation, color changes and nutrition quality parameters. *Ultrason. Sonochem.* **2017**, *36*, 50–58. [CrossRef]
24. Nowacka, M.; Wedzik, M. Effect of ultrasound treatment on microstructure, colour and carotenoid content in fresh and dried carrot tissue. *Appl. Acoust.* **2016**, *103*, 163–171. [CrossRef]
25. Ricce, C.; Rojas, M.L.; Miano, A.C.; Siche, R.; Augusto, P.E.D. Ultrasound pre-treatment enhances the carrot drying and rehydration. *Food Res. Int.* **2016**, *89*, 701–708. [CrossRef] [PubMed]
26. Feng, Y.; Yu, X.; Yagoub, A.E.A.; Xu, B.; Wu, L.; Zhou, C. Vacuum pretreatment coupled to ultrasound assisted osmotic dehydration as a novel method for garlic-slices dehydration. *Ultrason. Sonochem.* **2019**, *50*, 363–372. [CrossRef]
27. Nowacka, M.; Wiktor, A.; Anuszewska, A.; Dadan, M.; Rybak, K.; Witrowa-Rajchert, D. The application of unconventional technologies as pulsed electric field ultrasound and microwave-vacuum drying in the production of dried cranberry snacks. *Ultrason. Sonochem.* **2019**, *56*, 1–13. [CrossRef]
28. Ekezie, F.-G.C.; Sun, D.-W.; Han, Z.; Cheng, J.-H. Microwave-assisted food processing technologies for enhancing product quality and process efficiency: A review of recent developments. *Trends Food Sci Tech.* **2017**, *67*, 58–69. [CrossRef]
29. Fathi-Achachlouei, B.; Azadmard-Damirchi, S.; Zahedi, Y.; Shaddel, R. Microwave pretreatment as a promising strategy for increment of nutraceutical content and extraction yield of oil from milk thistle seed. *Ind. Crop. Prod.* **2019**, *128*, 527–533. [CrossRef]
30. Chahbani, A.; Fakhfakh, N.; Balti, M.A.; Mabrouk, M.; El-Hatmi, H.; Zouari, N.; Kechaou, N. Microwave drying effects on drying kinetics, bioactive compounds and antioxidant activity of green peas (*Pisum sativum* L.). *Food Biosci.* **2018**, *25*, 32–38. [CrossRef]
31. Mothibe, K.J.; Zhang, M.; Mujumdar, A.S.; Wang, Y.C.; Cheng, X. Effects of ultrasound and microwave pretreatments of apple before spouted bed drying on rate of dehydration and physical properties. *Dry. Technol.* **2014**, *32*, 1848–1856. [CrossRef]
32. Zhang, M.; Tang, J.; Mujumdar, A.; Wang, S. Trends is microwave-related drying of fruits and vegetables. *Trends Food Sci Technol.* **2006**, *17*, 524–534. [CrossRef]
33. Sabry, Z.A.; Bahlol, H.E.; El-Desouky, A.I.; Assous, M.T. Effect of microwave pre-treatment and drying methods on the carrot powder quality. *Middle East J. Appl. Sci.* **2016**, *6*, 349–356.
34. Brasiello, A.; Adiletta, G.; Russo, P.; Crescitelli, S.; Albanese, D.; Di Matteo, M. Mathematical modeling of eggplant drying: Shrinkage effect. *J. Food Eng.* **2013**, *114*, 99–105. [CrossRef]
35. Kowalski, S.J.; Mierzwa, D. Numerical analysis of drying kinetics for shrinkable products such as fruits and vegetables. *J. Food Eng.* **2013**, *114*, 522–529. [CrossRef]

36. Koç, B.; Eren, İ.; Kaymak-Ertekin, F. Modelling bulk density, porosity and shrinkage of quince during drying: The effect of drying method. *J. Food Eng.* **2008**, *85*, 340–349. [CrossRef]

37. Doymaz, İ.; İsmail, O. Drying characteristics of sweet cherry. *Food Bioprod. Process.* **2011**, *89*, 31–38. [CrossRef]

38. Adiletta, G.; Russo, P.; Senadeera, W.; Di Matteo, M. Drying characteristics and quality of grape under physical pretreatment. *J. Food Eng.* **2016**, *172*, 9–18. [CrossRef]

39. Johnson, A.C.; Ali Al Mukhaini, E.M. Drying studies on peach and strawberry slices. *Cogent Food Agric.* **2016**, *2*, 1141654. [CrossRef]

40. Senadeera, W.; Adiletta, G.; Önal, B.; Di Matteo, M.; Russo, P. Influence of different hot air drying temperatures on drying kinetics, shrinkage, and colour of persimmon slices. *Foods* **2020**, *9*, 101. [CrossRef]

41. AOAC. *Official Methods of Analysis*, 16th ed.; Association of Analytical Chemist: Arlington, VA, USA, 1998.

42. Fernandes, F.A.N.; Linhares, F.E.; Rodrigues, S. Ultrasound as pre-treatment for drying of pineapple. *Ultrason. Sonochem.* **2008**, *15*, 1045–1054. [CrossRef]

43. Oliveira, S.M.; Ramos, I.N.; Brandao, T.R.S.; Silva, C.L.M. Effect of air-drying temperature on the quality and bioactive characteristics of dried galega kale (*Brassica oleracea* L. Var. Acephala). *J. Food Process. Preserv* **2015**, 2485–2496. [CrossRef]

44. Jayes, W. The Sugar Engineers. Available online: http://www.sugartech.co.za/psychro (accessed on 1 July 2019).

45. Diamente, L.M.; Munro, P.A. Mathematical modelling of the thin layer solar drying of sweet potato slices. *Sol. Energy* **1993**, *51*, 271–276. [CrossRef]

46. Bala, B.K.; Woods, J.I. Simulation of the indirect natural-convection solar drying of rough rice. *Sol. Energy* **1994**, *53*, 259–266. [CrossRef]

47. Ramos, I.N.; Brandao, T.R.S.; Silva, C.L.M. Effect of pre-treatments on solar drying kinetics of red seedless. *Int. J. Food Stud.* **2014**, *3*, 239–247. [CrossRef]

48. Doymaz, I. Sun drying of figs: An experimental study. *J. Food Eng.* **2005**, *71*, 403–407. [CrossRef]

49. Yaldiz, O.; Ertekin, C.; Uzun, H.I. Mathematical modeling of thin layer solar drying of sultana grapes. *Energy* **2001**, *26*, 457–465. [CrossRef]

50. Mahmutoglu, T.; Emir, F.; Saygi, Y.B. Sun/solar drying of differently treated grapes and storage stability of dried grapes. *J. Food Eng.* **1996**, *29*, 289–300. [CrossRef]

51. Hsu, C.L.; Chen, W.; Wenga, Y.M.; Tsenga, C.Y. Chemical composition, physical properties, and antioxidant activities of yam flours as affected by different drying methods. *Food Chem.* **2003**, *83*, 85–92. [CrossRef]

52. Vega-Gálvez, A.; Ah-Hen, K.; Chacana, M.; Vergara, J.; Martínez-Monzó, J.; García-Segovia, P.; Lemus-Mondaca, R.; Di Scala, K. Effect of temperature and air velocity on drying kinetics, antioxidant capacity, total phenolic content, colour, texture and microstructure of apple (var. *Granny Smith*) slices. *Food Chem.* **2012**, *132*, 51–59. [CrossRef]

53. Salta, J.; Martins, A.; Santos, R.G.; Neng, N.R.; Nogueira, J.M.F.; Justino, J.; Rauter, A.P. Phenolic composition and antioxidant activity of Rocha pear and other pear cultivars–A comparative study. *J. Funct. Foods* **2010**, *2*, 153–157. [CrossRef]

54. Önal, B.; Adiletta, G.; Crescitelli, A.; Di Matteo, M.; Russo, P. Optimization of hot air drying temperature combined with pre-treatment to improve physico-chemical and nutritional quality of 'Annurca' apple. *Food Bioprod. Process.* **2019**, *115*, 87–99. [CrossRef]

55. Adiletta, G.; Petriccione, M.; Liguori, L.; Pizzolongo, F.; Romano, R.; Di Matteo, M. Study of pomological traits and physico-chemical quality of pomegranate (*Punica granatum* L.) genotypes grown in Italy. *Eur. Food Res. Technol.* **2018**, *244*, 1427–1438. [CrossRef]

56. Deng, L.Z.; Yang, X.H.; Mujumdar, A.S.; Zhao, J.H.; Wang, D.; Zhang, Q.; Wang, J.; Gao, Z.J.; Xiao, H.W. Red pepper (*Capsicum annuum* L.) drying: Effects of different drying methods on drying kinetics, physicochemical properties, antioxidant capacity, and microstructure. *Dry. Technol.* **2018**, *36*, 893–907. [CrossRef]

57. Wang, J.; Yang, X.H.; Mujumdar, A.S.; Fang, X.M.; Zhang, Q.; Zheng, Z.A.; Gao, Z.J.; Xiao, H.W. Effects of high humidity hot air impingement blanching (HHAIB) pre-treatment on the change of antioxidant capacity, the degradation kinetics of red pigment, ascorbic acid in dehydrated red peppers during storage. *Food Chem.* **2018**, *269*, 65–72. [CrossRef] [PubMed]

58. Box, G.E.P.; Hunter, W.G.; Hunter, J.S. *Statistics for Experiments an Introduction to Design, Data Analysis and Model Building*; John Wiley and Sons: New York, NY, USA, 1978.

59. Krokida, M.K.; Kiranoudis, C.T.; Maroulis, Z.B.; Marinos-Kouris, D. Effect of pre-treatment on color of dehydration products. *Dry. Technol.* **2000**, *18*, 1239–1250. [CrossRef]

60. Wojdylo, A.; Figiel, A.; Lech, K.; Nowicka, P.; Oszmiański, J. Effect of convective and vacuum-microwave drying on the bioactive compounds, color, and antioxidant capacity of sour cherries. *Food Bioprocess. Technol.* **2014**, *7*, 829–841. [CrossRef]

61. Horuz, E.; Bozkurt, H.; Karatas, H.; Maskan, M. Effect of hybrid (microwave-convectional) and convectional drying on drying kinetics, total phenolics, antioxidant capacity, vitamin C, color and rehydration capacity of sour cherries. *Food Chem.* **2017**, *230*, 295–305. [CrossRef]

62. Vadivambal, R.; Jayas, D.S. Changes in quality of microwave-treated agricultural products–A Review. *Biosyst. Eng.* **2007**, *98*, 1–16. [CrossRef]

63. Mothibe, K.J.; Zhang, M.; Nsor-atindana, J.; Wang, Y.-C. Use of ultrasound pre-treatment in drying of fruits: Drying rates, quality attributes and shelf life extension. *Dry. Technol.* **2011**, *29*, 1611–1621. [CrossRef]

64. Hayat, K.; Zhang, X.; Fraooq, U.; Abbas, S.; Jia, C.; Zhong, F.; Zhang, J. Efffect of microwave treatment on phenolic content and antioxidant activity of citrus mandarin pomace. *Food Chem.* **2010**, *123*, 423–429. [CrossRef]
65. Robinson, J.P.; Kingman, S.W.; Sanpe, C.E.; Shang, H.; Barronco, R.; Seeid, A. Separation of polyaromatic hydrocarbons from contamined soils using microwave heating. *Sep. Purif. Technol.* **2009**, *62*, 249–254. [CrossRef]
66. Amami, E.; Khezami, W.; Mezrigui, S.; Badwak, L.; Bejar, A.K.; Perez, C.T.; Kechaou, N. Effect of ultrasound-assisted osmotic dehydration pretreatment on the convective drying of strawberry. *Ultrason. Sonochem.* **2017**, *36*, 286–300. [CrossRef]
67. Wiktor, A.; Sledz, M.; Nowacka, M.; Rynak, K.; Vitrowa-Rajchert, D. The influence of immersion and contact ultrasound treatment on selected properties of apple tissue. *Appl. Acoust.* **2016**, *103*, 136–142. [CrossRef]
68. Di Scala, K.; Vega-Gálvez, A.; Uribe, E.; Oyanadel, R.; Miranda, M.; Veragara, J.; Quispe, I. Changes of quality characteristics of pepino fruit (*Solanum muricatum Ait*) during convective drying. *Int. J. Food Sci.* **2011**, *46*, 746–753. [CrossRef]
69. Ek, P.; Araújo, A.C.; Oliveira, S.M.; Ramos, I.N.; Brandão, T.R.S.; Silva, C.L.M. Assessment of nutritional quality and colour parameters of convective dried watercress(*Nasturtiumofficinale*). *J. Food Process. Preserv.* **2017**, *42*, 1–9.
70. Garau, M.C.; Simal, S.; Rosselló, C.; Femenia, A. Effect of air-drying temperature on physico-chemical properties of dietary fibre and antioxidant capacity of orange (*Citrus aurantium* v. Canoneta) by-products. *Food Chem.* **2007**, *104*, 1014–1024. [CrossRef]
71. Vega-Gálvez, A.; Di Scala, K.; Rodríguez, K.; Lemus-Mondaca, R.; Miranda, M.; López, J.; Perez-Won, M. Effect of air-drying temperature on physico-chemical properties, antioxidant capacity, colour and total phenolic content of red pepper (*Capcicum annuum*, L. var. Hungarian). *Food Chem.* **2009**, *117*, 647–653. [CrossRef]
72. Nicoli, M.C.; Anese, M.; Parpinel, M. Influence of processing on the antioxidant properties of fruit and vegetables. *Trends Foods Sci. Technol.* **1999**, *10*, 94–100. [CrossRef]
73. Kamiloğlu, S.; Capanoglu, E. Polyphenol content in figs (*Ficus carica* L.): Effect of sun drying. *Int. J. Food Prop.* **2015**, *18*, 521–535. [CrossRef]
74. Rahman, N.F.A.; Shamsudin, R.; Ismail, A.; Shah, N.N.A.K.; Varith, J. Effects of drying methods on total phenolic contents and antioxidant capacity of pomelo (*Citrus grandis* (L.) Osbeck) peels. *Innov. Food Sci. Emerg. Technol.* **2018**, *50*, 217–225. [CrossRef]
75. Doymaz, İ. Evaluation of some thin-layer drying models of persimmon slices (*Diospyros kaki* L.). *Energy Convers. Manag.* **2012**, *56*, 199–205. [CrossRef]
76. Corrêa, J.L.G.; Rasia, M.C.; Mulet, A.; Cárcel, J.A. Influence of ultrasound application on both the osmotic pre-treatment and subsequent convective drying of pineapple (*Ananas comosus*). *Innov. Food Sci. Emerg. Technol.* **2017**, *41*, 284–291. [CrossRef]
77. Nascimento, E.M.C.G.; Mulet, A.; Ascheri, J.L.R.; Carvalho, C.W.P.; Cárcel, J.A. Effects of high-intensity ultrasound on drying kinetics and antioxidant properties of passion fruit peel. *J. Food Eng.* **2016**, *170*, 108–118. [CrossRef]
78. Tao, Y.; Zhang, J.; Jiang, S.; Xu, Y.; Show, P.L.; Han, Y.; Ye, X.; Ye, M. Contacting ultrasound enhanced hot air convective drying of garlic slices: Mass transfer modelling and quality evaluation. *J. Food Eng.* **2018**, *235*, 79–88. [CrossRef]
79. Ju, H.-Y.; Law, C.-L.; Fang, X.-M.; Xiao, H.-W.; Liu, Y.-H.; Gao, Z.-J. Drying kinetics and evolution of the sample's core temperature and moisture distribution of yam slices (*Dioscorea alata* L.) during convective hot-air drying. *Dry. Technol.* **2016**, *34*, 1297–1306. [CrossRef]
80. Dehghannya, J.; Kadkhadaei, S.; Heshmati, M.K.; Ghanbarzadeh, B. Ultrasound-assisted intensification of a hybrid intermittent microwave-hot air drying process of potato: Quality aspect and energy consumption. *Ultrasonics* **2019**, *96*, 104–122. [CrossRef] [PubMed]

Article

Monitoring the Processing of Dry Fermented Sausages with a Portable NIRS Device

Alberto González-Mohino [1], Trinidad Pérez-Palacios [1], Teresa Antequera [1], Jorge Ruiz-Carrascal [1], Lary Souza Olegario [2] and Silvia Grassi [3,*]

[1] Meat and Meat Products Research Institute (IProCar), Food Technology, University of Extremadura, 10003 Cáceres, Spain; albertogj@unex.es (A.G.-M.); triny@unex.es (T.P.-P.); tantero@unex.es (T.A.); jruiz@unex.es (J.R.-C.)

[2] Department of Food Engineering, Technology Centre, Federal University of Paraiba, 58051-900 Joao Pessoa, Paraiba, Brazil; Laryolegario@hotmail.com

[3] Department of Food, Environmental, and Nutritional Sciences (DeFENS), Università degli Studi di Milano, via G. Celoria 2, 20133 Milan, Italy

* Correspondence: silvia.grassi@unimi.it; Tel.: +39-02-503-19179

Received: 6 August 2020; Accepted: 11 September 2020; Published: 14 September 2020

Abstract: This work studies the ability of a MicroNIR (VIAVI, Santa Rosa, CA) device to monitor the dry fermented sausage process with the use of multivariate data analysis. Thirty sausages were made and subjected to dry fermentation, which was divided into four main stages. Physicochemical (weight lost, pH, moisture content, water activity, color, hardness, and thiobarbiruric reactive substances analysis) and sensory (quantitative descriptive analysis) characterizations of samples on different steps of the ripening process were performed. Near-infrared (NIR) spectra (950–1650 nm) were taken throughout the process at three points of the samples. Physicochemical data were explored by distance to K-Nearest Neighbor (K-NN) cluster analysis, while NIR spectra were studied by partial least square–discriminant analysis; before these models, Principal Component Analysis (PCA) was performed in both databases. The results of multivariate data analysis showed the ability to monitor and classify the different stages of ripening process (mainly the fermentation and drying steps). This study showed that a portable NIR device (MicroNIR) is a nondestructive, simple, noninvasive, fast, and cost-effective tool with the ability to monitor the dry fermented sausage processing and to classify samples as a function of the stage, constituting a feasible decision method for sausages to progress to the following processing stage.

Keywords: dry-fermented sausages; near infrared spectroscopy; portable device; PLS-DA

1. Introduction

Dry-fermented meat products are one of the eldest and more remarkable groups of processed meats and constitute a key aspect in the identity, culture, and heritage of numerous regions. The great interest in traditional dry-fermented meats is specially remarkable in Europe, due not only to their great economic weight but also to their unique sensory features, which are a consequence of the raw material and the manufacturing process [1]. Traditional dry-fermented sausages are mostly manufactured with lean and fat from pork in small-scale production plants [2]. As other authors reported [3], most of these sausages are seasoned and processed with traditional manufacturing, the domestic environment being characterized by the limited degree of mechanization and final product control, which may bring higher heterogeneity to their quality. The implementation of control systems for the whole dry-fermented sausage process, ensuring the quality and safety of the product, would contribute to overcoming these issues. As most of traditional product industries are small-scale, such a control system should be simple, cheap, and easily implemented in fermenting-drying domestic or pilot-scale chambers.

Several analytical methods, such as physicochemical, chromatography, mass spectrometry, or sensory ones, have been shown to be able to collect accurate data from which the global quality of food products can be inferred. Nevertheless, these techniques are expensive, tedious, time- and solvent-consuming, and complex to use and require destruction of the sample.

Near-infrared (NIR) spectroscopy may well be an alternative to those other analysis methods, since it is cheaper than other instrumental techniques and allows for a nondestructive, simple, and fast analysis [4]. NIR spectroscopy combined with multivariate data analysis (chemometrics) have recently been demonstrated to be a solution of process analytical technology in the food industry [5]. Indeed, from NIR spectroscopy data, information about chemical, textural, and even microbiological compositions can be simultaneously inferred. Among the different chemometric methods that could be applied, Principal Component Analysis (PCA) and Partial Least Square-Discriminant Analysis (PLS-DA) became of interest, as they are able to take advantage of the structures in highly overlapping and colinear data [6].

Among the different commercially available NIR equipment, NIR handheld devices are a good alternative to benchtop instruments, being equally reliable but cheaper and faster and allowing in situ analyses. In fact, the advantages of NIR portable devices have been reported by other authors, highlighting cost reduction [7–9] as well as lower environmental impact [10] in comparison with benchtop ones. In addition, it is important to note that portable instruments are quite suitable for traditional product analysis, as reported in previous studies [11]. Portable NIR devices have been used to face different food-related issues, such as the determination of fish freshness [12] or the prediction of lycopene content in tomato [13]. Among different portable devices, MicroNIR is one of the most reliable due to its high resolution and broad spectral range. MicroNIR has been largely applied in several food matrices, including meat products [14], to assess dry cured ham quality parameters in dry cured ham [15], to monitor chicken meat authenticity [16], or to predict beef quality [17]. However, its use in monitoring the fermentation process in dry-fermented sausage has not been addressed yet.

Therefore, this work aims at developing a fast and noninvasive approach to monitor the process of traditional dry fermented sausages and to classify samples according to their processing stage by using a NIR spectroscopy handheld device coupled with multivariate data analysis.

2. Materials and Methods

2.1. Sausage Manufacturing

Twenty kilograms of pork lean and back fat (in proportion 4:1) were bought in a commercial supermarket (Mercadona, Caceres, Spain). Firstly, meat and fat were separately ground with a food grinding machine model PC-114 with a grinding plate of 4 mm (MAINCA, Equipamientos cárnicos S.L., Barcelona, España). Thereafter, the lean and fat were mixed by a mixing machine model RM-200 (MAINCA, Equipamientos cárnicos S.L., Barcelona, España) and added with the ingredients (salt (2.5% *w/w*), sucrose (0,75% *w/w*), garlic power (0,1%), spices (1% *w/w*), and sodium nitrite (100 ppm % *w/w*)). Subsequently, this mixture was stuffed into pork casings (35–40 mm diameter) of approximately 250 g each. Thus, 30 raw sausage samples were obtained. They were divided in two batches of 15 sausages (batch 1 and batch 2), which were subjected to the same fermentation and drying conditions but processed in different chambers.

The fermentation phase was carried out at 22–25 °C and 95% relative humidity (RH) for approximately 36 h until the pH reached 4.5. Thereafter, sausages were transferred into a drying-ripening chamber at 55 °C and 80% RH for 24 h (intense drying step) and continued drying at 15 °C for 60 h at an RH of 65%. The whole process took 120 h.

Figure 1 displays the processing conditions and the sampling. Sausages were analyzed by NIR spectroscopy at 0, 12, 24, 36, 48, 60, and 120 h. Moreover, three sausages for each processing step were used to perform destructive analyses to determine weight loss, water activity (aW), moisture content, and pH. The processing phases considered were the beginning of the processing (as raw material, RM), the end of fermentation (EF), and the intense drying (EID) stages. The rest of sausages ($n = 6$) finished the processing and were analyzed as a final product (FP). The determinations done in the FP batch were the same as the RW, EF, and EID ones but differed by adding instrumental color and texture, by lipid oxidation, and by sensory analysis.

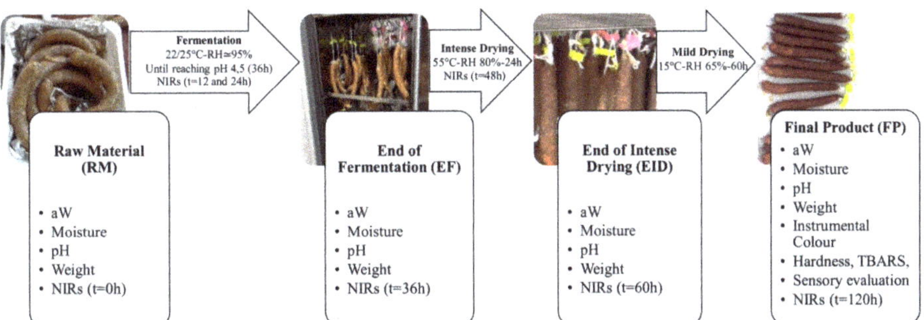

Figure 1. Scheme of the dry fermented sausage ripening process and analyses carried out at each stage. RH, relative humidity; aW, water activity; TBARS, thiobarbituric acid-reactive substances; NIRs, near infrared spectroscopy.

2.2. Physicochemical Analysis (FQA)

Measurement of pH was determined in three different locations of each sausage (to obtain a representative averaged pH) with a meat pH meter electrode probe model FC232D (HANNA Instruments S.L., Eibar, Spain) equipped with automatic temperature compensation. The pH meter was calibrated with commercial buffer solutions (Crison, Barcelona, Spain) at pH 4.0 and 7.0 prior to use.

Moisture content was determined by drying the samples (5 g) at 102 °C following the procedure of the official methods of Association of Official Agricultural Chemists (AOAC International reference method 935.29) [18].

Water activity (aW) was determined by a water activity measuring equipment (Lab Master-aw; NOVASINA AG, Lachen, Switzerland).

Instrumental color of the sausages was measured using a portable reflectance spectrophotometer (Konica Minolta CM-600d, Osaka, Japan) that was calibrated with a standard white calibration tile. The analysis was carried out according to the principles laid down by the Commission International d'Eclairage (CIE) [19]. The following color coordinates were determined: lightness (L*), redness (a*), and yellowness (b*).

Instrumental texture (hardness) was analyzed by a texturometer TA XT-2i Texture Analyser (Stable Micro Systems Ltd., Surrey, UK). For each sample, five cubes (1 cm^3) were obtained and analyzed. They were axially compressed to 50% of the original height with a flat plunger of 50 mm in diameter (P/50) at a crosshead speed of 2 mm $\times$ s^{-1} through a two-cycle sequence.

Lipid oxidation was measured by the thiobarbiruric reactive substances (TBARS) method, following the procedure described by Salih et al. [20] based on the concentration of malonaldehyde (MDA), and expressed as mg MDA/Kg sample.

These determinations were carried out in triplicate for each sample except for instrumental color and texture, which were analyzed in quintuplicate (due to the possible high variability of the samples and the laboratory error).

2.3. Sensory Analysis

Quantitative Descriptive Analysis (QDA) was carried out using 17 trained panelists (6 male and 11 female, age range 23–60 years). All of them were staff at the Meat and Meat Products Research

Institute (IProCar) of the University of Extremadura (Spain). Attributes evaluated by the panel were selected based on attributes reported in previous studies with similar products and taking into consideration the consensus reached by a focus group of 6 panelists. The following attributes were chosen: red color intensity and cohesiveness for appearance; hardness, juiciness, and chewiness for texture; and flavor intensity, saltiness, spicy flavor persistence, and hot-spicy flavor. A 10-cm unstructured scale was used for attributes scoring, and verbal anchors were fixed as "extremely low" to "extremely high" for all evaluated attributes. Samples (one slice per plate) were served on glass plates with a glass of mineral water and a piece of unsalted cracker to follow the rinsing protocol between samples. Evaluations took place in individual booths under white fluorescence light. The serving order of the samples was randomized according to the Williams Latin Square design. FIZZ software 2.20 C version (Biosystemes, Couternon, France, 2002) was used for collecting the scores.

2.4. MicroNIR Analysis

MicroNIR OnSite (VIAVI, Santa Rosa, CA) was used to analyze different sausages on the external surface of four different points equally spaced-out at each sampling time according to the scheme reported in Figure 2.

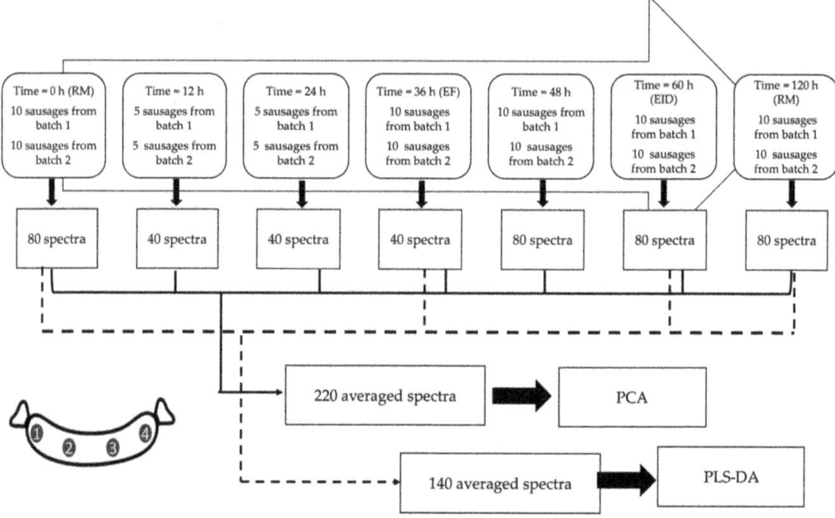

Figure 2. Scheme of the spectra acquisition procedure. RM, raw material; EF, the end of fermentation; end the intense drying (EID); FP, final product stages; PCA, Principal Component Analysis; and PLS-DA, Partial Least Square-Discriminant Analysis.

The MicroNIR spectral range was set to 950–1650 nm, with a 12.5 µs integration time and 200 scans, with a spectral bandwidth lower than 1.25% of center wavelength, typically 1% (e.g., at 1000 nm, the resolution is lower than 12.5 nm) and signal-to-noise ratio of 25,000.

Spectral acquisition was performed for 10 sausages selected from batch 1 and 10 from batch 2, every 12 h (i.e., at 0, 12, 24, 36, 48, and 60 h) and at the end of the process (120 h) for a total of seven sampling times and 440 spectra, as reported in Figure 2.

2.5. Data Analysis

Mean values and standard deviation were obtained from physicochemical and sensory data. Moreover, physicochemical results collected along the four stages of the process (weight losses, water activity, moisture, and pH) were explored by a Principal Component Analysis (PCA) to visualize the relationships among objects and variables with a biplot of scores and loadings. Thereafter, a clustering approach based on K-Nearest Neighbor's algorithm (K-NN) was applied. K-NN is a simple nonlinear classification approach based on Euclidean distance and not requiring any assumptions on the underlying data distribution [21], and it is able to solve complex classification issues [22]. The K-NN algorithm was applied to discriminate clusters according to the sausages' physicochemical characteristics to be later used as classes for the development of classification models based on NIR spectra.

Regarding MicroNIR, the four spectra collected for each sausage at each sampling time were averaged 2 by 2, thus merging the poles spectra and the ones collected closer to the longitudinal center. The averaged dataset (220 spectra × 125 wavelengths) was preprocessed to minimize the effect of noise and to enhance small but relevant spectral feature. Hence, the spectra dataset was transformed by smoothing (Savitzky-Golay, 3-wavelength gap size) followed by first derivative (Savitzky-Golay, 3-wavelength gap size and 2nd order polynomial) and mean center.

Data exploration by PCA lead to a reduction of the dataset due to outlier presence prior to developing a classification model by Partial Least Square Discriminant Analysis (PLS-DA). In order to do that, the 2-by-2 averaged spectra collected by MicroNIR for the sampling points corresponding to phases RM, EF, EID, and FP (140 spectra × 125 wavelengths) were split into a calibration set accounting for 66% of the data and a test set with the remaining 33% of samples. The PLS-DA method was selected to develop a classification model based on the NIR data according to the a priori classes defined by cluster analysis performed on the physicochemical data. A model was developed from the calibration dataset and internally validated by the Venetian blinds cross-validation procedure. Furthermore, the prediction capability was assessed by external validation using the test set, counting samples not used in the model development. According to Grassi et al. [21], PLS-DA was evaluated in all phases, i.e., calibration, cross-validation, and prediction, by two metrics, sensitivity (SENS) and specificity (SPEC), which are computed on the bases of four-factor (True Positive (TP), False Positive (FP), True Negative (TN), and False Negative (FN)). Thus, SENS describes the model capability to correctly recognize samples belonging to the considered class, whereas SPEC expresses the model capability to correctly reject samples belonging to all the other classes. Both metrics values are in a range from 0 (no correct prediction) to 1 (perfect classification).

Data analyses were performed under Matlab environment (R2017b, The Mathworks, Inc., Natick, MA, USA) eventually using the PLS toolbox v. 8.5 (Eigenvector Research, Inc., Manson, WA, USA) software package.

3. Results and Discussion

3.1. Physicochemical and Sensory Results

Table 1 shows, along the four stages, the mean averages and standards deviation of aW, moisture, pH, and weight loss of both dry-fermented sausages batches together, since no significant differences were found between them.

Table 1. Moisture, aW, pH, and weight loss results along the four stages of the process as mean values and standard deviation of batches 1 and 2.

Stage	Moisture (%)	aW	pH	Weight Loss (%)
Raw Material (RM)	60.2 ± 0.8 [1]	0.963 ± 0.01	5.72 ± 0.04	-
End of Fermentation (EF)	56.1 ± 1.7	0.961 ± 0.01	4.78 ± 0.11	10.2 ± 0.9
End of Intense Drying (EID)	41.9 ± 0.9	0.932 ± 0.02	5.17 ± 0.01	32.7 ± 1.7
Final Product (FP)	39.12 ± 0.8	0.875 ± 0.01	5.10 ± 0.07	35.8 ± 2.2

[1] Mean values and standard deviation of batches 1 and 2.

As expected, weight loss increased through the process, which is crucial for this type of product [23]. As a consequence, moisture and aW also followed a decreasing trend in their values from the first phase (RM) to the successive stages of the dry fermented process, as has been previously described [24,25]. Concerning pH, it also underwent an initial decrease between the RM and EF phases but, thereafter, increased during the drying stages, which is a common pattern for this type of product [25]. In the final product, TBARS (0.23 ± 0.01), instrumental color L* (53.8 ± 1.4), a* (21.7 ± 2.2), b* (25.1 ± 4.1), instrumental hardness (133.5 ± 6.8 kg), and QDA attributes (represented in Figure 3) were also analyzed in order to check whether these dry fermented sausages fulfill the sensory and technological requirements for this type of product. Comparing the found values with those reported by other authors, parameters such as aW (ranged from 0.84 to 0.87) and pH (ranged from 5.59 to 5.89) were similar to reported values [26,27]. Moreover, color intensity and saltiness values were comparable with those reported in these two studies when considering control traditional samples. Nevertheless, pH values in our sausages did not experience an increase after the fermentation phase as high as in other sausages [27,28]. Moisture (33.8%) and weight loss (33.0%) values in samples from the last stage of ripening obtained by other authors were also comparable with our results, which validates the dehydration process, crucial for guaranteeing the preservation of these kinds of meat products [29]. Regarding aW, this parameter is crucial for the safety of this kind of product, since values below 0.9 ensure a stable product at room temperature, limiting the growth of spoilage and pathogenic bacteria [30]. The obtained values in the final product (0.875 ± 0.01) ensure the safety of our sausages.

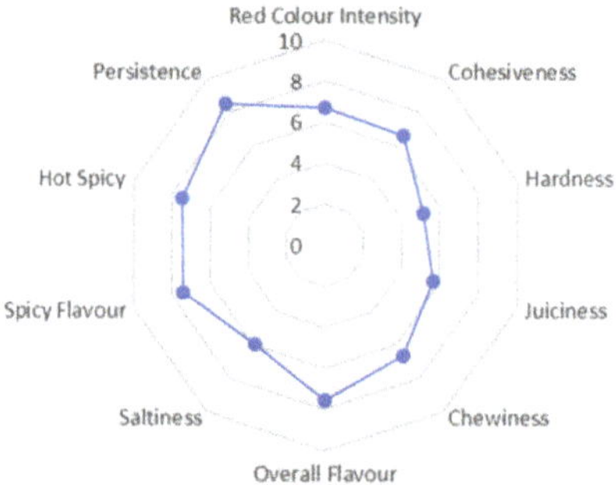

Figure 3. Results on sensory analysis from Quantitative Descriptive Analysis (QDA): the attributes are assessed within a 10-cm scale, from extremely low to extremely high.

If our traditional small-scale dry-fermented sausages are compared with commercial sausages, the results are coherent with those reported by Lopez et al. [1], who characterized ten commercial dry fermented sausages from a physicochemical and a sensory perspective, determining pH (ranged from 5.14 to 6.03), moisture (ranged from 28.75 to 48.70), and color parameters L* (ranged from 32.22 to 54.75), a* (ranged from 16.93 to 26.57), and b* (ranged from 6.42 to 15.91) of the commercial products.

Figure 4a displays the PCA biplot obtained for physicochemical data collected along the four steps of the dry-fermenting process for both batches, while Figure 4b displays the dendrogram obtained by cluster analysis through the K-nearest neighbor algorithm performed on the same dataset for both batches. The PCA biplot, defined by two first-principal components, accounts for 95.74% of the total variance (70.38 for PC1 and 25.36 for PC2). Both experimental batches were closely located for all stages of the processing. The RM group (t = 0 h, RM1 and RM2) is located in the I quadrant of the plot and correlates with pH and moisture content, both located in the same quadrant. Samples from the EF group (t = 36 h, EF1 and EF2) are situated in the IV quadrant, the same one where weight loss and aW are located. This location explains that PC2 differentiates samples according to changes mainly linked to pH, a parameter subjected to high variation during the fermentation process. Indeed, pH passes from an average of 5.7 to an average of 4.5 from RM to EF samples, respectively, thus distancing EF from the other samples and from the pH loadings. Samples from EID (t = 60, EID1 and EID2) and FP (t = 120, FP1 and FP2) are separated from RM and EF samples along PC1, showing highly negative PC1 scores. Their location is well explained by moisture content and weight variables, which, on the contrary, show high positive PC1 loadings. As a matter of fact, the drying process highly reduced moisture and weight of the sausages, especially in the intense drying step. Thus, EID and FP samples are characterized by similar PC1 scores but are quite different from RM and EF samples. The slight difference between EID and FP could be attributed to aW, for which the loadings call EID samples to slightly lower PC2 scores. PCA results on the physicochemical data could clearly separate three of the four stages of the dry fermented sausages process, though this analysis only constitutes a preliminary exploration. Further investigation by K-NN cluster clearly defined the existence of four groups according to the process phases at a K-NN distance of 1. In detail, Figure 4b reports the obtained dendrogram: RM1 and RM2 form a cluster differentiating from the other samples at 3 K-NN distance, the EF1 and EF2 cluster separates from the other samples at a distance of 2 K-NN, and EID1 and EID2 distinguish from FP1 and FP2 at a smaller distance (K-NN = 1), confirming the similarity observed by PCA.

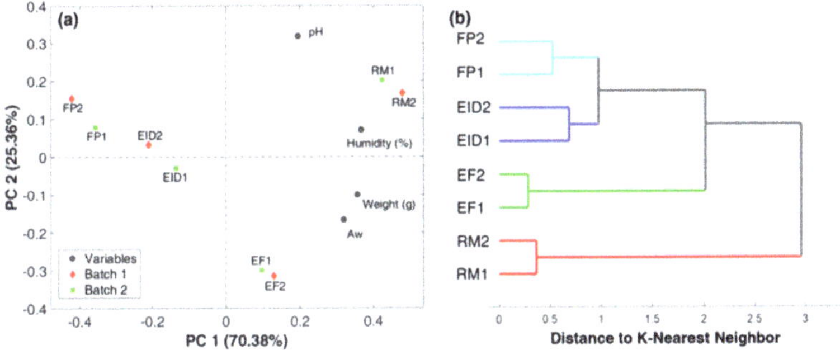

Figure 4. Physicochemical results: (**a**) biplot of Principal Component Analysis and (**b**) dendrogram of cluster analysis by K-nearest neighbor. Raw material, RM; end of fermentation, EF; end of intense drying, EID; and final product, FP.

3.2. Near Infrared Spectroscopy Results Exploration

NIR spectra acquired along the process are reported in Figure 5. The higher differences in the spectral absorptions are present around 930–1300 nm and 1150–1200 nm and from 1370 to 1650 nm. In these areas, absorption peaks of water are present: 979, 1200, and 1453 nm, corresponding to the first overtone of symmetric and asymmetric stretching, to a combination of stretching and bending, and to a combination of the stretching modes of OH bonds, respectively [31]. In particular, spectra acquired up to 48 h, i.e., before the drying phase, showed higher bands related to water absorption, whereas the loss of water during drying highly reduced the height of these bands. The reduction of absorption of the water bonds led to a better resolution of the shoulder present around 1150 nm, related to the second overtone of C–H [31] and possibly linked to the lipid fraction. The spectral changes are enhanced by transformation of the signal by smoothing and first derivative. In Figure 5b, it is possible to see how the spectra show a high variation along the considered range, discriminating the samples in two main groups, i.e., before and after the drying phases. Thus, it would be possible to establish the progress from one phase to the following visually according to spectra behavior. However, to better uncover the information hidden in the broad band characterizing the NIR spectra, it is necessary to use a multivariate approach.

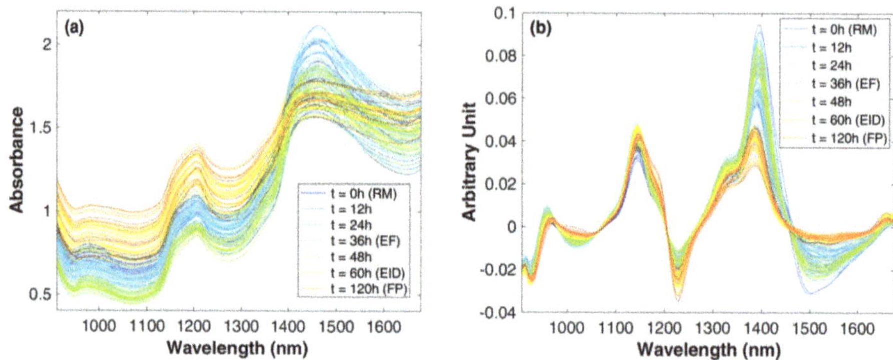

Figure 5. Near-infrared (NIR) spectra: (**a**) raw MicroNIR spectra and (**b**) MicroNIR spectra after first derivative transformation. Spectra are colored according to sampling time (h). Raw material, RM; end of fermentation, EF; end of intense drying, EID; and final product, FP.

Indeed, through PCA on the transformed spectral data, it was possible to unravel the relation between samples and variables. Figure 6a shows the PCA score plots defined by PC1 and PC3. First and third principal components accounted for 93.22% of the total variance (89.79 for PC1 and 3.43 for PC3). The sample distribution in the scores plot confirmed what was noticed in spectra visualization: the combination of PC1 and PC3 allowed for discrimination of samples before (from t = 0 to t = 48 h) and after the drying process (t ≥ 60 h).

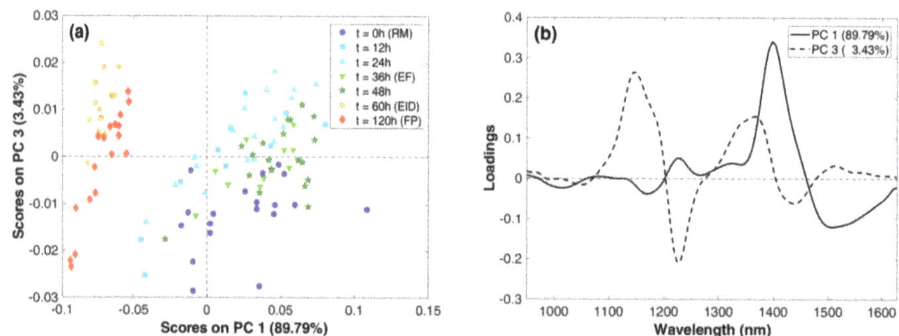

Figure 6. Principal Component Analysis on MicroNIR data: (**a**) score plots of PC1 vs. PC3 for samples colored according to sampling times and (**b**) loadings plot of PC1 and PC3.

As aforementioned, water content such as its activity is responsible for this effect, as observed in related loadings (Figure 6b). Indeed, the PC1 loadings highlighted the relevance of regions with maximums at 1224 and 1397 nm with positive effects and of regions with minimums at 1178 and 1497 nm with negative effects in the sample distribution. At the same extent, PC3 loadings are characterized by highly positive signal in the region with maximums at 1150 and 1360 nm and highly negative signal in the region with minimums at 1224 and 1435 nm. As other authors reported [32,33], NIR absorption is highly sensitive to variations in water content on meat, as occurred in our manufacturing process.

In the score plots (Figure 6a), a group of RM samples is distinguishable, which assumed negative PC3 scores, and it is mainly located in the IV quadrant. Samples collected between 12 and 48 h were more disperse, reproducing the inter-variability of the sausages along the process. Finally, spectra collected for samples undergoing the EID phase were well grouped in the II score quadrant (negative PC1 and positive P3), showing that the process absorbed the inter-variability of the sausages.

Furthermore, score distribution allowed the individuation of outliers, i.e., objects lying in an abnormal distance from observations of the same sampling time. This was the case for 4 samples for the RM and EID groups and 1 for the EF group.

3.3. PLS-DA Classification Models

The sample groups observed in the exploratory analysis did the groundwork for the development of classification models to discriminate samples by simple MicroNIR measures according to the process phase. For this purpose, the classes individuated by clustering analysis on the physicochemical results were used as a priori information (Y) to build PLS-DA classification models able to predict the ripening phase based on the MicroNIR data collected (X). The classification model was developed in calibration considering 88 of the collected spectra and validated both internally and externally by a prediction test set consisting of 44 spectra, i.e., by considering the 140 spectra collected for 20 sausages in phases RM, EF, EID, and FP net of the 8 outliers individuated by PCA. In Table 2, the metrics calculated for the model in cross-validation and prediction are reported. The PLS-DA model shows a successful classification ability as the RM, EID, and FP classes reached 1 for both sensitivity and specificity, proving the model capability to fully recognize samples belonging to the considered class (sensitivity = 1) together with the capability to correctly reject samples belonging to all other classes (specificity = 1). For example, reaching 1 for sensitivity means that all spectra belonging to the samples a priori identified as EID (ncal = 24 and npred = 12) were recognized by the PLS-DA model based on MicroNIR data as at the end of the intense drying process, as were the spectra belonging to RM, EF, and FP. The success of the classification model relies on the differences present in spectra. Indeed, the main differences observed in the spectral absorptions were present around 930–1300 nm and 1150–1200 nm and from 1370 to 1650 nm. In particular, changes were ascribed to peaks present at 979, 1200, and 1453 nm, which are linked to water absorption, the main compound changing during the process.

High specificity describes the perfect capability to correctly reject samples belonging to all other classes (specificity = 1). This was the case of the RM, EID, and FP classes, whereas for the EF cross-validation phase, it slowed to 0.98 as one spectrum, a priori defined as RM, was recognized by the classification model as EF. The misclassification could be linked to a higher moisture reduction on the pole points of one sausage at the RM phase, leading to a MicroNIR signal closer to the profile of an EF sample, i.e., with an higher absorption at peaks related to water presence (979, 1200, and 1453 nm). In any case, the prediction phase resulted in a perfect rejection of samples in all considered classes.

These results demonstrated that the data collected by the MicroNIR device allows for knowing the stage of the process, which may be useful to determine the passage from one stage to the next one.

Table 2. Results of PartialLleast Square–Discriminant analysis for sausage process stage identification: sensitivity (Sens) and specificity (Spec) values of models based on MicroNIR data. Raw material, RM; end of fermentation, EF; end of intense drying, EID; and final product, FP. In between brackets, the number of samples used for each single class for model calibration and prediction are given.

Validation Phase	N Sample	RM (24, 12)		EF (13, 6)		EID (24, 12)		FP (26, 14)	
		Sens	Spec	Sens	Spec	Sens	Spec	Sens	Spec
Cross-validation	88	1.00	1.00	1.00	0.98	1.00	1.00	1.00	1.00
Prediction	44	1.00	1.00	1.00	1.00	1.00	1.00	1.00	1.00

To our knowledge, no other work developed classification models to determine the ripening stage during the dry fermented sausage process. Previously, NIR spectroscopy has been demonstrated to be a valid tool for the evaluation of the ripening phase of salami, a typical European dry fermented sausages [29]. However, in their study, the sampling procedure was destructive, 9 slices were analyzed for each investigated salami, and a benchtop instrument was employed, resulting in a higher spectral range (800–2800 nm). The authors succeeded in describing the ripening evolution by fitting the PC1 scores obtained from spectral data elaboration as a function of ripening time; however, no supervised modelling has been developed. Likewise, Gaitán-Jurado et al. [34] proposed a Visible-NIR (400–2000 nm)-based approach for online quality control of the packed slices of chorizo and salchichón (Spanish dry fermented sausages). Besides that, in this study, slices were considered, highlighting the effect of different factors, mainly plastic turns around the sample and slice thickness, affecting the sampling procedure and, thus, the prediction models. However, these authors developed only an approach intended for final product quality control, not considering food processing control, i.e., not evaluating the potential of NIR to model phenomena occurring during ripening.

Furthermore, other authors [35] developed partial least squares (PLS) regression models for water activity and moisture content prediction from slices of two types of fermented sausages. These authors demonstrated that, by prediction of the two parameters, it is possible to control the drying process of fermented sausages by NIR spectroscopymodels. However, the procedure developed considered only two out of multiple parameters influencing sausage processing, just the drying process, and it relied on a destructive procedure, as it required product slicing.

4. Conclusions

In this work, the feasibility of using a portable NIR device to monitor and classify the ripening process of dry fermented sausages using multivariate data analysis has been demonstrated.

Physicochemical (destructive) determinations allowed for the classification of dry fermented sausages according to their stage of processing by means of K-NN cluster analysis. Likewise, the nondestructive technique of NIR spectroscopy also enabled this classification, based on the visual observation of the spectra and by applying chemometrics on absorbance results, achieving high sensitivity and specificity classification values with PLS-DA.

Thus, the use of a nondestructive, portable, noninvasive, fast, easy-to-use and cost-effective tool, such as the MicroNIR device of this study, may be implemented to monitor the industrial processing of dry fermented sausages. This appears as a very interesting quality-control monitoring strategy specially for small and medium-sized companies, which usually lack equipped laboratories and analytical facilities. The implementation of such a procedure would allow for control of the progress from one stage to another or even to identify sausages that have reached the end of processing without the need for any other type of analysis. Further studies on different types of sausage processing may be necessary before upscaling this monitoring strategy to an industrial setup.

Author Contributions: conceptualization, J.R.-C., T.A. and S.G.; methodology, A.G.-M., T.P.-P., L.S.O. and S.G.; software, A.G.-M. and S.G.; validation, S.G.; formal analysis, A.G.-M. and L.S.O.; investigation, T.P.-P., T.A. and J.R.-C.; resources, T.A., J.R.-C and T.P.-P.; writing—original draft preparation, A.G.-M. and S.G.; writing—review and editing, T.P.-P., T.A. and J.R.-C.; visualization, S.G. and J.R.-C.; project administration, T.A. and J.R.-C.; funding acquisition, T.A., T.P.-P. and J.R.-C. All authors have read and agreed to the published version of the manuscript.

Funding: This research received no external funding.

Conflicts of Interest: The authors declare no conflict of interest.

References

1. López, C.M.; Bru, E.; Vignolo, G.M.; Fadda, S. Main Factors Affecting the Consumer Acceptance of Argentinean Fermented Sausages. *J. Sens. Stud.* **2012**, *27*, 304–313. [CrossRef]

2. Parunović, N.; Savic, R.; Radovic, C. Qualitative properties of traditionally produced dry fermented sausages from meat of the autochthonous Mangalitsa pig breed. *IOP Conf. Ser. Earth Environ. Sci.* **2019**, *333*, 012035. [CrossRef]

3. Talon, R.; Lebert, I.; Lebert, A.; Leroy, S.; Garriga, M.; Aymerich, T.; Drosinos, E.; Zanardi, E.; Ianieri, A.; Fraqueza, M.J.; et al. Traditional dry fermented sausages produced in small-scale processing units in Mediterranean countries and Slovakia. 1: Microbial ecosystems of processing environments. *Meat Sci.* **2007**, *77*, 570–579. [CrossRef] [PubMed]

4. Giraudo, A.; Grassi, S.; Savorani, F.; Gavoci, G.; Casiraghi, E.; Geobaldo, F. Determination of the geographical origin of green coffee beans using NIR spectroscopy and multivariate data analysis. *Food Control.* **2019**, *99*, 137–145. [CrossRef]

5. Grassi, S.; Alamprese, C. Advances in NIR spectroscopy applied to process analytical technology in food industries. *Curr. Opin. Food Sci.* **2018**, *22*, 17–21. [CrossRef]

6. Feinberg, M. Multivariate calibration, by H. Martens and T. Naes. *TrAC Trends Anal. Chem.* **1990**, *9*, 419. [CrossRef]

7. Barthès, B.; Kouakoua, E.; Clairotte, M.; Lallemand, J.; Chapuis-Lardy, L.; Rabenarivo, M.; Roussel, S. Performance comparison between a miniaturized and a conventional near infrared reflectance (NIR) spectrometer for characterizing soil carbon and nitrogen. *Geoderma* **2019**, *338*, 422–429. [CrossRef]

8. Yan, H.; Siesler, H.W. Quantitative analysis of a pharmaceutical formulation: Performance comparison of different handheld near-infrared spectrometers. *J. Pharm. Biomed. Anal.* **2018**, *160*, 179–186. [CrossRef]

9. Serranti, S.; Bonifazi, G.; Gasbarrone, R. Kiwifruits ripening assessment by portable hyperspectral devices. *Sens. Agric. Food Quality Saf. X* **2018**, *10665*, 106650S. [CrossRef]

10. Casson, A.; Beghi, R.; Giovenzana, V.; Fiorindo, I.; Tugnolo, A.; Guidetti, R. Environmental advantages of visible and near infrared spectroscopy for the prediction of intact olive ripeness. *Biosyst. Eng.* **2020**, *189*, 1–10. [CrossRef]

11. Grassi, S.; Casiraghi, E.; Alamprese, C. Handheld NIR device: A non-targeted approach to assess authenticity of fish fillets and patties. *Food Chem.* **2018**, *243*, 382–388. [CrossRef] [PubMed]

12. Moon, E.J.; Kim, Y.; Xu, Y.; Na, Y.; Giaccia, A.J.; Lee, J.H. Evaluation of Salmon, Tuna, and Beef Freshness Using a Portable Spectrometer. *Sensors* **2020**, *20*, 4299. [CrossRef] [PubMed]

13. Goisser, S.; Wittmann, S.; Fernandes, M.; Mempel, H.; Ulrichs, C. Comparison of colorimeter and different portable food-scanners for non-destructive prediction of lycopene content in tomato fruit. *Postharvest Biol. Technol.* **2020**, *167*, 111232. [CrossRef]

14. Hassoun, A.; Måge, I.; Schmidt, W.F.; Temiz, H.T.; Li, L.; Kim, H.-Y.; Nilsen, H.; Biancolillo, A.; Aït-Kaddour, A.; Sikorski, M.; et al. Fraud in Animal Origin Food Products: Advances in Emerging Spectroscopic Detection Methods over the Past Five Years. *Foods* **2020**, *9*, 1069. [CrossRef]

15. Pérez-Santaescolástica, C.; Fraeye, I.; Barba, F.J.; Gómez, B.; Tomasevic, I.; Romero, A.; Moreno, A.; Toldrá, F.; Lorenzo, J.M. Application of non-invasive technologies in dry-cured ham: An overview. *Trends Food Sci. Technol.* **2019**, *86*, 360–374. [CrossRef]

16. Parastar, H.; Van Kollenburg, G.H.; Weesepoel, Y.; Doel, A.V.D.; Buydens, L.; Jansen, J. Integration of handheld NIR and machine learning to "Measure & Monitor" chicken meat authenticity. *Food Control.* **2020**, *112*, 107149. [CrossRef]

17. Savoia, S.; Albera, A.; Brugiapaglia, A.; Di Stasio, L.; Ferragina, A.; Cecchinato, A.; Bittante, G. Prediction of meat quality traits in the abattoir using portable and hand-held near-infrared spectrometers. *Meat Sci.* **2020**, *161*, 108017. [CrossRef]

18. Horwitz, W. *Official Methods of Analysis the of AOAC International*, 17th ed.; The Association of Official Analytical Chemists: Gaithersburg, MD, USA, 2000.

19. Billmeyer, F.W.; Saltzman, M. *Principles of Color Technology*, 2nd ed.; Wiley-Interscience: New York, NY, USA, 1981.

20. Salih, A.M.; Smith, D.M.; Price, J.F.; Dawson, L.E. Modified Extraction 2-Thiobarbituric Acid Method for Measuring Lipid Oxidation in Poultry. *Poult. Sci.* **1987**, *66*, 1483–1488. [CrossRef]

21. Grassi, S.; Benedetti, S.; Opizzio, M.; Di Nardo, E.; Buratti, S. Nardo Meat and Fish Freshness Assessment by a Portable and Simplified Electronic Nose System (Mastersense). *Sensors* **2019**, *19*, 3225. [CrossRef]

22. Oliveri, P.; Malegori, C.; Casale, M.; Tartacca, E.; Salvatori, G. An innovative multivariate strategy for HSI-NIR images to automatically detect defects in green coffee. *Talanta* **2019**, *199*, 270–276. [CrossRef]

23. Cavalheiro, C.P.; Ruiz-Capillas, C.; Herrero, A.M.; Jiménez-Colmenero, F.; Pintado, T.; Fries, L.L.M.; De Menezes, C.R. Effect of different strategies of Lactobacillus plantarum incorporation in chorizo sausages. *J. Sci. Food Agric.* **2019**, *99*, 6706–6712. [CrossRef] [PubMed]

24. Hu, Y.; Zhang, L.; Zhang, H.; Wang, Y.; Chen, Q.; Kong, B. Physicochemical properties and flavour profile of fermented dry sausages with a reduction of sodium chloride. *LWT* **2020**, *124*, 109061. [CrossRef]

25. Balamurugan, S.; Gemmell, C.; Lau, A.T.Y.; Arvaj, L.; Strange, P.; Gao, A.; Barbut, S. High pressure processing during drying of fermented sausages can enhance safety and reduce time required to produce a dry fermented product. *Food Control.* **2020**, *113*, 107224. [CrossRef]

26. Mendoza, E.; García, M.; Casas, C.; Selgas, M. Inulin as fat substitute in low fat, dry fermented sausages. *Meat Sci.* **2001**, *57*, 387–393. [CrossRef]

27. Talon, R.; Leroy, S.; Lebert, I.; Giammarinaro, P.; Chacornac, J.; Latorre-Moratalla, M.; Vidalcarou, C.; Zanardi, E.; Conter, M.; Lebecque, A. Safety improvement and preservation of typical sensory qualities of traditional dry fermented sausages using autochthonous starter cultures. *Int. J. Food Microbiol.* **2008**, *126*, 227–234. [CrossRef]

28. Aymerich, T.; Martin, B.; Garriga, M.; Vidal-Carou, M.; Bover-Cid, S.; Hugas, M. Safety properties and molecular strain typing of lactic acid bacteria from slightly fermented sausages. *J. Appl. Microbiol.* **2006**, *100*, 40–49. [CrossRef]

29. Alamprese, C.; Fongaro, L.; Casiraghi, E. Effect of fresh pork meat conditioning on quality characteristics of salami. *Meat Sci.* **2016**, *119*, 193–198. [CrossRef]

30. Taormina, P.J.; Sofos, J.N. Low-Water Activity Meat Products. In *The Microbiological Safety of Low Water Activity Foods and Spices*; Gurtler, J.B., Doyle, M.B., Kornacki, J.L., Eds.; Springer: New York, NY, USA, 2014.

31. Workman, J., Jr.; Weyer, L. *Practical Guide to Interpretive Near-Infrared Spectroscopy*, 1st ed.; CRC Press: Boca Raton, FL, USA, 2007.

32. Büning-Pfaue, H. Analysis of water in food by near infrared spectroscopy. *Food Chem.* **2003**, *82*, 107–115. [CrossRef]

33. Ishikawa, D.; Ueno, G.; Fujii, T. Estimation Method of Moisture Content at the Meat Surface During Drying Process by NIR Spectroscopy and Its Application for Monitoring of Water Activity. *Jpn. J. Food Eng.* **2017**, *18*, 135–143. [CrossRef]

34. Gaitán-Jurado, A.J.; Rincón, F.; Ortiz-Somovilla, V.; España-España, F. Detection of Factors Affecting the Acquisition of Visible-Near Infrared Fibre-Optic Probe Spectra of Commercial Meat Products. *J. Near Infrared Spectrosc.* **2008**, *16*, 111–119. [CrossRef]

35. Collell, C.; Gou, P.; Arnau, J.; Muñoz, I.; Comaposada, J. NIR technology for on-line determination of superficial aw and moisture content during the drying process of fermented sausages. *Food Chem.* **2012**, *135*, 1750–1755. [CrossRef] [PubMed]

Article

Influence of Electron Beam Irradiation on the Moisture and Properties of Freshly Harvested and Sun-Dried Rice

Lihong Pan [1,2,3,4], Jiali Xing [5], Xiaohu Luo [1,2,3,4,*], Yanan Li [1,2,3,4], Dongling Sun [6], Yuheng Zhai [1,2,3,4], Kai Yang [1,2,3,4] and Zhengxing Chen [1,2,3,4]

1   Key Laboratory of Carbohydrate Chemistry and Biotechnology, Ministry of Education, Jiangnan, Wuxi 214122, China; yelanplh@outlook.com (L.P.); lyn19881012ph@163.com (Y.L.); 6190112167@stu.jiangnan.edu.cn (Y.Z.); yangkai164@outlook.com (K.Y.); zxchen@jiangnan.edu.cn (Z.C.)
2   National Engineering Laboratory for Cereal Fermentation Technology, Jiangnan University, Wuxi 214122, China
3   Jiangsu Provincial Research Center for Bioactive Product Processing Technology, Jiangnan University, Wuxi 214122, China
4   Collaborative Innovation Center for Food Safety and Quality Control in Jiangsu Province, Jiangnan University, Wuxi 214122, China
5   Ningbo Institute for Food Control, Ningbo 315048, China; hellojiali77@gmail.com
6   Wuxi EL PONT Radiation Technology Co., Ltd., Wuxi 214151, China; sundl_1980@163.com
*   Correspondence: xh06326@gmail.com

Received: 15 July 2020; Accepted: 17 August 2020; Published: 19 August 2020

Abstract: Moisture content is an important factor that affects rice storage. Rice with high moisture (HM) content has superior taste but is difficult to store. In this study, low-dose electron beam irradiation (EBI) was used to study water distribution in newly harvested HM (15.03%) rice and dried rice (11.97%) via low-field nuclear magnetic resonance (LF-NMR). The gelatinization, texture and rheological properties of rice and the thermal and digestion properties of rice starch were determined. Results showed that low-dose EBI could change water distribution in rice mainly by affecting free water under low-moisture (LM) conditions and free water and bound water under HM conditions. HM rice showed smooth changes in gelatinization and rheological properties and softened textural properties. The swelling power and solubility index indicated that irradiation promoted the depolymerization of starch chains. Overall, low-dose EBI had little effect on the properties of rice. HM rice showed superior quality and taste, whereas LM rice exhibited superior nutritional quality. This work attempted to optimize the outcome of the EBI treatment of rice for storage purposes by analyzing its effects. It demonstrated that low-dose EBI was more effective and environmentally friendly than other techniques.

Keywords: Electron Beam Irradiation; rice; moisture; physicochemical properties

---

1. Introduction

Rice (*Oryza sativa* L.), the primary staple for more than half the world's population, is produced worldwide. Approximately 90% of rice is grown in Asia. From 2000 to 2013, the rice productivity of the U.S. increased by an estimated 29% or approximately 2.2% on an annual basis; this increase is second only to the increase in peanut productivity (at 3.5% annually) among major U.S. field crops [1]. China, as the world's largest rice producer and consumer, produced 199 million tons and consumed 203 million tons of rice between October 2018 and September 2019. Rice stocks reached 176 million tons in 2018. Rice consumption accounts for the largest grain ration, 81.96%, in China [2].

Considering the increasing levels of rice stocks, storing rice safely and securely has become a concern of researchers because improper storage easily results in considerable social and economic losses due to aging and insect infestation [3–5]. A good method for grain storage is necessary to maintain good rice quality in the face of massive rice stocks. Farmers generally use sunlight drying to reduce the moisture content of rice to extend storage time. However, although moisture reduction extends rice storage time, it will also reduce rice taste by increasing hardness [6,7]. Previous studies have shown that cooked rice grains have the best textural quality when they are stored with a moisture content of 15.70% [5]. People are increasingly seeking high-quality experiences along with improvement in life quality. Therefore, how to extend the storage time of rice while retaining its high-quality taste has become a thorny problem. Existing rice storage methods include natural drying, cryogenic storage, controlled-atmosphere storage, and chemical storage [8–10]. However, these methods have their drawbacks. Physical methods, such as natural drying and cryogenic storage, all require low rice moisture content. Although these methods can prolong rice storage time, achieving good taste quality is difficult. Controlled-atmosphere storage technology is a green environmental protection technology that is unsuitable for high-moisture (HM) grain because it causes alcohol accumulation and affects rice quality. Chemical methods can cause the accumulation of harmful substances and harm human health.

For several decades, food irradiation technology has been widely used to preserve food, such as fruit and meat [11,12]. In traditional food irradiation processing, the energy of gamma radiation produced by a radioisotope (CS-137 or Co-60) is transferred to irradiated food, killing the eggs and microorganisms that the food contains and thus extending shelf life [13]. Radionuclides have the disadvantage of continuous gamma ray emission, which can be a potential source of environmental hazards for operation staff and for some equipment. Electron beam irradiation (EBI), another food irradiation technique, is produced by an electron accelerator. In contrast to gamma irradiation, it can be switched off during off-duty hours without causing harm to personnel and equipment. The United Nations Food and Agriculture Organization suggested that the appropriate use of radiation for food decontamination is safe and that the irradiation of any food commodity with an overall average dose of up to 10 kGy presents no toxicological hazard and no special nutritional or microbiological problems [14].

Given that the EBI treatment of rice has no moisture requirement, newly harvested and non-dried rice can be irradiated. According to previous studies [5], an excellent texture and edible quality of rice are maintained by using this treatment. Our research team found that low-dose EBI has little effect on the quality of rice [15] while simultaneously extending storage time. In this study, EBI was applied to treat freshly harvested paddy rice and sun-dried rice, and the cooking quality and moisture migration of irradiated rice and the thermodynamics and digestion properties of the isolated starches were investigated. This work attempted to optimize the outcome of the EBI treatment of rice for storage purposes by analyzing the effects of this treatment method. It demonstrated that low-dose EBI was more effective and environmentally friendly than other methods.

2. Materials and Methods

2.1. Materials

Directly harvested paddy rice (15.03% moisture content) and sun-dried paddy rice (11.97% moisture content) were supplied by a peasant household (Nantong, Jiangsu, China). Pancreatic α-amylase (EC 3.2.1.1, 10 units/mg) and amyl glucosidase (EC 3.2.1.3, 3260 units/mL) were purchased from Megazyme (Wicklow, Ireland).

2.2. Preparation of EBI Paddy Rice

A total of 500 g of paddy rice was placed in a polyethylene bag and spread to a thickness of 1 cm. Paddy rice was irradiated at doses of 0, 1, 2, 3, or 4 kGy in separate batches by using an industrial electron accelerator operated at a dose rate of 2 kGy/s. The energy of accelerated electrons was 10 Mev, and the beam current was 1.0 mA with 1000 mm scan widths. After irradiation, all samples were stored in a desiccator at room temperature (~25 °C) for further analysis.

2.3. Low-Field Nuclear Magnetic Resonance Measurement

A MesoMR23-060V-I NMR spectrometer was used to determine moisture migration in irradiated rice (MesoMR23-060V-I, Niumag Co., LLC., Suzhou, China). To avoid the evaporation of water, 2 g of rice grains were weighed and rapidly placed into sample bottles that were then sealed and incubated at 25 °C for 30 min. At the same time, other rice grains were soaked in deionized water for 30 min. Surface moisture was wiped off from the rice grains, which were rapidly placed into sample bottles that were then sealed and incubated at 25 °C for 30 min to observe moisture migration. The Carr–Purcell–Meiboom–Gill pulse sequence scanning experiment was performed with the following parameters: number of sampling points TD = 14,992, number of repeated scans NS = 8, TW (ms) = 2000, NECH = 500, relaxation decay time DR = 3 s, with a 90° pulse of 7.52 ms, and 90°–180° pulse spacing of 14.48 ms. The inverse Laplace transformation of T2 curves was performed by using the associated software of the instrument.

2.4. Pasting Property of Rice

The pasting properties of rice flour were determined by using a Rapid Visco-Analyzer (RVA 4500, Perten Co. Ltd., Sydney, Australia). Rice paste (3.0 g dry basis, 25.0 g of deionized water) was held at 50 °C for 1 min and heated to 95 °C at a heating rate of 6 °C/min. The paste was held at 95 °C for 5 min and then cooled back to 50 °C at the same rate. Finally, the paste was held at 50 °C for 2 min. The peak, hold, and final viscosity values were obtained from viscograms.

2.5. Analysis of Rheological Properties

After the analysis of pasting properties, the corresponding dynamic and static rheological properties of each rice sample were determined as follows:

A rheometer (DHR-3 rheometer, Waters Co. Ltd., Milford, MA, USA) was used to analyze the viscoelastic properties of each rice sample. In brief, starch paste (approximately 4 mL) was loaded onto the lower plate. The upper parallel plate (40 mm diameter) was slowly lowered to reduce the gap between the two parallel plates to 1.0 mm prior to the run.

Dynamic viscoelasticity measurement: The test temperature was set at 25 °C, the scan strain was set to 1%, and the frequency was set from 0.1 Hz to 100 Hz to obtain the sample elastic modulus (G'), loss modulus (G''), and loss tangent (Tan δ = G''/G'). The changes in viscoelasticity during the sol-gel transition were indicated by the G' value.

Dynamic shear force recovery test: The temperature was set at 25 °C constantly, and the shearing rate ranged from 0.1 s^{-1} to 100 s^{-1}. The shear structure was recorded as the apparent viscosity.

2.6. Swelling Power and Solubiliy Index

Rice samples (0.5 g) were transferred into pre-weighed centrifuge tubes with 20 mL of distilled water. The rice flour suspensions were then incubated in a water bath for 30 min at 60 °C, 70 °C, 80 °C, and 90 °C with vortexing after every 5 min. After cooling the samples to room temperature, the tubes were centrifuged at 5000× g for 15 min. The supernatant was decanted in a preweighed aluminum specimen box. The weight gain of the centrifuge tubes was expressed as the swelling index. Moisture dishes were dried at 105 °C for 12 h and then cooled in a desiccator to room temperature. The weight gain of the moisture dishes was expressed as the solubility index.

2.7. Instrumental Texture Profile Analysis of Cooked Rice

The textural properties of the cooked rice were measured in accordance with the method of Li et al. [16] with some modifications. A total of 30 g of rice and 42 g of water were weighed and stewed in a rice cooker for 30 min. After the rice was cooled to room temperature, an 8.0 g subsample of cooked rice grains was weighed and placed in a single layer on the base plate and tested by using a TA.XTPlus Texture Analyzer (SMS Co. Ltd., Godalming, UK). A cylindrical probe (P/35) was applied with the following test speeds: pretest speed = 1 mm/s, test speed = 1 mm/s, and post-test speed = 1 mm/s. The strain was 60%, and the rice samples were compressed twice. Textural parameters, including hardness, cohesiveness, springiness, resilience, and chewiness, were calculated from the curves that were provided by the equipment.

2.8. Sensory Evaluation of Cooked Rice

The sensory analyses of cooked rice were performed by five trained panelists. The following sensory attributes were evaluated: taste, appearance, softness, flavor, and cold rice texture. The sensory evaluation of rice flours was conducted in accordance with the Chinese National Standard GB/T 15682-2008 [17]. Briefly, 10 g of rice was weighed out and rinsed twice with distilled water. Water was added to the sample at 1.3 times the sample volume. The sample was soaked at 25 °C for 30 min. Then, it was cooked in a steamer for 40 min and simmered for 20 min. The different samples were placed on a white porcelain plate. The reviewer tasted the sample while it was hot. Scoring was as follows: 25 points for taste, 20 points for flavor, 20 points for appearance, 30 points for softness, and 5 points for cold rice texture.

2.9. Isolation of Rice Starch

Rice starch was isolated via traditional alkaline extraction. Rice grains were ground and soaked in deionized water for 2 h and subsequently milled for 2 min by using a colloid mill. The mixture was subsequently centrifuged to remove water and then soaked in 0.2% sodium hydroxide solution for 48 h. The fluid was changed every 12 h. The filtrate was repassed through a 100-mesh sieve (pore size of 0.15 mm) and centrifuged at $6000 \times g$ for 10 min. Then, the upper gray matter (protein) was scraped off (repeated five times). The pH was adjusted to 7.0 by using dilute hydrochloric acid. This process was repeated five times to remove salt from the starch by using deionized water through centrifugation at $6000 \times g$ for 10 min. This process was followed by freeze–drying for 48 h. The obtained starch samples were stored in a desiccator for further analysis.

2.10. Differential Scanning Calorimetry

The thermal properties of starch samples were measured by using a differential scanning calorimeter (DSC3, Mettler Toledo Co. Ltd., Greifensee, Switzerland). Sample pretreatment was performed in reference to the protocol used by Zhou et al. [18] with some modifications. In a 40 μL capacity aluminum pan, distilled water was added to starch (3.0 mg) to obtain a starch suspension containing 70% water. The hermetically sealed pans were maintained at 4 °C for 12 h to ensure the equilibration of the starch and water before differential scanning calorimetry (DSC) analysis. The samples were scanned from 30 °C to 100 °C at a heating rate of 10 °C/min with an empty pan used as reference.

2.11. In Vitro Digestion

The digestibility of the starch samples was mainly determined in accordance with the method of Englyst and Wang [19,20] with some modifications. A 100 mg starch sample was added into a screw-cap tube and treated with 4 mL of digestive juice. The remaining steps were consistent with those of Wang's method.

2.12. Statistical Analysis

All data were processed by using an Origin 9.0 and presented as mean ± standard deviation (SD). Data were statistically analyzed via one-way analysis of variance and Duncan's multiple range test by using SPSS statistics 17.0. Differences were considered statistically significant at the 95% level ($p < 0.05$). Each sample was measured in triplicate.

3. Results and Discussion

3.1. Analysis of Water Status via Low-Field Nuclear Magnetic Resonance

Rice samples with low moisture (LM) content (11.97%) and high moisture (HM) content (15.03%) were subjected to low-field nuclear magnetic resonance (LF-NMR). The proton distributions in the control and irradiated rice samples were compared, and the results are presented in Figure 1. The control and irradiated rice samples were soaked simultaneously for 30 min for comparison, and the results are presented in Figure 2. The relative peak areas of the less- and more-mobile water fractions in each sample are displayed in Table S1. The T_2 relaxation time in the curves reflects the chemical environment encountered by protons in the sample, and the relaxation times of T_{21}, T_{22}, and T_{23} can be considered to be related to bound water, nonflowing water, and free water in paddy rice, respectively [21].

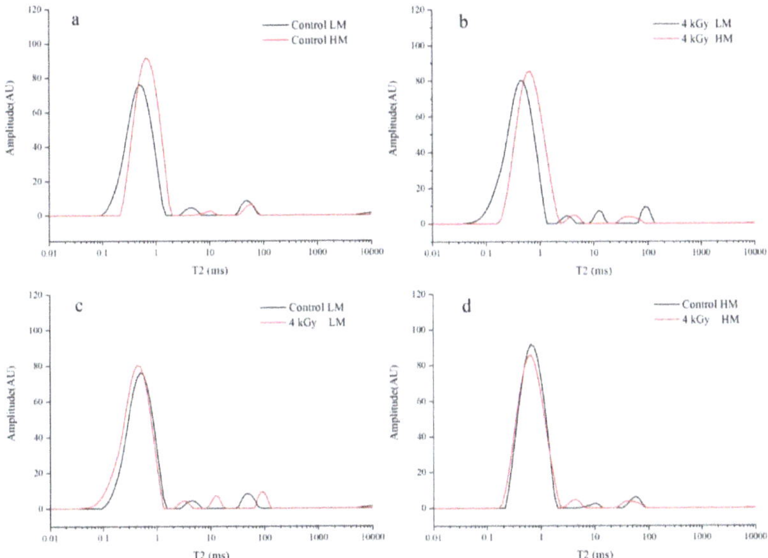

Figure 1. Water distribution in low moisture (LM) and high moisture (HM) rice samples under different irradiation doses. (**a**) Control LM and HM rice, (**b**) 4 kGy LM and HM rice, (**c**) Control and 4 kGy LM rice, (**d**) Control and 4 kGy HM rice.

The intensity in T_{21} (0.01–1 ms) was significantly different between freshly harvested rice and sun-dried rice. After irradiation, the bound water content of the two types of rice increased at the same time, and the free water content decreased. However, the proportion of the reduction shown by the free water content of LM rice was higher than that shown by the free water content of HM rice. In contrast to our study, a previous study showed that EBI treatment reduces the bound water content of egg white protein [22]. This discrepancy might be attributed to the different proteins present in eggs and rice. As shown in Figure 1a, the bound water content of freshly harvested rice was higher than that of sun-dried rice. During sun drying, part of the bound water was likely to have been converted into free water that evaporated under the influence of high temperature, leading to a reduction in the proportion of the bound water. After irradiation, the free water proportion of LM rice decreased more than that of the rice with HM content. LM rice mainly contained a large amount of free water, whereas HM rice contained a small amount of free water and a large amount of bound water. Irradiation mainly affected free water in LM rice but affected free water and bound water in HM rice. Therefore, free water in LM rice was greatly influenced by irradiation and mainly turned into free radicals. However, the rice with HM content contained more bound water than the rice with LM content. As free water was decomposed by irradiation, the electron beam also cleaved the double helixes of starches and proteins. This effect generated puncture pores on microcosmic surfaces and caused changes in the quantities of bound water and free water by changing their states through reducing hydrogen bonding such that the structure increased in flexibility and decreased in compactness. These effects thus changed water distribution.

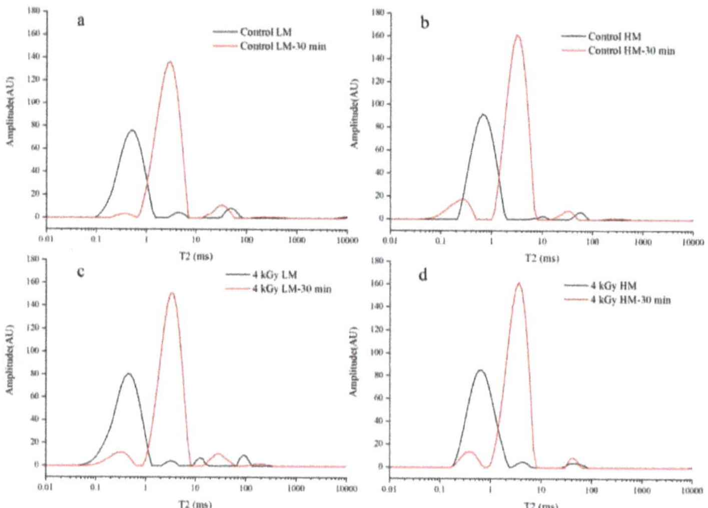

Figure 2. Water distribution in LM rice and HM rice samples after 30 min of soaking and treatment with different irradiation doses. (**a**) Control LM rice before and after 30 min soaking, (**b**) Control HM rice before and after 30 min soaking, (**c**) 4 kGy LM rice before and after 30 min soaking, (**d**) 4 kGy HM rice before and after 30 min soaking.

As shown in Figure 2 and Table S1, after 30 min of soaking, almost all the LM and HM rice showed a shortened T_{21} and prolonged T_{22}. The change in relaxation time (T_2) in LF-NMR can reflect the degree of freedom of water in the sample and the chemical environment encountered by the protons of the sample. A short T_2 (0.01–1 ms) is indicative of the tight binding degree between water molecules and matter and a low degree of freedom of water. It can be considered to be related to chemically bound water. A long T_2 is associated with weak binding force and a high degree of freedom, and it can be considered to be related to nonflowing water (1–10 ms) and free water (10–100 ms) [23]. Nonflowing water represents physically adsorbed water, i.e., water that is trapped or present in cellular structures and does not easily flow. Free water refers to water that can flow freely and that exhibits the lowest binding force and strongest mobility. Rice is generally soaked before cooking mainly to improve the speed and effect of gelatinization. This study aimed to explore whether irradiation and different water content systems could affect the water absorption efficiency of rice and thus ultimately affect the cooking quality of rice. As shown in Figure 2, irradiation after soaking in the LM state did not affect the value of T_{21}. By contrast, after soaking in the HM state, the T_{21} of irradiated rice moved to the right. This change indicated that the effect of irradiation tended to weaken the binding strength of organic matter and water. This result might be related to the effects of irradiation on free water and bound water in HM rice as mentioned above. In general, compared with that in rice that had not been exposed to the sun, the combination of water and protein starch in rice that had been exposed to the sun had lower strength, harder tissue, tighter networks, and lower water absorption rate. At the same time, irradiation had a positive effect on the water absorption rate of rice. Therefore, rice with a high water content and that had been treated with a specific dose of irradiation had the best water absorption rate and superior cooking quality.

3.2. Pasting Properties of Rice Samples

The RVA profiles of the rice flour samples are shown in Figure 3, and the parameters, including the peak viscosity (PV), holding viscosity (HV), final viscosity (FV), and pasting temperature (PT) of rice flour gelatinization are depicted in Table S2. Irradiation treatment resulted in a significant ($p < 0.05$) reduction in the PV of all the samples. Sultan et al. also reported a reduction in the HV of rice upon irradiation treatment [24]. Figure 3 shows that the PV, HV, and FV of LM rice were lower than those of HM rice. This result was consistent with the research results of Ali et al., who found that starches extruded under HM (22%) conditions have higher viscosity than those extruded under LM (14%) conditions [25]. The irradiation effect in the HM state is weaker than that in the LM state. Simultaneously, no significant difference in the PV of HM rice under 1 and 2 kGy ($p < 0.05$) irradiation was observed because the husks protected the rice grains from high irradiation doses. Electron beams destroy the interior structure of rice [26] and are likely to first act on water molecules [27]. Their action on a large number of water molecules results in the production of free radicals and a certain amount of thermal energy. Moreover, free radicals and energy cause the chemical bonds of biological macromolecules to break by acting on biological macromolecules. HM rice had more water molecules than LM rice. Most of the electron beam energy acted on water molecules, thus causing limited damage to starch macromolecules and small changes in viscosity.

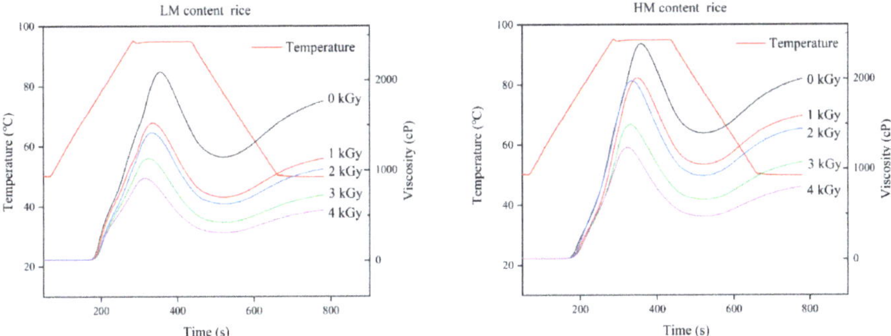

Figure 3. Rapid Visco-Analyzer (RVA) pasting profiles of LM and HM rice samples under different irradiation doses.

3.3. Rheological Properties

The rheological properties of natural starches and irradiated starches are depicted in Figure 4. The magnitudes of the elastic component, G′, and viscous component, G″ of the rice starch pastes increased with scanning frequency, and G′ was significantly higher than G″, which led to the absence of crossover between the two moduli over the range of applied frequencies, indicating that these parameters were all associated with weak gel properties [28]. This study revealed that both moduli of the gel prepared with irradiated rice samples were lower than those of the gel prepared with normal rice samples, whereas LM rice showed more reductions than HM rice. EBI treatment might have destroyed the continuous network structures and resulted in the changes in the internal structure of the rice kernel, thus weakening the gel network of the rice paste. These results were consistent with previously reported results [29]. The formation of the gel network in irradiated rice samples could be related to RVA results. Yu and Wang [30] stated that radiation can generate free radicals in starch macromolecules. These free radicals are capable of hydrolyzing chemical bonds and breaking large molecules into small dextrin fragments, destroying the gel network structure and consequently decreasing the apparent viscosity of irradiated rice flour. The apparent viscosity of HM rice flour changed more smoothly than that of LM rice flour likely because, given the large number of water molecules, irradiation energy first acted on water molecules instead of biomacromolecules.

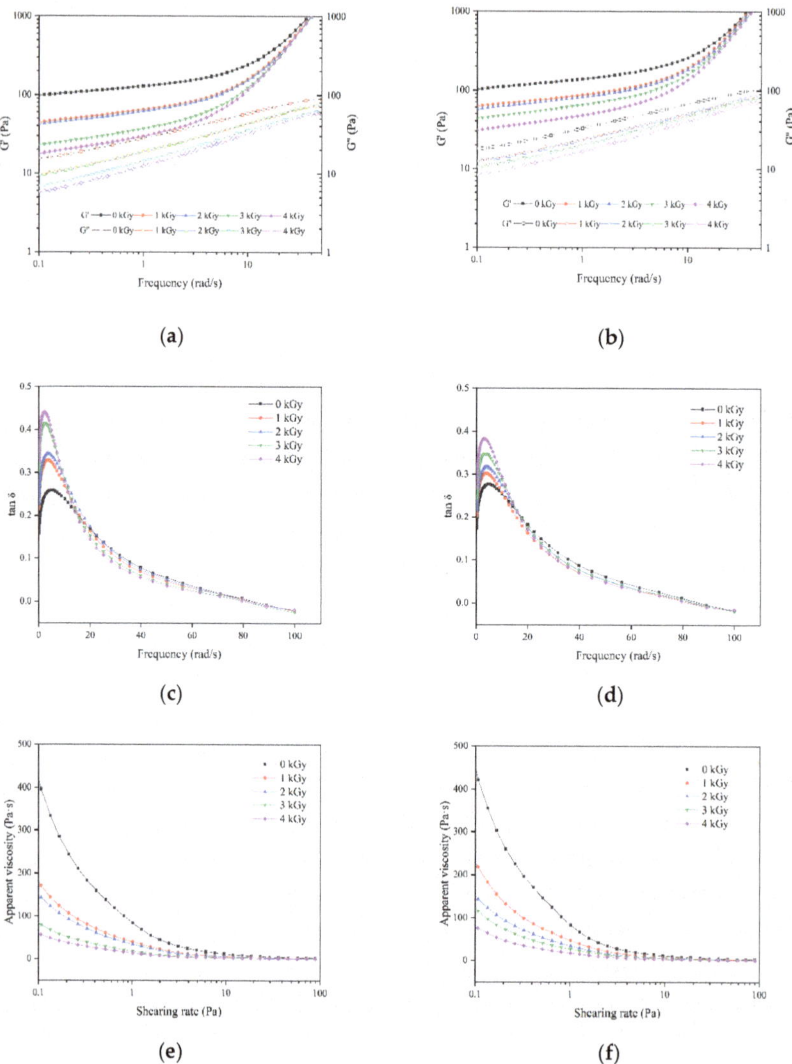

Figure 4. Changes in G′, G″ (**a,b**), tan δ (**c,d**), and apparent viscosity (**e,f**) of RSs under different electron beam irradiation (EBI) doses. (**a,c,e**) LM rice; (**b,d,f**) HM rice.

3.4. Swelling Power and Solubility Index

The swelling power and solubility index were assessed over the temperature range of 60 °C–90 °C and are presented in Table 1. The swelling power of all the rice flours increased significantly ($p < 0.05$) over the temperature range of 60 °C–90 °C. The values for LM rice ranged from 3.20 g/g to 14.37 g/g and those for HM rice ranged from 3.58 g/g to 13.40 g/g. The swelling power was observed to be a function of temperature, with the highest value observed at 90 °C [31]. In this study, the effect of irradiation dose on swelling power was not significant and only slightly increased this parameter.

In addition, the general swelling index of rice with HM content was slightly higher than that of rice with LM content. This difference was not significant. A previous study showed that the swelling power decreases because irradiation causes the depolymerization of amylopectin chains in starch molecules [32]. The swelling power slightly increases under a low irradiation dose. The slight opening of amylopectin chains under low-dose irradiation may have promoted water molecule entry, which leads to the increase in swelling power.

Table 1. Swelling power and solubility index of control and irradiated rice flours *.

Varieties	Dose	60 °C	70 °C	80 °C	90 °C
		SP (g/g)			
LM content	0	3.20 ± 0.04 [ap]	7.24 ± 0.06 [aq]	8.40 ± 0.01 [ar]	12.07 ± 0.04 [as]
	1	3.40 ± 0.02 [cp]	7.63 ± 0.12 [bq]	8.76 ± 0.10 [br]	12.69 ± 0.41 [abs]
	2	3.44 ± 0.02 [cp]	7.71 ± 0.04 [bcq]	8.85 ± 0.09 [br]	13.90 ± 0.37 [abs]
	3	3.28 ± 0.02 [bp]	7.78 ± 0.04 [bcq]	9.01 ± 0.12 [bcr]	14.10 ± 0.71 [abs]
	4	3.54 ± 0.01 [dp]	7.94 ± 0.06 [cq]	9.20 ± 0.00 [cr]	14.37 ± 0.87 [bs]
HM rice	0	3.58 ± 0.03 [ap]	7.28 ± 0.08 [aq]	8.20 ± 0.10 [ar]	11.33 ± 0.17 [as]
	1	4.12 ± 0.09 [bp]	7.24 ± 0.06 [aq]	8.33 ± 0.02 [ar]	11.94 ± 0.25 [abs]
	2	4.05 ± 0.11 [bp]	7.32 ± 0.08 [aq]	8.52 ± 0.32 [ar]	12.43 ± 0.64 [abcs]
	3	4.12 ± 0.04 [bp]	7.25 ± 0.01 [aq]	8.51 ± 0.08 [ar]	12.75 ± 0.13 [bcs]
	4	3.74 ± 0.19 [abp]	7.79 ± 0.23 [bq]	8.65 ± 0.09 [ar]	13.40 ± 0.02 [cs]
		SI (g/100 g)			
LM content	0	5.70 ± 0.06 [ap]	9.21 ± 0.14 [aq]	10.11 ± 0.13 [as]	10.44 ± 0.01 [ar]
	1	6.74 ± 0.34 [bp]	15.54 ± 0.03 [bq]	15.55 ± 0.32 [bq]	21.83 ± 0.85 [br]
	2	6.89 ± 0.32 [bcp]	16.45 ± 0.02 [cq]	16.79 ± 0.25 [bq]	23.32 ± 0.73 [br]
	3	7.79 ± 0.30 [cp]	19.56 ± 0.30 [dq]	19.77 ± 0.57 [cq]	29.48 ± 1.44 [cr]
	4	9.47 ± 0.14 [dp]	21.04 ± 0.11 [eq]	22.53 ± 0.25 [dr]	32.84 ± 1.34 [cs]
HM rice	0	5.11 ± 0.19 [ap]	7.40 ± 0.24 [aq]	8.81 ± 0.20 [as]	8.07 ± 0.06 [ar]
	1	6.35 ± 0.55 [bcp]	12.35 ± 0.09 [bq]	12.56 ± 0.42 [bq]	16.95 ± 0.26 [br]
	2	6.14 ± 0.10 [bp]	13.77 ± 0.07 [cq]	13.03 ± 0.78 [bq]	19.95 ± 0.91 [cr]
	3	7.31 ± 0.19 [cdp]	15.71 ± 0.20 [dq]	15.75 ± 0.37 [cq]	24.49 ± 0.28 [dr]
	4	7.83 ± 0.08 [dp]	18.15 ± 0.12 [eq]	18.93 ± 0.11 [dr]	29.89 ± 0.19 [es]

* Mean value ± SD with different superscript letters in the same column are significantly different ($p < 0.05$). SP = swelling power: SI = solubility index. The first letter represents vertical significance and the second letter represents horizontal significance.

The solubility index of all the rice flours increased slightly over the temperature range of 60 °C–90 °C. However, with respect to that of native flour, the solubility index of irradiated flours significantly increased ($p < 0.05$) over the range of 60 °C–90 °C. The solubility index of flours is largely attributed to the presence of soluble molecules, such as amylose, sugars, and albumins, in the flour [33]. The increased solubility of irradiated samples could be ascribed to the increased depolymerization of starch chains that may result in the enhanced hydration of particles [34].

3.5. Textural Properties of Cooked Rice

Texture profile analysis is one of the most important tests used to examine food texture. Table 2 shows the hardness, adhesiveness, springiness, cohesiveness, gumminess, chewiness, and resilience of cooked rice. The hardness of the rice irradiated at the dose of 2 kGy was higher than that of cooked nonirradiated rice and gradually increased from 4155.1 to 4266.5. The hardness of cooked rice is mainly affected by the molecular size of amylose and the ratio of amylose branches with DPs of 1000–2000. Moreover, the stickiness of cooked rice is negatively correlated with amylose content. Thus, the increased hardness of cooked irradiated rice (2 kGy) observed in this study might be attributed to the slight reduction in the molecular size of amylose. In general, rice with low water content was hard, whereas rice with high water content was soft. Irradiation reduced textural indexes, and texture was gradually destroyed with the increase in irradiation dose.

Table 2. Textural properties of cooked rice after EBI treatment at different doses *.

Dose/kGy		Hardness	Adhesiveness	Springiness	Cohesiveness	Gumminess	Chewiness	Resilience
LM rice	0	4155.1 ± 523.1 [a]	−1046.9 ± 148.4 [a]	0.72 ± 0.07 [a]	0.26 ± 0.01 [a]	1111.6 ± 168.6 [a]	833.6 ± 57.3 [ab]	0.08 ± 0.00 [a]
	1	3254.8 ± 486.7 [b]	−672.6 ± 96.6 [b]	0.60 ± 0.01 [b]	0.27 ± 0.03 [a]	899.5 ± 226.4 [ab]	537.2 ± 142.7 [b]	0.08 ± 0.01 [a]
	2	4266.5 ± 494.1 [a]	−710.3 ± 20.0 [b]	0.63 ± 0.03 [ab]	0.27 ± 0.02 [a]	1172.9 ± 186.8 [a]	731.5 ± 83.3 [a]	0.09 ± 0.01 [a]
	3	2793.0 ± 460.3 [b]	−441.1 ± 32.6 [c]	0.61 ± 0.04 [b]	0.25 ± 0.02 [a]	699.8 ± 118.5 [b]	422.7 ± 54.7 [b]	0.07 ± 0.01 [a]
	4	2520.5 ± 219.1 [b]	−477.7 ± 54.2 [c]	0.60 ± 0.00 [b]	0.26 ± 0.02 [a]	649.7 ± 73.2 [b]	395.6 ± 95.1 [b]	0.07 ± 0.00 [a]
HM rice	0	2476.0 ± 274.8 [a]	−602.8 ± 82.9 [a]	0.64 ± 0.02 [a]	0.25 ± 0.01 [ab]	608.7 ± 55.4 [a]	387.9 ± 35.1 [a]	0.06 ± 0.00 [a]
	1	2355.8 ± 81.3 [a]	−581.4 ± 36.1 [a]	0.58 ± 0.06 [ab]	0.25 ± 0.00 [ab]	592.0 ± 32.0 [a]	346.5 ± 47.9 [ab]	0.07 ± 0.00 [a]
	2	2074.3 ± 170.1 [ab]	−375.6 ± 40.8 [b]	0.53 ± 0.03 [bc]	0.25 ± 0.00 [ab]	517.1 ± 59.0 [ab]	273.3 ± 45.7 [bc]	0.08 ± 0.00 [b]
	3	1692.7 ± 147.4 [bc]	−327.0 ± 10.9 [b]	0.47 ± 0.02 [bc]	0.26 ± 0.01 [a]	438.5 ± 15.6 [b]	207.3 ± 14.7 [cd]	0.08 ± 0.00 [b]
	4	1427.8 ± 109.4 [c]	−311.3 ± 25.3 [b]	0.51 ± 0.05 [c]	0.24 ± 0.00 [b]	338.1 ± 18.1 [c]	173.7 ± 26.7 [d]	0.07 ± 0.00 [ab]

* Mean value ± SD with different superscript letters in the same column are significantly different ($p < 0.05$).

3.6. Sensory Properties of Cooked Rice

The sensory properties of cooked native and irradiated rice are presented in Figure 5. As seen in the figure below, HM rice had better sensory ratings than LM rice. Compared with that of nonirradiated rice, the aroma of irradiated HM rice decreased with the increase in irradiation dose and that of irradiated LM rice had almost disappeared. Compared with the control group, the HM rice irradiated at 1 and 2 kGy did not show significant changes in taste and continued to present light fragrance and sweet taste. LM rice did not have the same taste as HM rice. Unsatisfactory rice taste might be ascribed to the production of undesirable flavors as a result of the following: irradiation produces free radicals, which break rice starch molecules [27], expose hydrophobic side chains inside proteins [35], and partially oxidize protein and fat. In general, the quality and taste of HM rice were superior to those of LM rice.

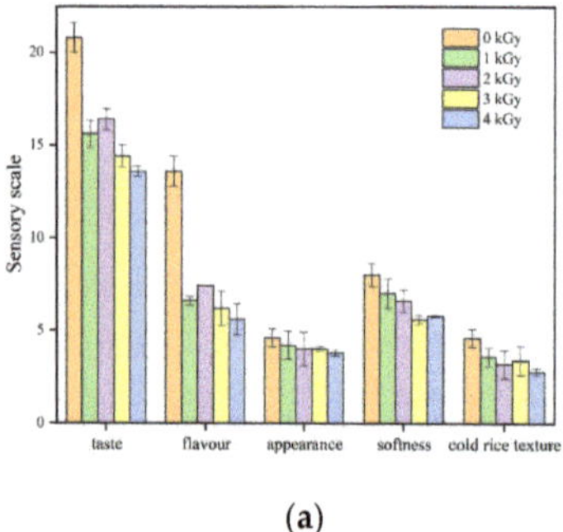

(a)

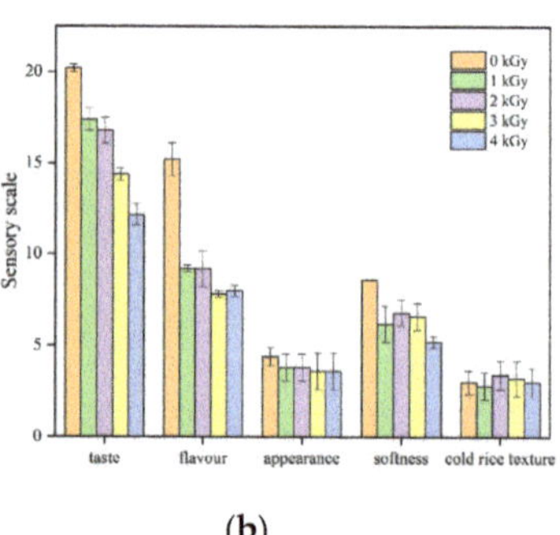

(b)

Figure 5. Sensory evaluation of cooked rice treated with various irradiation doses. (**a**) LM rice, (**b**) HM rice.

3.7. Thermal PROPERTIES

The DSC results for native and irradiated rice starches are presented in Table S3 and Figure 6, respectively. The figure shows that irradiation had a great influence on the thermodynamic properties of rice under LM condition but had little effect on HM rice. The heat absorption peak shifted to the left, and T_O, T_P, and T_C decreased in the LM state. Evan's theory [36] states that the degradation of the amorphous region of starch granules decreases the stability of the crystalline region and leads to the reduction in gelatinization temperature. Irradiation caused the degradation of starch molecules, thus reducing stability and resulting in the decrement in gelatinization temperature. Cooke et al. [37] pointed out that DSC results are more likely to show the change in crystal structure than the double helix structure of starch, indicating that irradiation has destroyed part of the double helix structure of starch.

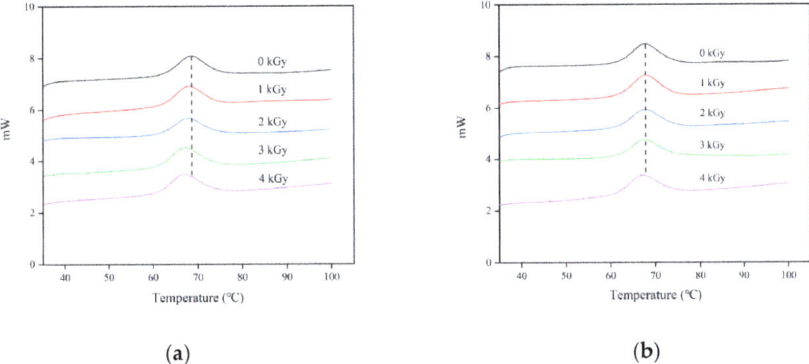

(**a**) (**b**)

Figure 6. Differential scanning calorimeter (DSC) thermograms of the starches isolated from rice grains treated with various irradiation doses. (**a**) LM rice, (**b**) HM rice.

3.8. In Vitro Digestion

The nutritional properties of the starches were evaluated via in vitro digestion. The rapidly digestible starches (RDS), slowly digestible starches (SDS), and resistant starches (RS) levels of native and irradiated rice are presented in Table 3. The results indicated that the digestive changes shown by all irradiated samples were not significant. In general, the RDS of rice samples with HM content were significantly higher than those of rice samples with LM content, whereas the RS of rice samples with HM content were significantly lower than those of rice samples with LM content. After consumption, RDS induces hyperglycemic response and insulin resistance, which easily cause several diet-related chronic diseases and metabolic syndromes; by contrast, RS is fermented only by microorganisms in the large intestine to produce short-chain fatty acids, which are beneficial to intestinal health [19,38]. Therefore, according to the results of this study, rice with LM content has better nutritional quality than rice with HM content.

Table 3. In vitro digestibility of the starches isolated from rice grains under various irradiation doses *.

Dose (kGy)		In Vitro Digestibility [#]		
		RDS (%)	SDS (%)	RS (%)
	0	28.21 ± 0.61 [a]	47.50 ± 0.73 [a]	24.29 ± 0.12 [ab]
	1	30.01 ± 0.49 [ab]	46.30 ± 0.73 [a]	23.69 ± 0.24 [ab]
LM rice	2	32.08 ± 1.22 [ab]	47.41 ± 0.85 [a]	20.51 ± 2.07 [a]
	3	28.38 ± 3.29 [a]	40.27 ± 0.73 [b]	31.35 ± 4.02 [c]
	4	33.97 ± 0.73 [b]	39.50 ± 2.56 [b]	26.53 ± 1.83 [bc]
	0	34.75 ± 0.85 [a]	47.93 ± 0.85 [a]	17.32 ± 0.00 [bc]
	1	35.44 ± 0.61 [a]	48.53 ± 0.49 [a]	16.03 ± 0.12 [b]
HM rice	2	36.77 ± 0.12 [a]	49.48 ± 0.37 [a]	14.05 ± 0.49 [a]
	3	37.33 ± 2.80 [a]	41.99 ± 3.41 [b]	20.68 ± 0.61 [c]
	4	36.98 ± 0.37 [a]	45.18 ± 1.58 [ab]	17.84 ± 1.22 [d]

* Mean value ± SD with different superscript letters in the same column are significantly different ($p < 0.05$). [#] RDS, SDS, and RS represent rapidly digestible, slowly digestible, and resistant starches, respectively.

4. Conclusions

The results of this study suggested that low-dose EBI treatment changed water distribution in newly harvested rice with HM (15.03%) and dried rice (11.97%). Radiation mainly affected the free water components of LM rice and the combined water and free water components of HM rice, as well as resulting in the destruction of biomolecules in both types of rice. EBI treatment reduced the viscosity of rice, and the change in LM rice was more obvious than that in HM rice. Rheological moduli also changed after irradiation, and the change in HM rice was gradual. The increase in swelling power after irradiation might be attributed to the slight opening of amylopectin chains under low-dose irradiation. This effect promoted the entry of water molecules. The increased solubility of irradiated samples could be ascribed to the increased depolymerization of starch chains that resulted in enhanced particle hydration. Low-dose irradiation had little effect on the gelatinization, texture, and rheological properties of most rice samples. The quality and taste of HM rice were superior, whereas the nutritional quality of LM rice was superior. This work was an attempt to optimize the outcome of the EBI treatment of rice for storage purposes by analyzing its effects. It showed that low-dose EBI was more effective and environmentally friendly than other methods.

Supplementary Materials: The following are available online at http://www.mdpi.com/2304-8158/9/9/1139/s1, Table S1: Differences in proton areas and peak times among rice samples with different moisture contents, Table S2: Pasting properties of LM and HM rice samples under various irradiation doses, Table S3: Thermal properties of flours prepared from rice grains treated with various irradiated doses.

Author Contributions: Conceptualization, L.P. and X.L.; formal analysis, L.P., J.X., X.L., Y.L., Y.Z. and K.Y.; investigation, L.P., X.L., Y.L. and D.S.; resources, D.S. and Z.C.; writing—original draft preparation, L.P. and J.X.; writing—review and editing, J.X. and X.L.; All authors have read and agreed to the published version of the manuscript.

Funding: This work was supported by the funding of the National Key Research and Development Program of China (2017YFD0401205), National Top Youth for Grain Industry (LQ2018302), and National First-Class Discipline Program of Food Science and Technology (JUFSTR20180203).

Conflicts of Interest: The authors declare no conflict of interest.

References

1.   ERS; USDA. Rice Overview. Available online: https://www.ers.usda.gov/topics/crops/rice/ (accessed on 3 June 2020).

2.   Network, China Industry Information. Available online: http://www.chyxx.com/industry/202004/850656.html (accessed on 10 April 2020).

3.   Keawpeng, I.; Venkatachalam, K. Effect of aging on changes in rice physical qualities. *Int. Food Res. J.* **2015**, *22*, 2180–2187.

4.   Srivastava, S.; Mishra, G.; Mishra, H.N. Fuzzy controller based E-nose classification of sitophilus oryzae infestation in stored rice grain. *Food Chem.* **2019**, *283*, 604–610. [CrossRef] [PubMed]

5.   Zhou, Z.; Robards, K.; Helliwell, S.; Blanchard, C. Ageing of stored rice: Changes in chemical and physical attributes. *J. Cereal Sci.* **2002**, *35*, 65–78. [CrossRef]

6.   Meullenet, J.F.; Marks, B.P.; Hankins, J.A.; Griffin, V.K.; Daniels, M.J. Sensory quality of cooked long-grain rice as affected by rough rice moisture content, storage temperature, and storage duration. *Cereal Chem.* **2000**, *77*, 259–263. [CrossRef]

7.   Tamaki, M.; Tashiro, T.; Ishikawa, M.; Ebata, M. Physicoecological studies on quality formation of rice kernel .4. effect of storage on eating quality of rice. *Jpn. J. Crop Sci.* **1993**, *62*, 540–546. [CrossRef]

8.   Wijayaratne, L.K.W.; Arthur, F.H.; Whyard, S. Methoprene and control of stored-product insects. *J. Stored Prod. Res.* **2018**, *76*, 161–169. [CrossRef]

9.   Richards, C.M.; Reilley, A.; Touchell, D.; Antolin, M.F.; Walters, C. Microsatellite primers for texas wild rice (Zizania Texana), and a preliminary test of the impact of cryogenic storage on allele frequency at these loci. *Conserv. Genet.* **2004**, *5*, 853–859. [CrossRef]

10.   Blumenthal, C.; Rawson, H.M.; McKenzie, E.; Gras, P.W.; Barlow, E.W.R.; Wrigley, C.W. Changes in wheat grain quality due to doubling the level of atmospheric Co2. *Cereal Chem.* **1996**, *73*, 762–766.

11.   Sendra, E.; Capellas, M.; Guamis, B.; Felipe, X.; MorMur, M.; Pla, R. Review: Food irradiation. general aspects. *Food Sci. Technol. Int.* **1996**, *2*, 1–11. [CrossRef]

12.   Al Khan, K.; Abrahem, M. Effect of irradiation on quality of spices. *Int. Food Res. J.* **2010**, *17*, 825–836.

13. Sung, W.C. Effect of gamma irradiation on rice and its food products. *Radiat. Phys. Chem.* **2005**, *73*, 224–228. [CrossRef]

14. Diehl, J.F. Food irradiation—Past, present and future. *Radiat. Phys. Chem.* **2002**, *63*, 211–215. [CrossRef]

15. Luo, X.H.; Li, Y.L.; Yang, D.; Xing, J.L.; Li, K.; Yang, M.; Wang, R.; Wang, L.; Zhang, Y.W.; Chen, Z.X. Effects of electron beam irradiation on storability of brown and milled rice. *J. Stored Prod. Res.* **2019**, *81*, 22–30. [CrossRef]

16. Li, H.Y.; Prakash, S.; Nicholson, T.M.; Fitzgerald, M.A.; Gilbert, R.G. The importance of amylose and amylopectin fine structure for textural properties of cooked rice grains. *Food Chem.* **2016**, *196*, 702–711. [CrossRef] [PubMed]

17. GAQS, Generally Administration of Quality Supervision. *Inspection of Grains and Oils-Method for Sensory Evaluation of Paddy or Rice Cooking and Eating Quality*; GB/T 15682-2008 [S]; GAQS: Beijing, China, 2008.

18. Zhou, X.; Ye, X.J.; He, J.; Wang, R.; Jin, Z.Y. Effects of electron beam irradiation on the properties of waxy maize starch and its films. *Int. J. Biol. Macromol.* **2020**, *151*, 239–246. [CrossRef]

19. Englyst, H.N.; Kingman, S.M.; Cummings, J.H. Classification and measurement of nutritionally important starch fractions. *Eur. J. Clin. Nutr.* **1992**, *46*, S33–S50.

20. Wang, R.; Zhang, T.Q.; He, J.; Zhang, H.; Zhou, X.; Wang, T.; Feng, W.; Chen, Z.X. Tailoring digestibility of starches by chain elongation using amylosucrase from neisseria polysaccharea via a zipper reaction mode. *J. Agric. Food Chem.* **2020**, *68*, 225–234. [CrossRef]

21. Zhou, X.X.; Liu, L.; Fu, P.C.; Lyu, F.; Zhang, J.Y.; Gu, S.Q.; Ding, Y.T. Effects of infrared radiation drying and heat pump drying combined with tempering on the quality of long-grain paddy rice. *Int. J. Food Sci. Technol.* **2018**, *53*, 2448–2456. [CrossRef]

22. Jin, Y.; Liang, R.; Liu, J.B.; Lin, S.Y.; Yu, Y.L.; Cheng, S. Effect of structure changes on hydrolysis degree, moisture state, and thermal denaturation of egg white protein treated by electron beam irradiation. *LWT Food Sci. Technol.* **2017**, *77*, 134–141. [CrossRef]

23. Sanchez-Alonso, I.; Moreno, P.; Careche, M. Low field nuclear magnetic resonance (Lf-Nmr) relaxometry in hake (Merluccius Merluccius L.) muscle after different freezing and storage conditions. *Food Chem.* **2014**, *153*, 250–257. [CrossRef]

24. Sultan, N.; Wani, I.A.; Masoodi, F.A. Moisture mediated effects of gamma-irradiation on physicochemical, functional, and antioxidant properties of pigmented brown rice (*Oryza Sativa*, L.) flour. *J. Cereal Sci.* **2018**, *79*, 399–407. [CrossRef]

25. Ali, S.; Singh, B.; Sharma, S. Impact of feed moisture on microstructure, crystallinity, pasting, physico-functional properties and in vitro digestibility of twin-screw extruded corn and potato starches. *Plant Foods Hum. Nutr.* **2019**, *74*, 474–480. [CrossRef] [PubMed]

26. Yu, Y.; Wang, J. Effect of gamma irradiation pre-treatment on drying characteristics and qualities of rice. *Radiat. Phys. Chem.* **2005**, *74*, 378–383. [CrossRef]

27. Bashir, K.; Aggarwal, M. Physicochemical, structural and functional properties of native and irradiated starch: A review. *J. Food Sci. Technol. Mysore* **2019**, *56*, 513–523. [CrossRef] [PubMed]

28. Yue, X.; Shang, W.; Liu, J.; Sun, H.; Strappe, P.; Zhou, Z. Analysis of the physiochemical properties of rice induced by postharvest yellowing during storage. *Food Chem.* **2020**, *306*, 125517.

29. Singh, S.; Singh, N.; Ezekiel, R.; Kaur, A. Effects of gamma-irradiation on the morphological, structural, thermal and rheological properties of potato starches. *Carbohydr. Polym.* **2011**, *83*, 1521–1528. [CrossRef]

30. Yu, Y.; Wang, J. Effect of gamma-ray irradiation on starch granule structure and physicochemical properties of rice. *Food Res. Int.* **2007**, *40*, 297–303. [CrossRef]

31. Sindhu, R.; Devi, A.; Khatkar, B.S. Physicochemical, thermal and structural properties of heat moisture treated common buckwheat starches. *J. Food Sci. Technol. Mysore* **2019**, *56*, 2480–2489. [CrossRef]

32. Liu, T.Y.; Ma, Y.; Xue, S.; Shi, J. Modifications of structure and physicochemical properties of maize starch by gamma-irradiation treatments. *LWT Food Sci. Technol.* **2012**, *46*, 156–163. [CrossRef]

33. Wani, I.A.; Wani, A.A.; Gani, A.; Muzzaffar, S.; Gul, M.K.; Masoodi, F.A.; Wani, T.A. Effect of gamma-irradiation on physico-chemical and functional properties of arrowhead (*Sagittaria Sagittifolia* L.) tuber flour. *Food Biosci.* **2015**, *11*, 23–32. [CrossRef]

34. Wu, D.X.; Shu, Q.Y.; Wang, Z.H.; Xia, Y.W. Effect of gamma irradiation on starch viscosity and physicochemical properties of different rice. *Radiat. Phys. Chem.* **2002**, *65*, 79–86. [CrossRef]

35. Bo, H.; Fan, D.-M.; Wusgal; Lian, H.-Z.; Zhang, H.; Chen, W. The radical production and oxidative properties of rice protein under microwave radiation. *Modern Food Sci. Technol.* **2015**, *4*, 151–156.

36. Evans, I.D.; Haisman, D.R. The effect of solutes on the gelatinization temperature range of potato starch. *Starch Staerke* **1982**, *34*, 224–231. [CrossRef]

37. Cooke, D.; Gidley, M.J. Loss of crystalline and molecular order during starch gelatinization—Origin of the enthalpic transition. *Carbohydr. Res.* **1992**, *227*, 103–112. [CrossRef]

38. Rahman, S.; Bird, A.; Regina, A.; Li, Z.; Ral, J.P.; McMaugh, S.; Topping, D.; Morell, M. Resistant starch in cereals: Exploiting genetic engineering and genetic variation. *J. Cereal Sci.* **2007**, *46*, 251–260. [CrossRef]

 *foods*

Article

Evolution of VOC and Sensory Characteristics of Stracciatella Cheese as Affected by Different Preservatives

Giuseppe Natrella, Graziana Difonzo, Maria Calasso, Giuseppe Costantino, Francesco Caponio and Michele Faccia *

Department of Soil, Plant and Food Sciences, University of Bari, Via Amendola 165/A, 70126 Bari, Italy; giuseppe.natrella@uniba.it (G.N.); graziana.difonzo@uniba.it (G.D.); maria.calasso@uniba.it (M.C.); giuseppe.costantino@uniba.it (G.C.); francesco.caponio@uniba.it (F.C.)
* Correspondence: michele.faccia@uniba.it; Tel./Fax: +39-080-5443012

Received: 7 September 2020; Accepted: 9 October 2020; Published: 12 October 2020

Abstract: Undesired volatile organic compounds (VOCs) can negatively affect the flavor of fresh food products; especially those characterized by a mild and delicate aroma. Finding connections between chemical and sensory analyses is a useful way to better understand the arising of off-flavors. A study was conducted on stracciatella; a traditional Italian cream cheese that is emerging on international markets. Samples were prepared by adding two different preservatives (alone or combined): sorbic acid and an olive leaf extract. Their influence on flavor preservation during refrigerated storage was investigated by chemical, microbiological and sensory analyses. A strong change of the VOC profile was ascertained after 8 days in the control cheese and in the sample added with leaf extract alone. The samples containing sorbic acid, alone or in combination with leaf extract, gave the best chemical and sensory results, demonstrating a significant shelf-life extension. In particular, these samples had lower concentrations of undesired metabolites, such as organic acids and volatiles responsible for off-flavor, and received better scores for odor and taste. Ex and Ex-So samples had significantly higher antioxidant activity than Ctr and So throughout the entire storage period, and the color parameter shows no differences among samples taken on the same day. The use of the olive leaf extract, at the concentration tested, seemed to be interesting only in the presence of sorbic acid due to possible synergic effect that mainly acted against Enterobacteriaceae.

Keywords: stracciatella cheese; volatile organic compounds; sensory characteristics; natural preservatives; cheese storage

---

1. Introduction

One of the decay's symptoms of an expiring food is its smell [1,2]. Thus, the study of the volatile organic compound (VOC) profile can be very useful for establishing the organoleptic status of the product. The onset of off-flavor can be caused by many factors, such as food handling, processing, spoilage microorganism and lipid oxidation [3–5]. Many researchers have investigated the VOC responsible of off-flavor and have tried to link them to specific sensory descriptors [5–10]. The fastest way to analyze these compounds is using SPME-GC/MS (Solid Phase Micro Extraction-Gas Chromatography/Mass Spectrometry). It is widely used among researchers for milk and dairy products [11–15] and other foods [16–18]. It is used also in the flavoromics approach [19,20], because it is capable of ppb detection level for a wide range of molecular weight. Such a kind of studies is very useful for fresh foods, which are preferred by modern consumers and are highly perishable. For cheeses, too, the consumer preference is increasingly orienting towards the fresh types, whose production volume in EU in 2019 was about 3.5 million tons (about 34% of total cheese manufactured) [21]. Several strategies have been proposed for slowing down their decaying process, including high pressure and x-ray treatments, modified atmosphere and active packaging and use of antimicrobial compounds [22–25]. Stracciatella is an Italian traditional fresh cream cheese that is emerging on the international market. It is made up of mixture of double cream and thin Mozzarella strands (manually shredded) [26]; it is commonly sold packaged in polyethylene trays heat-sealed with laminated film. Due to high perishability, it must be kept refrigerated until consumption but, being the signs of alteration only recognizable after tasting, a non-negligible part of the product sold in large-scale retail stores turns into waste. High perishability of stracciatella is mostly caused by fast fat oxidation and microbial growth, favored by high pH and moisture content (6.0–6.2 and 62–65%, respectively) [27]. Efforts for preserving shelf life and sensory characteristics of this cheese have reached poor results. Only little shelf-life extension was obtained by Gammariello et al. [28], by applying modified atmosphere packaging (MAP), and by Conte et al. [29] combining it with two antimicrobial compounds (lysozyme and EDTA). Dambrosio et al. [30] and Rea et al. [31] concluded that the principal obstacle lay in the manufacturing process that involves intense manipulation. Unfortunately, the stracciatella production process cannot be mechanized since it is impossible to reproduce the results obtained by the manual procedure (in particular the mozzarella shredding phase). The only strategy that can be used to keep longer the original sensory characteristic is the addition of preservatives. Unfortunately, only a few molecules are allowed by EU legislation in fresh dairy products, and the most commonly used is sorbic acid. This preservative is able to inhibit the growth of numerous microorganisms, depending on the types, species and strains, but also on substrate properties and environmental factors. According to the Codex Alimentarius Commission, the maximum concentration in fresh cheese (as total sorbates) is 1000 mg kg^{-1}. Even though sorbic acid has a GRAS (Generally Recognized As Safe) status, the dairy industries are asking for more natural preservatives. There is no information about recently studied aqueous extract obtained from olive leaf (OLE) in fresh cheeses. Its contribution on shelf life and chemical and sensory characteristics of stracciatella is still unknown. OLE has been recently tested for antimicrobial activity based on the high polyphenols content [32–34].

The aim of the present work was to study the relationships between the VOC profile, sensory features and microbiological characteristics of stracciatella cheese during storage, as affected by the addition of sorbic acid and OLE.

2. Materials and Methods

2.1. Preservatives Concentration and Sample Preparation

According to Ranieri et al. [35] OLE was produced at the laboratory scale, then its concentration to be added to cheese was chosen on the basis of both of the data reported in the literature and deriving from sensory preliminary trials. According to Caponio et al. [36] and Difonzo et al. [37], OLE added in an amount up to 1000 mg kg^{-1} was able to extend the shelf life of vegetable foods. The preliminary trials (data not shown) highlighted that OLE added at a concentration of 800 mg kg^{-1} negatively affected the sensory properties, causing an unpleasant odor and loss of freshness and frankness of flavor, while when present at a level of 400 mg kg^{-1} the typical aroma was maintained. A commercial food grade sorbic acid was supplied by Farmalabor (Canosa, Italy), and the level of addition was fixed at the maximum level (1000 mg kg^{-1}) allowed by European legislation in fresh dairy products [38].

Four types of stracciatella samples were prepared: 3 experimental, containing different types of antimicrobials (OLE, sorbic acid and a mixture of the two), plus a control (Table 1). The antimicrobials were added to 30% fat UHT (Ultra High Temperature) cream, then the cream was mixed with freshly prepared mozzarella strands (1:1 *w/w*) and gently homogenized at room temperature in a kneader (40 rounds per minute applied for 5 min). The obtained samples were stored at 8 °C in plastic trays mechanically sealed with laminated film. OLE, obtained as reported in Difonzo et al. [39], had a total phenol content of 151 mg gallic acid equivalent (GAE) g^{-1} and a value of 950 µmol Trolox equivalent (TE) g^{-1}. The samples were named Ex (sample with OLE), So (sample with sorbic acid), Ex-So (sample with OLE + sorbic acid) and Ctr (control). They were analyzed after 1 day and every 4 days from production until they resulted in being altered.

Table 1. Ingredients and preservatives concentration used to produce stracciatella cheese.

	CTR	Ex	So	Ex-So
Cream/mozzarella strands (*w/w* as *g*)	500/500	500/500	500/500	500/500
OLE (mg kg^{-1})	-	400	-	400
Sorbic acid (mg kg^{-1})	-	-	1000	1000

2.2. Chemical, Sensory and Microbiological Analyses

2.2.1. Volatile Organic Compounds (VOCs)

VOCs were extracted at 37 °C for 15 min as reported in a previous paper [40] after addition of 3-pentanone (81.3 ng) as an internal standard for semi-quantitation. A Triplus RSH autosampler was used, equipped with a divinylbenzene/carboxen/polydimethylsiloxane 50/30 mm SPME fiber assembly (Supelco, Bellefonte, PA, USA). The fiber was desorbed at 220 °C for 2 min in the injection port of the gas chromatograph operating in the splitless mode. The GC–MS analysis was performed using a Trace 1300 chromatograph equipped with a capillary column VF-WAX MS (60 m, 0.25 mm) and connected to mass spectrometer ISQ Series 3.2 SP1 (Thermo Scientific, Waltham, MA, USA). The operating conditions were: oven temperatures, 50 °C for 0.1 min then 13 °C min^{-1} up to 180 °C and 18 °C min^{-1} up to 220 °C with an isothermal for 1.5 min. The mass detector was set at 1700 V voltage; source temperature, 250 °C; ionization energy 70 eV and scan range 33–200 amu. Peak identification was done by means of Xcalibur V2.0 Qual Browse software (Thermo Fisher Scientific, Waltham, MA, USA) by matching with the reference mass spectra of the NIST (National Institute of Standards and Technology, Gaithersburg, MD, USA) library.

2.2.2. Antioxidant Activity and Oxidative Stability

The ABTS-TEAC (2,2′-azino-bis(3-ethylbenzothiazoline-6-sulfonic acid)/Trolox®-Equivalent Antioxidant Capacity) assay was carried out as described by Ranieri et al. [35] and the results were expressed as µmol TE g^{-1}. Induction time was determined by RapidOxy (Anton Paar Prove Tec GmbH, Blankenfelde-Mahlow, Germany) measuring the time needed for a 10% drop of the oxygen pressure. The set parameters were the following: T = 140 °C, P = 700 kPa [38].

2.2.3. Color

Colorimetric readings were carried out under D65 illuminant by using a spectro-colorimeter CM-700d (Konica Minolta Sensing, Osaka, Japan) equipped with a pulsed xenon lamp. The analysis was performed by placing the sample in a transparent quartz container. Lightness (*L**), redness (*a**, ±red-green) and yellowness (*b**, ±yellow-blue) were determined coordinates in the CIE (Commission Internationale de l'Éclairage) color space [41].

2.2.4. Organic Acids and pH

Organic acids were extracted as reported by Buffa et al. [42]. Separation was carried out on a Synergy Hydro RP column 80 Å, 4 µm, 250 mm × 4.6 mm (Phenomenex, Torrance, CA, USA), installed on a Waters HPLC composed of 600E pumps and a 996 diode array detector (Waters Corporation, Milford, CT, USA). Mobile phases were 0.1% orthophosphoric acid in water (eluent A) and acetonitrile (eluent B). The gradient was 0–18 min 100% A at 1 mL min^{-1} flow rate, then 18–18.3 min from 100% to 20% A; 18.3–19.5 min increasing flow rate to 1.4 mL min^{-1}, then 19.5–22.5 isocratic and 22.5–23 min from 20% to 100% A and 23–43 min isocratic. Detection was done at λ = 214 nm. A pH meter equipped with a dairy specific electrode (FC2020, Hanna instruments, Woonsocket, RI, USA) was used for pH measurement.

2.2.5. Sensory Analysis

Sensory analysis was assessed by a trained panel consisting of nine experts (4 women and 5 men, ranging in age from 25 to 60 years) selected according to the ISO 8586-1993 method. All of them were members of the Italian Association of Cheese Tasters (ONAF) and had attended a 20 h course about evaluation of cheese texture and flavor. Evaluation was carried by a quantitative descriptive analysis (QDA) and the samples were presented in a randomized and balanced way, in white disposable dishes coded by three-digit codes. The panelists described the cheese using sensory attributes chosen from the ONAF (Italian Organization of Cheese Tasters) vocabulary [43] and by scoring their intensity from 0 to 10. In the case of perception of an attribute not included in the vocabulary, it was only considered if the "weight percentage" (frequency of citations × perceived intensity) was more than 30%.

2.2.6. Cultivable Microbiota

Microbiological analyses were performed according to methods described in Minervini et al. [44], using culture media and supplements purchased from Oxoid (Oxoid Limited, Basingstoke, UK), starting from 20 g of stracciatella. The microbial groups counted were: total mesophilic aerobic (plate count agar at 30 °C for 48 h), presumptive mesophilic lactobacilli (MRS agar with 0.1 g L^{-1} cycloheximide at 30 °C for 48 h), presumptive mesophilic cocci (M17 agar with 0.1 g L^{-1} cycloheximide at 30 °C for 48 h), enterococci (Slanetz and Bartley agar at 37 °C for 48 h), staphylococci (Baird Parker agar supplemented with egg yolk tellurite at 37 °C for 48 h), Enterobacteriaceae (VRBGA at 37 °C for 24 h), *Pseudomonas* spp. (*Pseudomonas* agar with CFC supplement at 30 °C for 24 h), yeasts (wort agar supplemented with 0.1 g L^{-1} chloramphenicol at 30 °C for 48 h) and molds (potato dextrose agar supplemented with 0.1 g L^{-1} chloramphenicol at 25 °C for 5 days). The microbiological counts were confirmed by taking representative colonies for each medium, and checking them for morphology, motility, Gram staining reaction and catalase test.

2.3. Statistical Analysis

The data were statistically processed by one-way ANOVA followed by Tukey's HSD procedure at $p < 0.05$ using XLSTAT software (Addinsoft, NY, USA). Each sample was analyzed in triplicate. The data from the sensory analysis were also processed for generalized Procrustes analysis (GPA) with the same software.

3. Results and Discussion

3.1. VOC

Figure 1 shows the total amounts of volatile compounds during storage of the samples. As expected, low quantities were detected in the early days, ranging from 200 to 470 µg kg^{-1}. Until day 8, the differences among samples were not very relevant, even though they were sometimes statistically significant. Huge differences started after this time, with Ex and Ctr samples showing an increase to about 1800 µg kg^{-1}, then to about 2700 and 3600 µg kg^{-1} (at day 16), respectively. Differently, the concentrations in Ex-So and So remained almost constant during time. Increase in VOC formation during storage was expected, considering the typical chemical and microbiological characteristics of the product [27,30,31], and from these quantitative data it clearly appeared that a significant inhibition of their formation was associated to the presence of sorbic acid. In order to better understand the mechanisms of formation, the entire VOC data set was grouped into seven chemical classes (Table 2). The samples were rather similar at day 1 as to the most abundant classes (ketones, esters and aliphatic hydrocarbons), whereas some differences were found in some less represented groups such as aldehydes, alcohols and sulfur compounds, whose content is known to be influenced by manipulation (they are highly produced by spoilage microorganism). The main difference at T1 accounted on acids, which were more abundant in So and Ex-So, but it was mainly due to the presence of sorbic acid added as a preservative. As already observed under the quantitative point of view, a clear qualitative differentiation among samples started from day 8, and the pair Ctr/Ex and Ex-So/So behaving quite similarly until the end of storage. At T8 aldehydes and alcohols resulted to be higher in the pair Ctr/Ex, whereas esters and acids were more abundant in Ex-So/So. The control sample had the highest concentration of sulphur compounds and the lowest of aliphatic hydrocarbons. The higher concentrations of aldehydes and alcohols in Ctrl and Ex could be the consequence of two different phenomena: (i) lipid oxidation, which is a common source of aldehydes formation in fresh cheeses, from whose dehydrogenation the corresponding alcohols can successively derive [3,45] and (ii) microbial metabolism. At the end of storage (T16) almost all chemical groups were much higher in Ctr and Ex, whereas only sulphur compounds were more abundant in Ex-So. Finally, very low concentration of aliphatic hydrocarbons characterized the Ex-So sample.

As to the single VOC, 41 molecules were identified in the entire set of samples, the most abundant compounds detected at the end of the storage period are reported in Table 3. Acetaldehyde and 3-methylbutanal were found only in Ex and Ctr. Acetaldehyde can be formed throughout different pathways: ethanol oxidation should be a highly probable pathway, considering that ethanol concentration was very high in these samples (713.6 and 556.4 $\mu g\ kg^{-1}$ for Ex and Ctr respectively versus 7.8 and 4.2 $\mu g\ kg^{-1}$ for Ex-So and So, respectively). The mechanism based on phenolic compounds oxidation could also have played a role in Ex, considering the presence of polyphenols from the leaf extract [3]. 3-methylbutanal in cheese often arise from the conversion of the corresponding amino acid (isoleucine) by the LAB activity [46], and its concentration suggests a more marked microbial activity with respect to the samples without sorbic acid. This hypothesis was supported by the huge increase of ketones in Ctr and Ex, mainly connected to the formation of acetoin (1839.57 and 1512.48 $\mu g\ kg^{-1}$ respectively, versus 7.8 and 7.4 $\mu g\ kg^{-1}$ in So and Ex-So) that commonly derives from LAB and yeasts metabolism [47]. Ethyl acetate and ethyl butanoate increased much more in Ex, in connection with a higher presence of ethanol, which represents the limiting factor for their formation [48]. This aspect was probably connected to the absence of sorbic acid acting against yeasts. On the other hand, the presence of this preservative in So and Ex-So gave rise to the formation of ethyl sorbate. Among alcohols, besides ethanol, also 3-methyl-1-butanol greatly characterized the samples without sorbic acid, suggesting a higher microbial activity in them. In some food matrices such as fruit juices, 3-methyl-1-butanol and phenylethyl alcohol have been related to yeast metabolism [3]. The former was reported by Morales et al. [49] as an important component of volatilome of Enterobacteriaceae strains of dairy origin. Other volatiles that could be connected to faster microbial growth in Ctr and Ex samples were sulphur compounds (i.e., dimethyl sulfide) that are formed, for instance, by some Enterobacteriaceae species [7], and organic acids such as acetic, butanoic and hexanoic.

Table 2. Evolution of volatile compounds (μg kg^{-1}; mean values and standard deviation) during stracciatella cheese shelf life.

Parameter	T1				T4				T8				T12				T16			
	Ctr	Ex	So	Ex-So	Ctr	Ex	So	Ex-So	Ctr	Ex	So	Ex-So	Ctr	Ex	So	Ex-So	Ctr	Ex	So	Ex-So
Aldehydes	9.65[a] ± 1.55	8.00[a] ± 0.12	10.02[a] ± 1.99	0.78[b] ± 0.06	5.01[b] ± 2.42	5.19[b] ± 0.34	12.03[a] ± 2.15	5.82[b] ± 1.26	46.09[a] ± 20.77	39.94[a] ± 4.84	0.46[b] ± 0.02	0.52[b] ± 0.08	9.46[b] ± 1.05	36.58[a] ± 1.43	0.88[c] ± 0.25	1.03[c] ± 0.17	4.98[b] ± 0.54	20.89[a] ± 2.74	0.72[c] ± 0.3	0.83[c] ± 0.11
Ketones	57.04[a] ± 11.9	62.83[a] ± 0.92	67.84[a] ± 11.34	58.57[a] ± 4.4	58.36[b] ± 10.68	78.65[ab] ± 3.5	65.94[ab] ± 10.59	81.26[a] ± 2.19	105.58[a] ± 32.84	96.70[a] ± 1.86	80.20[a] ± 15.25	93.16[a] ± 2.97	1251.03[a] ± 64.69	865.58[b] ± 93.71	83.64[c] ± 7.9	85.64[c] ± 1.06	1894.70[a] ± 111.37	1603.86[b] ± 10.74	105.02[c] ± 7.19	82.11[d] ± 6.34
Esters	46.28[a] ± 18.4	62.21[a] ± 0.6	64.75[a] ± 18.6	61.29[a] ± 5.76	24.32[b] ± 7.23	58.56[a] ± 0.05	52.25[a] ± 7.23	57.46[a] ± 5.88	47.88[c] ± 11.75	77.24[b] ± 2.28	83.05[a] ± 1.55	86.36[a] ± 2.89	41.33d ± 3.68	178.18[a] ± 4.07	80.11[b] ± 0.06	75.04[c] ± 0.99	134.08[b] ± 7.24	187.41[a] ± 7.08	77.62[c] ± 14.99	103.54[c] ± 11.02
Alcohols	22.16[ab] ± 7.89	18.82[b] ± 4.0	27.75[a] ± 3.74	23.58[ab] ± 1.98	27.00[a] ± 2.92	13.68[c] ± 0.42	17.55[b] ± 1.53	17.36[b] ± 1.14	76.38[a] ± 25.79	115.65[a] ± 15.62	18.08[b] ± 0.52	18.16[b] ± 2.96	372.65[b] ± 21.01	629.53[a] ± 32.08	20.50[c] ± 5.29	30.49[c] ± 5.51	659.30[b] ± 60.15	1685.94[a] ± 76.51	17.51[d] ± 1.70	39.64[c] ± 8.6
Sulfur compounds	0.43[ab] ± 0.09	0.19[c] ± 0.14	0.52[a] ± 0.08	0.28[bc] ± 0.06	0.31[b] ± 0.03	0.41[a] ± 0.04	0.40[ab] ± 0.10	0.35[ab] ± 0.04	4.88[a] ± 1.13	0.51[b] ± 0.09	0.58[b] ± 0.02	0.64[b] ± 0.04	11.35[a] ± 3.38	0.54[b] ± 0.12	0.73[b] ± 0.07	0.80[b] ± 0.19	3.82[b] ± 0.03	3.90[b] ± 0.62	2.18[c] ± 0.15	7.51[a] ± 0.75
Acids	5.53[b] ± 1.89	3.29[b] ± 0.35	212.66[a] ± 33.12	275.92[a] ± 35.46	2.66[c] ± 0.76	4.42[b] ± 0.21	89.75[a] ± 16.42	85.54[a] ± 7.65	26.68[b] ± 19.02	17.95[b] ± 2.74	89.99[a] ± 8.93	114.94[a] ± 33.11	47.55[c] ± 10.53	80.32[bc] ± 26.1	164.21[a] ± 22.03	103.14[b] ± 4.07	126.86[a] ± 30.43	93.17[a] ± 6.01	139.67[a] ± 78.84	95.47[a] ± 8.14
Aliphatic hydrocarbons	59.35[a] ± 5.63	62.64[a] ± 4.3	61.53[a] ± 22.44	54.06[a] ± 5.32	0.00[c] ± 0.00	33.00[a] ± 0.63	43.37[a] ± 10.75	2.62[b] ± 0.17	1.35[b] ± 0.53	54.31[a] ± 15.58	82.17[a] ± 15.05	73.20[a] ± 10.46	17.59[b] ± 6.69	27.31[b] ± 4.39	80.26[a] ± 6.97	67.47[a] ± 7.15	14.20[a] ± 7.05	24.16[a] ± 6.65	33.80[a] ± 12.32	2.37[b] ± 0.03

T1, T4, T8, T12 and T16, stracciatella analyzed after 1, 4, 8, 12 and 16 days, respectively. Ctr, stracciatella control; Ex, stracciatella with olive leaf extract; So, stracciatella with sorbic acid; Ex-So, stracciatella with olive leaf extract and sorbic acid. Values in the rows with different superscripts at each sampling time differ at $p < 0.05$.

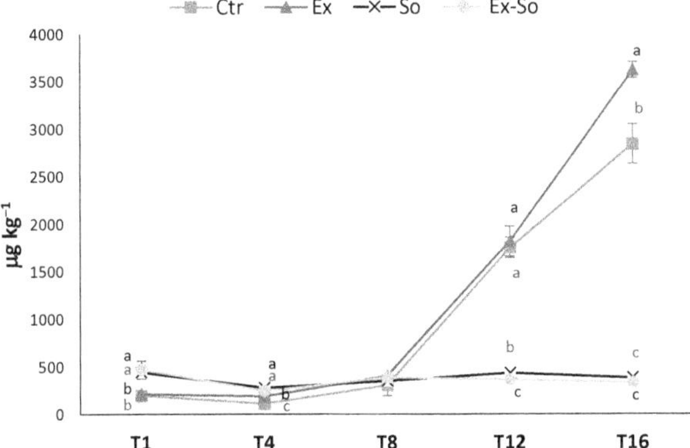

Figure 1. Total volatile compounds in stracciatella cheese during storage at 1 (T1), 4 (T4), 8 (T8), 12 (T12) and 16 (T16) days. Different letters above the lines indicate statistically different values ($p < 0.05$).

Table 3. Main volatile compounds ($\mu g\ kg^{-1}$) found at the end of shelf life (16 days) in stracciatella cheese (only compounds exceeding 1 μg are reported).

	Ctr	Ex	So	Ex-So	R
		Aldehydes			
Acetaldehyde	3.7 [b] ± 0.5	12.0 [a] ± 2.3	-	-	Ms,Std
3-methylbutanal	0.9 [b] ± 0.1	8.4 [a] ± 0.3	-	-	Ms,Std
		Ketones			
acetone	4.5 [b] ± 0.3	5.1 [b] ± 0.2	21.6 [a] ± 5.0	15.5 [a] ± 1.0	Ms,Std
2-butanone	15.9 [b] ± 0.9	55.0 [a] ± 2.4	19.4 [b] ± 2.8	19.0 [b] ± 0.8	Ms,Std
2-heptanone	19.7 [c] ± 0.6	22.4 [c] ± 0.8	49.9 [a] ± 4.1	32.9 [b] ± 2.5	Ms,Std
2-hydroxy-3-pentanone	11.8 [a] ± 1.6	5.1 [b] ± 0.6	-	-	Ms,Std
2-nonanone	2.4 [b] ± 0.2	2.8 [b] ± 0.2	5.0 [ab] ± 0.2	5.9 [a] ± 1.7	Ms,Std
acetoin	1839.5 [a] ± 109.0	1512.5 [b] ± 10.0	7.8 [c] ± 2.9	7.4 [c] ± 1.8	Ms,Std
		Esters			
ethyl acetate	37.5 [c] ± 0.8	107.5 [a] ± 5.8	68.3 [b] ± 13.8	81.8 [ab] ± 0.3	Ms,Std
butanoic acid, ethyl ester	93.5 [a] ± 7.7	75.0 [b] ± 1.3	2.8 [c] ± 0.6	13.0 [c] ± 1.3	Ms,Std
2-butenoic acid, ethyl ester (E)	0.5 [b] ± 0.1	1.4 [a] ± 0.1	0.4 [b] ± 0.0	-	Ms
hexanoic acid, ethyl ester	2.0 [a] ± 0.3	3.2 [a] ± 0.1	0.1 [b] ± 0.0	2.2 [a] ± 0.8	Ms,Std
sorbic acid, ethyl ester	-	-	6.0 [a] ± 0.5	6.5 [a] ± 1.9	Ms
		Alcohols			
ethanol	556.4 [b] ± 48.0	713.6 [a] ± 49.5	4.2 [c] ± 0.4	7.8 [c] ± 0.1	Ms,Std
1-propanol, 2 methyl-	9.7 [b] ± 0.6	76.3 [a] ± 1.7	3.2 [c] ± 0.3	-	Ms
1-butanol, 3-methyl	89.1 [b] ± 11.2	888.3 [a] ± 25.2	5.0 [c] ± 0.3	7.9 [c] ± 1.3	Ms,Std
1-pentanol	0.5 [b] ± 0.1	0.8 [b] ± 0.1	1.2 [b] ± 0.4	3.9 [a] ± 0.1	Ms,Std
4-methyl-2-hexanol	2.9 [b] ± 0.1	3.2 [b] ± 0.0	2.4 [b] ± 1.2	18.3 [a] ± 7.1	Ms
1-hexanol, 2-ethyl	0.4 [b] ± 0.1	0.9 [ab] ± 0.0	1.4 [a] ± 0.3	1.6 [a] ± 0.3	Ms
phenilethyl alcohol	0.1 [b] ± 0.0	2.7 [a] ± 0.1	0.1 [b] ± 0.0	0.1 [b] ± 0.0	Ms,Std
		Sulfur Compounds			
dimethyl sulfide	3.5 [b] ± 0.0	3.6 [b] ± 0.5	1.4 [c] ± 0.2	6.8 [a] ± 0.7	Ms,Std
		Acids			
acetic acid	19.6 [b] ± 8.4	52.2 [a] ± 5.4	3.4 [c] ± 0.2	10.1 [b] ± 3.1	Ms,Std
butanoic acid	43.9 [a] ± 12.2	14.6 [b] ± 0.3	5.8 [c] ± 3.4	5.9 [c] ± 1.5	Ms,Std
hexanoic acid	51.3 [a] ± 8.0	18.0 [b] ± 0.2	10.8 [c] ± 6.0	9.2 [c] ± 3.3	Ms,Std
octanoic acid	9.7 [a] ± 1.6	4.4 [b] ± 0.2	3.3 [b] ± 1.4	3.8 [b] ± 0.9	Ms,Std
propanoic acid, 2 methyl	0.1 [b] ± 0.0	1.4 [a] ± 0.0	-	-	Ms,Std
butanoic acid, 3-methyl	0.6 [b] ± 0.1	1.5 [a] ± 0.2	0.2 [bc] ± 0.0	-	Ms,Std
sorbic acid	-	-	115.9 [a] ± 68.3	66.3 [a] ± 0.7	Ms
		Aliphatic hydrocarbons			
1-heptene, 2,4-dimethyl-	14.2 [ab] ± 2.0	24.1 [ab] ± 6.7	38.8 [a] ± 12.3	2.3 [c] ± 0.1	Ms

Ctr, stracciatella control; Ex, stracciatella with olive leaf extract; So, stracciatella with sorbic acid; Ex-So, stracciatella with olive leaf extract and sorbic acid. R = identification method; Ms = mass spectrometer; Std = chemical standard. Values in the rows with different letters differ at $p < 0.05$.

3.2. Sensory Analysis and Possible Connections with VOC

Figure 2 shows the quantitative descriptive analysis (QDA) graphs divided into odor (A) and taste (B) perceptions. All fresh samples had almost the same olfactory characteristics until day 4, demonstrating the absence of any influence of the preservatives on the sensory profile. Starting from day 8, in perfect agreement with the evolution of VOC, a loss of the freshness characteristics was observed in Ctr and Ex. In particular, the overall intensity increased, and considering that stracciatella must present a mild and delicate aroma, it has to be considered a symptom of ongoing changes. This variation led to the loss of "frankness" defined as the characteristic aroma of the product; with time, undesired notes of sour milk, sourdough, banana and boiled cabbage were also perceived. These results matched well with the increase of some VOC compounds, such as butanoic acid, ethyl esters, ethanol, dimethyl sulphide and 3-methylbutanal. The same trend was also observed for Ex-So and So, but only starting from day 12. At day 16 Ctr and Ex samples were totally unacceptable, differently from Ex-So and So. This finding could be related to the strong increase of some VOC that can be considered as responsible of off-flavors. As to taste (B) no differences were observed among samples until day 8, then Ctr and Ex started to lose sweetness and to evidence bitterness and sourness; a slight increase in sourness was also perceive in So. The responsibility of these changes could be attributed to the formation of free hydrophobic amino acids deriving from proteolysis and/or increase of organic acids. At day 12, Ex and Ctr were rejected for excessive sourness and sweetness decreased in Ex-So and So, but they were still judged as acceptable. At day 16 also these latter samples were rejected for bad smell (they were not tasted). Figure 3 shows the GPA plot based on the dataset of odor sensory analysis. Here it is possible to observe the distance among samples on a bidimensional plot. The two dimensions account for 98.80% of the total variance. Most of the differences (93.08%) can be explained on F1, where So and Ex-So were positioned on the left side of the plot, correlated to fresh milk and frankness perceptions. Ex and Ctr were on the right side of the plot, very far between them and related to ripe fruit and sour yogurt respectively, and in both cases also with sourdough, sour milk and odor intensity. This information allows us to make connections between the VOC profile and sensory analyses: So and Ex-So resulted to be quite similar, while Ex and Ctr were far from them and strongly affected by off-flavors. Thus, using sorbic acid or a mixture of the two preservatives seems to not interfere negatively on the product.

The potentially involved molecules in stracciatella flavor were hypothesized by calculating the odor active value (OAV) of VOC according to Qian et al. [50], on the basis of the odor thresholds in air taken from the literature. The VOC having OAV > 1 are shown in Table 4. Ethyl-acetate exceeded a value of 1 in all samples at any time, and should be one of the key odorants in this cheese. It presents an ethereal and green note and was found at the highest level in Ex, where also its precursors, ethanol and acetic acid, were at the highest level than all samples. Hexanal (green grass odor) should only play a role in fresh cheese, since OAV > 1 was only found at day 1. Surprisingly, ethanol was only present at day 1 with OAV = 1.3 in So and Ex-So, but the reason was unclear. As expected, it was not newly formed in these samples due to the inhibition of yeasts, whereas it strongly increased in Ctr and Ex, reaching a maximum OAV value of 69.6 and 89.2 at day 16 respectively. Other two potentially aroma-active esters, ethyl butanoate and ethyl hexanoate responsible of banana and apple notes [51], increased faster in Ctr and Ex. In these two samples, also acetoin (buttery and creamy odor) could contribute to the aroma, whereas in So and Ex-So it remained under the perception threshold. At day 8, also 3-methylbutanal (ethereal and fruity odor, but acrid when present at high concentration) exceeded a value of 1 in Ex and Ctr.

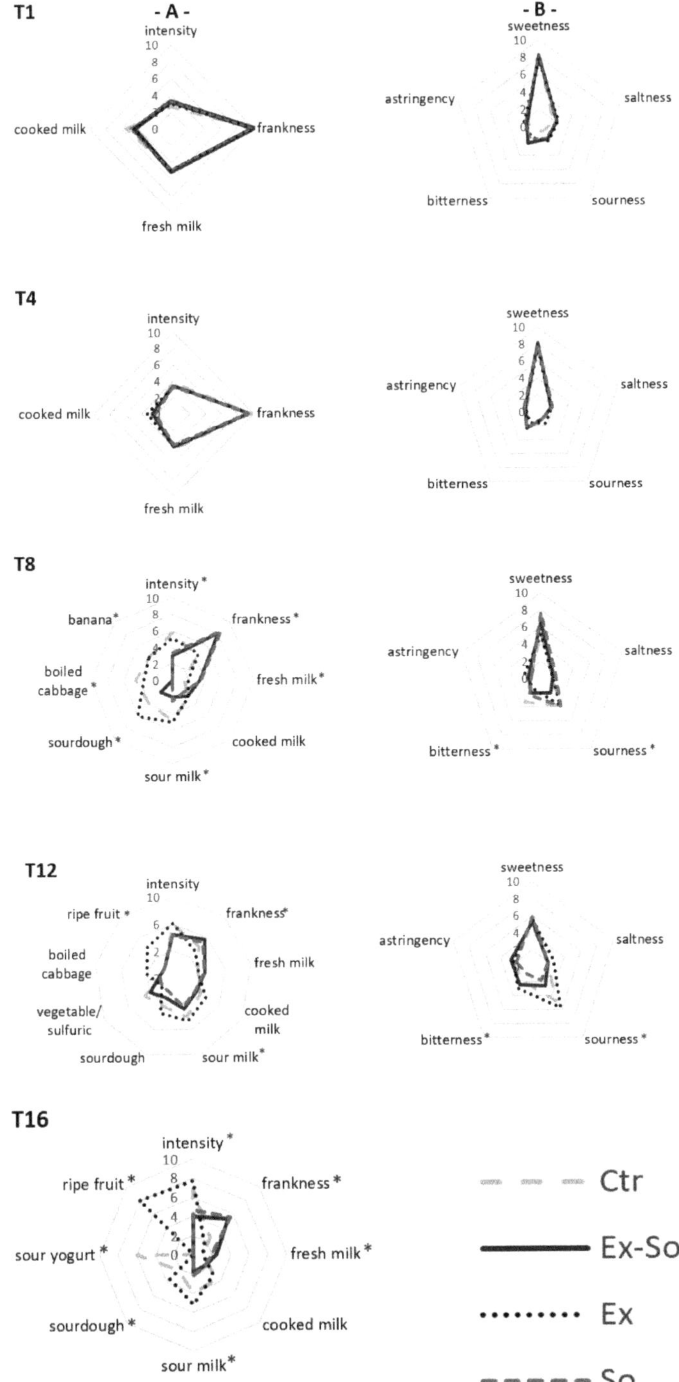

Figure 2. Quantitative descriptive analysis (QDA) of stracciatella odor (**A**) and taste perceptions (**B**). * = statistically different values ($p < 0.05$). Sampling time is the same as Figure 1.

Table 4. Volatile compounds with odor active value (OAV) > 1 during stracciatella cheese shelf life.

Volatile Compound	Ctr					Ex					So					Ex-So					Odor Description
	T1	T4	T8	T12	T16	T1	T4	T8	T12	T16	T1	T4	T8	T12	T16	T1	T4	T8	T12	T16	
3-Methylbutanal	-	4.5	228.2	14.2	4.5	-	-	197.2	180.5	42.3	-	-	-	-	-	-	-	-	-	-	Powerful penetrating, fusty, acrid, apple-like
Hexanal	2.0	-	-	-	-	1.6	1.0	-	-	-	2.0	2.5	-	-	-	-	1.2	-	-	-	Green, fatty, fruity
Nonanal	-	-	-	-	-	-	-	-	-	-	-	-	-	-	-	-	-	-	1.0	-	Sweet, orange, orange peel
2-Nonanone	-	-	1.2	-	-	-	-	-	1.1	-	-	-	-	-	1.0	-	-	-	-	1.2	Fresh sweet, weedy, earthy, herbal
Acetoin	-	-	-	1.5	2.3	-	-	-	1.0	1.9	-	-	-	-	-	-	-	-	-	-	Intense buttery, creamy
Ethyl acetate	9.3	4.9	8.3	5.1	7.5	12.4	11.7	12.0	27.1	21.5	12.1	10.1	15.2	14.3	13.7	11.5	10.8	15.5	13.2	16.4	Ethereal, fruity, green
Butanoic acid, ethyl ester	-	-	45.4	149.8	935.0	-	-	131.2	302.1	749.8	-	-	28.7	31.9	27.9	-	-	45.2	38.9	130.1	Pineapple
Hexanoic acid, ethyl ester	-	-	3.8	1.3	6.8	-	-	7.6	36.6	10.8	-	-	1.3	1.4	-	-	-	2.1	5.5	7.2	Floral, fruity, apple, banana, pineapple
Ethanol	-	1.0	2.4	35.1	69.6	-	-	5.6	29.5	89.2	1.3	-	-	-	-	1.3	-	-	-	1.0	Pleasant, weak, ethereal, vinous
1-Butanol, 3-methyl	-	-	-	1.2	1.3	-	-	-	5.3	12.5	-	-	-	-	-	-	-	-	-	-	Banana, alcohol, fruity
Phenilethyl alcohol	1.4	-	-	-	-	-	-	-	-	2.3	1.0	-	-	-	-	-	-	-	-	-	Characteristic rose-like
Dimethyl sulfide	-	-	14.3	36.6	11.8	-	-	-	-	11.9	-	-	-	-	4.7	-	-	-	-	22.7	Unpleasant odor of wild radish, cabbage-like

Ctr, stracciatella control; Ex, stracciatella with olive leaf extract; So, stracciatella with sorbic acid; Ex-So, stracciatella with olive leaf extract and sorbic acid. T1, T4, T8, T12 and T16, stracciatella analyzed after 1, 4, 8, 12 and 16 days, respectively.

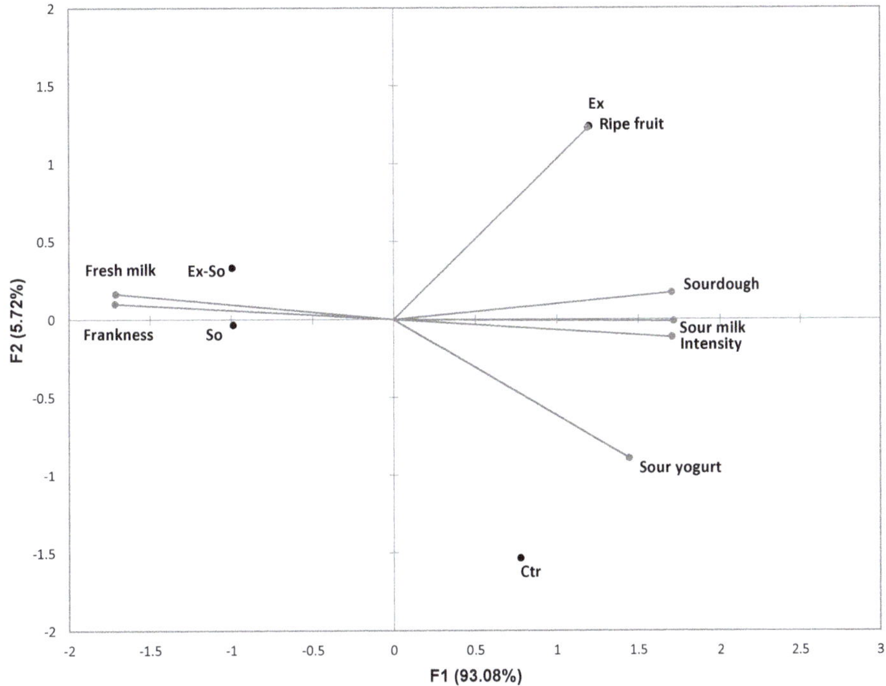

Figure 3. Generalized Procrustes analysis (GPA) of the dataset of the odor sensory analysis.

3.3. Chemical and Microbial Analyses

3.3.1. Antioxidant Activity and Color

As shown in Figure 4A, Ex and Ex-So samples had significantly higher antioxidant activity than Ctr and So throughout the entire storage period. Similar results were reported in studies on different food matrices [38,52]. The oxidative stability test carried out by RapidOxy gave different results (Figure 4B). No significant differences between samples were found at day 1, the So samples had the lowest induction time values after 4 and 8 days, and the Ex-So samples showed a higher value than Ctr at day 16, in accordance with antioxidant activity. These results could be due to the OLE effect, but the synergy with sorbic acid and interaction with the food matrix cannot be excluded. Overall, the results indicated that OLE, both added alone and together with sorbic acid, exerted an antioxidant activity in cheese. Different authors showed the antioxidant effect exerted by OLE in dairy products and according to literature the antioxidant power was lower in Ctr in a range of 40–80% and decreased during storage [53,54]. From this outcome it can be hypothesized that the higher presence of aldehydes in the VOC profile of OLE samples was mostly due to microbial activity rather than to fat oxidation.

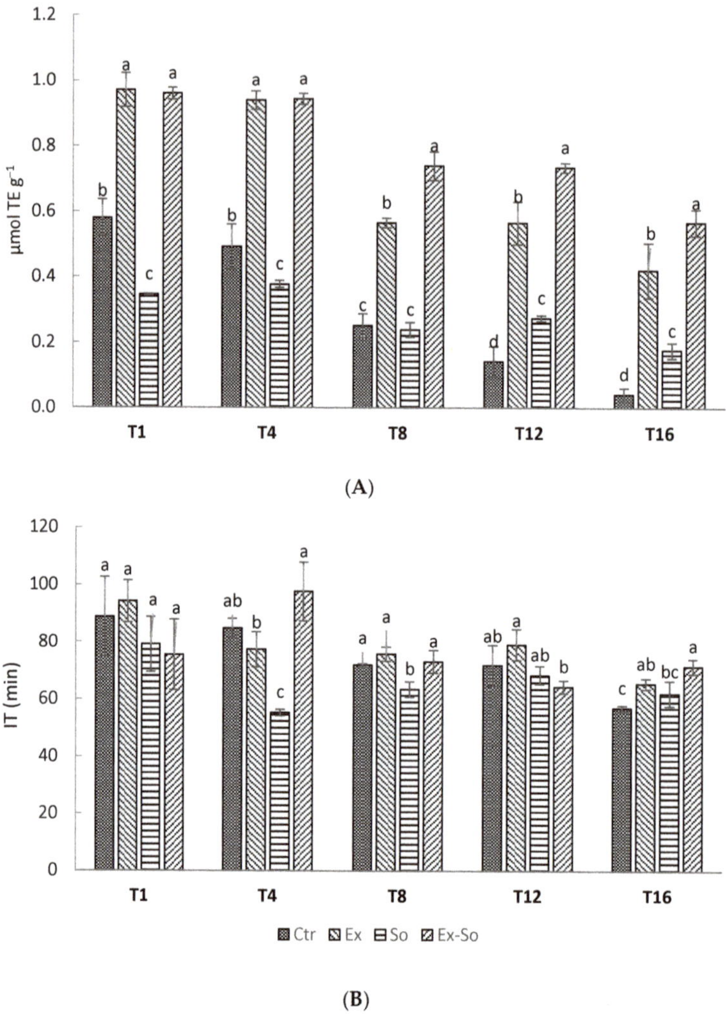

(A)

(B)

Figure 4. Antioxidant activity evaluation by means of ABTS-TEAC (2,2'-azino-bis(3-ethylbenzothiazoline-6-sulfonic acid)/Trolox®-Equivalent Antioxidant Capacity) (**A**) and oxidative stability measurement by RapidOxy (**B**). Sampling time is the same as Figure 1. Different letters at each sample indicate statistically different values ($p < 0.05$). Abbreviations: IT, Induction time.

As to color (Table 5), the values of some indices changed with time, but no significant differences were observed among samples taken at the same day. L^* index decreased, in accordance with other authors [31,36], and according to García-Pérez et al. [55] it should be connected to the acidification of the product. Differently, the a^* index slightly increased during storage, whereas b^* index showed a non-defined trend. The increase of a^* value could originate from polymerization of polyphenols and related tendency to browning [56].

Table 5. Color parameters (mean values and standard deviation) during stracciatella cheese shelf life.

Parameter	T1				T4				T8				T12				T16			
	Ctr	Ex	So	Ex-So	Ctr	Ex	So	Ex-So	Ctr	Ex	So	Ex-So	Ctr	Ex	So	Ex-So	Ctr	Ex	So	Ex-So
L^*	86.54 [a] ± 11.84	92.34 [a] ± 7.21	93.29 [a] ± 2.00	95.35 [a] ± 4.89	95.27 [a] ± 3.87	94.12 [a] ± 4.95	95.28 [a] ± 3.00	97.9 [a] ± 1.28	95.3 [a] ± 1.67	95.16 [a] ± 1.12	98.32 [a] ± 1.32	93.84 [a] ± 1.98	93.28 [a] ± 0.15	90.64 [a] ± 0.86	93.58 [a] ± 2.77	92.96 [a] ± 3.87	90.17 [a] ± 4.41	88.70 [a] ± 4.91	90.77 [a] ± 1.31	77.27 [b] ± 2.55
a^*	−0.21 [a] ± 0.13	−0.38 [a] ± 0.12	−0.16 [a] ± 0.03	−0.3 [a] ± 0.03	−0.08 [a] ± 0.17	−0.23 [a] ± 0.10	−0.14 [a] ± 0.01	−0.38 [a] ± 0.08	−0.06 [a] ± 0.20	−0.28 [a] ± 0.12	−0.12 [a] ± 0.01	−0.25 [a] ± 0.18	0.10 [a] ± 0.05	0.06 [a] ± 0.09	0.02 [a] ± 0.04	−0.05 [a] ± 0.10	0.21 [a] ± 0.05	0.10 [b] ± 0.10	0.02 [b] ± 0.01	0.26 [a] ± 0.20
b^*	8.70 [a] ± 0.49	8.67 [a] ± 0.10	8.36 [a] ± 0.21	8.93 [a] ± 0.57	7.52 [a] ± 0.12	8.51 [a] ± 0.39	9.08 [a] ± 0.91	8.16 [a] ± 0.69	8.84 [a] ± 0.16	10.0 [a] ± 0.17	8.31 [b] ± 0.46	8.79 [b] ± 0.21	8.79 [b] ± 0.06	9.97 [a] ± 0.43	9.08 [a] ± 0.99	9.28 [a] ± 0.52	8.67 [a] ± 0.10	9.17 [a] ± 0.50	9.17 [a] ± 0.25	7.61 [b] ± 0.76

T1, T4, T8, T12 and T16, stracciatella analyzed after 1, 4, 8, 12 and 16 days, respectively. Ctr, stracciatella control; Ex, stracciatella with olive leaf extract; So, stracciatella with sorbic acid; Ex-So, stracciatella with olive leaf extract and sorbic acid. Values in the rows with different superscripts at each sampling time differ at $p < 0.05$.

3.3.2. Organic Acids and pH

The changes in organic acids in dairy products are highly connected to microbial activities. In bottled milk and non-fermented products, as is stracciatella, they are undesired since they are responsible of off-flavors. From Figure 5, it can be observed that changes started early in Ctr and Ex samples: being a fermentable substrate, citric acid decreased from 0.91 and 0.93 mg kg^{-1} to 0.09 and 0.26 mg kg^{-1}, respectively; lactic acid increased from 0.12 and 0.13 mg kg^{-1} to 0.65 and to 1.05; acetic acid increased from 0.01 to 0.11 and to 0.24 mg kg^{-1}. There is no information concerning organic acids content of stracciatella cheese, but looking at other dairy products it is possible to confirm our results. In fact, citric acid is often involved in the Krebs cycle, and this justifies its reduction. Moreover, the increase in other organic acids (i.e., acetic and citric) is explained by higher microbial activities during storage [57–59]. According to Adda et al. [60] citric acid is used as a substrate by the microorganism to produce pyruvic and acetic acid, this finding clearly suits with our results. Differently, only slight variations were observed in the other samples containing sorbic acid or a mixture of the two preservatives. These results matched well with the decreasing trend of pH that went from 6.34 to 5.89 in Ctr and from 6.29 to 5.83 in Ex; in contrast, pH remained almost unchanged in So and Ex-So. The lower values in these latter samples found already at day 1 (6.02 and 6.14, respectively) were caused by the acidic properties of sorbic acid (pKa 4.76). These findings highlight a possible antimicrobial effect of olive leaf extract in synergy with sorbic acid.

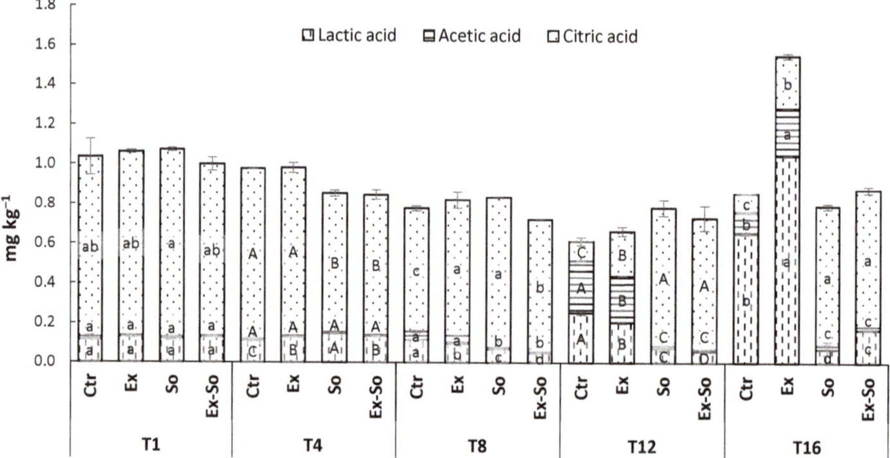

Figure 5. Trend of organic acids in stracciatella cheese during storage. Sampling time is the same as Figure 1. Different letters at each sample time for each acid indicate statistically different values ($p < 0.05$).

3.3.3. Cultivable Microbiota

The evolution of cultivable microbiota during cheese storage matched well with the results of the chemical and sensory analyses. It changed with different magnitudes in samples during storage, depending on the storage time and microbial group (Figure 6). The cell densities of mesophilic aerobic microorganisms ranged from about 4.1 (Ex and Ex-So) to 4.81 (Ctr) log CFU g^{-1} after 1 day of storage. At the end of shelf life, the values were higher than 7.0 log CFU g^{-1} in all samples, without significant differences ($p > 0.05$) between experimental and control. However, preservatives exerted an effect during storage, since the highest cell load was reached more rapidly (at 8 days) in Ctr than in Ex and Ex-So samples (at 16 days). Thus, the probability to find earlier VOC responsible for off-flavor in these samples was higher. Lactic acid bacteria (LAB) in stracciatella are considered altering agents, just like in bottled milk. The average cell densities of mesophilic lactobacilli and cocci at day 1 were 4.7 and 5.1 log CFU g^{-1}, respectively, while enterococci were about 2.6 log CFU g^{-1}, without significant differences among samples ($p > 0.05$). Lactobacilli and cocci remained almost constant during storage in So or in Ex-So samples, whereas their number increased up to 1 log cycle in Ctr and Ex cheeses at 16 days. Additionally, enterococci increased up to 1 log cycle at the end of storage, but only in Ctr (about 3.6 log CFU g^{-1}), whereas it decreased in all other samples. These findings suggest that OLE was effective in controlling enterococci when used alone, whereas it only had an effect on the other two LAB groups when associated to sorbic acid. Our results corroborate the findings of Roila et al. [61] and Servili et al. [62], who verified that the inhibitory effects of a polyphenol extract from an olive oil byproduct against lactic acid bacteria in mozzarella and a fermented milk beverage were scarce and dose-dependent. According to Hurtado et al. [63], some LAB strains can degrade oleuropein and metabolize specific compounds of OLE. In addition, the antimicrobial activity of the leaf extract in dairy products may be partially inhibited by the chemical interaction between phenolic hydroxyl groups and milk proteins [64]. *Pseudomonas* spp. are the most feared spoilage bacteria in fresh cheeses, since they can replicate at refrigeration conditions [65,66]. After 1 day of storage, the two cheeses containing OLE showed a lower ($p < 0.05$) cell density of *Pseudomonas* spp. than control. Successively, the counts increased in all samples reaching the maximum cell densities of about 6.3 log CFU g^{-1}. Despite the effect observed at the early stage, OLE did not work against these bacteria, both alone and in the presence of sorbic acid. Inefficacy could be due to two main reasons: (1) OLE concentration was too low to have an effect. In fact, antimicrobial activity of olive polyphenols against Gram-negative bacteria, such as *Escherichia coli* and *Pseudomonas*, was reported to be dose-dependent [61,67] and (2) pH of the cheese (6.0–6.2) was too high for allowing efficacy of sorbic acid: in cottage cheese it acted against *Pseudomonas* at pH 4.6–5.1 [68]. As for Enterobacteriaceae, after the first day of storage significant differences were observed between Ctr (3.8 log CFU g^{-1}) and So (3.7 log CFU g^{-1}) compared to Ex (1.7 log CFU g^{-1}) and Ex-So (<1 log CFU g^{-1}) samples. Successively, they disappeared more rapidly in the cheeses added with antimicrobials. This finding is of particular relevance, because Enterobacteriaceae are involved in spoilage phenomena in the dairy product. In accordance with VOC results, Enterobacteriaceae are known as producers of some VOC as 3-methyl-1-butanol, found only in samples without sorbic acid [49,69]. Staphylococci and molds were not detected, differently from yeasts. Overall, effectiveness of OLE in controlling yeasts was ascertained at early phases (day 4), successively it had an effect only in the presence of sorbic acid, confirming the usefulness of sorbic acid in controlling mold and yeast.

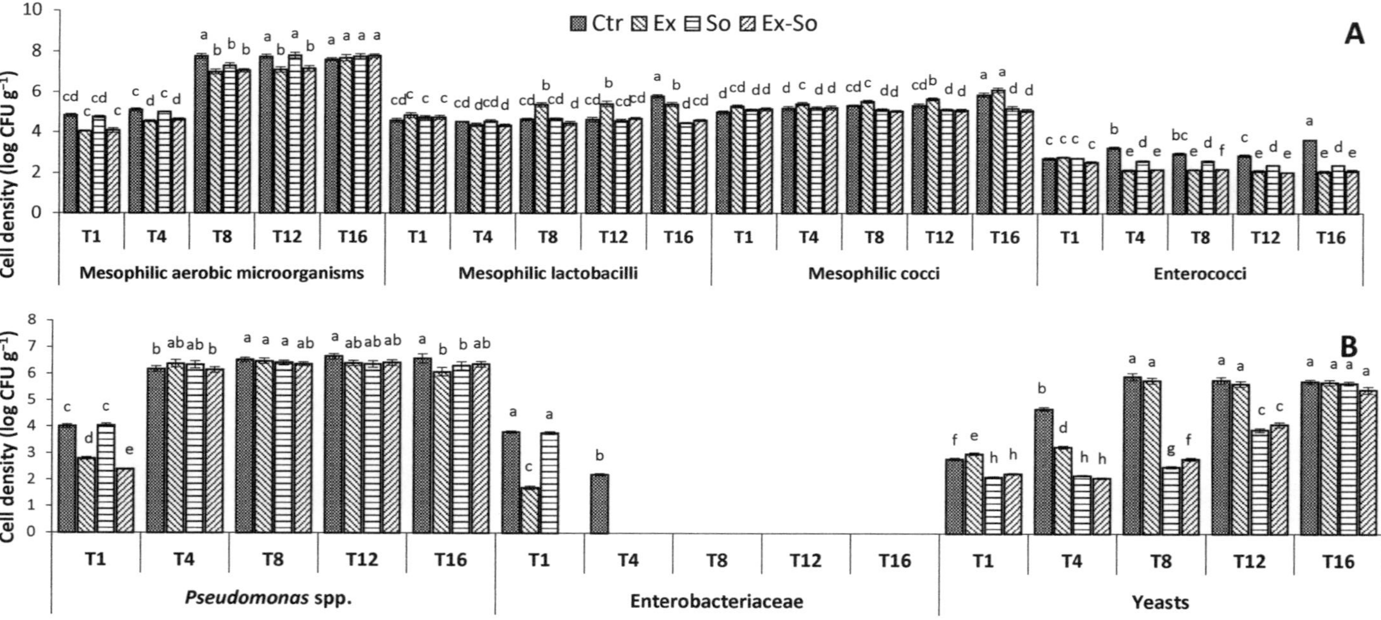

Figure 6. Cell numbers (log CFU g^{-1}, average values of 3 triplicates ± SD) of mesophilic aerobic microorganisms, presumptive mesophilic lactobacilli, mesophilic cocci and enterococci (**A**), *Pseudomonas* spp., Enterobacteriaceae and yeasts (**B**) in stracciatella samples. Ctr = control; Ex = added with olive leaf extract; So = added with sorbic acid; Ex-So = added with both sorbic acid and olive leaf extract. Sampling time is the same as Figure 1. Different letters for each microbial group indicate significant differences at $p < 0.05$.

4. Conclusions

The results of the present work pointed out the importance of relating the VOC profile and sensory results as a decay's symptom in fresh cheeses. In particular, VOC analysis could be a fundamental tool for predicting the shelf life of stracciatella, together with the classic microbiological and sensory assays. In addition, it could help to monitor the formation of some key volatiles responsible for an off-flavor and prevent rejection of the product by consumers in large-scale retails. The study allowed us to suitably evaluate the influence of the preservatives on the sensory characteristics. The samples added with sorbic acid or a combination of sorbic acid and olive leaf extract gave the best result as to the evolution of the VOC profile. It was reflected in better flavor preservation during storage, allowing us to better preserve both stracciatella's delicate aroma and taste. The color parameter results did not show any relevant difference among products taken at the same day, so the addition of preservatives does not influence the product aspect. Although stracciatella OLE-enriched samples reached higher antioxidant activity than Ctr and So, such performances were mostly connected to the well-known antimicrobial effect of sorbic acid, also further to a possible synergic effect with OLE, in particular against *Enterococcus* spp. and Enterobacteriaceae. Possibly better results might be reached also against other spoilage microorganisms by increasing the extract concentration, but the negative impact on the sensory characteristics appears to be a main limit. Nevertheless, the possible synergic effect could help producers to reduce the concentration of chemical preservatives. A further synergic effect could be obtained by slight lowering of pH of the cheese, compatibly with maintaining the flavor acceptable.

Author Contributions: G.N., G.D., F.C. and M.F. edited the 'Antioxidant activity and color', 'Organic acids and VOC' and 'Sensory characteristics' sections, whereas M.C. and G.C. edited the 'Cultivable microbiota' section. All authors have contributed in writing the original draft, reviewing and editing the manuscript. All authors have read and agreed to the published version of the manuscript.

Funding: This research was funded by the AGER 2 Project, grant no. 2016-0105.

Conflicts of Interest: The authors declare that they have no known competing financial interests or personal relationships that could have appeared to influence the work reported in this paper.

References

1. Ghosh, P.R.; Fawcett, D.; Sharma, S.B.; Poinern, G.E.J. Progress towards sustainable utilisation and management of food wastes in the global economy. *Int. J. Food Sci.* **2016**, *2016*. [CrossRef] [PubMed]

2. Djekic, I.; Miloradovic, Z.; Djekic, S.; Tomasevic, I. Household food waste in Serbia—Attitudes, quantities and global warming potential. *J. Clean. Prod.* **2019**, *229*, 44–52. [CrossRef]

3. Tylewicz, U.; Inchingolo, R.; Rodriguez-Estrada, M.T. *Food Aroma Compounds*; Elsevier Inc.: Cambridge, MA, USA, 2017; ISBN 9780128052570.

4. Shahidi, F. Indicators for evaluation of lipid oxidation and off-flavor development in food. *Dev. Food Sci.* **1998**, *40*, 55–68. [CrossRef]

5. Wilkes, J.G.; Conte, E.D.; Kim, Y.; Holcomb, M.; Sutherland, J.B.; Miller, D.W. Sample preparation for the analysis of flavors and off-flavors in foods. *J. Chromatogr. A* **2000**, *880*, 3–33. [CrossRef]

6. Wilson, A.D.; Oberle, C.S.; Oberle, D.F. Detection of off-flavor in catfish using a conducting polymer electronic-nose technology. *Sensors* **2013**, *13*, 15968–15984. [CrossRef]

7. Wang, S.; Chen, H.; Sun, B. Recent progress in food flavor analysis using gas chromatography-ion mobility spectrometry (GC-IMS). *Food Chem.* **2020**, *315*, 126158. [CrossRef]

8. Vazquez-Landaverde, P.A.; Velazquez, G.; Torres, J.A.; Qian, M.C. Quantitative determination of thermally derived off-flavor compounds in milk using solid-phase microextraction and gas chromatography. *J. Dairy Sci.* **2005**, *88*, 3764–3772. [CrossRef]

9. Natrella, G.; Faccia, M.; Lorenzo, J.M.; De Palo, P.; Gambacorta, G. Short communication: Sensory characteristics and volatile organic compound profile of high-moisture mozzarella made by traditional and direct acidification technology. *J. Dairy Sci.* **2020**, *103*. [CrossRef]

10. Ercan, D.; Korel, F.; Karagül Yüceer, Y.; Kınık, Ö. Physicochemical, textural, volatile, and sensory profiles of traditional Sepet cheese. *J. Dairy Sci.* **2011**, *94*, 4300–4312. [CrossRef]

11. Wolf, I.V.; Perotti, M.C.; Zalazar, C.A. Composition and volatile profiles of commercial Argentinean blue cheeses. *J. Sci. Food Agric.* **2011**, *91*, 385–393. [CrossRef]

12. Cheng, H. Volatile flavor compounds in yogurt: A review. *Crit. Rev. Food Sci. Nutr.* **2010**, *50*, 938–950. [CrossRef] [PubMed]

13. Karatapanis, A.E.; Badeka, A.V.; Riganakos, K.A.; Savvaidis, I.N.; Kontominas, M.G. Changes in flavour volatiles of whole pasteurized milk as affected by packaging material and storage time. *Int. Dairy J.* **2006**, *16*, 750–761. [CrossRef]

14. D'Incecco, P.; Limbo, S.; Hogenboom, J.; Rosi, V.; Gobbi, S.; Pellegrino, L. Impact of extending hard-cheese ripening: A multiparameter characterization of Parmigiano reggiano cheese ripened up to 50 months. *Foods* **2020**, *9*, 268. [CrossRef] [PubMed]

15. Natrella, G.; Gambacorta, G.; Faccia, M. Volatile organic compounds throughout the manufacturing process of Mozzarella di Gioia del Colle PDO cheese. *Czech J. Food Sci.* **2020**, *38*, 215–222. [CrossRef]

16. Munekata, P.E.S.; Domínguez, R.; Franco, D.; Bermúdez, R.; Trindade, M.A.; Lorenzo, J.M. Effect of natural antioxidants in Spanish salchichón elaborated with encapsulated n-3 long chain fatty acids in konjac glucomannan matrix. *Meat Sci.* **2017**, *124*, 54–60. [CrossRef]

17. Angerosa, F.; Servili, M.; Selvaggini, R.; Taticchi, A.; Esposto, S.; Montedoro, G. Volatile compounds in virgin olive oil: Occurrence and their relationship with the quality. *J. Chromatogr. A* **2004**, *1054*, 17–31. [CrossRef]

18. Lorenzo, J.M.; Carballo, J.; Franco, D. Effect of the inclusion of chestnut in the finishing diet on volatile compounds of dry-cured ham from celta pig breed. *J. Integr. Agric.* **2013**, *12*, 2002–2012. [CrossRef]

19. Tian, H.; Xu, X.; Chen, C.; Yu, H. Flavoromics approach to identifying the key aroma compounds in traditional Chinese milk fan. *J. Dairy Sci.* **2019**. [CrossRef]

20. Gracka, A.; Jeleń, H.H.; Majcher, M.; Siger, A.; Kaczmarek, A. Flavoromics approach in monitoring changes in volatile compounds of virgin rapeseed oil caused by seed roasting. *J. Chromatogr. A* **2016**, *1428*, 292–304. [CrossRef]

21. CLAL. EU-28: Dairy Sector. Available online: https://www.clal.it/en/?section=quadro_europa (accessed on 30 September 2020).

22. Altieri, C.; Scrocco, C.; Sinigaglia, M.; Del Nobile, M.A. Use of chitosan to prolong mozzarella cheese shelf life. *J. Dairy Sci.* **2005**, *88*, 2683–2688. [CrossRef]

23. Evert-Arriagada, K.; Hernández-Herrero, M.M.; Juan, B.; Guamis, B.; Trujillo, A.J. Effect of high pressure on fresh cheese shelf-life. *J. Food Eng.* **2012**, *110*, 248–253. [CrossRef]

24. Ricciardi, E.F.; Lacivita, V.; Conte, A.; Chiaravalle, E.; Zambrini, A.V.; Del Nobile, M.A. X-ray irradiation as a valid technique to prolong food shelf life: The case of ricotta cheese. *Int. Dairy J.* **2019**, *99*, 104547. [CrossRef]

25. Di Pierro, P.; Sorrentino, A.; Mariniello, L.; Giosafatto, C.V.L.; Porta, R. Chitosan/whey protein film as active coating to extend Ricotta cheese shelf-life. *LWT Food Sci. Technol.* **2011**, *44*, 2324–2327. [CrossRef]

26. Gammariello, D.; Conte, A.; Attanasio, M.; Del Nobile, M.A. A study on the synergy of modified atmosphere packaging and chitosan on stracciatella shelf life. *J. Food Process Eng.* **2011**, *34*, 1394–1407. [CrossRef]

27. Trani, A.; Gambacorta, G.; Gomes, T.F.; Loizzo, P.; Cassone, A.; Faccia, M. Production and characterisation of reduced-fat and PUFA-enriched Burrata cheese. *J. Dairy Res.* **2016**, *83*, 236–241. [CrossRef] [PubMed]

28. Gammariello, D.; Conte, A.; Di Giulio, S.; Attanasio, M.; Del Nobile, M.A. Shelf life of Stracciatella cheese under modified-atmosphere packaging. *J. Dairy Sci.* **2009**, *92*, 483–490. [CrossRef]

29. Conte, A.; Brescia, I.; Del Nobile, M.A. Lysozyme/EDTA disodium salt and modified-atmosphere packaging to prolong the shelf life of burrata cheese. *J. Dairy Sci.* **2011**, *94*, 5289–5297. [CrossRef]

30. Dambrosio, A.; Quaglia, N.C.; Saracino, M.; Malcangi, M.; Montagna, C.; Quinto, M.; Lorusso, V.; Normanno, G. Microbiological quality of burrata cheese produced in Puglia region: Southern Italy. *J. Food Prot.* **2013**, *76*, 1981–1984. [CrossRef]

31. Rea, S.; Marino, L.; Stocchi, R.; Branciari, R.; Loschi, A.R.; Miraglia, D.; Ranucci, D. Differences in chemical, physical and microbiological characteristics of Italian Burrata cheeses made in artisanal and industrial plants of the Apulia region. *Ital. J. Food Saf.* **2016**, *5*. [CrossRef]

32. Tavakoli, H.; Hosseini, O.; Jafari, S.M.; Katouzian, I. Evaluation of physicochemical and antioxidant properties of yogurt enriched by olive leaf phenolics within nanoliposomes. *J. Agric. Food Chem.* **2018**, *66*, 9231–9240. [CrossRef]

33. Palmeri, R.; Parafati, L.; Trippa, D.; Siracusa, L.; Arena, E.; Restuccia, C.; Fallico, B. Addition of olive leaf extract (OLE) for producing fortified fresh pasteurized milk with an extended shelf life. *Antioxidants* **2019**, *8*, 255. [CrossRef] [PubMed]

34. Roila, R.; Valiani, A.; Ranucci, D.; Ortenzi, R.; Servili, M.; Veneziani, G.; Branciari, R. Antimicrobial efficacy of a polyphenolic extract from olive oil by-product against "Fior di latte" cheese spoilage bacteria. *Int. J. Food Microbiol.* **2019**, *295*, 49–53. [CrossRef] [PubMed]

35. Ranieri, M.; Di Mise, A.; Difonzo, G.; Centrone, M.; Venneri, M.; Pellegrino, T.; Russo, A.; Mastrodonato, M.; Caponio, F.; Valenti, G.; et al. Green olive leaf extract (OLE) provides cytoprotection in renal cells exposed to low doses of cadmium. *PLoS ONE* **2019**, *14*, e0214159. [CrossRef] [PubMed]

36. Caponio, F.; Difonzo, G.; Calasso, M.; Cosmai, L.; De Angelis, M. Effects of olive leaf extract addition on fermentative and oxidative processes of table olives and their nutritional properties. *Food Res. Int.* **2019**, *116*, 1306–1317. [CrossRef] [PubMed]

37. Difonzo, G.; Squeo, G.; Calasso, M.; Pasqualone, A.; Caponio, F. Physico-chemical, microbiological and sensory evaluation of ready-to-use vegetable pâté added with olive leaf extract. *Foods* **2019**, *8*, 138. [CrossRef]

38. European Commission. Commission regulation (EU) No 1129/2011 of 11 November 2011 amending Annex II to Regulation (EC) No 1333/2008 of the European Parliament and of the Council by establishing Union list of food additives. *Off. J. Eur. Union* **2011**, *295*, 1–177.

39. Difonzo, G.; Pasqualone, A.; Silletti, R.; Cosmai, L.; Summo, C.; Paradiso, V.M.; Caponio, F. Use of olive leaf extract to reduce lipid oxidation of baked snacks. *Food Res. Int.* **2018**, *108*, 48–56. [CrossRef]

40. Faccia, M.; Trani, A.; Natrella, G.; Gambacorta, G. Short communication: Chemical-sensory and volatile compound characterization of ricotta forte, a traditional fermented whey cheese. *J. Dairy Sci.* **2018**, *101*, 5751–5757. [CrossRef]

41. CIE, International Commission on Illumination. *Survey of Reference Materials for Testing the Performance of Spectrophotometers and Colorimeters*; Publication CIE No. 114.1; Central Bureau of the CIE: Vienna, Austria, 1994.

42. Buffa, M.; Guamis, B.; Saldo, J.; Trujillo, A.J. Changes in organic acids during ripening of cheeses made from raw, pasteurized or high-pressure-treated goats' milk. *LWT Food Sci. Technol.* **2004**, *37*, 247–253. [CrossRef]

43. Gambera, A. Metodica di assaggio dei formaggi. In *L'assaggio dei Formaggi*, 2nd ed.; Gambera, A., Ed.; Comunecazione Snc: Bra, Italy, 2017; pp. 9–23.

44. Minervini, F.; Conte, A.; Del Nobile, M.A.; Gobbetti, M.; De Angelis, M. Dietary fibers and protective lactobacilli drive burrata cheese microbiome. *Appl. Environ. Microbiol.* **2017**, *83*, e01494-17. [CrossRef]

45. McSweeney, P.L.; Fox, P.F. *Advanced Dairy Chemistry*, 4th ed.; Springer: Boston, MA, USA, 2003. [CrossRef]

46. Fröhlich-Wyder, M.T.; Bisig, W.; Guggisberg, D.; Jakob, E.; Turgay, M.; Wechsler, D. Cheeses with propionic acid fermentation. In *Cheese: Chemistry, Physics and Microbiology*, 4th ed.; McSweeney, P.L.H., Fox, P.F., Cotter, P.D., Everett, D.W., Eds.; Elsevier Academic Press: Amsterdam, The Netherlands, 2017; Volume 2, pp. 889–910. [CrossRef]

47. Natrella, G.; Gambacorta, G.; De Palo, P.; Lorenzo, J.M.; Faccia, M. Evolution of volatile compounds from milk to curd during manufacturing of mozzarella. *Mljekarstvo* **2020**, *70*. [CrossRef]

48. Ziino, M.; Condurso, C.; Romeo, V.; Giuffrida, D.; Verzera, A. Characterization of "Provola dei Nebrodi", a typical Sicilian cheese, by volatiles analysis using SPME-GC/MS. *Int. Dairy J.* **2005**, *15*, 585–593. [CrossRef]

49. Morales, P.; Feliu, I.; Fernández-García, E.; Nuñez, M. Volatile compounds produced in cheese by Enterobacteriaceae strains of dairy origin. *J. Food Prot.* **2004**, *67*, 567–573. [CrossRef] [PubMed]

50. Qian, M.; Reineccius, G.A. Quantification of aroma compounds in parmigiano reggiano cheese by a dynamic headspace gas chromatography-mass spectrometry technique and calculation of odor activity value. *J. Dairy Sci.* **2003**, *86*, 770–776. [CrossRef]

51. Wolf, I.V.; Perotti, M.C.; Bernal, S.M.; Zalazar, C.A. Study of the chemical composition, proteolysis, lipolysis and volatile compounds profile of commercial Reggianito Argentino cheese: Characterization of Reggianito Argentino cheese. *Food Res. Int.* **2010**, *43*, 1204–1211. [CrossRef]

52. Nunes, M.A.; Pimentel, F.B.; Costa, A.S.G.; Alves, R.C.; Oliveira, M.B.P.P. Olive by-products for functional and food applications: Challenging opportunities to face environmental constraints. *Innov. Food Sci. Emerg. Technol.* **2016**, *35*, 139–148. [CrossRef]

53. Peker, H.; Arslan, S. Effect of olive leaf extract on the quality of low fat apricot yogurt. *J. Food Process. Preserv.* **2017**, *41*, e13107. [CrossRef]

54. Cho, W.Y.; Kim, D.H.; Lee, H.J.; Yeon, S.J.; Lee, C.H. Quality characteristic and antioxidant activity of yogurt containing olive leaf hot water extract. *CyTA J. Food* **2020**, *18*, 43–50. [CrossRef]

55. García-Pérez, F.J.; Lario, Y.; Fernández-López, J.; Sayas, E.; Pérez-Alvarez, J.A.; Sendra, E. Effect of orange fiber addition on yogurt color during fermentation and cold storage. *Color Res. Appl.* **2005**, *30*, 457–463. [CrossRef]

56. Farrag, A.F.; Bayoumi, H.M.; Ibrahim, W.A.; El-Sheikh, M.M.; Eissa, H.A. Characteristics of white soft cheese fortified with hibiscus soft drink as antimicrobial and hypertension treatment. *Science* **2017**, *12*, 122–129. [CrossRef]

57. Zeppa, G.; Rolle, L. A study on organic acid, sugar and ketone contents in typical piedmont cheeses. *Ital. J. Food Sci.* **2008**, *20*, 127–139.

58. Akalin, A.S.; Gönç, S.; Akbaş, Y. Variation in organic acids content during ripening of pickled white cheese. *J. Dairy Sci.* **2002**, *85*, 1670–1676. [CrossRef]

59. Manolaki, P.; Katsiari, M.C.; Alichanidis, E. Effect of a commercial adjunct culture on organic acid contents of low-fat Feta-type cheese. *Food Chem.* **2006**, *98*, 658–663. [CrossRef]

60. Adda, J.; Gripon, J.C.; Vassal, L. The chemistry of flavour and texture generation in cheese. *Food Chem.* **1982**, *9*, 115–129. [CrossRef]

61. Roila, R.; Branciari, R.; Ranucci, D.; Ortenzi, R.; Urbani, S.; Servili, M.; Valiani, A. Antimicrobial activity of olive mill wastewater extract against Pseudomonas fluorescens isolated from mozzarella cheese. *Ital. J. Food Saf.* **2016**, *5*, 111–115. [CrossRef] [PubMed]

62. Servili, M.; Rizzello, C.G.; Taticchi, A.; Esposto, S.; Urbani, S.; Mazzacane, F.; Di Maio, I.; Selvaggini, R.; Gobbetti, M.; Di Cagno, R. Functional milk beverage fortified with phenolic compounds extracted from olive vegetation water, and fermented with functional lactic acid bacteria. *Int. J. Food Microbiol.* **2011**, *147*, 45–52. [CrossRef]

63. Hurtado, A.; Reguant, C.; Bordons, A.; Rozès, N. Lactic acid bacteria from fermented table olives. *Food Microbiol.* **2012**, *31*, 1–8. [CrossRef]

64. Von Staszewski, M.; Pilosof, A.M.R.; Jagus, R.J. Antioxidant and antimicrobial performance of different Argentinean green tea varieties as affected by whey proteins. *Food Chem.* **2011**, *125*, 186–192. [CrossRef]

65. Liao, C.H. Pseudomonas and related genera. In *Food Spoilage Microorganisms*, 1st ed.; Blackburn, C.W., Ed.; Woodhead Publishing: Boca Raton, FL, USA, 2006; pp. 507–540. [CrossRef]

66. Faccia, M.; Gambacorta, G.; Natrella, G.; Caponio, F. Shelf life extension of Italian mozzarella by use of calcium lactate buffered brine. *Food Control* **2019**, *100*, 287–291. [CrossRef]

67. Markín, D.; Duek, L.; Berdícevsky, I. In vitro antimicrobial activity of olive leaves. *Mycoses* **2003**, *46*, 132–136. [CrossRef]

68. Brocklehurst, T.F.; Lund, B.M. Microbiological changes in cottage cheese varieties during storage at +7 °C. *Food Microbiol.* **1985**, *2*, 207–233. [CrossRef]

69. Lucera, A.; Mastromatteo, M.; Conte, A.; Zambrini, A.V.; Faccia, M.; Del Nobile, M.A. Effect of active coating on microbiological and sensory properties of fresh mozzarella cheese. *Food Packag. Shelf Life* **2014**, *1*, 25–29. [CrossRef]

 *foods*

Communication

Heat- and Ultrasound-Assisted Aqueous Extraction of Soluble Carbohydrates and Phenolics from Carob Kibbles of Variable Size and Source Material

Chrystalla Antoniou [1], Angelos Kyratzis [1], Youssef Rouphael [2], Stelios Stylianou [3] and Marios C. Kyriacou [1,*]

[1] Agricultural Research Institute, P.O. Box 22016, 1516 Nicosia, Cyprus; chrystalla.antoniou@ari.gov.cy (C.A.); a.kyratzis@ari.gov.cy (A.K.)

[2] Department of Agricultural Sciences, University of Naples Federico II, 80055 Portici, Italy; youssef.rouphael@unina.it

[3] Mellona, P.O. Box 54097-CY3720, 4151 Limassol, Cyprus; stelios@mellona.com.cy

* Correspondence: m.kyriacou@ari.gov.cy; Tel.: +357-22403221

Received: 27 August 2020; Accepted: 22 September 2020; Published: 25 September 2020

Abstract: Aqueous extraction of carob kibbles is the fundamental step in the production of carob juice and carob molasses. Improving the theoretical yield in sugars during organic solvent-free aqueous extraction is of prime interest to the food industry. Collateral extraction of phenolics, however, must be monitored as it influences the sensory and functional profile of carob juice. We presently examined the impact of source material, kibble size, temperature, and duration on the efficiency of extracting sugars and phenolics aqueously by conventional heat-assisted (HAE) and ultrasound-assisted (UAE) methods. Source material was the most influential factor determining the concentration of phenolics extracted by either method. Source material also influenced the relative proportions of sucrose, glucose, and fructose, which may impact the perceived sweetness of the juice. Kibble size (medium size M = 9–13 mm; powder size P = 1–4 mm) was more influential with UAE than HAE for both sugars and phenolics but was rendered less influential with prolonged UAE duration. Increasing HAE temperature (from 25 °C to 75 °C) favored the extraction of phenolics over sugars; however, prolonging extraction at 25 °C improved sugar yield without excessive yield in phenolics. Disproportionate extraction of phenolics over sugars limits the use of heat-assisted extraction to improve sugar yield in carob juice production and may shift the product's sensory profile toward astringency. Prolonged extraction at near ambient temperature can, however, improve sugar yield, keeping collateral extraction of phenolics low. Ultrasound agitation constitutes an effective means of extracting sugars from powder-size kibbles. Industrial application of both methodologies depends on the targeted functional and sensory properties of carob juice.

Keywords: carob kibbles; carob juice; aqueous extraction; sugars; phenolics

1. Introduction

Carob is an evergreen species (*Ceratonia siliqua* L.) of the *Fabaceae* family. It is naturally self-propagated and cultivated in Cyprus as well as other countries of the Mediterranean basin, preferring mild and dry habitats [1]. In the last 5–10 years, carob cultivation has attracted renewed interest from growers driven by growing global demand for carob-based food products. According to the Food and Agriculture Organization statistics, the leader in the production of carob in 2018 was Portugal with 41,734 tons, followed by Italy, Morocco, Turkey, Greece, Cyprus, Algeria, Spain, and Lebanon [2]. The low-input cultivation practices required for carob production in combination with the rich bioactive constitution of its fruit renders carob a sustainable crop warranting further investigation.

The carob fruit is a pod consisting of about 90% pulp and 10% seeds by weight at full maturity [3]. Carob seeds are in fact considered the most valuable part of the pods exploited industrially for the production of carob bean gum (locust bean gum, LBG), a widely used natural food additive [4]. Separation of the seeds from the pod by milling also delivers carob pulp kibbles as a low-cost by-product. Carob pulp kibbles have been used primarily as livestock feed, as cacao substitute in the confectionery and beverage industries, and as a regional culinary ingredient mostly in the Mediterranean basin [5,6]. In recent times, increasing attention has been given to products of carob pulp kibbles such as carob powder, fiber, juice, and molasses used by the food industry for developing a wide range of health-promoting or niche food products, including gluten-free ones [1,6]. Many studies have demonstrated the health-promoting effects of these products, notably their contribution to the prevention of colon cancer and hepatocellular carcinoma, wide antioxidant properties, reduction of diarrheal symptoms, and lowering of Low Density Lipoprotein cholesterol, as well as antidiabetic effects [7–10].

Carob kibbles are typically characterized by high sugar content (48–56% d.w.), crude fiber (around 40% d.w.), protein (2–7% d.w.), ash (2–3% d.w.), total polyphenols (1.4–2.0% d.w.), and low fat content (0.5–1% d.w.) [1,7,11]. Aqueous extraction of carob kibbles is performed conventionally at ambient conditions to deliver a juice rich in water-soluble sugars and phenolics [12]. Carob juice is a natural energy drink, widely consumed in Egypt, shown to have comparable sensory properties to grape juice, but higher phenolic content [13]. Carob juice can blend with other fruit juices to increase the nutraceutical value of the product due to the antioxidant and antimicrobial properties of polyphenols present in the carob juice [13–15]. The extraction of juice high in sugars is the first step in the industrial production of molasses (carob syrup) which is a concentrate of carob juice (65–80 °Brix) produced by slow simmering, without the addition of sugars or other additives [16]. Molasses is a nutritious, energy-rich, and healthy product that can be directly consumed or used as a functional ingredient alternative to sugar in the food and pharmaceutical industries [16,17]. The high sugar content of molasses renders it suitable for flavoring yogurt [18], while it is also used as a nutrient medium for *Aspergillus niger*, *Lactobacillus casei*, and the yeast *Saccharomyces cerevisiae* to produce citric acid and fermentation-derived lactic acid and bioethanol [1].

Several studies have been conducted to characterize the chemical composition of carob kibbles and to investigate the nutritional value and beneficial properties of kibble products [1,7,19–22]. To the best of our knowledge, however, very few works have examined the efficiency of organic solvent-free and time-efficient methodologies in extracting sugar-rich natural juice from carob kibbles, with the outlook of being applied on an industrial scale [19,23]. While the principal interest in such methods lies mainly in their sugar extraction efficiency, their yield in phenolic compounds must also be monitored as they may influence the oxidative stability and potential nutraceutical value but also the sensory profile of the products [13,24]. In this respect, the present short study examined the compositional differences in terms of soluble carbohydrates and phenolics of carob juice produced using two aqueous and environmentally friendly methods easy to apply at industrial scale extraction: 1. aqueous extraction under gentle agitation (conventional method—HAE) and 2. ultrasonic-assisted extraction (UAE). The HAE is traditionally applied by the industry for producing carob molasses. UAE, on the other hand, was examined briefly as a potential means of accelerating the extraction process. With each methodology, variable extraction parameters were appraised, including source material, kibble size, extraction duration, and temperature, in order to assess their comparative effect on the efficiency of extracting sugars and phenolics.

2. Materials and Methods

2.1. Plant Material

Fully mature carob pods were harvested from two principal phenotypes of local carob (*Ceratonia siliqua* L.) landrace: Lefkaritiki (LF) and Mavroteratsia (MV). These two phenotypes produce carobs of slightly different morphology although they are genetically proximate [25]. Carobs of Mavroteratsia were collected from a unitary source at low altitude (15 m) in the coastal zone (Zygi), while carobs of Lefkaritiki were collected from an inland unitary source at higher altitude (510 m) (Lefkara). As phenotype is the result of genotype-environment interaction, for the purposes of this study, we use the term "source material" to identify the lower-altitude material of Mavroteratsia and the higher-altitude material of Lefkaritiki.

2.2. Carob Juice Extraction

Carob pods were washed under tap water to remove debris, rinsed with deionized water, and patted dry. Then the pods were coarsely ground in a Vita Prep 3 blender (Vita-Mix Corp., Cleveland, OH, USA) operated at low speed and deseeded. The carob pulp was further ground to deliver kibbles of medium size (M = 9–13 mm) and powder size (P = 1–4 mm). Both sizes were placed in 50 mL falcon tubes and covered in distilled water (c. kibbles $\approx$ 15.3 g/H_2O $\approx$ 26.1 g) to obtain a kibbles/water mass ratio of near 0.6. Extraction was performed either by a conventional heat-assisted method (HAE) or an ultrasonic-assisted method (UA). The HAE extraction was performed by incubating the tubes in an OLS200 Grant Instruments water bath (Cambridgeshire, UK) under gentle agitation (80 rpm) in a range of temperatures (25, 50, 75 °C) and durations (80, 160, 240, 320 min). The UA extraction was performed by placing the tubes in an S40H Elma ultrasonic water bath (Elma Schmidbauer GmbH, Singen, Germany) operated at 340 Watt power and 50/60 Hz frequency and thermostated at 40 °C for 30 or 60 min. Suspensions from both extraction methods were then strained through organza and centrifuged for 15 min at 19,341× g. Clear supernatants were used to determine nonstructural carbohydrates and total phenolics. All treatments were replicated three times.

2.3. Total Phenolics Content

The total phenolic content (TPC) of the aqueous extracts was determined according to the method of Singleton et al. (1999) [26] with slight modifications, previously described by Kyriacou et al. (2016) [27]. Quantification was performed on a Jasco V-550 UV-vis spectrophotometer (Jasco Corp., Tokyo, Japan) against linear calibration with external gallic acid standards over the range of 50–500 mg L^{-1}, yielding a regression coefficient R^2 > 0.99. The TPC of the extract was expressed in gallic acid equivalents (GAE) mg L^{-1}.

2.4. Soluble Carbohydrates Content

For the analysis of water-soluble carbohydrates extract clarification was performed using Carrez Clarification Kit (Sigma-Aldrich, St. Louis, MO, USA). Separation and quantification of nonstructural carbohydrates (glucose, fructose, and sucrose) were accomplished by liquid chromatography on an Agilent HPLC system (Agilent Technologies, Santa Clara, CA, USA) equipped with a 1200 Series quaternary pump and a 1260 Series Refractive Index detector operated by Chem-Station software. Injection volume was 20 µL and separation was performed on a Waters 4.6 × 250 mm carbohydrate column (Waters, Milford, MA, USA) at 35 °C using an acetonitrile:water (82:18) mobile phase at a flow rate of 1.5 mL min^{-1}. Quantification was performed against fructose, glucose, and sucrose calibrating curves established using six external standard concentrations (0.2–2.0 g 100 mL^{-1}) with a coefficient of determination (R^2) greater than 0.999 and expressed as g 100 mL^{-1}.

2.5. Reagents and Standards

Gallic acid, fructose, Folin Ciocalteu's Phenol reagent, and sodium carbonate were purchased from Sigma Aldrich (Steinheim, Germany). Glucose and methanol were purchased from Merck (Darmstadt, Germany). Sucrose was purchased from Fluka AG (Buchs, Switzerland) and acetonitrile was purchased form Supelco (Bellefonte, PA, USA).

2.6. Statistical Analysis

All analysis was performed using SPSS statistical package (IBM, SPSS, Chicago, IL, USA, ver. 25). Data were subjected to analysis of variance (ANOVA) and the percentage of total variance accounted for by the main effects and their interactions is presented. Mean comparisons were performed by Tukey's b test. Two-tailed t test was used to compare mean values of treatments between the two extraction methodologies. Regression analysis on mean values was employed for profiling change in total phenolics content with extraction time at different temperatures for the two sources of material. Regression analysis was performed regardless of kibble size due to the very low percentage of variance explained by this factor.

3. Results and Discussion

Sugars and phenolics are the key components shaping the sensory and functional profile of carob juice. A moderate presence of phenolics may improve the stability and antioxidant capacity of the juice without imparting an astringent flavor at the expense of sweetness [14,19,27]. Carob pulp is high in sugars, chiefly sucrose (up to 52 g 100 g^{-1} d.w), fructose (1.8–12.5 g 100 g^{-1} d.w), and glucose (1.8–10.2 g 100 g^{-1} d.w), the relative content of which contributes differentially to perceived juice sweetness [7]. Cultivar and harvest maturity constitute factors that may substantially affect fruit sugar content and influence extraction efficiency [27,28]. In terms of phenolic constitution, carob pods contain mainly gallic acid, hydrolysable and condensed tannins, flavonol glycosides, and traces of isoflavonoids, the concentrations of which may also be influenced by cultivar and physiological stage of maturity at harvest [7,15,29,30].

Aside from the extraction of carob juice for consumption as beverage, the extraction of juice high in sugars also constitutes the fundamental step for industrial production of molasses (carob syrup), which is a concentrate of carob juice (65–80° Brix) produced by slow simmering, without the addition of sugars or other additives [16]. Several methodologies were previously examined concerning the extraction of sugars and phenolics from carob kibbles [22,23,30]; however, the parameters influencing the efficiency of aqueous extraction have received little attention despite the widespread application of aqueous extraction by the carob industry [23]. The efficiency of aqueously extracting sugars and phenolics from carob kibbles was presently assessed for heat-assisted and ultrasound-assisted extraction methodologies. The relative effect of key parameters (source material, temperature, duration, kibble size) on extraction efficiency was assessed through their relative contribution to the total variance of sugars and phenolics concentrations in the obtained carob juice (Tables 1 and 2).

Table 1. Percentage of variance explained by main effects and mean comparisons for fructose (FRU), glucose (GLU), sucrose (SUC), total sugars, and phenolics obtained by conventional extraction (HAE) using two source materials (LF: Lefkaritiki; MV: Mavroteratsia) and two kibble sizes (M: medium-size kibbles; P: powder-size kibbles) in a range of extraction temperatures (25, 50, and 75 °C) and durations (80, 160, 240, 320 min). GAE: gallic acid equivalents. Interaction plots of main effects on sugars and phenolics are presented separately in Figure 1A,B.

Source of Variation	Fructose (g 100 mL^{-1})	Glucose (g 100 mL^{-1})	Sucrose (g 100 mL^{-1})	Total Sugars (g 100 mL^{-1})	Phenolics (mg L^{-1} GAE)
Source material			*Means*		
LF	2.6b	1.7b	11.3a	15.6a	2595a
MV	3.4a	2.5a	8.2b	14.0b	715b
Kibble size					
M	3.0a	2.1	10.4a	15.5a	1573b
P	2.9b	2.1	9.1b	14.1b	1737a
Temperature					
25	2.8b	2.1ab	9.7	14.5b	830c
50	3.1a	2.10a	9.8	15.0a	1341b
75	3.0a	2.0b	9.9	14.9ab	2844a

Table 1. *Cont.*

Source of Variation	Fructose (g 100 mL^{-1})	Glucose (g 100 mL^{-1})	Sucrose (g 100 mL^{-1})	Total Sugars (g 100 mL^{-1})	Phenolics (mg L^{-1} GAE)
Time					
80	2.7b	1.8b	9.2c	13.8b	1127d
160	2.9b	1.9b	10.6a	15.4a	1608c
240	3.1a	2.2a	9.8b	15.0a	1840b
320	3.2a	2.3a	9.5bc	15.0a	2148a
			Percentage of Variance		
Source materials	42.5 ***	50.3 ***	39.9 ***	12.9 ***	37.4 ***
Kibble size	0.8 *	0	6.3 ***	9.6 ***	0.3 **
Temperature	4.0 ***	1.0 *	0.1	0.8	30.0 ***
Time	13.8 ***	9.6 ***	4.5 ***	7.8 ***	5.3 ***

* Significant effect at the 0.05 level; ** significant effect at the 0.01 level; *** significant effect at the 0.001 level. Data represent means of three biological replicates. Means followed by different lowercase letters within each column denote significant difference at $p < 0.05$.

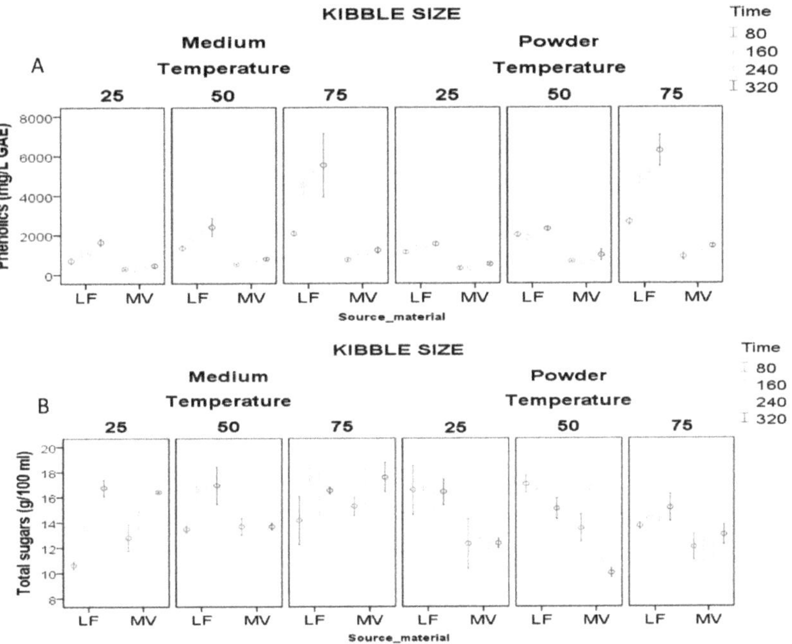

Figure 1. Interaction plots for total phenolics and sugars obtained by heat-assisted extraction (HAE; (**A**,**B**)) of different source materials (LF: Lefkaritiki; MV: Mavroteratsia) and kibble sizes (medium and powder). HAE extraction varied in time (80, 160, 240, 320 min) and temperature (25, 50, and 75 °C). Data points represent means of three replicates with standard error bars.

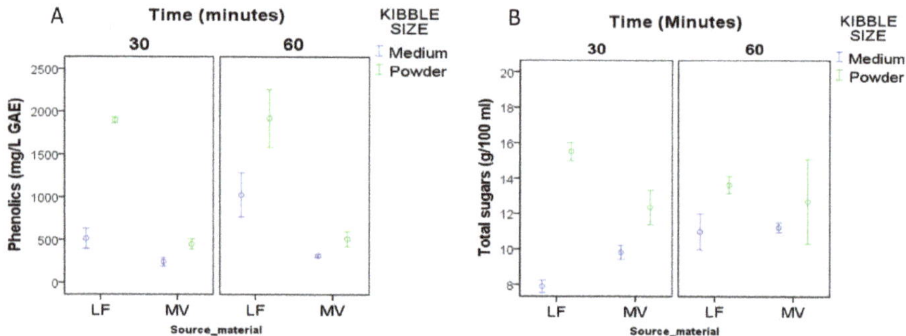

Figure 2. Interaction plots for total phenolics and sugars obtained by ultrasound-assisted extraction (UAE; (**A**,**B**)) of different source materials (LF: Lefkaritiki; MV: Mavroteratsia) and kibble sizes (medium and powder). UAE extraction varied in extraction time (30 and 60 min). Data points represent means of three replicates with standard error bars.

Table 2. Percentage of variance explained by main effects and mean comparisons for fructose (FRU), glucose (GLU), sucrose (SUC), total sugars, and phenolics obtained by ultrasound-assisted extraction (UAE) using two source materials (LF: Lefkaritiki; MV: Mavroteratsia) and two kibble sizes (M: medium-size kibbles; P: powder-size kibbles) under two extraction times (30 and 60 min). Interaction plots of main effects on phenolics and sugars are presented separately in Figure 2A,B, respectively.

Source of Variation	Fructose (g 100 mL⁻¹)	Glucose (g 100 mL⁻¹)	Sucrose (g 100 mL⁻¹)	Total Sugars (g 100 mL⁻¹)	Phenolics (mg L⁻¹ GAE)
Source material			*Means*		
LF	1.8b	1.4b	8.8a	12.0	1338.8a
MV	2.5a	1.9a	7.1b	11.5	379.9b
Kibble size					
M	1.9b	1.6	6.5b	10.0b	523.9b
P	2.5a	1.7	9.3a	13.6a	1194.9a
Time					
30	2.1	1.7	7.6b	11.4	775.7b
60	2.3	1.7	8.2a	12.1	943.1a
			Percentage of Variance (%)		
Source material	40.1 ***	52.0 ***	18.7 ***	1.1	53.8 ***
Kibble size	33.2 ***	4.6	48.2 ***	59.7 ***	26.3 ***
Time	2.3	0.0	2.1 *	2.6	1.6 **

* Significant effect at the 0.05 level; ** significant effect at the 0.01 level; *** significant effect at the 0.001 level. Data represent means of three biological replicates. Means followed by different lowercase letters within each column denote significant difference at $p < 0.05$.

In the conventional heat-assisted method (HAE), variation in phenolics concentration was determined mainly by source material and temperature, while variation in total sugars concentration was not affected by temperature (Table 1). Variation in phenolics content foremost and sugars secondarily was observed in previous studies and attributed to genetic and environmental factors, harvest maturity, and postharvest storage [11,31–33]. Lefkaritiki delivered a juice higher in both sugars and phenolics than Mavroteratsia. The current results indicate that the source material can significantly affect the sugar and phenolics yield potential of aqueous extracts. This effect was more pronounced on phenolics than sugars, with Lefkaritiki yielding 3.6-fold higher phenolics concentration and 1.1-fold higher total sugars concentration than Mavroteratsia (Table 1). These phenotypes of the local carob landrace are genetically similar and bear limited differences in pod morphology. They were sourced, however, from different environments, with Lefkaritiki collected at 510 m altitude and Mavroteratsia near sea level. It has been previously established for arboricultural species as well as seasonal crops that phenolic content is highly modulated by abiotic stress conditions such as water stress, salinity, light intensity, and heat stress [34,35]. Given that carob constitutes an underutilized tree crop largely limited to cultivation in marginal lands of highly alkaline, calcareous, and infertile soils, where it subsists strictly under rainfed conditions, the impact of agro-environment on the phenolic content of the pod is unsurprising. Further research is therefore warranted to investigate the significance of growth environment on the actual pod content and ultimately juice extract concentrations of sugars and especially phenolics, variation in which can significantly influence the biological activity of carob kibble products [32].

Kibble size had limited effect overall on the extraction of phenolics and moderate effect on sugars (Table 1). Similarly, a previous study demonstrated no effect of kibble size on the phytochemical composition of carob juice [13]. Significantly higher phenolics concentration was obtained with each incremental increase of extraction temperature and time (Table 1; Figure 1A). The most pronounced incremental increase in phenolics concentration was observed with both kibble sizes of Lefkaritiki when extraction temperature was raised to 75 °C and extraction time to 160 min (Figure 1A). Regression analysis highlighted the increase in phenolic concentration with progressive extraction time at each temperature (Supplementary Figure S1). The highest rate of increase in phenolic concentration over extraction time was observed for both source materials at 75 °C, yielding strong regressions (LF: R^2 = 0.848, p < 0.001 and MV: R^2 = 0.820, p = 0.002). Lefkaritiki was more responsive than Mavroteratsia, as shown by the slopes of the regressions (13.8 vs. 2.1, respectively). This observation is in accordance with previous findings that higher extraction temperatures yield higher phenolic concentrations [12,36]. Temperature and time effects on phenolic concentration may be partly explained by effective solubilization of condensed tannins abounding in carob pulp [19].

Extraction temperature overall had marginal effects on juice fructose and glucose concentrations and no effect on sucrose and total sugars concentrations (Table 1). Total sugar yield increased with time only for middle-sized kibbles extracted at 25 °C. It is worth noting that at 25 °C, the extraction of total sugars from middle-sized kibbles of Lefkaritiki was particularly increased when extraction was extended to 240 min, without eliciting an excessive concomitant extraction of phenolics (Figure 1B). According to Roseiro et al. (1991), the yield of carob carbohydrates aqueously extracted at ambient temperature (20 °C) increased with the extraction time, reaching a steady state after 6 h [12]. At higher than ambient temperatures, prolonging the extraction time had no apparent effect on sugars concentration while it increased the phenolics yield, thereby adding astringency to the juice owing in part to the presence of high molecular weight tannins [29,37,38]. High levels of tannins render carob juice more bitter in taste and metallic in aroma compared to grape juice [13]. On the other hand, phenolics and particularly hydrolysable tannins have gained substantial attention in the food and pharmaceutical industries as antioxidant and antimicrobial agents that improve the shelf life of a product [1,14]. Raising extraction temperature up to 75 °C for periods longer than 160 min has been presently shown effective in spiking the phenolic content of carob juice to be used as a stabilizing agent in juice blending.

Kibble size was the most influential factor for the UAE extraction of sugars (Table 2). It also explained substantial variation in phenolics. The duration of sonication (30 vs. 60 min) had nonsignificant and minimal impact on the concentrations of sugars and phenolics in the sonicated aqueous extracts, respectively (Table 2; Figure 2A,B). However, the effect of kibble size on both sugars and phenolics extraction decreased when sonication time increased from 30 min to 60 min (Figure 2A,B). As opposed to gentle agitation (HAE described above), sonication facilitated more efficient extraction from powdered than medium-sized kibbles (Figure 2A,B). This might be explained by the larger surface-to-volume ratio of powdered kibbles which were effectively kept in suspension by constant sonication, whereas in the case of gentle agitation, solvent circulation through powdered kibbles was impeded by sedimentation of the solid phase. This is an observation of practical significance for the carob industry since powder-sized kibbles are a major low-cost by-product of industrial carob milling, which inevitably ends up as animal fodder or source material for the production of carob juice and molasses.

Comparing the two extraction methodologies under similar temperature–time conditions by t-Test analysis, the HAE at 50 °C for 80 min (14.4 g 100 mL^{-1}; 1150.3 mg L^{-1} GAE) demonstrated significantly higher total sugars and phenolics yield (P-sugars < 0.001, P-phenolics < 0.01) compared to UAE at 40 °C for 60 min (12.1 g 100 mL^{-1}; 943.2 mg L^{-1} GAE). The lower phenolics concentration in the juice facilitated by UAE was therefore outweighed by the lower sugar yield obtained. Nonetheless, both methodologies can be applied potentially for industrial production of carob juice depending on the targeted functional and sensory properties of the extracted juice.

4. Conclusions

The results of this study provide an assessment of key parameters that affect the efficiency of aqueously extracting sugars and phenolics from carob kibbles. The source material was most influential in determining the concentration of phenolics in carob juice for both the heat-assisted and ultrasound-assisted extraction; however, the importance of the carob material's environment of origin warrants further investigation in this respect. The source material also influenced the relative proportions of sucrose, glucose, and fructose in the juice, which bear an impact on the perceived sweetness of the latter. Of the extraction parameters examined, kibble size proved more influential in UAE than HAE for both sugars and phenolics by facilitating better suspension and preventing sedimentation of powder-sized particles. Kibble size was less influential as UAE duration increased. In the case of HAE, increasing extraction temperature favored the yield in phenolics over sugars, thus shifting the sensory profile of the juice toward astringency. However, prolonging the extraction time at 25 °C improved sugar yield without excessive extraction of phenolics.

Supplementary Materials: The following are available online at http://www.mdpi.com/2304-8158/9/10/1364/s1, Figure S1: Regression of total phenolics concentration in carob juice with extraction time (60, 180, 240, and 320 min) using heat-assisted conventional method over a range of temperatures (25, 50, 75 °C) for the two source materials (LF: Lefkaritiki; MV: Mavroteratsia).

Author Contributions: Conceptualization, M.C.K., C.A., and S.S.; methodology, M.C.K. and C.A.; formal analysis, M.C.K. and C.A.; resources, M.C.K. and Y.R.; data curation, C.A. and A.K.; writing—original draft preparation, M.C.K., C.A., and A.K.; writing—review and editing, M.C.K., C.A., A.K., Y.R., and S.S.; supervision, M.C.K. and Y.R.; project administration, M.C.K. and A.K. All authors have read and agreed to the published version of the manuscript.

Funding: This research was co-funded by the European Regional Development Fund and the Republic of Cyprus through the Cyprus Research and Innovation Foundation (Project: BlackGold INTEGRATED /0916/0019).

Conflicts of Interest: The authors declare no conflict of interest.

References

1. Nasar-Abbas, S.M.; Huma, Z.; Vu, T.-H.; Khan, M.K.; Esbenshade, H.; Jayasena, V. Carob Kibble: A Bioactive-Rich Food Ingredient. *Compr. Rev. Food Sci. Food Saf.* **2016**, *15*, 63–72. [CrossRef]

2. FAOSTAT Food and Agriculture Organization of the United Nations, Final 2018 Data. Available online: http://faostat.fao.org/site/567/DesktopDefault.aspx#ancor (accessed on 6 May 2020).

3. Wursch, P.; Del Vedovo, S.; Rosset, J.; Smiley, M. The tannin granules from ripe carob pod. *Lebensm. Wiss. Technol.* **1984**, *17*, 351–354.

4. Bouzouita, N.; Khaldi, A.; Zgoulli, S.; Chebil, L.; Chekki, R.; Chaabouni, M.; Thonart, P. The analysis of crude and purified locust bean gum: A comparison of samples from different carob tree populations in Tunisia. *Food Chem.* **2007**, *101*, 1508–1515. [CrossRef]

5. Kotrotsios, N.; Christaki, E.; Bonos, E.; Florou-Paneri, P. Carobs in productive animal nutrition. *J. Hellen. Vet. Med. Soc.* **2017**, *62*, 48–57. [CrossRef]

6. Yousif, A.K.; Alghzawi, H. Processing and characterization of carob powder. *Food Chem.* **2000**, *69*, 283–287. [CrossRef]

7. Goulas, V.; Stylos, E.; Chatziathanasiadou, M.V.; Mavromoustakos, T.; Tzakos, A.G. Functional Components of Carob Fruit: Linking the Chemical and Biological Space. *Int. J. Mol. Sci.* **2016**, *17*, 1875. [CrossRef]

8. Youssef, M.; Kamal, E.; El-Manfaloty, A.M.; Moshera, M.; Ali, H.M. Assessment of proximate chemical composition, nutritional status, fatty acid composition and phenolic compounds of carob (*Ceratonia siliqua* L.). *Food Pub. Health* **2013**, *3*, 304–308.

9. Zunft, H.J.F.; Lüder, W.; Harde, A.; Haber, B.; Graubaum, H.J.; Koebnick, C.; Grünwald, J. Carob pulp preparation rich in insoluble fibre lowers total and LDL cholesterol in hypercholesterolemic patients. *Eur. J. Nutr.* **2003**, *42*, 235–242. [CrossRef]

10. Theophilou, I.C.; Neophytou, C.M.; Constantinou, A.I. Carob and its Components in the Management of Gastrointestinal Disorders. *J. Hepatol Gastroenterol* **2017**, *1*, 005.

11. Khlifa, M.; Bahloul, A.; Kitane, S. Determination of chemical composition of carob pod (*Ceratonia siliqua* L.) and its morphological study. *J. Mater. Environ. Sci.* **2013**, *4*, 348–353.

12. Roseiro, J.C.; Gírio, F.M.; Collaço, M.A. Yield improvements in carob sugar extraction. *Process. Biochem.* **1991**, *26*, 179–182. [CrossRef]

13. Rababah, T.M.; Al-u'datt, M.; Ereifej, K.; Almajwal, A.; Al-Mahasneh, M.; Brewer, S.; Alsheyab, F.; Yang, W. Chemical, functional and sensory properties of carob juice. *J. Food Qual.* **2013**, *36*, 238–244. [CrossRef]

14. Buzzini, P.; Arapitsas, P.; Goretti, M.; Branda, E.; Turchetti, B.; Pinelli, P.; Ieri, F.; Romani, A. Antimicrobial and antiviral activity of hydrolysable tannins. *Mini Rev. Med. Chem.* **2008**, *8*, 1179–1187. [CrossRef] [PubMed]

15. Owen, R.W.; Haubner, R.; Hull, W.E.; Erben, G.; Spiegelhalder, B.; Bartsch, H.; Haber, B. Isolation and structure elucidation of the major individual polyphenols in carob fibre. *Food Chem.Toxicol.* **2003**, *41*, 1727–1738. [CrossRef]

16. Tounsi, L.; Karra, S.; Kechaou, H.; Kechaou, N. Processing, physico-chemical and functional properties of carob molasses and powders. *J. Food Meas. Charact.* **2017**, *11*, 1440–1448. [CrossRef]

17. Tounsi, L.; Ghazala, I.; Kechaou, N. Physicochemical and phytochemical properties of Tunisian carob molasses. *J. Food Meas. Charact.* **2019**, *14*, 20–30. [CrossRef]

18. Atasoy, A.F. The effects of carob juice concentrates on the properties of yoghurt. *Int. J. Dairy Technol.* **2009**, *62*, 228–233. [CrossRef]

19. Kumazawa, S.; Taniguchi, M.; Suzuki, Y.; Shimura, M.; Kwon, M.-S.; Nakayama, T. Antioxidant activity of polyphenols in carob pods. *J. Agric. Food Chem.* **2002**, *50*, 373–377. [CrossRef]

20. El Bouzdoudi, B.; El Ansari, Z.N.; Mangalagiu, I.; Mantu, D.; Badoc, A.; Lamarti, A. Determination of Polyphenols Content in Carob Pulp from Wild and Domesticated Moroccan Trees. *Am. J. Plant Sci.* **2016**, *7*, 1937–1951. [CrossRef]

21. Rtibi, K.; Selmi, S.; Grami, D.; Amri, M.; Eto, B.; El-Benna, J.; Sebai, H.; Marzouki, L. Chemical constituents and pharmacological actions of carob pods and leaves (*Ceratonia siliqua* L.) on the gastrointestinal tract: A review. *Biomed. Pharmacother.* **2017**, *93*, 522–528. [CrossRef]

22. Papaefstathiou, E.; Agapiou, A.; Giannopoulos, S.; Kokkinofta, R. Nutritional characterization of carobs and traditional carob products. *Food Sci. Nutr.* **2018**, *6*, 2151–2161. [CrossRef] [PubMed]

23. Almanasrah, M.; Roseiro, L.B.; Bogel-Lukasik, R.; Carvalheiro, F.; Brazinha, C.; Crespo, J.; Kallioinen, M.; Mänttäri, M.; Duarte, L.C. Selective recovery of phenolic compounds and carbohydrates from carob kibbles using water-based extraction. *Ind. Crops Prod.* **2015**, *70*, 443–450. [CrossRef]

24. Shahidi, F.; Naczk, M. *Phenolics in Food and Nutraceuticals*; CRC Press: Boca Raton, FL, USA, 2003.

25. Kyratzis, A.; Antoniou, C.; Papayiannis, L.C.; Graziani, G.; Rouphael, Y.; Kyriacou, M.C. Carob pod morphology and physicochemical composition of indigenous Cyprus germplasm are modulated by genotype, agro-environmental zone and growing season. 2020; to be submitted.

26. Singleton, V.L.; Orthofer, R.; Lamuela-Raventós, R.M. Analysis of total phenols and other oxidation substrates and antioxidants by means of folin-ciocalteu reagent. In *Methods in Enzymology*; Elsevier: Amsterdam, The Netherlands, 1999; Volume 299, pp. 152–178.

27. Kyriacou, M.C.; Emmanouilidou, M.G.; Soteriou, G.A. Asynchronous ripening behavior of cactus pear (Opuntia ficus-indica) cultivars with respect to physicochemical and physiological attributes. *Food Chem.* **2016**, *211*, 598–607. [CrossRef]

28. Kader, A.A. Flavor quality of fruits and vegetables. *J. Sci. Food Agric.* **2008**, *88*, 1863–1868. [CrossRef]

29. Papagiannopoulos, M.; Wollseifen, H.R.; Mellenthin, A.; Haber, B.; Galensa, R. Identification and quantification of polyphenols in Carob Fruits (*Ceratonia siliqua* L.) and derived products by HPLC-UV-ESI/MS. *J. Agric. Food Chem.* **2004**, *52*, 3784–3791. [CrossRef] [PubMed]

30. Stavrou, I.J.; Christou, A.; Kapnissi-Christodoulou, C.P. Polyphenols in carobs: A review on their composition, antioxidant capacity and cytotoxic effects, and health impact. *Food Chem.* **2018**, *269*, 355–374. [CrossRef]

31. Fidan, H.; Petkova, N.; Sapoundzhieva, T.; Abanoz, E.I. Carbohydrate Content in Bulgarian and Turkish Carob Pods and Their Products. *CBU Int. Conf. Proc.* **2016**, *4*, 796–802. [CrossRef]

32. Custodio, L.; Fernandes, E.; Escapa, A.L.; Fajardo, A.; Aligue, R.; Albericio, F.; Neng, N.R.; Nogueira, J.M.; Romano, A. Antioxidant and cytotoxic activities of carob tree fruit pulps are strongly influenced by gender and cultivar. *J. Agric. Food Chem.* **2011**, *59*, 7005–7012. [CrossRef]

33. Biner, B.; Gubbuk, H.; Karhan, M.; Aksu, M.; Pekmezci, M. Sugar profiles of the pods of cultivated and wild types of carob bean (*Ceratonia siliqua* L.) in Turkey. *Food Chem.* **2007**, *100*, 1453–1455. [CrossRef]

34. Nasrabadi, M.; Ramezanian, A.; Eshghi, S.; Kamgar-Haghighi, A.A.; Vazifeshenas, M.R.; Valero, D. Biochemical changes and winter hardiness in pomegranate (*Punica granatum* L.) trees grown under deficit irrigation. *Sci. Hortic.* **2019**, *251*, 39–47. [CrossRef]

35. Tavarini, S.; Gil, M.I.; Tomas-Barberan, F.A.; Buendia, B.; Remorini, D.; Massai, R.; Degl'Innocenti, E.; Guidi, L. Effects of water stress and rootstocks on fruit phenolic composition and physical/chemical quality in Suncrest peach. *Ann. Appl. Biol.* **2011**, *158*, 226–233. [CrossRef]

36. Turhan, I.; Tetik, N.; Aksu, M.; Karhan, M.; Certel, M. Liquid–solid extraction of soluble solids and total phenolic compounds of carob bean (*Ceratonia siliqua* L.). *J. Food Process. Eng.* **2006**, *29*, 498–507. [CrossRef]

37. Roseiro, L.B.; Duarte, L.C.; Oliveira, D.L.; Roque, R.; Bernardo-Gil, M.G.; Martins, A.I.; Sepúlveda, C.; Almeida, J.; Meireles, M.; Gírio, F.M.; et al. Supercritical, ultrasound and conventional extracts from carob (*Ceratonia siliqua* L.) biomass: Effect on the phenolic profile and antiproliferative activity. *Ind. Crops Prod.* **2013**, *47*, 132–138. [CrossRef]

38. Ayaz, F.A.; Torun, H.; Ayaz, S.; Correia, P.J.; Alaiz, M.; Sanz, C.; Gruz, J.; Strnad, M. Determination of chemical composition of anatolian carob pod (*Ceratonia siliqua* L.): Sugars, amino and organic acids, minerals and phenolic compounds. *J. Food Qual.* **2007**, *30*, 1040–1055. [CrossRef]

Article

Optimization of the Effect of Pineapple By-Products Enhanced in Bromelain by Hydrostatic Pressure on the Texture and Overall Quality of Silverside Beef Cut

Diana I. Santos [1], Maria João Fraqueza [2,*], Hugo Pissarra [2], Jorge A. Saraiva [3], António A. Vicente [4] and Margarida Moldão-Martins [1]

[1]   LEAF, Linking Landscape, Environment, Agriculture and Food, School of Agriculture, University of Lisbon, Tapada da Ajuda, 1349-017 Lisbon, Portugal; dianaisasantos@isa.ulisboa.pt (D.I.S.); mmoldao@isa.ulisboa.pt (M.M.-M.)
[2]   CIISA, Centre for Interdisciplinary Research in Animal Health, Faculty of Veterinary Medicine, University of Lisbon, Av. da Universidade Técnica, Pólo Universitário, Alto da Ajuda, 1300-477 Lisbon, Portugal; hpissarra@fmv.ulisboa.pt
[3]   LAQV-REQUIMTE, Department of Chemistry, University of Aveiro, 3810-193 Aveiro, Portugal; jorgesaraiva@ua.pt
[4]   CEB, Centro de Engenharia Biológica, Departamento de Engenharia Biológica, Universidade de Minho, Campus de Gualtar, 4710–057 Braga, Portugal; avicente@deb.uminho.pt
*   Correspondence: mjoaofraqueza@fmv.ulisboa.pt; Tel.: +351-213-652-889

Received: 28 September 2020; Accepted: 23 November 2020; Published: 26 November 2020

Abstract: Dehydrated pineapple by-products enriched in bromelain using a hydrostatic pressure treatment (225 MPa, 8.5 min) were added in marinades to improve beef properties. The steaks from the silverside cut (2 ± 0.5 cm thickness and weight 270 ± 50 g), characterized as harder and cheaper, were immersed in marinades that were added to dehydrated and pressurized pineapple by-products that corresponded to a bromelain concentration of 0–20 mg tyrosine, 100 g^{-1} meat, and 0–24 h time, according to the central composite factorial design matrix. Samples were characterized in terms of marination yield, pH, color, and histology. Subsequently, samples were cooked in a water-bath (80 °C, 15 min), stabilized (4 °C, 24 h), and measured for cooking loss, pH, color, hardness, and histology. Marinades (12–24 h) and bromelain concentration (10–20 mg tyrosine.100 g^{-1} meat) reduced pH and hardness, increased marination yield, and resulted in a lighter color. Although refrigeration was not an optimal temperature for bromelain activity, meat hardness decreased (41%). Thus, the use of pineapple by-products in brine allowed for the valorization of lower commercial value steak cuts.

Keywords: pineapple by-products; hydrostatic pressure; bromelain; enzyme activity; marinade; meat; texture

---

1. Introduction

Bromelain is a proteolytic enzyme present in the *Bromeliaceae* family to which pineapple (*Ananas comosus* L.) belongs. Bromelain is also found in pineapple by-products (cores, peels, and leaves), though in lower amounts than in stems and fruits [1,2]. The minimally processed industry produces a large amount ~50% (*w/w*) of pineapple by-products [3]. These by-products remain metabolically active and respond to postharvest abiotic stresses that induce biosynthesis and accumulation on secondary metabolites. Hydrostatic pressure treatments (50–250 MPa), as abiotic stress, activate cellular processes [4]. Pineapple by-products are rich in compounds of interest (i.e., bromelain) that can be enhanced by the application of abiotic stresses via hydrostatic pressure, avoiding extraction technologies. Extraction usually involves large quantities of volatile and flammable organic solvents with negative impacts on the environment and economy. The application of a green technology could value by-products with potential application in the food industry. On the other hand, consumers demand processes without chemicals. Therefore, the food industry's potential valorization of the pineapple by-products enriched in bromelain as a novel food ingredient allows for the possibility of consuming compounds that offer health benefits and reduce environmental impact [5].

The marinating process is a method that improves the sensory properties of meat. Marinating consists of immersing or injecting meat with a solution that may contain several ingredients, i.e., water, salts, vinegar, lemon juice, wine, soy sauce, essential oils, tenderizers, herbs, spices, and organic acids with the aim of flavoring and tenderizing meat or meat products [6]. Marinades are important in muscles rich in connective tissue and tenderizing properties offer means of commercial improvement [7,8]. Marinades develop an extra succulent texture on the surface of meat and reduce water loss during cooking. The soaking in marinades increases the tenderness of meat due to the increase in volume induced by the pH of the muscle fibers and/or connective tissue, faster proteolytic weakening of the muscular structure, and greater collagen solubilization after cooking [9].

The proteolytic enzymes present in fruits can degrade myofibrillar proteins and collagen in meat and, consequently, have beneficial effects on tenderness. Meat tenderized with proteolytic enzymes improves consumer acceptability, although the improvement in tenderization can be related with a decrease in yield [10]. The positive effect of proteolytic enzymes obtained from fruits on meat tenderness has been reported by some authors [11,12]. Cysteine proteases such as bromelain have been studied in relation to meat tenderization [13]. The amino acid structure of meat is crucial for the softening effect by bromelain because fragmentation occurs in lysine, alanine, tyrosine, and glycine [14]. Bromelain affects the structure of the myosin and actin filaments in myofibrillar proteins. Proteolytic enzymes increase the rate of myofibril fragmentation in meat and interrupt the intramuscular connective tissue structure [15,16]. Bromelain is active against collagen proteins [12]. Bromelain has maximum enzymatic activity in the temperature range of 37–70 °C [17]. At refrigeration temperatures, the enzyme has low activity, although it is active at 0 °C [18].

The USDA Food Safety Inspection Service currently considered the enzymes bromelain, papain, ficin, *Bacillus* proteases, and aspartic proteases to be safe [12]. Proteolytic enzymes extracted from plants, namely papain, bromelain, actinidin, zingibain, and ficin have been widely used as meat tenderizers in several countries [13,19].

The present work aimed to optimize the effects of marinade with pineapple by-products (enriched in bromelain through treatment with hydrostatic pressure) at different concentrations and contact time on the sensory characteristics (particularly the hardness) of steaks from silverside cuts considered to have a lower commercial value due to its sensory characteristics.

2. Materials and Methods

2.1. Meat Samples Collecting and Preparation

The beef cuts silverside (lot 332199042) were from animals that were born, raised, slaughtered, deboned, cut, and vacuum packaged in Poland according to EU standard commercial practices. Further, they were transported under refrigeration (0–4 °C) conditions to Portugal where they were again labelled (lot 194948). Two independent beef cuts (silverside cut constituted mainly by *M. gluteobiceps and M. semitendinosus*) from different bovines (Piece 1: 4715 g and Piece 2: 4225 g) were supplied by a central logistic of hypermarkets in Santarém, Portugal, and transported under refrigerated conditions (0–4 °C) to the Faculty of Veterinary Medicine, University of Lisbon. The assay was performed 10 days after the animals slaughtering. Each beef cut was sliced with a sharp knife into 16 steaks with 2 ± 0.5 cm thickness and a weight of 270 ± 50 g.

2.2. Production of Bromelain Powder from Pineapple By-Products

Pineapple (*Ananas comosus* L.) by-products were provided by the company Campotec S.A. located in Torres Vedras, west center of Portugal. The core pineapples (~104.5 × 30 mm) were stored under refrigeration (5 ± 1 °C) approximately 3 h prior to packaging in PA/PE-90 (Alempack - Embalagens Flexíveis, Elvas, Portugal) that were vacuum sealed (85% of vacuum).

An abiotic stress in packaged by-products (pineapple core) was applied using a pilot-scale high pressure hydrostatic equipment (Hiperbaric 55, Burgos, Spain) with a 55 L vessel, according to the conditions optimized in a previous work (225 MPa during 8.5 min) [20]. The core pineapple samples after pressurization were stored at 5 ± 1 °C for 24 h. After storage time all samples were frozen at −80 °C and lyophilized (Coolsafe Superior Touch, Scanvac, Denmark) during 7 days at −94 °C and 0.2 to 0.09 mbar. Further, 300 g samples were ground in a food processor (Bimby Thermomix TM31, Vorwerk Thermomix, Cloyes-sur-le-Loir, France) at a speed of 10,200 rpm for 25 s to obtain the powder sample. The enzymatic activity of bromelain was quantified in the pineapple powder core sample and showed 9.33 ± 0.29 mg tyrosine. g^{-1} dry matter according to Chakraborty (2014) with some modifications [21].

2.3. Marinating

2.3.1. Optimization of Pineapple By-Products Enhanced in Bromelain Application on Marinated Steaks

Response surface methodology (RSM) constructed on a two-variable central composite rotatable design (CCRD) was used as a function of two factors [22]. The experimental data through a stepwise multiple regression analysis using StatisticaTM v.8 Software (StatSoft Inc., Tulsa, OK, USA, 2007) was fitted to a second-order polynomial equation to predict each dependent variable (Y). The three-dimensional response surface designs as a function of independent variables (bromelain concentration (X_1; mg tyrosine.100 g^{-1} meat) and time (X_2; min)). b_0 is the interception and b_i, b_j, and b_{ij} (i,j = 1,2) are the linear, quadratic, and interaction coefficients, respectively, that are described by the second order polynomial models, using decoded variables, as follows Equation (1).

$$Y_i = b_0 + b_1X_1 + b_2X_2 + b_{11}X_1^2 + b_{22}X_2^2 + b_{12}X_1X_2 \tag{1}$$

A brine was prepared by dissolving 1% (*w/w*) of coarse salt (purified sea salt for seasoning and cooking, Pingo Doce: Lot AL73527697) in water and stored at 4 ± 1 °C for 3 h to stabilize. Each steak was placed in a container (23 × 16 × 5.5 cm) and the brine was added.

The pineapple powder core (0 to 20 mg tyrosine.100 g^{-1} meat) was suspended in 400 mL of brine according to the experimental design matrix in Table 1. The brine mixture with pineapple powder core was then added to the container with the meat, and the meat remained immersed at 4 ± 1 °C for the time (0 to 24 h) defined by the experimental design for each sample.

The samples were weighed, the color and pH of the steaks was determined, and the pH of the brine were evaluated before marinating. After the marinade, the samples were weighed again, the color and pH of the steaks was determined, and the pH of the brine were evaluated, and the sample was collected for the histological analysis. Subsequently, the steaks were packaged in PA/PE 90 μm (polyamide (PA) 20 μm, polyethylene (PE) 70 μm provided by Luis Sanchez and Filhos Lda., Sintra, Portugal) that were vacuum sealed (85% vacuum - Henkovac, quality vacuum systems, De Brand, Netherlands). The samples were prepared in duplicate for each treatment of the experimental design.

Table 1. Experimental design with coded and decoded values of independent variables (bromelain concentration and time of contact) for steak samples and dependent variables: pH, marination yield, cooking yield, color, and hardness. (C) means the central point.

| Run | Coded Factors | | Decoded Factors | | pH | | | Marination Yield (%) | Cooking Yield (%) | Colour | | | | | | | | | Hardness Steak (N) |
| | Time | Concentration | Time (h) | Concentration (mg Tyrosine. 100 g^{-1} Meat) | Brine | Marinated Steak | Cooked Steak | | | Initial | | | Marinated Steak | | | Cooked Steak | | | |
										L*	a*	b*	L*	a*	b*	L*	a*	b*	
8	0.00000	1.41421	12	20	4.69 ± 0.18	5.05 ± 0.01	5.83 ± 0.03	2.38 ± 0.37	69.31 ± 0.76	38.48 ± 1.12	23.57 ± 1.08	7.23 ± 0.76	51.24 ± 1.99	11.72 ± 1.35	4.27 ± 1.43	56.67 ± 1.48	7.72 ± 0.69	8.75 ± 0.68	32.54 ± 3.91
9 (C)	0.00000	0.00000	12	10	4.81 ± 0.04	5.11 ± 0.02	5.84 ± 0.03	2.68 ± 0.29	67.70 ± 0.25	38.20 ± 1.20	24.50 ± 1.28	7.48 ± 1.15	48.89 ± 2.23	12.92 ± 1.35	3.73 ± 0.60	56.80 ± 1.53	7.33 ± 0.47	8.83 ± 0.72	35.50 ± 4.70
2	−1.00000	1.00000	3.5	17	4.31 ± 0.03	5.29 ± 0.10	5.65 ± 0.03	1.48 ± 0.08	68.08 ± 0.49	38.27 ± 1.92	25.05 ± 1.58	7.17 ± 1.10	45.08 ± 1.41	15.01 ± 0.76	3.75 ± 0.54	54.51 ± 0.90	7.85 ± 0.29	8.82 ± 0.51	38.36 ± 3.19
4	1.00000	1.00000	20.5	17	4.60 ± 0.06	4.99 ± 0.05	5.73 ± 0.01	1.97 ± 0.46	68.23 ± 1.95	38.40 ± 0.82	24.37 ± 1.19	7.34 ± 0.72	49.90 ± 3.13	11.94 ± 1.58	5.93 ± 1.12	58.53 ± 0.78	6.83 ± 0.41	8.77 ± 0.32	30.35 ± 2.50
10 (C)	0.00000	0.00000	12	10	4.93 ± 0.01	5.18 ± 0.01	5.84 ± 0.03	3.53 ± 0.87	66.27 ± 0.05	38.91 ± 1.63	24.27 ± 2.18	7.33 ± 1.59	54.11 ± 4.02	11.27 ± 0.78	4.35 ± 1.91	55.90 ± 4.86	8.20 ± 1.84	10.01 ± 0.42	35.47 ± 4.40
7	0.00000	−1.41421	12	0	5.61 ± 0.05	5.39 ± 0.07	5.99 ± 0.03	3.58 ± 0.16	69.64 ± 0.84	39.23 ± 1.28	24.88 ± 1.37	7.86 ± 1.22	51.22 ± 2.30	12.72 ± 1.29	4.26 ± 0.40	55.98 ± 1.27	8.44 ± 0.69	10.73 ± 0.44	48.14 ± 4.99
5	−1.41421	0.00000	0	10	4.14 ± 0.03	5.34 ± 0.06	5.77 ± 0.03	0.07 ± 0.09	71.18 ± 1.98	37.52 ± 1.08	23.57 ± 0.68	5.87 ± 0.96	39.24 ± 0.51	21.21 ± 1.47	5.75 ± 0.72	50.65 ± 1.68	10.32 ± 1.33	9.57 ± 0.39	41.30 ± 8.58
6	1.41421	0.00000	24	10	4.86 ± 0.05	5.07 ± 0.06	5.81 ± 0.05	4.19 ± 0.42	63.13 ± 1.92	38.54 ± 1.66	25.71 ± 2.97	8.26 ± 2.55	52.11 ± 3.80	11.47 ± 0.68	5.18 ± 1.52	59.63 ± 0.93	6.42 ± 0.37	9.18 ± 0.32	36.16 ± 6.81
11 (C)	0.00000	0.00000	12	10	4.88 ± 0.01	5.12 ± 0.09	5.80 ± 0.04	2.95 ± 0.63	64.33 ± 0.95	39.30 ± 2.40	25.42 ± 2.59	8.46 ± 1.61	49.70 ± 6.45	13.55 ± 1.72	5.31 ± 1.51	59.45 ± 0.85	6.72 ± 0.26	9.42 ± 0.46	37.14 ± 11.02
12 (C)	0.00000	0.00000	12	10	4.89 ± 0.16	5.14 ± 0.02	5.81 ± 0.06	3.54 ± 0.74	66.44 ± 0.46	39.93 ± 0.90	25.25 ± 2.88	8.75 ± 1.63	53.60 ± 0.39	11.05 ± 1.92	4.60 ± 1.10	59.45 ± 1.93	6.54 ± 0.91	8.59 ± 0.65	37.03 ± 8.04
1	−1.00000	−1.00000	3.5	3	5.16 ± 0.03	5.34 ± 0.04	5.77 ± 0.08	2.01 ± 0.12	67.21 ± 0.29	39.61 ± 1.66	24.34 ± 1.63	7.22 ± 1.72	46.45 ± 0.51	16.03 ± 0.91	4.34 ± 0.89	55.56 ± 1.00	7.72 ± 0.44	9.80 ± 0.56	43.10 ± 7.60
3	1.00000	−1.00000	20.5	3	5.34 ± 0.06	5.21 ± 0.03	5.86 ± 0.01	3.97 ± 0.51	67.47 ± 0.76	38.64 ± 1.17	25.31 ± 1.46	7.73 ± 1.33	50.34 ± 3.80	11.93 ± 1.13	4.74 ± 0.97	58.60 ± 0.73	7.41 ± 0.94	10.25 ± 0.54	38.02 ± 7.83

2.3.2. Brine Effect on Marinated Steaks

In order to study the effect of adding brine, marinades were performed without adding pineapple powder over time (0–24 h). In the control sample, the steak was placed in the container and the brine 1% (*w/w*) was not added (0 h). Brines 1% (*w/w*) without the addition of pineapple powder core were added to the steak to assess the effect of the brine on steak over time (3.5 h, 12 h, 20.5 h, and 24 h).

2.4. Cooking

The cooking process was carried out in a water bath set at 80 ± 1 °C, in which the vacuum-packed meat (4 ± 1 °C) was added. When the temperature of the water bath achieved 80 ± 1 °C again, the cooking process started and lasted for 15 min in order to reach 75 ± 1 °C at the coldest point (center of the steaks). Subsequently, the steaks were placed in a water and ice bath for 30 to 40 min until reaching 4 ± 1 °C in the slowest cooling point (center of the steak) [23]. The cooking temperature and the cooling of the steaks were monitored by thermocouple (*Ellab* a-s model CTF 9008, Copenhagen, Denmark). After cooking and cooling, the samples were stored for about 24 h at 4 °C to stabilize the characteristics of the steaks. Subsequently, the samples were weighed again, the color and pH of the steaks were evaluated, the texture was determined, and samples were collected for histology.

2.5. Analytical Methods

Each steak sample subjected to each treatment was used to analyze the margination yield, cooking yield, pH, color, texture, and histology. The standard deviation in the results indicates the variation between the two pieces of meat and the repetitions of the analytical method.

2.5.1. pH Measurement

The pH values were measured with a digital pH meter (Hanna Instruments, Woonsocket, RI, USA). The pH meter was calibrated with pH 7 and 4 buffers before pH determination. The stainless-steel blade was fitted to the solid sample probe to facilitate penetration into the meat. Cutting the meat allowed direct contact between the probe and the sample, so that the pH was measured directly without further sample preparation. The initial pH of the steaks was measured 2 h after opening the vacuum bag and the initial pH of the brine was measured 3 h after preparing the solution. The pH of the steaks and brine was measured again after the marinating time of each sample. After cooking (24 h), the pH of the steaks was determined three times on each sample [24].

2.5.2. Measurement of Marination Yield

The initial weight of the steaks was determined immediately after cutting using an analytical balance (Sartorius A200S, Göttingen, Germany). After the marinade time, the steaks were removed from the brine, dried with paper towels to remove superficial water (blotted drying), and weighed again. The marination yield (MY) was calculated by the difference between raw and marinated weights as following: marination yield (%) = ((weight of marinated steak − weight of raw steak)/weight of raw steak) × 100.

2.5.3. Measurement of Cooking Yield

The meat samples were weighed just before cooking. The steaks after 24 h of stabilization at 4 °C after cooking were removed from the bag, dried with paper towels to remove excess surface moisture, and reweighed using an analytical balance. The cooking yield was calculated by the difference between raw and cooked weights as following: Cooking yield (%) = (weight of cooked steak/weight of raw marinated steak) × 100 [23].

2.5.4. Color Measurement

The color was measured on the surface of steaks with a Minolta CR 300 (Konica Minolta, Osaka, Japan) using the L*, a*, b* coordinates (CIELab color system). The measurement Minolta CR 300 uses diffuse illumination/0° viewing angle and aperture sizes of 8 mm. The calibration was performed with a white ceramic reference (standard illuminant D65). The initial color of the steaks was measured 2 h after opening the vacuum bag of the piece of meat, after the marinating time associated with each sample and 24 h after cooking the steaks. The color was measured 3 times on each sample and each value resulted from the arithmetic mean of three measurements. The results are presented in terms of variation (Δ) in relation to the initial color for marinated samples. In the case of cooked samples, the color variation (Δ) was calculated in relation to the unmarinated cooked sample.

2.5.5. Mechanical Texture Measurement

The texture properties of the steaks were analyzed with a texture analyzer TA. XT plus (Stable Micro Systems, Surrey, UK) equipped with a Warner–Bratzler V-shaped blade (0.9 mm thickness and a triangular aperture of 60°). The sample was cut into small pieces of known size with parallelepipedal form (2 × 1 × 1 cm). The texture analyzer was calibrated with a 5 kg load cell. The steaks were cut perpendicular to the longitudinal muscle fibers. The Warner–Bratzler blade was pressed down (compression test mode) at a constant pre-test and test speed of 5 mm.s^{-1}, and post-test speed of 10 mm.s^{-1} through the sample and the maximum shear force (Newton) were recorded. The texture of the steaks was determined twenty times on each sample [23,25].

2.5.6. Histological Technique

For histological analysis, a small portion of a steak (about 2 × 2 × 1 cm) was collected from each treatment's sample (marinade and cooking) and control (raw steak). These samples were placed individually in plastic cups with 10% neutral formalin until the histological processing took place. The sample cuts must be fine to facilitate the fixing of the formol (thickness not exceeding 1 cm). These samples were fixed in 10% neutral buffered formalin and submitted for histological routine processing technique (fixation in formalin, embedding in paraffin, and staining with hematoxylin and eosin). The slides were observed under an optical microscope, and photographs of the histological sections were made with an Olympus DP21 digital camera [26].

2.5.7. Statistical Analysis and Model Fitting

The experimental data was statistically evaluated through the StatisticaTM v.8 Software (StatSoft Inc., 2007, Tulsa, OK, USA) from Statsoft (StatSoft Inc., Tulsa, OK, USA) [27]. The multiple regression analysis was fitted to a second-order polynomial equation to predict each dependent variable (pH, marination yield, cooking yield, color, and hardness of steaks). The three-dimensional response surface designs are described by the second order polynomial models, using decoded variables, as a function of independent variables bromelain concentration on the brine (mg tyrosine.100 g^{-1} meat) and contact time (h). The adequacy of the model to fit the experimental data was verified by the analysis of variance (ANOVA) and coefficient of determination (R^2) and adjusted R^2 (Adj-R^2) [22].

For the tested parameters, obtained results were compared to those of the same contact times of the experimental design. Tukey's honest significant difference (HSD) test was used to determine the significant differences among means for different treatments. The accepted level of significant differences was $p < 0.05$.

3. Results

3.1. Optimization of Pineapple By-Products Enhanced in Bromelain Application on Marinated Steaks

3.1.1. pH Value

The addition of greater amounts of bromelain to the brine promoted a decrease in pH, making the brine more acidic (the initial pH of steak was 5.44), as can be seen in Figure 1a. The surface response methodology described very well the pH change that occurs over time at the different concentrations of bromelain introduced in the marinade. The model presented a very good fit, with $R^2 = 0.97$ and Adj-$R^2 = 0.94$ and the adequacy of the second-order polynomial model (Table 2) as confirmed by the not significant lack of fit ($p > 0.05$). According to the model, the variables time and concentration of bromelain significantly influence the marinade pH values. Marinades with higher amounts of enzyme showed lower pH values (pH = 4.31–4.69). The lowest enzyme concentrations showed higher pH values (pH = 5.49–5.67) regardless of the marinating time (Figure 1a). Marinades with bromelain concentrations greater than 10 mg tyrosine.100 g^{-1} meat had pH values below 5.

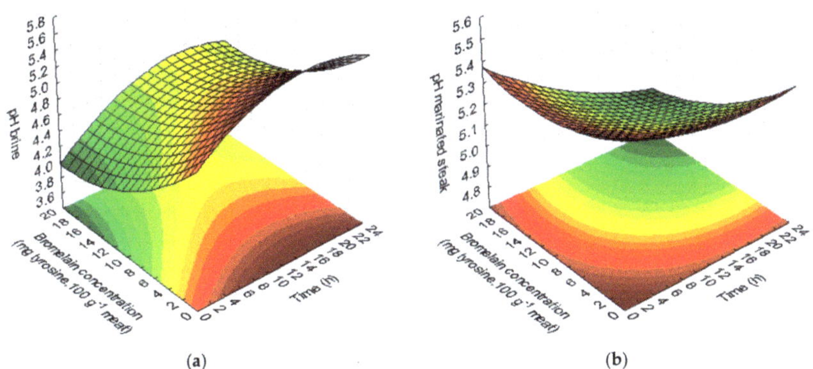

(a)   (b)

Figure 1. Effect of marinades on pH as a function of bromelain concentration (mg tyrosine.100 g^{-1} meat) and time (h). Response surfaces fitted to pH of brine (**a**) and pH of marinated steak (**b**).

The quadratic model described the behavior of pH in steaks after marinade (Table 2) with good fit ($R^2 = 0.95$ and Adj-$R^2 = 0.92$). Both bromelain concentration and immersion time had a linear negative significant effect. When the bromelain quantities were smaller, but the immersion time is longer (greater than or equal to 12 h), the steak pH decreased (Figure 1b). Samples with no added enzyme (0 mg tyrosine.100 g^{-1} meat) with small amounts (3 mg tyrosine.100 g^{-1} meat) for a short time (3.5 h) or 10 mg tyrosine.100 g^{-1} meat for a very short time (0 h), showed significantly higher pH values ($p < 0.05$) than steak samples marinated with higher concentrations of bromelain enzyme (17–20 mg tyrosine.100 g^{-1} meat). The pH values of treated steaks range from 4.99 to 5.44, with the highest value corresponding to the initial pH of the steak (without marinade). The lowest pH values were obtained for marinated steaks with 17 mg tyrosine.100 g^{-1} meat for 3.5 h (pH = 4.31) and 20.5 h (pH = 4.60) ($p < 0.05$).

Table 2. Regression coefficients of second-order polynomial equations for each response variable: pH of brine, marinated steak, cooked steak; marination yield (MY); cooking yield (CY); color of marinade and cooked steak; and hardness (H).

Parameter		Equation	R^2	R^2 adj	Lack of Fit
pH	Brine	$pH = 5.111^* + 0.077\ t^* - 0.002\ t^{2*} - 0.115\ C^* + 0.003\ C^{2*} + 0.0005\ tC$	0.97	0.94	Not significant
	Marinated steak	$pH = 5.476^* - 0.016\ t^* + 0.0004\ t^2 - 0.021\ C^* + 0.0008\ C^{2*} - 0.0007\ tC$	0.95	0.92	Not significant
	Cooked steak	$pH = 5.820^* + 0.017\ t - 0.0005\ t^2 - 0.016\ C + 0.0004\ C^2 - 0.00006\ tC$	0.68	0.42	Significant
Marination yield (MY)		$MY = 0.275 + 0.375\ t^* - 0.008\ t^2 + 0.059\ C - 0.003\ C^2 - 0.006\ tC$	0.86	0.75	Not significant
Cooking yield (CY)		$CY = 66.180^* - 2.746\ t + 0.706\ t^2 + 0.290\ C + 2.969\ C^2 - 0.053\ tC$	0.55	0.18	Not significant
Color	Marinated steak	$\Delta L = 1.858 + 1.267\ t^* - 0.036\ t^{2*} + 0.125\ C - 0.005\ C^2 - 0.0007\ tC$	0.84	0.71	Not significant
		$\Delta a = -2.836 - 1.123\ t^* + 0.027\ t^{2*} - 0.185\ C + 0.002\ C^2 + 0.011\ tC$	0.86	0.74	Not significant
		$\Delta b = -0.484 - 0.421\ t^* + 0.012\ t^2 - 0.094\ C + 0.001\ C^2 + 0.009\ tC$	0.57	0.21	Not significant
	Cooked steak	$\Delta L = 12.630^* + 0.507\ t^* - 0.009\ t^2 + 0.246\ C - 0.009\ C^2 - 0.0006\ tC$	0.85	0.73	Not significant
		$\Delta a = -13.691^* - 0.330\ t + 0.006\ t^2 - 0.232\ C + 0.010\ C^2 + 0.004\ tC$	0.53	0.14	Not significant
		$\Delta b = 4.468^* - 0.223\ t + 0.007\ t^2 - 0.240\ C + 0.009\ C^2 - 0.0007\ tC$	0.53	0.13	Not significant
Hardness steak (H)		$H = 48.916^* - 0.398\ t + 0.009\ t^2 - 1.064\ C^* + 0.030\ C^2 - 0.012\ tC$	0.89	0.80	Not significant

*: Parameter significantly affecting the response variable, $p < 0.05$.

The quadratic model did not adjust to the pH values of the steaks after cooking. The pH values fluctuated between 5.65 and 5.99.

3.1.2. Marination Yield

The quadratic model generated for the marination yield (MY) was significant in fitting the experimental data to a 95% confidence level. The R^2 and Adj-R^2 values were high (Table 2) and indicated a good fit to the data. In addition, the suitability of second-order polynomial models was validated by the lack of fit non-significant ($p > 0.05$). Steak samples had a MY of 0.07% to 4.19%. Only the linear positive significant effect of immerging time was observed. Figure 2 shows that longer times (12–24 h) increase the MY of the steak twice. Longer times (20.5–24 h) and low or intermediate bromelain concentrations (0–10 mg tyrosine.100 g^{-1} meat) increased steak MY, while shorter times (0–3.5 h) and intermediate or high concentrations (10–17 mg tyrosine.100 g^{-1} meat) of enzymes decreased MY.

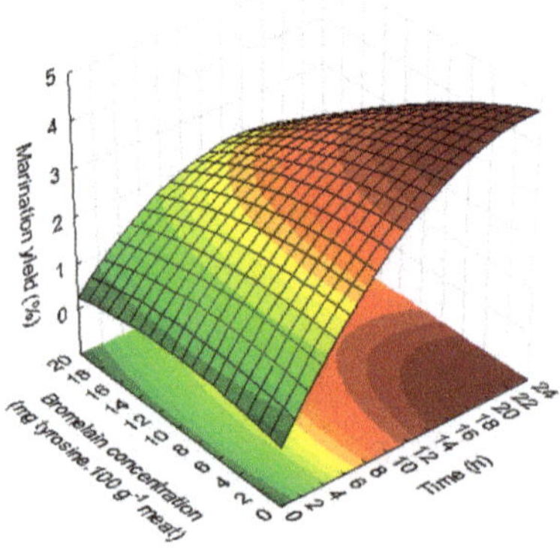

Figure 2. Effect of marinades on marination yield of the steaks as a function of marinades: bromelain concentration (mg tyrosine.100 g^{-1} meat) and time (h).

3.1.3. Cooking Yield

The generated model did not have a good fit and the study variables (bromelain concentration and time) did not have a significant influence on cooking performance (Table 2).

3.1.4. Color

The obtained models for the color parameters show that the color is significantly influenced by the immersion time in brine and that the concentration of bromelain does not have any significant effect on this parameter. The effect of the time during which meat was in the brine on ΔL^* was more relevant than the effect of bromelain concentration (Table 2).

The values presented in response surface methodology are related to the difference between the marinated samples and initial values of color. The quadratic models generated for the values of L for both marinated and cooked steaks were significant in fitting the experimental data to a 95% confidence level, and the R^2 and Adj-R^2 values were high (Table 2). The suitability of second-order polynomial models was validated by the lack of fit non-significant ($p > 0.05$). With respect to the L^* parameter of the marinated steaks and cooked steaks, linear positive and quadratic negative significant effects ($p < 0.05$) of the marinating time were observed.

The second-order polynomial model generated for ΔL^* shows that higher ΔL^* values for marinade steaks were observed between 12–22 h, regardless of the bromelain concentration (Figure 3a).

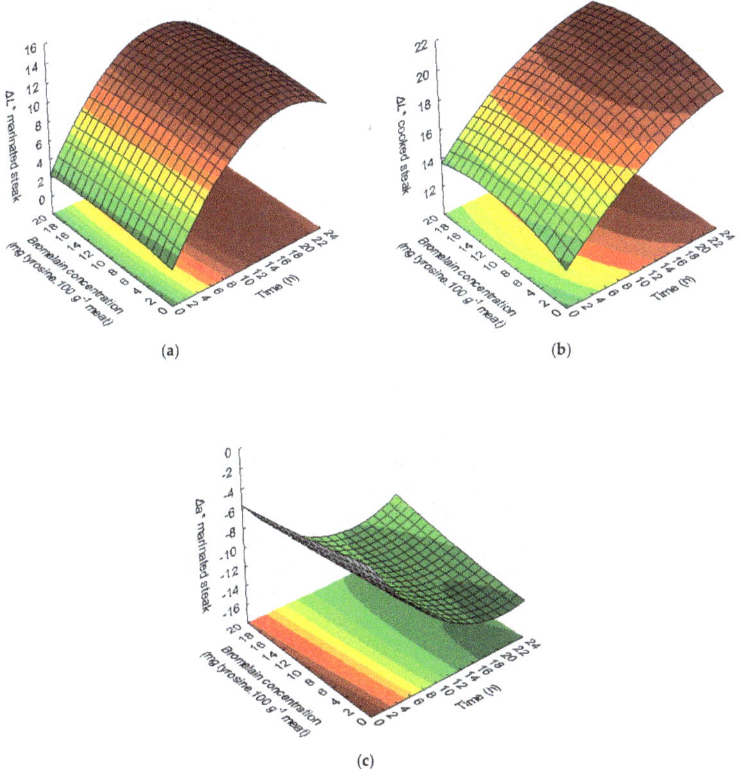

Figure 3. Response surfaces fitted for color parameters in marinated steaks: parameter ΔL^* (**a**), parameter Δa^* (**c**), and parameter ΔL^* in cooked steaks (**b**).

Figure 3b shows that an increase in the value of parameter ΔL* of steaks after cooking was observed after 15 h of marinade, regardless of the concentration of bromelain. Time is the significant variable in the model, while the bromelain concentration has no significance in the model. It was observed that longer times (>12 h), with or without the addition of the enzyme, made the steaks lighter/paler after cooking ($p < 0.05$).

The color parameter a* (Table 1) showed significantly higher values (21.21) for bromelain concentrations of 10 mg tyrosine.100 g^{-1} meat, with shorter marinade times (0 h). The longest marinades (24 h) with the addition of bromelain enzyme (10 mg tyrosine.100 g^{-1} meat) had the lowest value of parameter a* (11.47). After cooking, steaks without marinade had a* value of 10.16 and the longest marinade (24 h) with addition of enzyme (10 mg tyrosine.100 g^{-1} meat) induced the lowest a* value on steak (6.42).

The model obtained for parameter Δa* of the marinated steak presents a good fit to the data (R^2 = 0.86 and Adj-R^2 = 0.74) and a not significant lack of fit (Table 2). Figure 3c shows that the highest values of Δa* were obtained for shorter times (significant variable), regardless of the concentration of bromelain in the marinade. Parameter Δa* emphasizes the loss of meat pigments, becoming less red due to the loss of myoglobin.

The quadratic models obtained for color parameters Δa* and Δb* after steak cooking have insufficient adjustment to the data (R^2 < 0.75) and therefore are not shown. Cooked samples with shorter marinating time showed lower Δa* values and, as for ΔL* values, the variable that influenced the results was time.

The quadratic model generated for the color parameter Δb* of the marinated steak did not fit and did not present a well-defined tendency (R^2 < 0.75). The color values obtained for the marinated steaks (3.26–6.32) did not show significant differences ($p > 0.05$) in color parameter b*. After cooking, the color values for parameter b* (8.75–11.99) also showed non-significant differences (Table 1). The steak sample with shortest time of contact with the marinade presented the highest a* and the lowest L*, in other words, they presented a color similar to the non-marinated steak (raw).

3.1.5. Texture (Hardness)

The quadratic model obtained for hardness after cooking have adjustment of the data range (R^2 > 0.75). The model showed a tendency of hardness to decrease for higher marinating times and higher concentrations of bromelain (Figure 4). Steaks that were not marinated had a significantly higher hardness ($p < 0.05$) than marinated steaks. Samples with a higher concentration of bromelain (17 mg and 20 mg tyrosine.100 g^{-1} meat) showed a tendency towards less hard steaks, although longer marinades (>12 h) with lower bromelain concentrations (3 and 10 mg tyrosine.100 g^{-1} meat) also reduced the hardness (Table 1).

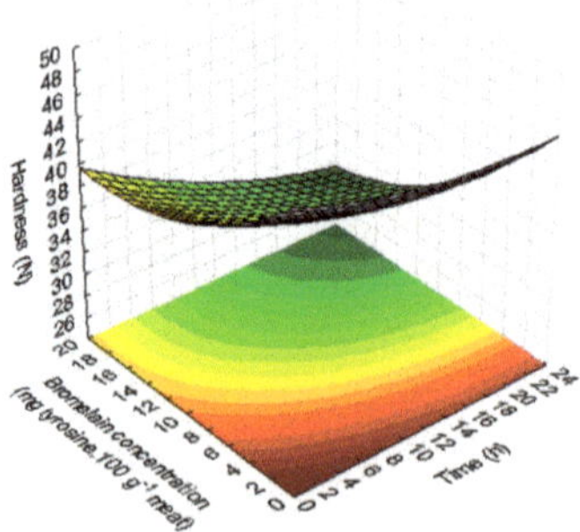

Figure 4. Response surface fitted for hardness as a function of marinades (bromelain concentration (mg tyrosine.100 g^{-1} meat) and time (h)).

3.1.6. Evaluation of Histological Technique

The images obtained through the histological technique can be seen in the Table 3. Skeletal muscle fibers appear to be more affected by marinating time than by the concentration of bromelain. The skeletal muscle fibers suffered a slight degradation for the marinade with short times (3.5 h) and a moderate degradation for higher marinade times (12 h), although the concentration of bromelain was lower (10 mg tyrosine. 100 g^{-1} meat) in the marinade lasting 12 h than in the 3.5 h (17 mg tyrosine. 100 g^{-1} meat) marinade.

The cooked sample showed severe disorganization in the skeletal muscle fibers. These structural changes in skeletal muscle fibers occurred due to heat treatment.

The intermuscular collagen fibers presented a moderate degree of degradation, regardless of the marinade time and the concentration of the bromelain enzyme. The complete disorganization of the inter-muscular collagen fibers was observed for the cooked samples and not for the raw sample, with the same marinating conditions. It can be concluded that the disorganization of the inter-muscular collagen fibers is an effect of the heat treatment and not an effect of the conditions of the marinade.

Muscle fibers appeared swollen in the sample with the highest concentration of bromelain enzyme in the marinade, while for lower concentrations in the marinade the muscle fibers remain generally preserved.

Table 3. Histological technique images collected through optical microscope with digital camera as a function of marinades and cooking: bromelain concentration (mg tyrosine.100 g^{-1} meat) and time (h).

Sample Processing	Concentration (mg Tyrosine. 100 g^{-1} Meat)	Time (h)	Microphotography	Qualitative Description
Raw	10	12		On this microphotograph is visible a moderate degree of disorganization of skeletal muscle fibers and a moderate degree of disorganization of the inter-muscular collagen fibers (H&E. Barr 200 µm).
	17	3.5		On this microphotograph is visible a slight degree of disorganization of skeletal muscle fibers. (H&E. Barr 200 µm).

Table 3. *Cont.*

Sample Processing	Concentration (mg Tyrosine. 100 g^{-1} Meat)	Time (h)	Microphotography	Qualitative Description
		20.5		On this microphotography is visible a moderate degree of disorganization of the inter-muscular collagen fibers and muscle fibers are generally preserved (H&E. Barr 200 µm).
	20	12		On this microphotography is visible a moderate degree of disorganization of the inter-muscular collagen fibers and muscle fibers are swollen (H&E. Barr 200 µm).
Cooked		12		On this microphotograph is visible a severe degree of disorganization of skeletal muscle fibers and a complete disorganization of the inter-muscular collagen fibers (H&E. Barr 200 µm).

3.2. Brine Effect on Marinated Steaks

The initial brine (without addition of enzymes) showed a higher pH (7.42) and significantly different ($p < 0.05$) from the remaining brines (Figure 5a). The time (3.5 h, 12 h, 20.5 h, 24 h) showed a significant effect ($p < 0.05$) on lowering the pH of marinated steak and the pH of brine. We observed an increase of the cooked steak pH compared to the initial steak and brine pH (Figure 5a).

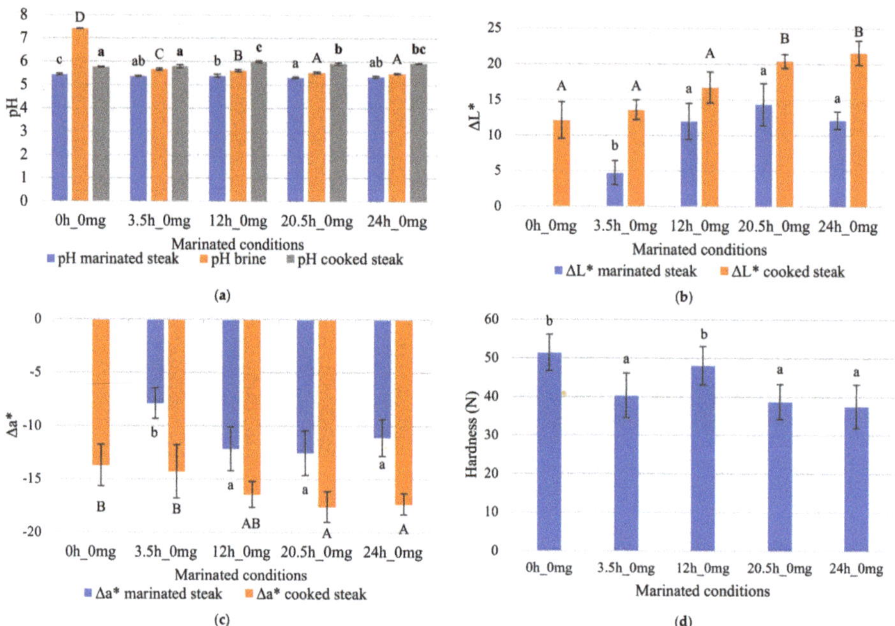

Figure 5. Study of the effect of adding brine on marinated steaks: (**a**) pH, (**b**) ΔL* color, (**c**) Δa* color, and (**d**) hardness. Error bars represent ± standard deviation (pH and color: $n = 3$ and hardness: $n = 20$). pH: Different letters express significant differences between pH marinated steak samples (lower case letters); between pH brine samples (upper case letters); and between pH cooked steak samples (bold lower-case letters). Color: Different letters express significant differences between color marinated steak samples (lower case letters) or between color cooked steak samples (upper case letters).

A 24 h marinating time did not significantly influence MY (2.11–3.58%) when steaks were immersed just in brine.

The steak that was not marinated (control sample) showed a higher cooking yield (82.04%) than the marinated samples. The contacting time between steak and brine without bromelain enzyme imparted a reduction in cooking yield of 8.53% to 12.40% (69.64–73.51%).

The steaks without marinade showed the lowest value of L* (38.14), while the highest value of L* (52.45) was obtained for samples marinated for 20.5 h without adding enzymes, thus verifying the significant effect ($p < 0.05$) of the marinade time (Figure 5b). The L* parameter of the marinated steaks (Table 2) was significantly influenced by the marinating time ($p < 0.05$). After cooking, the untreated steak presented the lowest L* value (50.26) while samples marinated for 24 h showed the highest L* values (58.66) for a bromelain concentration of 0 mg tyrosine.100 g^{-1} meat. The color parameter ΔL* in marinated steak samples was significantly higher for samples with marinating time equal to or greater than 12 h. In cooked samples, the parameter ΔL* was significantly higher for marinades 20.5 h and 24 h, compared to shorter times (Figure 5b).

The marinated steak samples showed significantly lower Δa* values for marinades greater than 12 h and the cooked steak samples showed lower values for marinating times greater than 20.5 h (Figure 5c).

The immersion of steaks in brine seems to promote a reduction in hardness. Samples with marinade times 3.5 h, 20.5 h, and 24 h showed significantly lower values than samples without marinade (Figure 5d). The brine was advantageous in reducing the firmness of the steaks.

4. Discussion

The pineapple by-product is acidic and lowers the pH of the meat, which consequently increases the marination yield and lowers cooking yield. The pH of the brine significantly reduced to pH values below 5, with the increase in the concentration of bromelain, and for this reason the pH of meat was decreased with consequent denaturation of proteins, including possibly myoglobin. These changes on meat pH may have implications for its structure with influence on MY and cooking yield.

The enzymatic activity of bromelain in steak tenderization, besides being affected by immersion time and bromelain concentrations, is also affected by marinade pH and temperature [28,29]. The pH was reported to decrease in steak, chicken, and squid samples, after marinades with bromelain were at room temperature [30].

A slight change in pH can affect meat stability, and the pH promotes myoglobin oxidation. The pH value of meat greatly impacts MY, tenderness, and juiciness. The pH value of steak meat decreased after treatment with commercial pineapple stem bromelain and the greatest reduction was observed for steak meat treated with purified core bromelain [31].

Besides the pH of meat, the enzymatic action of bromelain on the protein structure can influence the marination yield. The fibers in the muscle are broken and lose their ability to retain water. Higher enzyme concentrations reduce the marination yield and, consequently, the meat loses water. MY is a crucial quality characteristic for sensory attributes, but also particularly important economically. The meat's ability to retain its own water is an important feature of meat quality and contributes to its juiciness, flavor, and color [32]. The longer the steaks are in the brine, the greater the water absorption since the salt can promote protein solubilization and adsorb water. The addition of salt increased the MY of the meat, with a high MY due to a low interaction between actin and myosin [33]. The reduction in pH also influences the structure of myofibrillar proteins (principally myosin and actin) [34,35].

The salt (sodium chloride) changes the isoelectric point of the meat proteins to lower pH values and, consequently, changes the properties of the meat tissues, thus increasing water binding [36]. Salt increases the solubility of meat proteins and also increases ionic strength, which influences the increase of MY [37,38]. Sodium chloride (0 to 2%) applied to steak decreases the meat drip and increases the marination yield [38]. The marination yield is related to the myofibrillar protein, namely the proteins and structures that bind and entrap water. Myofibrillar proteins and myofibrils and muscle cells can entrap water, but this depends on the influence of pH, ionic strength, and oxidation [39]. MY is related to water structure and ion bonding. The binding of sodium and chloride ions in proteins and the salt-protein interactions in the water structure may explain the effect of salt on water retention through the selective binding of chloride ions in the hydrophobic regions of the myosin filaments [40]. The hydration force consists of binding the ion to the surface of the protein and forming the water around the ion creates a repulsive barrier [41].

Other authors have also referred the behavior of steak under bromelain marinade observed in this study. The reduction in MY depended on the influence of bromelain on the denaturation of myofibrillar proteins through the hydrolysis of tissue fibers [30,42].

The MY tendency can maintain, decrease, or increase. Some studies using enzyme treatments reported that the increases were not significant, while others described significant increases in MY in buffalo meat, steak, and turkey meat [43–45]. Other studies registered MY reductions in steak, chicken, and squid [32]. The reduction in the pH of meat from egg-laying chickens also influenced MY [46]. A study carried out in the post mortem of pork meat also observed the relationship between the decrease in pH, the mobility of myowater and related to the MY of the meat [35]. Another study demonstrated that the application of NaCl in steak increased the water retention capacity in the meat, which also reduced losses during cooking [47].

A higher enzyme concentration and a longer marinating time influenced the loss of cooking yield because the enzymatic action allowed for the alteration of cellular structure and reduced the capacity to retain water; on the other hand, the marinade with low salt concentrations allowed for MY to increase. The incorporation of NaCl (0.7% or 1.2% *w/w*) and cooking at temperatures (60–65 °C) increased the MY of muscles (*M. gluteobiceps*) with reduced weight loss during cooking. Meat treated with salt (NaCl) showed an increase in protein solubilization that may be related to the elimination of myofibrillar proteins and act as an impediment to the extraction of myosin. The softening of the protein structure and the conformational changes of myofibrillar proteins would allow water to be retained inside the tissue [48]. The elimination of water increased with an increase in cooking temperature and the contraction of myofibrillar proteins due to the increase in temperature for the period of warming, which influenced the results of cooking loss [49].

Meat quality and cooking yield are an industry concern, with repercussions also in meat attributes that affect consumer satisfaction and depend on the MY and meat pH. Cooking losses are reflected in meat weight loss. The results of this study were consistent with the results found by other authors. Weight loss due to cooking increased in samples marinated that contained pineapple juice over marinating time in egg-laying chicken samples [46]. The bromelain powder produced from pineapple increased the cooking loss of steak round cuts [15].

The cooking yield of steak, chicken, and squid samples treated with bromelain was lower than the control samples and it was also found that the cooking performance decreased with the increase of the bromelain extract [30]. In the present study, it was seen that the addition of bromelain promoted a decrease in the pH value of the meat, which caused less water retention and, consequently, lower cooking yields.

The analysis of the color values obtained in these assays indicates that the time the meat is in the brine is more relevant than the concentration of bromelain added to brine, in the color parameter ΔL. The steaks becoming paler can be explained by the solubilization of myoglobin in the brine during the marinade and, consequently, the concentration of myoglobin in steaks decreases. The changes observed in parameter Δa emphasize the loss of pigments in the color of the meat dipped in the brine.

Steak samples marinated with higher concentrations of bromelain have a lower pH and a lighter color with less red color. A study with bovine *Longissimus dorsi* muscles observed that the addition of citric acid in a brine at 2% NaCl (*w/v*) caused steaks to become lighter and less red in color than the brine at 2% NaCl (*w/v*). One of the reasons for this is that the lower pH in the citric acid solution influences the denaturation of muscle proteins and causes a greater reflection of light. The effects of marinades on luminosity and redness may result from effects on pH or ionic strength. The low pH also favors the oxidation of myoglobin [50].

A study with *semitendinosus* steak evaluated the purged liquid and found that the samples treated with enzyme injection (pancreatin) had a larger volume of liquid and had a darker red color in purged fluid. The author suggested that the weight lost in these samples signified water-soluble sarcoplasmic proteins that contained myoglobin. The steaks treated with enzymes showed a lighter color, probably due to loss of myoglobin [51].

Another study reported a decrease in luminosity (L* value) and an increase in redness (a* value) in the steak meat cut samples treated by the dipping method with purified bromelain reconstituted in distilled water compared to the control sample. These results were the opposite of the results obtained in this study, but the author mentions possible explanations for the observed color changes. Reducing the amount of metmyoglobin to myoglobin can improve color stability, since oxidation of myoglobin maintains the red color of meat. Lipid oxidation and pigment oxidation in steak are also associated with meat discoloration [31]. Bromelain being reconstituted only in distilled water and without adding sodium chloride may be an explanation for the differences observed. Moreover, the facts that the meat was put in contact with marinade for only 2 h at 25 °C with purified bromelain could explain why the meat was not paler. Furthermore, the initial pH meat value (6.1) was higher than the values reported in the present study (pH = 5.44) and, consequently, the behavior of the meat may not be the same.

The marinade with pineapple juice showed changes in color parameters in chicken meat after cooking [46]. Changes in the color of the meat can be caused by heat treatment [52]. The increase in temperature during cooking makes the meat paler (increases the L* value) and reduces the red color (decreases the a* value) [53].

The results obtained by other authors agreed with the results obtained in this study. The values of L * and b * increased for steak round cuts samples treated with bromelain powder [15].

Many studies of marinated meat concluded that marinades containing pineapple juice had the greatest impact on meat tenderness and caused the most intense structural and textural alterations, although the attractiveness of the meat worsened [8,9]. Bromelain has been extensively studied by several authors as meat tenderizer for various animals such as steak, mutton, chicken, and pork [30,54,55]. Pineapple peel bromelain was an efficient softener for the steak, chicken, and squid [30]. The samples of steak, chicken, and squid muscles treated with bromelain extract showed lower firmness (shear force values decreased) with an increase of the amount of bromelain [30]. Pineapple juice marinade has also improved the tenderness of various parts of spent chicken meat and increased consumer acceptability. The author suggests marinating as a technique to increase the economic value of spent chicken meat [46].

The results obtained in this study showed a reduction in hardness between 6% and 41%. Several authors have observed a decrease in the Warner–Bratzler shear force of beef between 7% and 21% in beef meat with the application of several proteases, including bromelain [18,31]. In another study, the action of bromelain on meat tissue also increased tenderness. The steak meat samples was incubated at room temperature (25 ± 3 °C) for 2 h and treated with purified core bromelain showed a greater reduction (52.1%) in the Warner–Bratzler shear force than samples treated with commercial stem bromelain (26.8%), although both samples showed a reduction in the Warner–Bratzler shear force compared to the control samples [31]. In the present study, the beef meat was marinated with the bromelain present in the pineapple core by-products without extraction or purification. These marinades took place at refrigerated temperatures (5 ± 1 °C) for longer periods of time (0–24 h) and showed a reduction maximum of 41% in Warner–Bratzler shear force compared to the control samples. Although the refrigeration temperature does not approach the optimum temperature of the bromelain activity, the meat's hardness showed a considerable reduction.

Resistance depends on the quantity of intramuscular connective tissue, intramuscular fat, and sarcomere length. The action of proteolytic enzymes on myofibrillar proteins promotes a decrease in meat firmness due to the breakdown of these proteins into small peptides or proteins with low molecular weight [30,56]. Toughness is instigated by enhanced cross-linking in connective tissue [57]. The integrity of the structure of connective tissues and myofibrils can influence meat softness [12].

Inactivation of bromelain (75 ± 1 °C) may not be complete with medium or rare degrees of cooking; there is the possibility of high residual enzymatic activity after cooking, which can cause over-tenderization [13].

The optimum conditions for bromelain to have maximum activity are temperatures in the range 37–70 °C [17]. Meat quality legislation does not allow storage at temperatures above 7 °C [58]. Although the marinade process was at a temperature below the favorable conditions for the enzyme (4 °C), bromelain showed advantageous results in tenderizing the steaks.

Histology allows the detection of specific tissue components and architectural changes promoted by distinct types of meat processing [59]. Immersion treatments change the structure of steak meat tissue, and the distance between cells increases with decreasing pH. The analysis of histological images showed that the swelling of the meat tissue is due to the increase in the extracellular space or to the swelling of the meat fibers [50]. Another study with bovine muscle notes that the microstructure undergoes profound changes during acidification, but that it regains normal structure when the pH is adjusted to the common pH of raw bovine muscle. This disorganization of muscle proteins and the absorption of water promote a decrease in shear strength and an increase in steak meat tenderization [28]. Steak *Longissimus lumborum* muscle samples immersed in the sonicated papain solution also presented larger intracellular spaces and cell wall disruption, the most evident changes being in the surface layer of the meat [26].

The marinating and cooking processes did not improve the yield since the increased water retention (4%) during marinade was lost during cooking (19%), presenting total losses of 15%. The hydration of the meat is due to the infiltration of the marinade in extracellular spaces [60]. The qualitative changes promoted by cooking suggest gradual modifications in the denaturation of collagen fibers, increasing softness, in a study with female carabeef (buffalo) meat. Higher temperature and longer cooking time reduced the shear force value and collagen content, and increased pH, cooking loss, collagen solubility, and tenderness values [61]. Comparable results were found for bovine Semitendinosus muscle increased when roasted at 80–90 °C and soluble collagen amount in bovine meat roasted to 80 °C [62].

5. Conclusions

The addition of dehydrated pineapple by-products enriched in bromelain via high hydrostatic pressures to marinades induced positive modifications on meat characteristics. Pineapple by-products were acidic and lower the pH of meat. The lower pH of steak during cooking induced a greater precipitation of proteins with greater loss of yield in cooking. The color of the steaks was more influenced by marinating time than the concentration of bromelain. Long marinades and the addition of bromelain decreased the steaks' hardness. Skeletal muscle fibers appeared to be more affected by marinating time than by the concentration of bromelain. Severe structural changes occurred due to heat treatment. The intermuscular collagen fibers presented a moderate degree of degradation, regardless of the marinade conditions.

Longer marinades (12–24 h) with a higher concentration of bromelain (10–20 mg tyrosine.100 g^{-1} meat) reduced pH (8%) and hardness (41%) of the steaks, increased marination yield (4%), and turned them lighter in color (L* marinated steak increased 38%) compared to samples without marinade. The pineapple by-products are a source of bromelain and, thus, this work contributes both to the recovery of an agro-industrial waste and to the economic valorization of steak pieces with less commercial value due to their hardness.

Author Contributions: Conceptualization, D.I.S., M.J.F., and M.M.-M.; methodology, D.I.S., M.J.F., H.P., and M.M.-M.; formal analysis, D.I.S. and H.P.; investigation, D.I.S., M.J.F., H.P., and M.M.-M.; writing—original draft preparation, D.I.S., M.J.F., and H.P.; writing—review and editing, D.I.S., M.J.F., H.P., J.A.S., A.A.V., and M.M.-M.; supervision M.J.F., J.A.S., A.A.V., and M.M.-M.; funding acquisition M.J.F., J.A.S., A.A.V., and M.M.-M. All authors have read and agreed to the published version of the manuscript.

Funding: The first author acknowledges the financial support from Fundação para a Ciência e a Tecnologia (FCT), Portugal, through a doctoral fellowship (SFRH/BD/109124/2015). This work was supported by the national funding of FCT, under the scope of the strategic funding to the research units LEAF, Linking Landscape, Environment, Agriculture and Food, School of Agriculture, University of Lisbon, 1349-017 Lisbon (UIDP/04129/2020), CIISA, Interdisciplinary Centre of Research in Animal Health, Faculty of Veterinary Medicine, University of Lisbon, 1300-477, Lisbon (UIDB/00276/2020) FMV and PDR2020-1.0.1-FEADER-031359, funded by the European Regional Development Fund (ERDF)), QOPNA (UID/QUI/00062/2019), LAQV-REQUIMTE, Department of Chemistry, University of Aveiro, 3810-193 Aveiro (UIDB/50006/2020) and CEB, Centre of Biological Engineering, Department of Biological Engineering, University of Minho, 4710–057 Braga (UIDP/04469/2019) through national funds and where applicable co-financed by the FEDER, within the PT2020 Partnership Agreement.

Acknowledgments: The authors acknowledge their industrial partner Campotec, located in Torres Vedras, for supplying the pineapple by-products used in this study. M.J.F. is a member of network Red CYTED HEALTHY MEAT (119RT0568).

Conflicts of Interest: The authors declare no conflict of interest.

References

1. Umesh Hebbar, H.; Sumana, B.; Raghavarao, K.S.M.S. Use of reverse micellar systems for the extraction and purification of bromelain from pineapple wastes. *Bioresour. Technol.* **2008**, *99*, 4896–4902. [CrossRef]
2. Ketnawa, S.; Chaiwut, P.; Rawdkuen, S. Extraction of bromelain from pineapple peels. *Food Sci. Technol. Int.* **2011**, *17*, 395–402. [CrossRef] [PubMed]
3. Ketnawa, S.; Chaiwut, P.; Rawdkuen, S. Pineapple wastes: A potential source for bromelain extraction. *Food Bioprod. Process.* **2012**, *90*, 385–391. [CrossRef]
4. Jacobo-Velázquez, D.A.; del Rosario Cuéllar-Villarreal, M.; Welti-Chanes, J.; Cisneros-Zevallos, L.; Ramos-Parra, P.A.; Hernández-Brenes, C. Nonthermal processing technologies as elicitors to induce the biosynthesis and accumulation of nutraceuticals in plant foods. *Trends Food Sci. Technol.* **2017**, *60*, 80–87.
5. Campos, D.A.; Gómez-García, R.; Vilas-Boas, A.A.; Madureira, A.R.; Pintado, M.M. Management of fruit industrial by-products—a case study on circular economy approach. *Molecules* **2020**, *25*, 320. [CrossRef]
6. Pathania, A.; McKee, S.R.; Bilgili, S.F.; Singh, M. Antimicrobial activity of commercial marinades against multiple strains of *Salmonella* spp. *Int. J. Food Microbiol.* **2010**, *139*, 214–217. [CrossRef]
7. Burke, R.M.; Monahan, F.J. The tenderisation of shin beef using a citrus juice marinade. *Meat Sci.* **2003**, *63*, 161–168. [CrossRef]
8. Zochowska-Kujawska, J.; Lachowicz, K.; Sobczak, M. The tenderisation of wild boar meat using a calcium chloride, kefir, wine and pineapple marinade. *Electron. J. Pol. Agric. Univ. Ser. Food Sci. Technol.* **2010**, *13*, 2.
9. Zochowska-Kujawska, J.; Lachowicz, K.; Sobczak, M. Effects of fibre type and kefir, wine lemon, and pineapple marinades on texture and sensory properties of wild boar and deer longissimus muscle. *Meat Sci.* **2012**, *92*, 675–680. [CrossRef]
10. Botinestean, C.; Gomez, C.; Nian, Y.; Auty, M.A.E.; Kerry, J.P.; Hamill, R.M. Possibilities for developing texture-modified beef steaks suitable for older consumers using fruit-derived proteolytic enzymes. *J. Texture Stud.* **2018**, *49*, 256–261. [CrossRef]
11. Zainal, S.; Nadzirah, K.Z.; Noriham, A.; Normah, I. Optimisation of beef tenderisation treated with bromelain using response surface methodology (RSM). *Agric. Sci.* **2013**, *04*, 65–72. [CrossRef]
12. Calkins, C.R.; Sullivan, G. *Adding Enzymes to Improve Beef Tenderness*; National Cattlemen's Beef Association: Centennial, CO, USA, 2007; pp. 1–5.
13. Bekhit, A.A.; Hopkins, D.L.; Geesink, G.; Bekhit, A.A.; Franks, P. Exogenous Proteases for Meat Tenderization. *Crit. Rev. Food Sci. Nutr.* **2014**, *54*, 1012–1031. [CrossRef] [PubMed]
14. Buyukyavuz, A. Effect of Bromelain on Duck. In *All Theses*; Clemson University: Clemson, SC, USA, 2014; 68p.
15. Nadzirah, K.Z.; Zainal, S.; Noriham, A.; Normah, I. Application of bromelain powder produced from pineapple crowns in tenderising beef round cuts. *Int. Food Res. J.* **2016**, *23*, 1590–1599.
16. Gerelt, B.; Ikeuchi, Y.; Suzuki, A. Meat tenderization by proteolytic enzymes after osmotic dehydration. *Meat Sci.* **2000**, *56*, 311–318. [CrossRef]
17. Manzoor, Z.; Nawaz, A.; Mukhtar, H.; Haq, I. Bromelain: Methods of Extraction, Purification and Therapeutic Applications. *Braz. Arch. Biol. Technol.* **2016**, *59*. [CrossRef]
18. Sullivan, G.A.; Calkins, C.R. Application of exogenous enzymes to beef muscle of high and low-connective tissue. *Meat Sci.* **2010**, *85*, 730–734. [CrossRef]
19. Qihe, C.; Guoqing, H.; Yingchun, J.; Hui, N. Effects of elastase from a Bacillus strain on the tenderization of beef meat. *Food Chem.* **2006**, *98*, 624–629. [CrossRef]
20. Santos, D.I.; Pinto, C.A.; Corrêa-Filho, L.C.; Saraiva, J.A.; Vicente, A.A.; Moldão-Martins, M. Effect of moderate hydrostatic pressures on the enzymatic activity and bioactive composition of pineapple by-products. *J. Food Process Eng.* **2020**. [CrossRef]
21. Chakraborty, S.; Rao, P.S.; Mishra, H.N. Effect of pH on Enzyme inactivation kinetics in high-pressure processed pineapple (*Ananas comosus* L.) puree using response surface methodology. *Food Bioprocess Technol.* **2014**, *7*, 3629–3645. [CrossRef]
22. Montgomery, D.C. *Design and Analysis of Experiments*, 9th ed.; John Wiley & Sons, Inc.: Hoboken, NJ, USA, 2017; ISBN 9781119113478.
23. Honikel, K.O. Reference methods for the assessment of physical characteristics of meat. *Meat Sci.* **1998**, *49*, 447–457. [CrossRef]

24. *ISO 2917:1999 ISO 2917:1999(en) Meat and Meat Products — Measurement of pH (Reference Method)*; International Organization for Standardization: Geneva, Switzerland, 1999.

25. James, B.J.; Yang, S.W. Effect of cooking method on the toughness of bovine M. Semitendinosus. *Int. J. Food Eng.* **2012**, *8*. [CrossRef]

26. Barekat, S.; Soltanizadeh, N. Effects of Ultrasound on Microstructure and Enzyme Penetration in Beef Longissimus lumborum Muscle. *Food Bioprocess Technol.* **2018**, *11*, 680–693. [CrossRef]

27. *StatSoft STATISTICA*; Data Analysis Software System; TIBCO Software: Palo Alto, CA, USA, 2007.

28. Ke, S.; Huang, Y.; Decker, E.A.; Hultin, H.O. Impact of citric acid on the tenderness, microstructure and oxidative stability of beef muscle. *Meat Sci.* **2009**, *82*, 113–118. [CrossRef] [PubMed]

29. Ketnawa, S.; Sai-Ut, S.; Theppakorn, T.; Chaiwut, P.; Rawdkuen, S. Partitioning of bromelain from pineapple peel (Nang Lae cultv.) by aqueous two phase system. *Asian J. Food Agro-Ind.* **2009**, *2*, 457–468.

30. Ketnawa, S.; Rawdkuen, S. Application of Bromelain Extract for Muscle Foods Tenderization. *Food Nutr. Sci.* **2011**, *2*, 393–401. [CrossRef]

31. Chaurasiya, R.S.; Sakhare, P.Z.; Bhaskar, N.; Hebbar, H.U. Efficacy of reverse micellar extracted fruit bromelain in meat tenderization. *J. Food Sci. Technol.* **2015**, *52*, 3870–3880. [CrossRef]

32. Gokoglu, N.; Yerlikaya, P.; Ucak, I.; Yatmaz, H.A. Effect of bromelain and papain enzymes addition on physicochemical and textural properties of squid (*Loligo vulgaris*). *J. Food Meas. Charact.* **2017**, *11*, 347–353. [CrossRef]

33. Honikel, K.O.; Hamm, R. Measurement of water-holding capacity and juiciness. In *Quality Attributes and their Measurement in Meat, Poultry and Fish Products*; Springer: Berlin/Heidelberg, Germany, 1994; pp. 125–161.

34. Honikel, K.O.; Kim, C.J.; Hamm, R.; Roncales, P. Sarcomere shortening of prerigor muscles and its influence on drip loss. *Meat Sci.* **1986**, *16*, 267–282. [CrossRef]

35. Pearce, K.L.; Rosenvold, K.; Andersen, H.J.; Hopkins, D.L. Water distribution and mobility in meat during the conversion of muscle to meat and ageing and the impacts on fresh meat quality attributes—A review. *Meat Sci.* **2011**, *89*, 111–124. [CrossRef]

36. Hamm, R. Biochemistry Of Meat Hydration. *Adv. Food Res.* **1961**, *10*, 355–463.

37. Hamm, R. The influence of pH on the protein net charge in the myofibrillar system. *Reciprocal Meat Conf. Proc.* **1994**, *47*, 5–9.

38. Medyński, A.; Pospiech, E.; Kniat, R. Effect of various concentrations of lactic acid and sodium chloride on selected physico-chemical meat traits. *Meat Sci.* **2000**, *55*, 285–290. [CrossRef]

39. Huff-Lonergan, E.; Lonergan, S.M. Mechanisms of water-holding capacity of meat: The role of postmortem biochemical and structural changes. In *Proceedings of the Meat Science*; Elsevier: Amsterdam, The Netherlands, 2005; Volume 71, pp. 194–204.

40. Puolanne, E.; Halonen, M. Theoretical aspects of water-holding in meat. *Meat Sci.* **2010**, *86*, 151–165. [CrossRef] [PubMed]

41. Grigsby, J.J.; Blanch, H.W.; Prausnitz, J.M. Cloud-point temperatures for lysozyme in electrolyte solutions: Effect of salt type, salt concentration and pH. *Biophys. Chem.* **2001**, *91*, 231–243. [CrossRef]

42. Murphy, R.Y.; Marks, B.P. Effect of meat temperature on proteins, texture, and cook loss for ground chicken breast patties. *Poult. Sci.* **2000**, *79*, 99–104. [CrossRef]

43. Naveena, B.M.; Mendiratta, S.K.; Anjaneyulu, A.S.R. Tenderization of buffalo meat using plant proteases from Cucumis trigonus Roxb (Kachri) and Zingiber officinale roscoe (*Ginger rhizome*). *Meat Sci.* **2004**, *68*, 363–369. [CrossRef]

44. Karakaya, M.; Ockerman, W.H. The effect of NaCl-K2HPO4, some plant enzymes and oils on the emulsion and water holding capacities in beef. *GIDA* **2002**.

45. Doneva, M.; Miteva, D.; Dyankova, S.; Nacheva, I.; Metodieva, P.; Dimov, K. Efficiency of plant proteases bromelain and papain on turkey meat tenderness. *Biotechnol. Anim. Husb.* **2015**, *31*, 407–413. [CrossRef]

46. Kadıoğlu, P.; Karakaya, M.; Unal, K.; Babaoğlu, A.S. Technological and textural properties of spent chicken breast, drumstick and thigh meats as affected by marinating with pineapple fruit juice. *Br. Poult. Sci.* **2019**, *60*, 381–387. [CrossRef]

47. Aktaş, N.; Aksu, M.I.; Kaya, M. The Influence of Marination with Different Salt Concentrations on the Tenderness, Water Holding Capacity and Bound Water Content of Beef. *Turk. J. Vet. Anim. Sci.* **2003**, *27*, 1207–1211.

48. Vaudagna, S.R.; Pazos, A.A.; Guidi, S.M.; Sanchez, G.; Carp, D.J.; Gonzalez, C.B. Effect of salt addition on sous vide cooked whole beef muscles from Argentina. *Meat Sci.* **2008**, *79*, 470–482. [CrossRef] [PubMed]

49. Sánchez del Pulgar, J.; Gázquez, A.; Ruiz-Carrascal, J. Physico-chemical, textural and structural characteristics of sous-vide cooked pork cheeks as affected by vacuum, cooking temperature, and cooking time. *Meat Sci.* **2012**, *90*, 828–835. [CrossRef] [PubMed]

50. Sharedeh, D.; Gatellier, P.; Astruc, T.; Daudin, J.D. Effects of pH and NaCl levels in a beef marinade on physicochemical states of lipids and proteins and on tissue microstructure. *Meat Sci.* **2015**, *110*, 24–31. [CrossRef] [PubMed]

51. Janz, J.A.M.; Pietrasik, Z.; Aalhus, J.L.; Shand, P.J. The effects of enzyme and phosphate injections on the quality of beef semitendinosus. *Can. J. Anim. Sci.* **2005**, *85*, 327–334. [CrossRef]

52. Fletcher, D.L.; Qiao, M.; Smith, D.P. The relationship of raw broiler breast meat color and pH to cooked meat color and pH. *Poult. Sci.* **2000**, *79*, 784–788. [CrossRef]

53. Sen, A.R.; Naveena, B.M.; Muthukumar, M.; Vaithiyanathan, S. Colour, myoglobin denaturation and storage stability of raw and cooked mutton chops at different end point cooking temperature. *J. Food Sci. Technol.* **2014**, *51*, 970–975. [CrossRef]

54. Bille, P.G.; Taapopi, M.S. Effects of two commercial meat tenderizers on different cuts of goat's meat in Namibia. *Afr. J. Food, Agric. Nutr. Dev.* **2008**, *8*, 417–426.

55. Ieowsakulrat, S.; Theerasamran, P.; Chokpitinun, C.; Pinitglang, S. Tenderization of pork steak with fruit bromelain from pineapple monitored by sensory evaluation and texturometer. *Warasan Witthayasat Kaset* **2011**.

56. Kemp, C.M.; Sensky, P.L.; Bardsley, R.G.; Buttery, P.J.; Parr, T. Tenderness—An enzymatic view. *Meat Sci.* **2010**, *84*, 248–256. [CrossRef]

57. Archile-Contreras, A.C.; Cha, M.C.; Mandell, I.B.; Miller, S.P.; Purslow, P.P. Vitamins e and C May increase collagen turnover by intramuscular fibroblasts. Potential for improved meat quality. *J. Agric. Food Chem.* **2011**, *59*, 608–614. [CrossRef]

58. Regulamento (CE), n.o 853/2004 REGULAMENTO (CE) n.o 853/2004 do PARLAMENTO EUROPEU E DO CONSELHO de 29 de Abril de 2004 que estabelece regras específicas de higiene aplicáveis aos géneros alimentícios de origem animal. *J. Of. União Eur.* **2004**, 61.

59. Ghisleni, G.; Stella, S.; Radaelli, E.; Mattiello, S.; Scanziani, E. Qualitative evaluation of tortellini meat filling by histology and image analysis. *Int. J. Food Sci. Technol.* **2010**, *45*, 265–270. [CrossRef]

60. Liu, Z.; Xiong, Y.L.; Chen, J. Morphological examinations of oxidatively stressed pork muscle and myofibrils upon salt marination and cooking to elucidate the water-binding potential. *J. Agric. Food Chem.* **2011**, *59*, 13026–13034. [CrossRef] [PubMed]

61. Vasanthi, C.; Venkataramanujam, V.; Dushyanthan, K. Effect of cooking temperature and time on the physico-chemical, histological and sensory properties of female carabeef (buffalo) meat. *Meat Sci.* **2007**, *76*, 274–280. [CrossRef]

62. Palka, K. The influence of post-mortem ageing and roasting on the microstructure, texture and collagen solubility of bovine semitendinosus muscle. *Meat Sci.* **2003**, *64*, 191–198. [CrossRef]

Article

Sensory Quality Evaluation of Superheated Steam-Treated Chicken Leg and Breast Meats with a Combination of Marination and Hot Smoking

Woo-Hee Cho [1] and Jae-Suk Choi [1,2,*]

[1] Seafood Research Center, Industry-Academic Cooperation Foundation, Silla University, Busan 46958, Korea; ftrnd3@silla.ac.kr
[2] Department of Food Biotechnology, College of Medical and Life Sciences, Silla University, Busan 46958, Korea
* Correspondence: jsc1008@silla.ac.kr; Tel.: +82-51-248-7789

Abstract: As the sensory qualities of meat processed using methods such as superheated steam, marination, and hot smoking have not been examined, this study analyzed the sensory quality of chicken meats (leg, breast) and its chemical correlation by determining optimal processing conditions (superheated steam treatment, marination, and hot smoking). Chicken meats were defrosted using room temperature, running tap water, or high-frequency defroster. Marinated meats with herbal extract solution were treated with superheated steam and then hot smoked with wood sawdust; sensory evaluations were performed at each processing step. The products were analyzed for fatty acids and nutrients, along with storage tests under different conditions. High-frequency defrosting showed the lowest drip loss and thawing time compared to other methods. Bay leaves and oak wood were selected as the best sub-materials for higher sensory scores. Optimal superheated steam conditions showed higher overall acceptance (8.86, 8.71) and were set as follows; leg meat (225 °C; 12 min 20 s), breast meat (223 °C; 8 min 40 s). The final meat products possessed good nutritional composition and no severe sensory spoilages were detected during storage despite microbial and chemical degradations. Thus, regular sensory evaluations at each processing step and storage condition were effective for developing superior chicken meat products.

Keywords: chicken meat; sensory evaluation; superheated steam; marination; hot smoking; storage effect

Citation: Cho, W.-H.; Choi, J.-S. Sensory Quality Evaluation of Superheated Steam-Treated Chicken Leg and Breast Meats with a Combination of Marination and Hot Smoking. *Foods* **2021**, *10*, 1924. https://doi.org/10.3390/foods10081924

Academic Editors: Sidonia Martinez and Javier Carballo

Received: 13 July 2021
Accepted: 17 August 2021
Published: 19 August 2021

Publisher's Note: MDPI stays neutral with regard to jurisdictional claims in published maps and institutional affiliations.

1. Introduction

Chicken meat is a representative major food ingredient that provides good sensory quality as well as nutritional composition and is used almost globally except in certain religions such as Buddhism. Specifically, chicken meat is high in protein and low in fat; in particular, it does not contain trans-fats [1]. Recently, several chicken meats have been processed to prepare completely cooked products to be served as home meal replacements so that consumers can simply consume them regardless of place and time. For example, smoked chicken meat, chicken sausage, and fried chicken are the commonly marketed chicken meat products. For developing high-quality food products, it is very important to establish the optimal processing conditions for maximizing the initial qualities of products, considering the different times required for delivering to the actual consumer. Processes involving high-frequency thawing are reported to minimize the disadvantages of un-optimized thawing methods that typically result in larger ice crystals [2]. Superheated steam treatment results in good sensory improvement like moist texture and decreases both the processing time and microbial activity [3]. Further, marination improves the meat attributes by masking odor [4], and hot smoking provides desirable sensory properties such as flavor and aroma to foods [5] as well as inhibits microorganisms [6].

Several quality analysis methods including microbial, physicochemical, and sensory analyses have been used to evaluate food quality. Among these properties, sensory properties have been recently measured using instruments such as color meter, electronic nose, sodium meter, and texture analyzer. Nevertheless, intuitive sensory evaluation by human senses is remarkably effective, as all sensory parameters should be evaluated comprehensively rather than individually. In reality, numerous studies regarding sensory characteristics have used universal properties such as overall acceptance, which consider all indexes (e.g., appearance, odor, taste, and texture) simultaneously [7].

So far, few studies have examined the sensory qualities of processed meat simultaneously treated with methods such as superheated steam, marination, and hot smoking. Therefore, in this study different processing methods were applied to develop high-quality chicken meat (leg and breast) products; methods including high-frequency thawing, superheated steam treatment, marination, and hot smoking were combined based on previous results. Further, the present study focused on the effect of storage on the qualities of the processed chicken leg and breast, and investigated chemical mechanism related to its sensory results, along with the nutritional composition analysis including fatty acids and basic nutrients.

2. Materials and Methods

2.1. Sample Preparation

Chicken leg and breast (Harim Corp., Iksan, Korea) meats were obtained raw from Busan Poultry Cooperative (Busan, Korea) and processed based on the development scheme outlined in Figure 1. Each sample was grouped by similar size [leg; 115–135 (126.4 ± 9.8) mm (W), 54–58 (55.9 ± 1.8, 53.5 ± 2.1) mm (H, T), and breast; 130–140 (136.3 ± 4.2) mm (W), 50–60 (55.7 ± 4.6) mm (H), 34–38 (36.1 ± 1.8) mm (T)] and weight [leg; 90–100 (97.9 ± 5.3) g, and breast; 135–155 (144.2 ± 8.2) g], respectively. All chicken meats were stored at freezing temperature (-18 ± 3 °C) and defrosted using the following thawing methods: room temperature (RT; 15 ± 1 °C in an incubator as per HACCP standards; JSMI-04C, JC Research, Gongju-si, Korea), under running tap water (RW; 23 ± 1 °C), and with high-frequency defrosting (HFD; 27 MHz and 11 kW input power; TEMPERTRON FRT-10, Yamamoto Vinita Co. Ltd., Osaka, Japan).

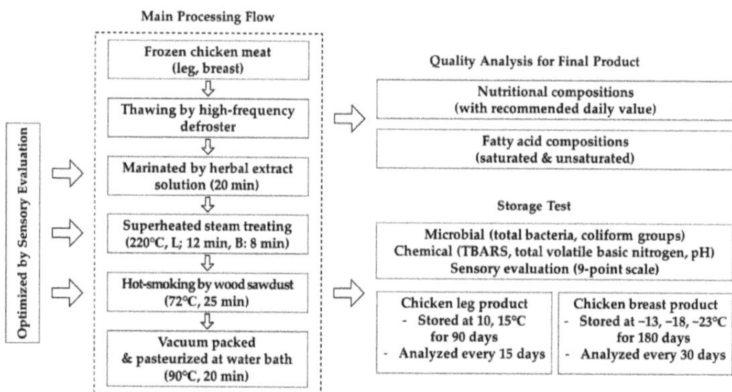

Figure 1. Schematic representation of the development of chicken leg and breast products: L, leg and B, breast.

All herbs (bay leaves, coriander powder, fennel whole, thyme whole, cumin seeds, basil whole, basil powder, and star anise) used for the hot water extract as marination solution were purchased from Solpyo Foods (Gyeonggido, Namyangjusi, Korea) except for sea buckthorn fruit powder, which was obtained from Hub-in-Korea (Gyeonggido, Gimposi, Korea). All marination solutions were prepared by boiling in hot water (100 °C) for 20 min with packaging in a non-woven fabric filter bag at 3% (w/v). The hot water extract solutions were used after being cooled to RT. Oak, apple, chestnut, walnut, and cherry wood sawdust sticks for hot smoking treatment were purchased from Shinsei (Shinsei Sangyo Co., Tokyo, Japan).

2.2. Drip Loss Measurement

Drip loss was measured by comparing the weights of the frozen and defrosted unit samples with each thawing method. All frozen samples and plastic bags (17.7 cm × 18.8 cm; Ziploc for frozen, SC Johnson Korea Ltd., Seoul, Korea) were individually weighed using an electronic scale (MW-2N, CAS Ltd., Yangju-si, Korea), packed at as low-in-air conditions as possible, and defrosted at the same indoor temperature (20 ± 1 °C). After being fully defrosted, the weight of each plastic bag was measured without the defrosted chicken part, which was held as a drip. The drip loss was expressed as the percentage of the initial weight.

Drip loss (%) = [(plastic bag weight after defrosting (g) − original plastic bag weight (g))/frozensample weight (g)] × 100.

2.3. Preparation for Treatment

For marination, the defrosted chicken leg and breast meat were immersed in herbal extract solutions containing additional 5% (w/v) saline for 20 min and were dried on a perforated stainless steel rack at RT for 10 min. Marinated chicken samples were heated using a superheated steam roaster (DFC-560A-2R/L, Naomoto Corporation, Osaka, Japan) under the meat-specific heating condition and treated with wood stick smoke using a hot smoker (Braii Smoker, Bradley, Canada) at 72 °C (Combustion temperature; 250–350 °C) for 25 min considering the conditions of Tirtawijaya et al. [8] and Jeffe et al. [9]. The processed samples were then cooled at RT until the core temperature decreased to below 50 °C and then packaged in vacuum plastic bags. The final products were pasteurized for 20 min in a water bath at 90 °C.

2.4. Experiment for Optimization of Superheated Steam Treatment

Superheated steam treatment conditions for chicken leg and breast were optimized using response surface methodology (RSM). Table 1. shows each heating condition for the central composite design (CCD) including independent variables and coded ranges (−1.414, −1, 0, +1, +1.414). The treating temperature (X_1) and time (X_2) for both chicken parts were set as respective independent variables. The response surface models of each treatment were derived with the response results, which were obtained with different code combinations, using MINITAB 18 (Minitab Inc., State College, PA, USA). As the response result, the dependent variable (overall acceptance) was measured by a 9-point scale evaluation for sensory analysis. The model was represented as a function of the independent variables using the following quadratic Formula (1):

$$Y = \beta_0 + \sum_{i=1}^{k} \beta_i X_i + \sum_{i=1}^{k} \beta_{ii} X_i^2 + \sum_{i=1}^{k-1} \sum_{j=2}^{k} \beta_{ij} X_i X_j \tag{1}$$

where β_0, β_i, β_{ii}, and β_{ij} are the regression coefficients for intercept, linear, quadratic, and interaction terms of the model, respectively. The optimal conditions for superheated steam treatment to chicken leg and breast were obtained through statistical evaluation of the model by analysis of variance (ANOVA) and their actual validation.

Table 1. Independent variables, codes, and actual levels for optimizing the superheated treatment of chicken products.

Independent Variables	Symbol	Unit	Meats	Range Level				
				−1.414	−1	0	+1	+1.414
Temperature	X_1	°C	Leg	192	200	220	240	248
			Breast	192	200	220	240	248
Time	X_2	min	Leg	11.1	11.5	12.5	13.5	13.9
			Breast	7.1	7.5	8.5	9.5	9.9

2.5. Sensory Evaluation

Sensory evaluation (appearance, odor, taste, texture, and overall acceptance) of all samples was done by 21 trained panelists belonging to the Industry-Academic Foundation at Silla University (Busan, Korea) based on a 9-point scale. The result was rated with the following score standards: 9 (best quality), 5 (acceptable limit), and 1 (worst quality).

2.6. Microbial Analysis

The microbial quality of the processed product was analyzed by testing the total bacteria count (TBC) and total coliform group (TCG) parameters, respectively. The respective tests were measured in triplicate in accordance with the Association of Official Analytical Chemists (AOAC) 990.12 [10] and 991.14 [11] using rehydratable dry-film media (Aerobic Count Plates and E. coli/Coliform Count plates; 3M, Saint Paul, MN, USA).

2.7. Thiobarbituric Acid Reactive Substances

Thiobarbituric Acid Reactive Substances (TBARS) testing was conducted according to the methods described by Yildiz [12] and Mohibbullah et al. [13]. Each sample homogenized with 20% trichloroacetic acid (TCA) in 2M phosphoric acid solution was filtered and incubated with 0.005M thiobarbituric acid for 30 min in a water bath at 95 °C. The sample was cooled to RT and absorbance was measured at 530 nm using a SPECTROstar Nano microplate reader (BMG Labtech, Ortenberg, Germany).

2.8. Total Volatile Basic Nitrogen

Total volatile basic nitrogen (TVBN) was measured using the Conway microdiffusion method based on the procedure described by the Ministry of Food and Drug Safety (MFDS), Korea [14]. The sample solution dispensed in a Conway chamber was incubated at 37 °C for 90 min. To quantify TVBN, 0.01 N NaOH was titrated with reactive substances (0.01 N H_2SO_4) in the chamber, mixed with Brunswick reagent beforehand.

2.9. Hydrogen Ion Concentration (pH)

The pH values of processed chicken leg and breast were measured using an OHAUS Starter 3100 pH meter coupled with a glass electrode (Ohaus, Seoul, Korea) complying with the method described by the MFDS [15].

2.10. Analysis of Nutritional Quality

Nutritional compositions (moisture, ash, salinity, calories, sodium, carbohydrates, sugars, dietary fiber, crude fat, trans fat, saturated fat, cholesterol, crude protein, calcium, iron, potassium, and vitamin D) of the processed chicken leg and breast were quantitatively analyzed in line with AOAC 925.09, 923.03, 979.09, 962.09, and 923.05 [16].

2.11. Fatty Acid Analysis

For analyzing the fatty acid compositions of the processed chicken leg and breast, a gas chromatograph (GC) (GC-2010 Plus, Shimadzu Corp., Kyoto, Japan) equipped with a flame ionization detector (FID) was used as described in the AOAC 963.22 [17]. After lipid extraction with ether treatment and methylation, fatty acids were separated using the GC-MS column (100 m $\times$ 0.25 mm $\times$ 0.25 µm; Supelco Inc., Bellefonte, PA, USA) at an oven temperature of 240 °C. The fatty acid values were computed by comparing retention times with standard components.

2.12. Storage Quality Analysis

To evaluate the shift in the storage quality of processed chicken leg and breast, the microbial (TBC and TCG), chemical (TBARS, TVBN, and pH), and sensory (appearance, odor, taste, texture, and overall acceptance) indexes were analyzed at different storage conditions (leg; at 10 °C and 15 °C for 90 days, breast; at -13 °C, -18 °C, and -23 °C for 180 days) based on the commonly marketed temperature and shelf-life periods for the specific meat groups [18].

2.13. Statistical Analysis

Except for the optimization of superheated steam treatment, all experimental values were measured in triplicate and were analyzed by one-way ANOVA and *t*-tests using SPSS version 23.0 (IBM Corp., Armonk, NY, USA). Statistical significance was set at a *p*-value < 0.05.

3. Results and Discussion

3.1. Effect of High-Frequency Defrosting on Variation in Drip Loss

The drip loss results of defrosted chicken leg and breast using different thawing methods are shown in Figure 2. The drip loss of defrosted chicken leg indicated that samples thawed with RW and HFD prepared samples showed significantly lower drip loss compared to those thawed at RT, but showed no significant difference between each other. A significantly lower drip loss was observed in the only chicken breast sample that was defrosted using HFD, followed by RT and RW. Overall, the HFD method had a significantly higher effect on decreasing drip loss variation in each chicken part compared to the RT treatment. Furthermore, the chicken breast also showed a significantly effective decrease in drip loss when defrosted by HFD compared to the other methods studied ($p < 0.10$).

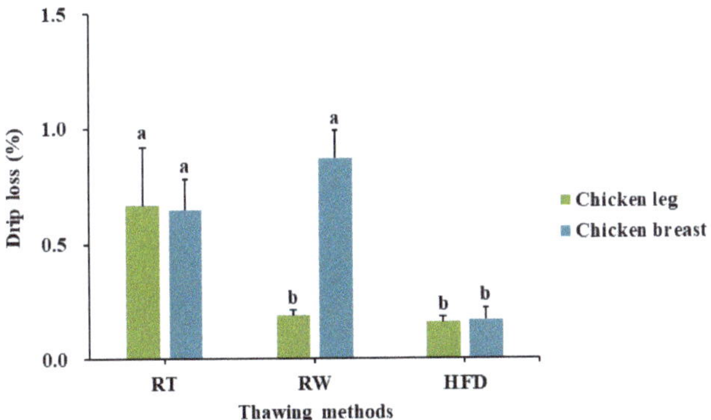

Figure 2. Comparison of drip loss percentage in defrosted samples using different thawing methods (RT, room temperature; RW, running tap water; HFD, high-frequency defroster). Values are mean $\pm$ SE. Different letters (a,b) in each column indicate significant differences among the means by Tukey's test ($p < 0.10$).

In terms of thawing time, each method required different lengths of time for defrosting the samples completely, although they showed similar results by parts: RT (10 h), RW (120 min), and HFD (20 min). Generally, thawing under running tap water was faster than under still air such as at room temperature due to a higher coefficient of heat transfer [19]. In HFD, defrosting by a high-frequency electrode is accelerated because of molecular friction in the muscle cells by a high frequency electrode, which differs from the traditional defrosting methods using thermal conduction. Specifically, the RW and HFD methods exhibited a similar drip loss level in the chicken leg part; however, it was more effective to use HFD considering the thawing time. In contrast, HFD-treated chicken breast samples showed better improvement in terms of both thawing time and drip loss level compared to those treated with RT and RW.

HFD appeared to decrease the drip loss level of both meats more than the other method. In other words, HFD enabled the majority of water inside the foods to be absorbed into the tissue and shortened the time for maximum ice crystal formation using internal heating and balanced melting [20].

The short thawing time and low drip loss level exhibited by HFD can improve quality parameters such as color or hardness as reported in a previous study [21] that examined the same thawing conditions (RT, RW, HFD). Thus, the HFD method was proved to be an optimal thawing method for the preparation of each raw material (chicken leg and breast) considering both the thawing time and drip loss. It can also be effective for maintaining good product quality and minimizing the variables that negatively affect the subsequent sensory evaluation steps.

3.2. Optimization of Superheated Steam Treatment of Chicken Leg and Breast

Table 2 shows the response results (Y_L and Y_B; Overall acceptance) from different samples treated with the already defined superheated steam treatment conditions (X_1; Temperature, X_2; Time) for chicken leg and breast. Based on these results, the regression equations for the response model were respectively computed as follows:

$$Y_L = 8.573 + 0.2084\,X_1 - 0.1196\,X_2 - 0.549\,X_1X_1 - 0.501\,X_2X_2 - 0.380\,X_1X_2 \qquad (2)$$

$$Y_B = 8.540 + 0.351\,X_1 + 0.230\,X_2 - 1.083\,X_1X_1 - 0.570\,X_2X_2 - 0.093\,X_1X_2 \qquad (3)$$

Table 2. Central composite design of independent variables and response of dependent variables during optimization of superheated treatment of chicken products: X_1, temperature; X_2, time; Y_L, overall acceptance of chicken leg; Y_B, overall acceptance of chicken breast.

Run No.	Coded Values		Chicken Leg			Chicken Breast		
			Actual Values		Responses	Actual Values		Responses
	X_1	X_2	X_1	X_2	Y_L	X_1	X_2	Y_B
1	−1	−1	200	11.5	6.95	200	7.5	6.10
2	1	−1	240	11.5	7.90	240	7.5	6.57
3	−1	1	200	13.5	7.67	200	9.5	7.14
4	1	1	240	13.5	7.10	240	9.5	7.24
5	−1.41	0	192	12.5	7.14	192	8.5	5.71
6	+1.41	0	248	12.5	8.05	248	8.5	7.29
7	0	−1.41	220	11.1	8.00	220	7.1	7.48
8	0	+1.41	220	13.9	7.38	220	9.9	7.57
9	0	0	220	12.5	8.38	220	8.5	8.81
10	0	0	220	12.5	8.67	220	8.5	8.33
11	0	0	220	12.5	8.67	220	8.5	8.48

The analyzed coefficients with a significant effect on each response and their correlations are presented in Table 3. Both the R^2 and p-values of the respective models (Y_L and Y_B) satisfied the common recommendation (>0.8) in the previous studies [22] and the statistical standards (<0.05), respectively. Lack of fits revealed inappropriate correlations or the inclusion of considerable factors such as interaction and quadratics, which were also fulfilled the standard (>0.05) in this study. This also indicates that the respective models were suitable for deriving the optimal conditions of superheated steam treatment for chicken leg and breast products.

Table 3. Analysis of variance for the response of dependent variables during optimization of superheated treatment of chicken products: X_1, temperature; X_2, time; Y_L, overall acceptance of chicken leg; Y_B, overall acceptance of chicken breast.

Responses	*p*-Value				
	Model	Linear (X_1, X_2)	Quadratic (X_1X_1, X_2X_2)	Interaction (X_1X_2)	Lack of Fit
Y_L $(R^2 = 0.908)$	>0.013	0.123	>0.006	>0.035	0.23
Y_B $(R^2 = 0.903)$	>0.014	0.096	>0.004	0.682	0.193

Considering the coefficient affecting the responses by the models, both quadratics along with the interaction of Y_L showed a significant influence on the results ($p < 0.05$), whereas the linear effects were not valid ($p > 0.05$). Figure 3a,b show the visualized response model for superheated steam treatment of chicken leg and breast, respectively, in a three-dimensional form. The graphs formed a convex curve as they were set to closely central conditions on account of the quadratic impact.

Subsequently, the models showed the maximum response values (overall acceptance) of 8.61 and 8.59 individually. In relation to the optimal range of factors (temperature and time), each model appeared to show almost the same temperature (about 220 °C) whereas the time differed (about 12.5 and 8.5 min for leg and breast, respectively). In general, bone-in meat requires a longer heating time to cook compared to de-boned parts as the bone contributes to slowing heat transfer. Similarly, the cooking time using oven was about two times higher in bone-in pork than in bone-less one in the study of Zilmmermann [23]. In the present study, the chicken leg and breast treated at temperatures lower than 220 °C for less than 12.5 and 8.5 min were not fully cooked and had a few pink spots. However, higher temperatures and longer treatment times resulted in overcooked products, with burned surfaces.

As described in Table 4, the optimal superheated steam treatment conditions for chicken leg and breast were completely determined by the method of maximizing the model responses to each coded variable and validating them with actual experimental results. The models exported 8.61 and 8.59 of the predicted responses (P) when the factors were set to the optimum conditions, respectively; therefore, the experimental samples of chicken leg and breast were respectively prepared at 225 °C for 12 min 20 s and at 223 °C for 8 min 40 s to practically set the equipment. The obtained values (E) were 8.86 and 8.71, and E/P values indicating the error level of prediction were calculated as 1.03 and 1.01 for chicken leg and breast with a high desirability, respectively (>0.9). Further, these completed conditions were continually employed in the next processing steps in this study.

These results were found to be different from those of a previous study [3] in which the sensory evaluation results show that, the best superheated steam-treating conditions of chicken breast fillet are high steam temperature (350 °C) and short heating time (6 min). However, in the present study, the conditions were optimized with lower temperatures and longer times than those in the previous study, and resulted from different sample sizes. In addition, sliced samples were used in the previous study. Therefore, it indicates that the different optimal superheated steam-treating conditions for processing chicken breast also could be confirmed depending on such differences. On the other hand, no studies on superheated steam-treated chicken leg have been described, thus highlighting the importance of the present work.

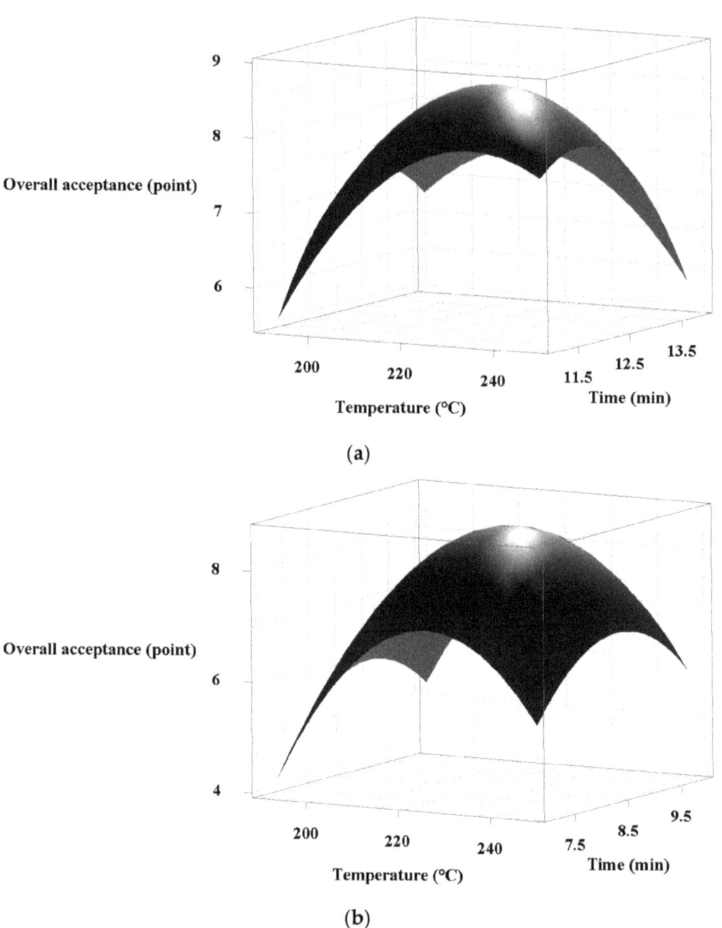

Figure 3. Three-dimensional response surface plots of superheated steam treated chicken products for overall acceptance with respect to temperature and time of roasting: (**a**) chicken leg, (**b**) chicken breast.

Table 4. Optimum conditions for superheated steam-treated chicken products using response surface methodology.

Responses	Optimum Conditions		Predicted Value (p)	Experimental Value (E)	E/P	Desirability
	X_1	X_2				
Y_L	+0.27 (225.43 °C)	−0.21 (12.29 min)	8.61	8.86 ± 0.14	1.03	0.97
Y_B	+0.16 (223.14 °C)	+0.19 (8.69 min)	8.59	8.71 ± 0.18	1.01	0.93

Values are mean ± SE.

3.3. Effect of Marination with Herbal Extract Solutions on Sensory Evaluation

The results of sensory evaluation for determining the optimal herbal extract solution are represented in Figure 4. Herein, bay leaves were established as the optimal herb for marinating both chicken meats and similar score patterns were obtained for both. Specifically, the groups with significantly improved overall acceptance score compared to the control among chicken leg meats were those with bay leaves- and sea buckthorn fruit powder-treatment. In addition, the scores of chicken breast groups were significantly increased in star anise-treated groups, even above those of the other two groups ($p < 0.05$). Interestingly, for chicken breast in particular, the odor score in every marinated group was significantly higher compared to the control one as opposed to the leg. However, no group showed significant variations in appearance and texture indexes.

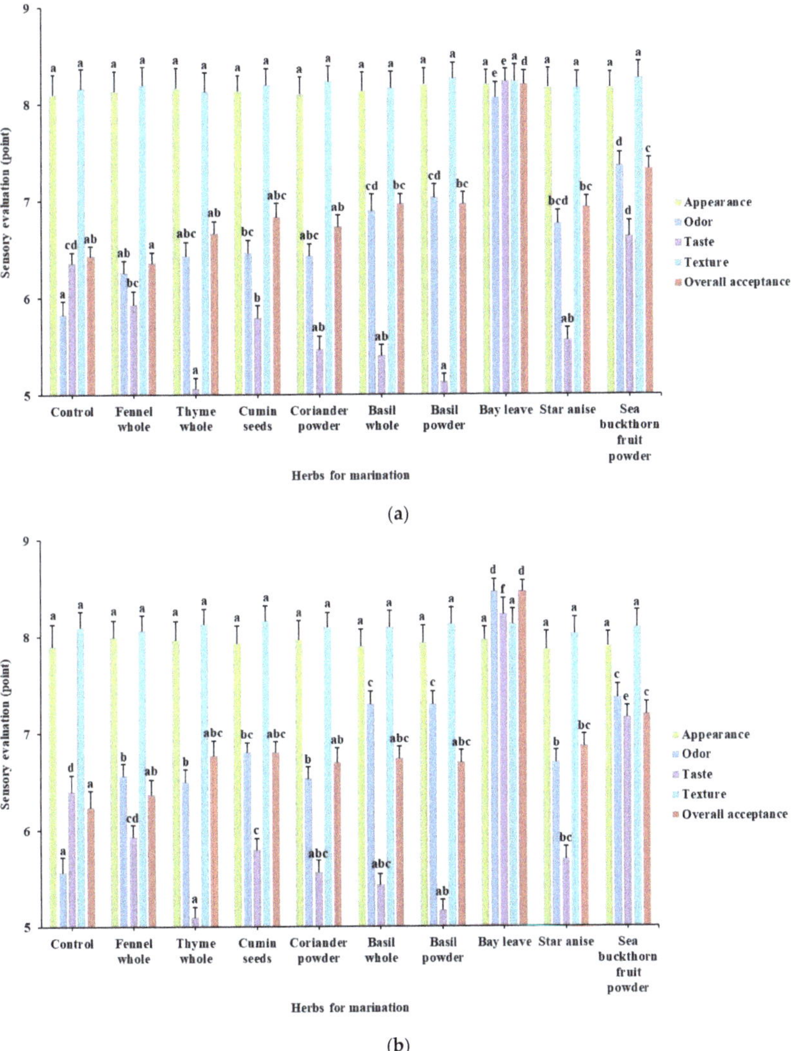

(a)

(b)

Figure 4. Sensory evaluation of marinated chicken leg and breast with different herbal extract solutions: (**a**), chicken leg; (**b**), chicken breast. Values are mean ± SE. Different letters (a–f) in the respective colored column indicate significant differences among the means by Tukey's test ($p < 0.05$).

These results are consistent with those of previous studies about the relationship between herbs and sensory changes. Kurup et al. [24] showed that among the selected herbs and spices, bay leaves, thyme, and coriander have the properties of deodorizing/masking. The study also reported that a harsh and bitter taste from spices was considered to have resulted from the presence of alkaloids, glycosides, and organic and inorganic salts. Lee and An [25] reported that a traditional beef dish with basil added showed decreased taste scores in sensory tests compared to the control group.

3.4. Hot Smoking Treatment

The superheated steam-treated and marinated chicken meats were then hot smoked using different wood sawdust as the final processing step, and the results are shown in Figure 5. The results indicated that oak wood had the highest overall score among the sawdust sources used for both meats, and its values were significantly higher than those of other groups including the control group ($p < 0.05$). Further, all the groups did not show a significant difference in the appearance and texture, similar to the marination results. Interestingly, except for the oakwood-treated group, the overall acceptance, odor and taste scores of other groups (apple, chestnut, walnut, and cherry) were lower compared to the control group. In particular, apple, walnut, and cherry wood smoke-treated groups were rated to have significantly reduced points among the sources used ($p < 0.05$).

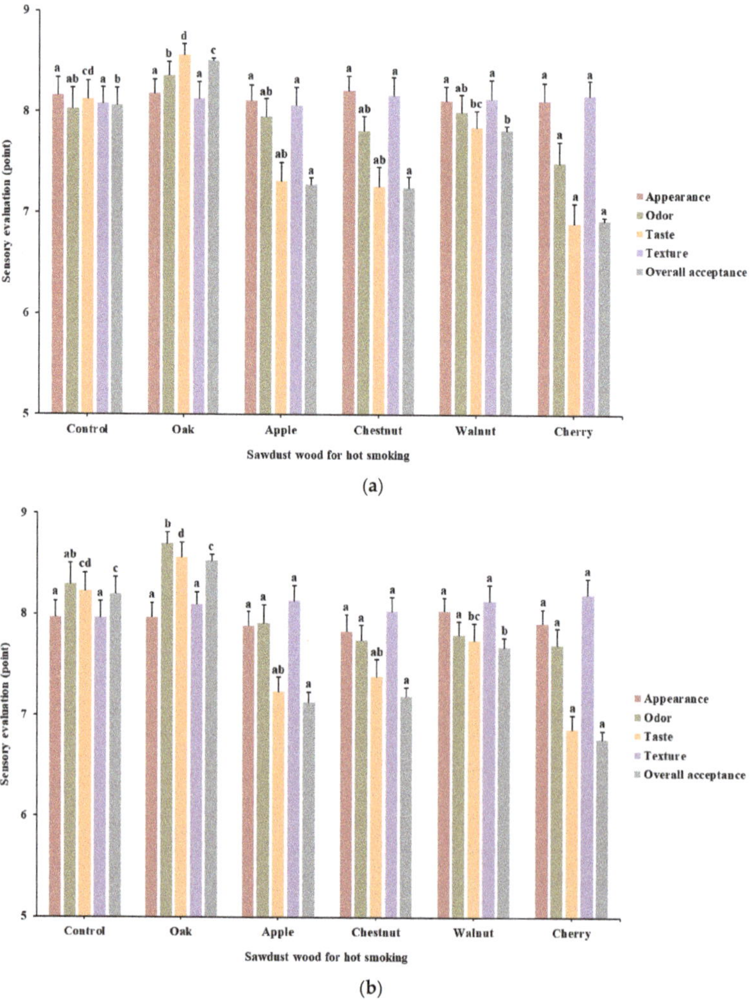

Figure 5. Sensory evaluation of hot-smoked chicken leg and breast with different sawdust: (**a**), chicken leg; (**b**), chicken breast. Values are mean ± SE. Different letters (a–d) in the respective colored column indicate significant differences among the means by Tukey's test ($p < 0.05$).

These different score changes between smoke wood types might be attributed to the different amounts of flavoring contents. Migdał et al. [26] reported that the fruit wood, especially that of apple trees, is rich in hemicellulose, and Ratnani and Widiyanto [27] reported that a fruity and sweet flavor substances such as furfural and furans are produced from the decomposition of hemicellulose. However, in the present study, the chicken meat products were not harmonized with hot-smoking using fruit wood in the described conditions, because it was assumed that the high fruit aroma and taste would cover the essential flavors of chicken meats.

3.5. Fatty Acid Analysis

The fatty acid composition profiles of processed chicken leg and breast are shown in Table 5. The results indicated that the processed chicken leg possessed outstandingly higher total fatty acid amount (12.37 g/100 g) than the chicken breast (3.57 g/100 g). The different total fatty acid amounts were derived from the nutritional characteristics of each muscle. Among saturated fatty acids (SFA), palmitic acid in both the leg and breast meat accounted for over two thirds of the fatty acid groups at 3.04 g (/100 g) and 0.88 g (/100 g), respectively. Further, oleic acid was the most abundant fatty acid (5.33 g/100 g and 1.44 g/100 g) among monounsaturated fatty acids (MUFA), as well as total fatty acids, in both the leg and breast part. Among polyunsaturated fatty acids (PUFA), linoleic acid (1.87 g/100 g and 0.52 g/100 g) was the most abundant in both meats. Kralik et al. [28] reported that the dominant SFAs in chicken fat were palmitic and stearic acids whereas unsaturated fatty acids mainly include oleic, linoleic, and arachidonic acids.

Table 5. Fatty acid composition of processed chicken leg and breast.

Fatty Acids	Shorthand	Chicken Leg (100 g)	Chicken Breast (100 g)
Caprylic acid	C8:0	0.00	-
Capric acid	C10:0	0.00	0.00
Lauric acid	C12:0	0.00	0.00
Myristic acid	C14:0	0.07	0.03
Pentadecanoic acid	C15:0	0.02	0.03
Palmitic acid	C16:0	3.04	0.88
Magaric acid	C17:0	0.01	0.01
Stearic acid	C18:0	0.76	0.23
Arachidic acid	C20:0	0.01	0.00
Heneicosylic acid	C21:0	0.01	0.00
Behenic acid	C22:0	0.01	0.00
$\sum$ SFA [1]		3.94	1.18
Myristoleic acid	C14:1	0.02	0.01
Palmitoleic acid	C16:1	0.93	0.24
Magaoleic acid	C17:1	0.01	0.01
Oleic acid	C18:1	5.30	1.44
Eicosenoic acid	C20:1	0.01	0.00
Eicosadienoic acid	C20:2	0.02	0.01
Erucic acid	C22:1	0.02	0.01
$\sum$ MUFA [2]		6.33	1.73
Linoleic acid	C18:2 *n*-6	1.87	0.52
γ-Linolenic acid	C18:3 *n*-6	0.02	0.01
Dihomo γ-Linolenic acid	C20:3 *n*-6	0.00	0.00
Arachidonic acid	C20:4 *n*-6	0.07	0.04
$\sum$ *n*-6		1.96	0.57
Linolenic acid	C18:3 *n*-3	0.13	0.04
Eicosatrienoic acid	C20:3 *n*-3	0.00	0.00
Eicosapentaenoic acid (EPA)	C20:5 *n*-3	0.00	0.01

Table 5. *Cont.*

Fatty Acids	Shorthand	Chicken Leg (100 g)	Chicken Breast (100 g)
Docosapentaenoic acid (DPA)	C22:5 *n*-3	0.01	0.01
Docosahexaenoic acid (DHA)	C22:6 *n*-3	0.00	0.03
$\sum$ *n*-3		0.15	0.09
$\sum$ PUFA [3]		2.10	0.66
Total fatty acid (g)		12.37	3.57
PUFA/SFA		0.53	0.56
n-6/*n*-3		13.07	6.33

[1] SFA: saturated fatty acid. [2] MUFA: monounsaturated fatty acid. [3] PUFA: polyunsaturated fatty acid.

Wood et al. [29] mentioned that meat firmness is affected by the melting point of fatty acids in foods, indicating that melting point is decreased with increasing 18C fatty acids such as oleic acid. Similarly, oleic acid increases the perception of juiciness and meat-like flavor [30]. In this study, the processed chicken meats were shown to have good texture and flavor through the sensory evaluation of both leg and breast meat, receiving more softening evaluations in the leg compared to the breast. Another previous study reported that the chicken meat having different fatty acid compositions showed different sensory evaluation results in both the leg and breast [31].

3.6. Nutritional Value

The nutritional compositions of processed chicken leg and breast are presented in Table 6. Comparing both products with each other, the chicken leg product showed higher amounts of calories, fats, and cholesterol compared to the chicken breast. Chicken thighs including leg are known to be fattier than the breast and are also moister in comparison [32]. Koh et al. [33] studied the nutrition of chicken depending on the meats and showed that the highest fat amount was included in the wing part, followed by the thigh and breast. Additionally, unlike livestock meat such as beef and lamb, there is no trans-fat in chicken meat [1].

Table 6. Nutritional values of processed chicken leg and breast.

Parameters		Chicken Leg (100 g)		Chicken Breast (100 g)		Daily Values [1]
		Amount	% DV [2]	Amount	% DV [2]	
Ash	g	1.31	-	1.50	-	-
Calories	cal	241.46	-	149.26	-	-
Sodium	g	0.22	9.6	0.24	10.4	2.3
Carbohydrate	g	0.65	0.2	1.18	0.4	275
Sugar	g	0.62	1.2	1.00	2.0	50
Dietary fiber	g	1.46	5.2	2.16	7.7	28
Crude fat	g	12.46	16.0	3.58	4.6	78
Trans fat	g	-	-	-	-	2
Saturated fat	g	3.94	19.7	1.18	5.9	20
Cholesterol	mg	138.63	46.2	75.53	25.2	300
Crude protein	g	25.83	51.7	28.08	56.2	50
Vitamin D	µg	-	-	-	-	20
Potassium	g	0.28	6.0	0.32	6.8	4.7
Iron	mg	5.27	29.3	3.18	17.7	18
Calcium	g	0.15	11.5	0.08	6.2	1.3

[1] According to Nutrition Facts Labeling Requirements, US Food and Drug Administration. [2] The % daily value (DV) indicates how much a nutrient in a serving of food contributes to the daily diet. Normally, 2000 calories a day is used in general nutrition advice. Reprinted from US Food and Drug Administration (2020) [34].

Table 6 shows the respective daily value percentage of each processed product based on the recommended nutrition facts by the FDA. Herein, chicken breast provides a higher amount of protein amount than chicken leg considering unit calories. In case of sodium levels, both meats were measured to keep about 10% DV; however, they revealed different salty tastes compared to each other in actual evaluation, as the saltiness of chicken leg tended to be more concentrated on the surface rather than the inside and the panelists reported feeling a higher salty taste with their first bite. Richter [35] reported that the skin disrupts salt penetration into the meat. According to Tompkin [36], dark meat contains more iron than white meat. In reality, among the nutrition lists, iron content was higher in the processed chicken leg part than in the breast part.

3.7. Effect of Storage Conditions on Processed Chicken Meats

Different chicken meat products treated with optimized superheated steam conditions and the best combination of herbal marination and hot-smoking were packaged and tested for each marketed storage condition to investigate the effect of storage on sensory qualities along with microbial and physicochemical parameters.

3.7.1. Sensory Qualities

Tables 7 and 8 show the sensory results for processed chicken legs and breasts, respectively. At different storage temperatures on chicken leg croups, the parameters that changed the fastest in chicken leg groups were odor at 10 °C and taste at 15 °C, respectively. Both of these were the most significantly decreased during the storage period ($p < 0.05$). The sensory panelists reported that the chicken leg showed slightly unusual and stale smell as storage time increased, and these were more serious in samples stored at 15 °C than in those stored at 10 °C. Li et al. [37] reported that the production of off-flavor and odor in meat products is attributed to volatile compounds owing to the metabolites produced by spoilage microorganisms during storage. These could be strongly related to changed TBC and TVBN levels as shown in subsequent results.

Table 7. Periodical values of sensory evaluation of processed chicken leg stored at different temperatures (10 °C, and 15 °C) for 90 days.

Temp	Day	Appearance	Odor	Taste	Texture	Overall Acceptance
10 °C	0	8.48 [a] ± 0.15	8.29 [a] ± 0.16	8.52 [a] ± 0.13	8.24 [a] ± 0.15	8.52 [a] ± 0.11
	15	8.43 [a] ± 0.13	8.19 [a,b] ± 0.15	8.48 [a] ± 0.13	8.24 [a] ± 0.14	8.62 [a] ± 0.13
	30	8.33 [a] ± 0.16	8.05 [a,b] ± 0.20	7.95 [a,b] ± 0.15	8.43 [a] ± 0.16	8.38 [a] ± 0.15
	45	8.24 [a] ± 0.14	7.86 [a,b,c] ± 0.16	7.90 [a,b] ± 0.14	8.05 [a,b] ± 0.18	8.14 [a] ± 0.14
	60	8.10 [a] ± 0.12	7.48 [b,c] ± 0.18	7.86 [a,b] ± 0.26	8.00 [a,b] ± 0.20	8.10 [a] ± 0.19
	75	7.19 [b] ± 0.13	6.71 [d] ± 0.20	7.33 [b] ± 0.21	7.29 [c] ± 0.16	7.33 [b] ± 0.14
	90	7.19 [b] ± 0.16	7.24 [c,d] ± 0.21	7.48 [b] ± 0.18	7.48 [b,c] ± 0.19	7.48 [b] ± 0.13
15 °C	0	8.48 [a] ± 0.15	8.29 [a] ± 0.16	8.52 [a] ± 0.13	8.24 [a] ± 0.15	8.52 [a] ± 0.11
	15	8.29 [a] ± 0.12	8.14 [a] ± 0.13	8.10 [a,b] ± 0.10	8.29 [a] ± 0.14	8.29 [a] ± 0.12
	30	8.19 [a] ± 0.11	8.05 [a] ± 0.19	7.90 [a,b] ± 0.15	7.81 [ab] ± 0.16	8.19 [a] ± 0.13
	45	8.10 [a] ± 0.12	8.05 [a] ± 0.15	7.71 [b,c] ± 0.14	8.00 [a] ± 0.15	8.05 [a] ± 0.13
	60	7.33 [b] ± 0.17	6.86 [b] ± 0.16	7.14 [c] ± 0.19	7.24 [b] ± 0.14	7.29 [b] ± 0.14
	75	7.14 [b] ± 0.13	6.90 [b] ± 0.17	7.10 [c] ± 0.17	7.24 [b] ± 0.15	7.05 [b] ± 0.13
	90	7.05 [b] ± 0.15	6.90 [b] ± 0.18	7.19 [c] ± 0.19	7.24 [b] ± 0.12	7.19 [b] ± 0.15

Values are mean ± SE. Different letters (a–d) in each column indicate significant differences among means by Tukey's test ($p < 0.05$).

Table 8. Periodical values of sensory evaluation in processed chicken breast stored at different temperatures (-13 °C, -18 °C, and -23 °C) for 180 days.

Temp	Day	Appearance	Odor	Taste	Texture	Overall Acceptance
-13 °C	0	8.48 [a] ± 0.13	8.14 [a] ± 0.13	8.33 [a] ± 0.14	8.29 [a] ± 0.16	8.48 [a] ± 0.13
	30	8.43 [a,b] ± 0.11	8.10 [a] ± 0.12	8.19 [a,b] ± 0.13	8.05 [a] ± 0.15	8.10 [a,b] ± 0.15
	60	8.10 [a,b,c] ± 0.18	7.71 [a,b] ± 0.17	7.95 [a,b] ± 0.16	8.14 [a] ± 0.17	8.24 [a,b] ± 0.15
	90	8.10 [a,b,c] ± 0.18	7.90 [a,b] ± 0.14	8.05 [a,b] ± 0.19	7.86 [a] ± 0.17	8.19 [a,b] ± 0.15
	120	7.76 [a,b,c] ± 0.19	7.43 [b] ± 0.11	7.76 [a,b] ± 0.15	7.76 [a] ± 0.15	8.05 [a,b] ± 0.13
	150	7.67 [b,c] ± 0.21	7.67 [a,b] ± 0.19	7.90 [a,b] ± 0.17	7.57 [a] ± 0.20	7.81 [b] ± 0.16
	180	7.62 [c] ± 0.22	7.57 [a,b] ± 0.16	7.57 [b] ± 0.13	7.57 [a] ± 0.20	7.76 [b] ± 0.15
-18 °C	0	8.48 [a] ± 0.13	8.14 [a] ± 0.13	8.33 [a] ± 0.14	8.29 [a] ± 0.16	8.48 [a] ± 0.13
	30	8.24 [a,b] ± 0.14	8.10 [a] ± 0.12	8.14 [a] ± 0.16	8.24 [a] ± 0.14	8.43 [a] ± 0.15
	60	8.48 [a] ± 0.11	8.00 [a] ± 0.14	8.10 [a] ± 0.18	8.05 [a] ± 0.16	8.33 [a] ± 0.13
	90	8.29 [a,b] ± 0.17	7.81 [a,b] ± 0.15	8.10 [a] ± 0.18	8.19 [a] ± 0.18	8.38 [a] ± 0.15
	120	7.62 [b] ± 0.19	7.29 [b] ± 0.17	7.62 [a] ± 0.16	7.62 [a] ± 0.18	7.90 [a] ± 0.15
	150	8.05 [a,b] ± 0.20	8.00 [a] ± 0.18	7.62 [a] ± 0.18	7.81 [a] ± 0.21	8.14 [a] ± 0.19
	180	8.10 [a,b] ± 0.17	7.90 [a,b] ± 0.15	7.95 [a] ± 0.21	7.90 [a] ± 0.18	8.05 [a] ± 0.16
-23 °C	0	8.48 [a] ± 0.13	8.14 [a] ± 0.13	8.33 [a] ± 0.14	8.29 [a] ± 0.16	8.48 [a] ± 0.13
	30	8.24 [a] ± 0.15	8.10 [a] ± 0.17	8.10 [a,b] ± 0.14	8.19 [a] ± 0.18	8.38 [a] ± 0.15
	60	8.19 [a] ± 0.18	8.00 [a] ± 0.17	8.00 [a,b] ± 0.18	7.86 [a] ± 0.16	7.95 [a] ± 0.15
	90	7.76 [a] ± 0.19	7.71 [a] ± 0.20	8.00 [a,b] ± 0.15	7.81 [a] ± 0.20	8.14 [a] ± 0.19
	120	7.90 [a] ± 0.17	7.76 [a] ± 0.19	7.52 [b] ± 0.19	7.57 [a] ± 0.20	7.86 [a] ± 0.16
	150	7.76 [a] ± 0.17	7.67 [a] ± 0.16	7.76 [a,b] ± 0.19	7.71 [a] ± 0.14	7.90 [a] ± 0.15
	180	7.86 [a] ± 0.19	7.71 [a] ± 0.20	8.00 [a,b] ± 0.18	7.57 [a] ± 0.16	7.95 [a] ± 0.18

Values are mean ± SE. Different letters (a–c) in each column indicate significant differences among means by Tukey's test ($p < 0.05$).

Chicken legs stored at both 10 °C and 15 °C showed a significant score decrease in appearance comparing the initial and final day groups ($p < 0.05$). This change was mainly investigated as the result of a color change to a bluish hue according to panelist comments. Katiyo et al. [38] reported that the formation of deoxymyoglobin could affect the change in meat to blue color. In this study, the vacuum packaging formed an oxygen-free condition, which was considered to cause such a change.

When it comes to the significant changes in the texture of processed chicken products during storage periods, fat content plays an important role in determining the texture and tenderness for meat. Kilcast and Lewis [39] said that the crystals surrounded by liquid are formed inside lipid cells at chilling temperatures. In a study by Pande and Akoh [40], larger crystals and softer texture quality or mouthfeel was reported in the final product. Likewise, the chill-storage chicken legs in this study appeared to show a similar tendency with such studies, indicating that the fattier meat of chicken leg became over-soft and inelastic as time passed during the periodical sensory evaluation.

In contrast to chicken leg products, in Table 8. there were few significant sensory degradations in the processed chicken breast during the storage periods. In particular, the samples stored at -23 °C were rated to have almost no significant score changes in all sensory parameters during storage periods, whereas a slight exterior degradation was measured in samples stored at -13 °C, which would be confirmed as a result of the color change to a bluish hue. The samples stored at -13 °C were also found to show significantly decreased taste after the storage period compared to other groups. Different frozen storage conditions could play a crucial role in the taste of foods when defrosted because of dripping loss. Specifically, slow freezing makes the water crystals inside food larger as they slowly pass through a maximum ice crystal formation zone. When they are defrosted for serving and re-heated, they produce a drip that includes important nutrition and tasty compounds. According to Olsson et al. [41], drips decrease the quality of the flesh because of the loss of such desirable components.

3.7.2. Microbial Qualities

The present study analyzed the values of TBC and TCG for processed chicken legs and breasts during each storage condition as shown in Table 9. TBC values were significantly increased for chicken leg stored at 15 °C than for chicken leg stored at 10 °C; similarly, processed chicken breast exhibited higher TBC at a higher storage temperature than at other lower storage temperatures. However, in the whole period, TBC contents were gradually increased in the leg meat products, whereas those of breast meat fluctuated but were maintained between 1.0–1.5 Log CFU/g for 180 days. The latter case could be explained by the inhibition of microorganisms resulting from the oxygen-free condition of vacuum packaging and frozen storage conditions [42]. Overall, no TCG was detected in both meat products for all storage periods and the TBC values did not exceed the acceptable limit (<5 Log CFU/g). This result might be strongly affected by superheated steam treatment at high temperatures, which is consistent with a similar study [43] in which superheated steam-treated chicken skin showed a greater decrease in the number of *Listeria innocua* (CLIP 20595) compared to the sample without superheated steam treatment. Moreover, hot-smoking treatment seems to cause chemically induced inhibition by phenolic compounds as well as thermal inhibition using smoke. Heiszler et al. [44] reported that the increasing surface temperature caused by smoking treatment led to phenol and formaldehyde deposition, which then improved bacteriostatic and bactericidal effects.

Table 9. Periodical values in the total bacterial count (TBC; Log CFU/g) of processed chicken products stored at different temperatures for different storage days (90 and 180).

Day	Chicken Leg		Day	Chicken Breast		
	10 °C	15 °C		−13 °C	−18 °C	−23 °C
0	2.53 [a] ± 0.04	2.53 [a,b] ± 0.04	0	ND	ND	ND
15	1.78 [a] ± 0.36	2.29 [a] ± 0.02	30	1.48 [a] ± 0.44	1.00 ± 0.00	ND
30	2.02 [a] ± 0.30	2.43 [a] ± 0.08	60	1.30 ± 0.00	1.00 ± 0.00	1.40 ± 0.00
45	1.84 [a] ± 0.45	2.73 [b,c] ± 0.04	90	1.09 [a] ± 0.12	1.15 [a] ± 0.21	1.20 [a] ± 0.28
60	2.11 [a] ± 0.25	2.77 [b,c] ± 0.10	120	1.39 [a] ± 0.30	1.35 [a] ± 0.07	1.24 [a] ± 0.34
75	2.14 [a] ± 0.47	2.88 [c] ± 0.12	150	1.36 [a] ± 0.22	1.15 [a] ± 0.21	1.16 [a] ± 0.28
90	2.84 [a] ± 0.17	3.43 [d] ± 0.04	180	1.62 [a] ± 0.28	1.30 [a] ± 0.30	1.24 [a] ± 0.34

Values are mean ± SE. Different letters (a–d) in each column indicate significant differences among the means by Tukey's test ($p < 0.05$).

3.7.3. Physicochemical Qualities

Figure 6 shows the periodical values in the chemical responses in the processed chicken leg and breast at each storage condition. The values of TBARS for each product are included in Figure 6a. Both processed meats showed change trends similar to those of microbial testing, demonstrating that the TBARS values in chicken leg tended to be steadily increased whereas those in chicken breast were irregularly increased. These values were significantly increased in the final storage period compared to the initial periods ($p < 0.05$).

Domínguez et al. [45] reported that the acceptable limit of TBARS value is 2–2.5 MDA/kg, and suggested that meat and meat products do not become rancid within this standard limit. Likewise, our sensory panelists actually perceived neither severe bitterness nor rancidity, which was the corresponding result, considering that the experimental values remained low.

Figure 6b shows the TVBN levels of processed chicken meats during the storage periods. Comparing both chicken meats at each final storage period, the leg product showed 20.53 ± 0.12 and 24.15 ± 0.70 mg% at 10 °C and 15 °C, whereas the breast product showed 12.13 ± 2.64, 11.78 ± 2.63, and 14.23 ± 0.51 mg% at −13 °C, −18 °C, and −23 °C, respectively. Several previous studies and government food institutes have established the permissible limit of TVBN for fresh meat as follows: 15 mg% [46,47], 20 mg% [48]. Thus, the present study confirmed that chicken legs stored for less than 75 days did not exceed the recommended limit whereas all chicken breast samples satisfied the standard limits.

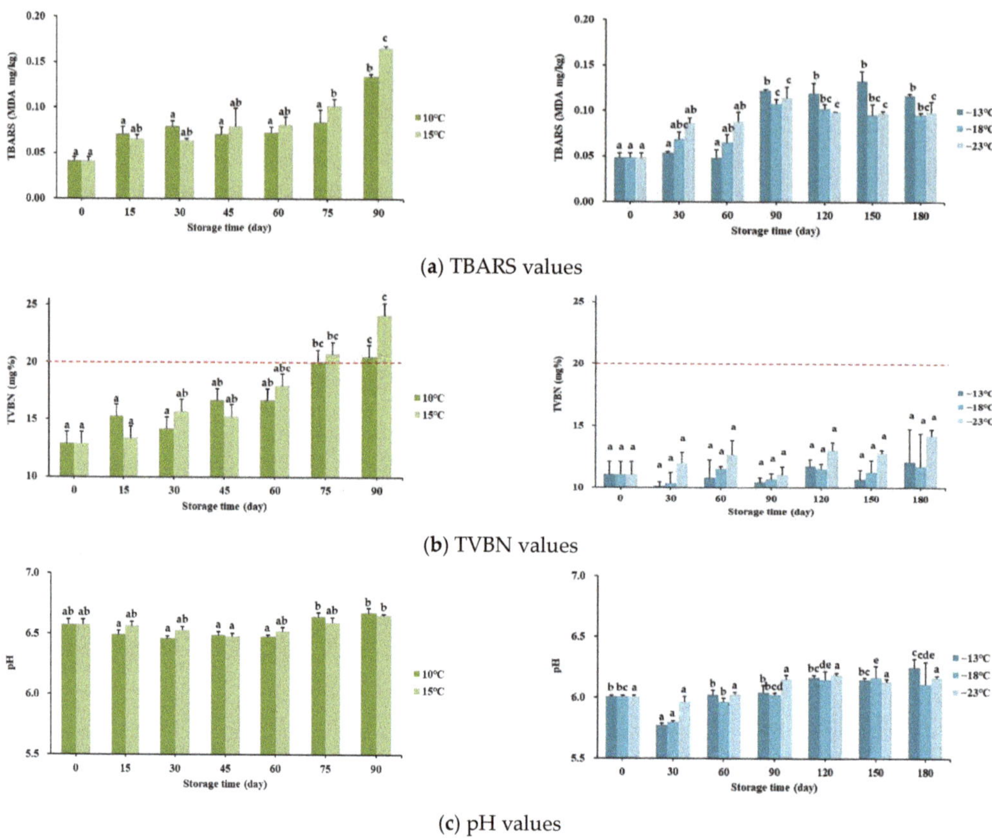

(**a**) TBARS values

(**b**) TVBN values

(**c**) pH values

Figure 6. Periodical values in the chemical properties (TBARS, TVBN, pH) of processed chicken products stored at different temperatures for storage days (90 and 180). Values are mean ± SE. Different letters (a–e) indicate significant differences among means by Tukey's test ($p < 0.05$); ▮▮ values for chicken leg, ▮▮▮ values for chicken breast.

Related to muscle protein decomposition, higher TVBN values could have an important sensory meaning in the following changes. First, the higher TVBN contributes to the off-flavor of meat owing to the rise in byproducts such as acid compounds and mineral nitrogen [49]. Second, higher TVBN values provokes a rotten and ammonium odor from foods [50]. These sensory responses were similarly observed in the actual sensory test.

As the final indicator in this experiment, pH values cross-reflect chemical reaction by other spoilage factors and are shown in Figure 6c. In this study, we observed that processed chicken leg meat had higher pH values than chicken breast meat. In detail, the former showed pH 6.4–6.7 whereas the latter showed pH 5.7–6.3. From the previous studies on cooked chicken meats, the following results have been assembled. First, chicken thigh has higher pH levels by about 0.5 compared to breast meat [51]. Second, superheated steam-treated chicken breast showed about pH 5.8 in the studies by Choi et al. [52] and Chun et al. [53]. Our results were mostly consistent with these studies.

4. Conclusions

With the purpose of developing preferable chicken products, the present study examined the effects of high frequency thawing, determined the optimal superheated steam treatment conditions and sub-materials (marination herb and smoke wood) and compared the storage effect of the final products on the subsequent sensory evaluation by well-trained panelists. Superheated steam treatments for leg and breast meat were optimized at 225 °C for 12 min 20 s and at 223 °C for 8 min 40 s, respectively, showing the best sensory qualities. For better flavor and taste, bay leaf extract was employed for marination of each meat and oak wood was selected as the best hot-smoking sawdust. The final products possessed excellent nutritional composition with balanced fatty acids between PUFA and SFA. Furthermore, both processed chicken leg and breast showed well-maintained sensory qualities scoring over 7 points at each storage condition; however, microbial and chemical degradation of the leg meat product was observed when stored under chilling conditions. Thus, the regular and diversified sensory evaluations related to the chemical correlation at each processing step and storage condition were effective for develop superior chicken leg and breast products.

Author Contributions: Conceptualization, W.-H.C. and J.-S.C.; methodology, J.-S.C.; software, W.-H.C.; validation, W.-H.C. and J.-S.C.; formal analysis, W.-H.C.; investigation, W.-H.C., J.-S.C.; data curation, W.-H.C., J.-S.C.; writing—original draft preparation, W.-H.C.; writing—review and editing, J.-S.C.; visualization, W.-H.C.; supervision, J.-S.C.; project administration, J.-S.C.; funding acquisition, J.-S.C. Both authors have read and agreed to the published version of the manuscript.

Funding: This research was supported by the Ministry of Trade, Industry, and Energy (MOTIE), Korea Institute for Advancement of Technology (KIAT) For Regional Program Evaluation (IRPE) through the Encouragement Program for The Social and Economic Innovation Growth (project number P0013031).

Institutional Review Board Statement: The study was conducted according to the guidelines of the Declaration of Helsinki, and approved by the Institutional Review Board of Silla University (protocol code 1041449-202006-HR-007).

Informed Consent Statement: Informed consent was obtained from all subjects involved in the study.

Data Availability Statement: Data supporting reported results are available upon request.

Acknowledgments: Our sincere thanks to C.-Y.J. for providing the resources for research and special thanks to N.-H.K. and J.-W.L. for technical support.

Conflicts of Interest: The authors declare no conflict of interest for this article.

References

1. Ismail, I.; Joo, S.-T. Poultry Meat Quality in Relation to Muscle Growth and Muscle Fiber Characteristics. *Food Sci. Anim. Resour.* **2017**, *37*, 873–883. [CrossRef]
2. Jung, S.B.; Kim, T.H.; Son, T.Y.; Yu, E.S.; Shin, J.Y.; Jung, J.Y.; Hwang, J.W.; Yang, J.Y. A Study on Development of the High Frequency Thawing Machine. *J. Korean Soc. Ind. Converg.* **2018**, *21*, 301–307. [CrossRef]
3. Oh, J.-H.; Yoon, S.; Choi, Y. The Effect of Superheated Steam Cooking Condition on Physico-Chemical and Sensory Characteristics of Chicken Breast Fillets. *Korean J. Food Cook. Sci.* **2014**, *30*, 317–324. [CrossRef]
4. Lunde, K.; Egelandsdal, B.; Choinski, J.; Mielnik, M.; Flåtten, A.; Kubberød, E. Marinating as a technology to shift sensory thresholds in ready-to-eat entire male pork meat. *Meat Sci.* **2008**, *80*, 1264–1272. [CrossRef] [PubMed]
5. Milly, P.J. Antimicrobial Properties of Liquid Smoke Fractions. Doctoral Dissertation, University of Georgia, Athens, GA, USA, December 2003.
6. Asita, A.O.; Campbell, I.A. Anti-microbial activity of smoke from different woods. *Lett. Appl. Microbiol.* **1990**, *10*, 93–95. [CrossRef]
7. Pereira, P.A.P.; De Salvia, M.A.; Carneiro, J.D.D.S. Sensory perception of Brazilian petit suisse cheese by a consumer panel using three-way internal and external preference maps. *MOJ Food Process. Technol.* **2020**, *8*, 109–112. [CrossRef]
8. Tirtawijaya, G.; Park, Y.; Won, N.E.; Kim, H.; An, J.H.; Jeon, J.; Park, S.; Yoon, S.; Sohn, J.H.; Kim, J.; et al. Effect of steaming and hot smoking treatment combination on the quality characteristics of hagfish (*Myxine glutinosa*). *J. Food Process. Preserv.* **2020**, *44*, 14694. [CrossRef]
9. Jaffe, T.R.; Wang, H.; Chambers, E. Determination of a lexicon for the sensory flavor attributes of smoked food products. *J. Sens. Stud.* **2017**, *32*, e12262. [CrossRef]
10. AOAC International. Method of Analysis–Official methods 990.12. In *International Official Methods of Analysis Official Methods*, 18th ed.; Association of Official Analytical Chemists: Arlington, VA, USA, 2002.
11. AOAC International. Method of Analysis–Official methods 991.14. In *International Official Methods of Analysis Official Methods*, 18th ed.; Association of Official Analytical Chemists: Arlington, VA, USA, 2002.
12. Yildiz, P.O. Effect of Essential Oils and Packaging on Hot Smoked Rainbow Trout during Storage. *J. Food Process. Preserv.* **2014**, *39*, 806–815. [CrossRef]

13. Mohibbullah; Won, N.E.; Jeon, J.-H.; An, J.H.; Park, Y.; Kim, H.; Bashir, K.M.I.; Park, S.-M.; Kim, Y.S.; Yoon, S.-J.; et al. Effect of superheated steam roasting with hot smoking treatment on improving physicochemical properties of the adductor muscle of pen shell (*Atrina pectinate*). *Food Sci. Nutr.* **2018**, *6*, 1317–1327. [CrossRef]

14. Ministry of Food and Drug Safety (MFDS). *8th General Analysis Method. Foodcode (Sik-Poom-Gong-Jeon)*; MFDS: Cheongju, Korea, 2020. Available online: https://www.foodsafetykorea.go.kr/foodcode/01_03.jsp?idx=11142 (accessed on 23 June 2021).

15. Ministry of Food and Drug Safety (MFDS). *4th General Analysis Method. Food additives code (Sik-Poom-Cheom-Ga-Mul-Gong-Jeon)*; MFDS: Cheongju, Korea, 2020. Available online: https://www.foodsafetykorea.go.kr/foodcode/04_03.jsp?idx=878 (accessed on 23 June 2021).

16. AOAC International. Method of Analysis–Official methods 925.09, 923.03, 979.09, 962.09, and 923.05. In *International Official Methods of Analysis Official Methods*, 17th ed.; Association of Official Analytical Chemists: Washington, DC, USA, 2000.

17. AOAC International. Method of Analysis–Official methods 963.22. In *International Official Methods of Analysis Official Methods*, 16th ed.; Association of Official Analytical Chemists: Arlington, VA, USA, 1995.

18. Ministry of Food and Drug Safety (MFDS). *The Guidelines for Shelf-life Experiment of Food, Livestock, and Health & Functional Food*, 5th ed.; MFDS: Cheongju, Korea, 2020. Available online: https://www.mfds.go.kr/brd/m_1060/view.do?seq=14285&srchFr= &srchTo=&srchWord=&srchTp=&itm_seq_1=0&itm_seq_2=0&multi_itm_seq=0&company_cd=&company_nm=&page=1 (accessed on 12 August 2021).

19. Yu, L.-H.; Lee, E.-S.; Chen, H.-S.; Jeong, J.-Y.; Choi, Y.-S.; Lim, D.-G.; Kim, C.-J. Comparison of Physicochemical Characteristics of Hot-boned Chicken Breast and Leg Muscles during Storage at 20 °C. *Food Sci. Anim. Resour.* **2011**, *31*, 676–683. [CrossRef]

20. Jung, S.B.; Seo, T.R.; Jung, H.J.; Kim, B.K.; Cho, Y.J. Effect of High Frequency Thawing and General Thawing Methods on the Quality of Frozen Mackerel, Alaska pollack, Japanese Spanish mackerel, and Yellow croaker. *J. FISHRIES Mar. Sci. Educ.* **2016**, *28*, 1152–1158. [CrossRef]

21. Park, D.H.; Lee, S.; Kim, H.Y.; Seo, J.-H.; Lee, J.; Kim, S.; Shin, H.; Choi, M.-J. The Effect of Packaging, Freezing Method, and Thawing Method on the Quality Properties of Blanched *Colocasia esculenta* (L.) Schott Stem. *Food Eng. Prog.* **2016**, *20*, 135–142. [CrossRef]

22. Joglekar, A.M.; May, A.T. Product excellence through design of experiments. *Cereal Foods World* **1987**, *32*, 857–868.

23. Zimmermann, W.J. An Approach to Safe Microwave Cooking of Pork Roasts Containing Trichinella spiralis. *J. Food Sci.* **1983**, *48*, 1715–1718. [CrossRef]

24. Kurup, A.H.; Deotale, S.; Rawson, A.; Patras, A. Thermal Processing of Herbs and Spices. *Herbs, Spices Med. Plants.* Available online: https://onlinelibrary.wiley.com/doi/abs/10.1002/9781119036685.ch1 (accessed on 1 June 2021). [CrossRef]

25. Lee, J.H.; An, L.H. Effects of Herbs and Green Tea on the Sensory and the Antioxidative Qualities of Beef-Yukwonjeon. *J. of the East Asian Soc. Diet. Life* **2007**, *17*, 808–815.

26. Migdał, W.; Zając, M.; Węsierska, E.; Tkaczewska, J.; Kulawik, P. PROCEEDINGS. Available online: http://istocar.bg.ac.rs/wp-content/uploads/2015/12/Proceedings-2015-2.pdf (accessed on 24 June 2021).

27. Ratnani, R.D. Widiyanto A Review of Pyrolisis of Eceng Gondok (Water hyacinth) for Liquid Smoke. In Proceedings of the E3S Web of Conferences, Semarang, Indonesia, 14–15 August 2018; EDP Sciences: Les Ulis, France, 18 August 2018; Volume 73, p. 05010.

28. Kralik, G.; Kralik, Z.; Grčević, M.; Hanžek, D. Quality of Chicken Meat. *Anim. Husb. Nutr.* **2018**, *63*. [CrossRef]

29. Wood, J.D.; Enser, M.; Fisher, A.V.; Nute, G.R.; Whittington, F.M.; Richardson, R.I. Effects of diets on fatty acids and meat quality. *Options Méditerr. Ser. A* **2003**, *67*, 133–141.

30. Smith, S.B. Marbling and Its Nutritional Impact on Risk Factors for Cardiovascular Disease. *Food Sci. Anim. Resour.* **2016**, *36*, 435–444. [CrossRef]

31. Kim, Y.H.; Min, J.S.; Hwang, G.S.; Lee, S.O.; Kim, I.S.; Bark, H.I.; Lee, M.H. Fatty acids composition and sensory characteristics of the commercial native chicken meat. *Korean J. Food Sci. Technol.* **1999**, *31*, 964–970.

32. Northcutt, J.; A Foegeding, E.; Edens, F. Water-Holding Properties of Thermally Preconditioned Chicken Breast and Leg Meat. *Poult. Sci.* **1994**, *73*, 308–316. [CrossRef]

33. Koh, H.-Y.; Yu, I.-J. Nutritional Analysis of Chicken Parts. *J. Korean Soc. Food Sci. Nutr.* **2015**, *44*, 1028–1034. [CrossRef]

34. Industry Resources on the Changes to the Nutrition Facts Label. Available online: https://www.fda.gov/food/food-labeling-nutrition/industry-resources-changes-nutrition-facts-label (accessed on 25 June 2021).

35. Richter, S. Investigating Ice slurry's Perceived Mechanical Abrasive Quality to Increase Pathogen Reduction on Poultry during Immersion Chilling. Doctoral Dissertation, Georgia Institute of Technology, Atlanta, GA, USA, 26 April 2018.

36. Tompkin, R.B.; Christiansen, L.N.; Shaparis, A.B. Iron and the antibotulinal efficacy of nitrite. *Appl. Environ. Microbiol.* **1979**, *37*, 351–353. [CrossRef] [PubMed]

37. Li, X.; Zhu, J.; Li, C.; Ye, H.; Wang, Z.; Wu, X.; Xu, B. Evolution of Volatile Compounds and Spoilage Bacteria in Smoked Bacon during Refrigeration Using an E-Nose and GC-MS Combined with Partial Least Squares Regression. *Molecules* **2018**, *23*, 3286. [CrossRef] [PubMed]

38. Katiyo, W.; De Kock, R.; Coorey, R.; Buys, E.M. Sensory implications of chicken meat spoilage in relation to microbial and physicochemical characteristics during refrigerated storage. *LWT* **2020**, *128*, 109468. [CrossRef]

39. Kilcast, D.; Lewis, D.F. Structure and texture—their importance in food quality. *Nutr. Bull.* **1990**, *15*, 103–113. [CrossRef]

40. Pande, G.; Akoh, C.C. Enzymatic modification of lipids fortrans-free margarine. *Lipid Technol.* **2013**, *25*, 31–33. [CrossRef]

41. Olsson, G.B.; Ofstad, R.; Lødemel, J.B.; Olsen, R.L. Changes in water-holding capacity of halibut muscle during cold storage. *LWT* **2003**, *36*, 771–778. [CrossRef]

42. Mathew, R.; Jaganathan, D.; Anandakumar, S. Effect of vacuum packaging method on shelf life of chicken. *Imp. J. Interdiscip. Res.* **2016**, *2*, 1859–1866.

43. Kondjoyan, A.; Portanguen, S. Effect of superheated steam on the inactivation of Listeria innocua surface-inoculated onto chicken skin. *J. Food Eng.* **2008**, *87*, 162–171. [CrossRef]

44. Heiszler, M.G.; Kraft, A.A.; Rey, C.R.; Rust, R.E. Effect of time and temperature of smoking on microorganisms on frankfurters. *J. Food Sci.* **1972**, *37*, 845–849. [CrossRef]

45. Domínguez, R.; Pateiro, M.; Gagaoua, M.; Barba, F.J.; Zhang, W.; Lorenzo, J.M. A Comprehensive Review on Lipid Oxidation in Meat and Meat Products. *Antioxidants* **2019**, *8*, 429. [CrossRef] [PubMed]

46. Tang, X.; Yu, Z. Rapid evaluation of chicken meat freshness using gas sensor array and signal analysis considering total volatile basic nitrogen. *Int. J. Food Prop.* **2020**, *23*, 297–305. [CrossRef]

47. Holman, B.W.; Bekhit, A.E.-D.A.; Waller, M.; Bailes, K.L.; Kerr, M.J.; Hopkins, D.L. The association between total volatile basic nitrogen (TVB-N) concentration and other biomarkers of quality and spoilage for vacuum packaged beef. *Meat Sci.* **2021**, *179*, 108551. [CrossRef] [PubMed]

48. Ministry of Food and Drug Safety (MFDS). *5th Standards and Specifications by Food. Foodcode (Sik-Poom-Gong-Jeon)*; MFDS: Cheongju, Korea, 2020. Available online: https://www.foodsafetykorea.go.kr/foodcode/01_03.jsp?idx=54 (accessed on 3 July 2021).

49. Effect of Methodology on the Determination of Total Volatile Basic Nitrogen as an Index of Quality of Meat and Fish. *Egypt. J. Food Saf.* **2013**, *1*, 23–34. [CrossRef]

50. Altissimi, S.; Mercuri, M.L.; Framboas, M.; Tommasino, M.; Pelli, S.; Benedetti, F.; Di Bella, S.; Haouet, N. Indicators of protein spoilage in fresh and defrosted crustaceans and cephalopods stored in domestic condition. *Ital. J. Food Saf.* **2017**, *6*, 6921. [CrossRef]

51. Choe, J.; Kim, H.-Y. Physicochemical characteristics of breast and thigh meats from old broiler breeder hen and old laying hen and their effects on quality properties of pressed ham. *Poult. Sci.* **2020**, *99*, 2230–2235. [CrossRef]

52. Choi, Y.-S.; Hwang, K.-E.; Jeong, T.-J.; Kim, Y.-B.; Jeon, K.-H.; Kim, E.-M.; Sung, J.-M.; Kim, H.-W.; Kim, C.-J. Comparative Study on the Effects of Boiling, Steaming, Grilling, Microwaving and Superheated Steaming on Quality Characteristics of Marinated Chicken Steak. *Food Sci. Anim. Resour.* **2016**, *36*, 1–7. [CrossRef]

53. Chun, J.-Y.; Kwon, B.-G.; Lee, S.-H.; Min, S.-G.; Hong, G.-P. Studies on Physico-chemical Properties of Chicken Meat Cooked in Electric Oven Combined with Superheated Steam. *Food Sci. Anim. Resour.* **2013**, *33*, 103–108. [CrossRef]

Article

Effect of Different Processing Methods on Quality, Structure, Oxidative Properties and Water Distribution Properties of Fish Meat-Based Snacks

Asad Nawaz [1,2], Enpeng Li [1,2,*], Ibrahim Khalifa [3], Noman Walayat [4], Jianhua Liu [4], Sana Irshad [5], Anam Zahra [6], Shakeel Ahmed [7], Mario Juan Simirgiotis [7], Mirian Pateiro [8] and José M. Lorenzo [8,9]

[1] Key Laboratory of Plant Functional Genomics of the Ministry of Education/Jiangsu Key Laboratory of Crop Genomics and Molecular Breeding, College of Agriculture, Yangzhou University, Yangzhou 225009, China; 007298@yzu.edu.cn
[2] Co-Innovation Center for Modern Production Technology of Grain Crops, Yangzhou University, Yangzhou 225009, China
[3] Food Technology Department, Faculty of Agriculture, Benha University, Moshtohor, Benha 13736, Egypt; Ibrahiem.khalifa@fagr.bu.edu.eg
[4] College of Food Science and Technology, Zhejiang University of Technology, Hangzhou 310014, China; nomanrai66@zjut.edu.cn (N.W.); Jhliu@zjut.edu.cn (J.L.)
[5] School of Environmental Studies, China University of Geo Sciences, Wuhan 430074, China; sanairshad55@gmail.com
[6] Islamabad Campus, University Institute of Diet and Nutritional Sciences, University of Lahore, Islambad 45750, Pakistan; Anam.zahra@dnsc.uol.edu.pk
[7] Campus Isla Teja, Instituto de Farmacia, Facultad de Ciencias, Universidad Austral de Chile, Valdivia 5090000, Chile; Shakeel1177@uach.cl (S.A.); mario.simirgiotis@uach.cl (M.J.S.)
[8] Centro Tecnológico de la Carne de Galicia, Avd. Galicia No. 4, Parque Tecnológico de Galicia, San Cibrao das Viñas, 32900 Ourense, Spain; mirianpateiro@ceteca.net (M.P.); jmlorenzo@ceteca.net (J.M.L.)
[9] Área de Tecnología de los Alimentos, Facultad de Ciencias de Ourense, Universidad de Vigo, 32004 Ourense, Spain
* Correspondence: lep@yzu.edu.cn

Citation: Nawaz, A.; Li, E.; Khalifa, I.; Walayat, N.; Liu, J.; Irshad, S.; Zahra, A.; Ahmed, S.; Simirgiotis, M.J.; Pateiro, M.; et al. Effect of Different Processing Methods on Quality, Structure, Oxidative Properties and Water Distribution Properties of Fish Meat-Based Snacks. *Foods* **2021**, *10*, 2467. https://doi.org/10.3390/foods10102467

Academic Editor: Joongho Kwon

Received: 4 September 2021
Accepted: 12 October 2021
Published: 15 October 2021

Publisher's Note: MDPI stays neutral with regard to jurisdictional claims in published maps and institutional affiliations.

Abstract: Snack foods are consumed around to globe due to their high nutrition, taste and versatility; however, the effects of various processing methods on quality, structure and oxidative properties are scare in the literature. This study aims to evaluate the effect of various processing methods (frying, baking and microwave cooking) on quality, structure, pasting, water distribution and protein oxidative properties of fish meat-based snacks. The results showed that the frying method induced a significantly ($p < 0.05$) higher expansion than baking and microwave methods. Texture in terms of hardness was attributed to the rapid loss of water from muscle fiber, which resulted in compact structure and the increased hardness in microwave cooking, whereas in frying, due to excessive expansion, the hardness decreased. The pasting properties were significantly higher in baking, indicating the sufficient swelling of starch granules, while low in microwave suggest the rapid heating, which degraded the starch molecules and disruption of hydrogen bonds as well as glycosidic linkage and weakening of granules integrity. The water movement assessed by Low Field Nuclear Magnetic Resonance (LF-NMR) showed that frying had less tight and immobilized water, whereas microwave and baking had high amounts of tight and immobilized water, attributing to the proper starch-protein interaction within matrix, which was also evidenced by scanning electron microscopy (SEM) analysis. The protein oxidation was significantly ($p < 0.05$) higher in frying compared to baking and microwave cooking. The findings suggest the endorsement of baking and microwave cooking for a quality, safe and healthy snacks.

Keywords: cooking methods; fish meat snacks; LF-NMR; SEM; protein oxidation

221

1. Introduction

Snack foods are popular around the globe due to their balanced nutrition, highly bioavailable bioactive compounds, specific aroma and taste as well as cost effectiveness [1]. Traditionally, these are made from wheat flour and frying methods, which lead to some undesirable consequences, such as low nutrition, high oil content [2], allergens to gluten protein [3] and poor water-holding capacity [4]. The role of processing methods has always been neglected or less perceivable influence, but these are of high importance in terms of loss of nutrition, structure, safety and physiochemical properties [5], and most recently alternation at molecular levels [6]. Thus, there is need to find some other alternatives of wheat flour as well as to find some other processing methods, which will not only improve the nutrition, but also improve the physiochemical properties. Thus, studying various processing methods and their impact on structure, quality and oxidative properties is inevitable.

Processing methods have a critical role in the quality and structure of food especially the mode of heat transfer, such as conduction (transfer of heat via intervening matter), convection (heat transfer by fluid) and radiation (heat transfer through space) [7]. Heat disrupts the native conformation of polypeptides, which hold the molecules together, resulting the increase in thermal motion, loss of secondary and tertiary structure [6] and rupturing the intermolecular forces, e.g., electrostatic or non-polar interaction and disulfide bonds [8]. Further, once proteins are unfolded in response of thermal processing, aggregation of proteins occur and scramble the disulfide bonding, resulting in side chain modifications [9], which cross link with other polypeptides that are exposed to oxidation. In addition, processing methods have significant influence in terms of structural modification [6], textural changes [10], water distribution [11], microstructural changes [12] and even oxidative changes [13]. On the other hand, the heating of starch granules results in the loss of crystalline structure as well as gelatinization, which changes the structure and composition of starch granules. Thus, studying these structural and physio-chemical modifications in the starch–protein matrix is foreseeable for a safe and healthy product, especially for snack foods that are consumed after every meal.

Fish meat has recently drawn a considerable attention of food industrialist due to its highly bioavailable protein [14], essential amino acids [15] and the fact that it is suitable for all religions in terms of religious concepts [16]. Since fish meat production has increased in the world, its use in processing food has received immense attention and has been used in combination with various other snacks foods, such as cassava starch [3], wheat flour [17], fish bone [2], potato powder [5] dietary fiber [18] and most recently the oxidation of protein, which, owing to various cooking methods and its relationship with the physiochemical properties of the end product has received considerable attention.

There have been many studies in the literature that focus on the nutrition, quality or the use of fish meat in snacks. However, there is no such report that narrates the effect of processing methods on the structure, quality and physiochemical properties of fish meat snacks. Therefore, the key aim of this study was to assess the effect of different processing methods (frying, baking and microwave heating) on structure, quality, and physiochemical properties of fish meat-based snacks. Snacks based on wheat flour were prepared by the partial replacement of fish meat and cooked with various different processing methods. The physiochemical properties, such as expansion, water hydration capacity and texture, were analyzed in order to assess the effect of different processing methods. Moreover, pasting properties were determined in order to predict the gelling behavior in different processing methods. In addition, water distribution properties were assessed in all processing methods using low-field nuclear magnetic resonance analysis. The study will provide useful information for the best use of processing method and open new window for the feasibility of different methods.

2. Materials and Methods

2.1. Materials

Grass carp (*Ctenopharyngodon Idella*, 1.5 ± 0.2 kg) in a deceased form was purchased from local supermarket of Yangzhou, Jiangsu, China. The composition of fish was as follows: moisture 77.5%, protein 18.5%, ash 0.9% and fat 2.35%. Wheat flour with 12% protein, 1.5% fat and 0.42% ash was acquired from Jinsha River industry Co., Ltd., Chengdu, Sichuan, China. Salt and cooking oil (sunflower oil, Arwana brand) were purchased from local supermarket. Potato powder (moisture 13%, protein 7.9% and fat 1%) was purchased from Zhang Jiakou Co., Ltd., Zhangjiakou, Hubei, China. All other chemicals used in this study were of analytical grade and were acquired from Sigma-Aldrich, Inc. (Natick, MA, USA), and were used without further purification.

2.2. Preparation of Snacks

Snacks were prepared according to the formulation of our previous study [14] with slight modifications. Briefly, wheat flour (70 g), potato powder (20 g), minced fish meat (10 g), tricalcium phosphate 0.03% w/w and salt (1.5%, w/w) were added in a mixing bowl, while 22 mL of water (32% w/w of wheat flour) was added into it and kneaded manually for 4 min until a uniform dough was obtained. The kneaded dough was covered by polythene sheet and kept at room temperature for proper hydration. After this, hydrated dough was rolled by a noodle-making machine (Model: FKM-180, Yongkang Electrical appliances, Jinhua, Zhejiang, China) roller in order to make a uniform sheet with a thickness of 3 mm. This uniform sheet was cut into square pieces of 3.5 × 3.5 cm, which were further dried in hot air oven for 2 h at 45 °C. The dried snacks were subjected to various cooking methods: frying (165 °C for 35–45 s) by a thermostatically controlled electric fryer (PFE8, IME, Salzburg, Austria), baking (220 °C for 6 min in a baking oven) was done in thermostatically controlled oven (Model: HGB-20D, Rudong Jiahua Food Machinery, Co., Ltd., Nantong, Jiangsu, China) and microwave cooking (1000 W for 3 min for each side) using a microwave oven (NE-1037, Panasonic, Osaka, Japan). The prepared snacks were packed in high-density polyethylene bags for further analysis. For each cooking method, cooking was performed in triplicates and for each batch, 1 kg of snacks were prepared with three replicates (3 kg for each cooking method).

2.3. Expansion of Snacks

The expansion of snacks was measured following the methodology of Nawaz, Xiong [19] with modifications. This was measured before and after cooking. For each cooking method, six snacks from each treatment were marked with permanent ink marker, and thickness (before and after cooking) was measured from a center point using digital Vernier caliper, having a minimum resolution of 0.01 mm. All values reported in results are mean of six replicates and were calculated by following formula:

$$\text{Expansion (mm)} = \text{Thickness before cooking} - \text{Thickness after cooking}$$

2.4. Physicochemical Properties

Water hydration capacity (WHC) was calculated by following the study of Nawaz, Xiong [20]. All values reported in the results are mean values of triplicates. Color parameters were measured following the methodology of Nawaz, Xiong [14] using Minolta CR-300 Colorimeter (Osaka, Japan). These were measured by making a powder of each treatment by grinding and then packing into transparent polyethylene bags. Lightness (L *), redness (a *) and yellowness (b *) were calculated in triplicates for each treatment. Texture was calculated following the methodology of Nawaz, Xiong [20] using a three bending point (HDP/3PB) probe by means of TA-XT2 plus texture analyzer (stable micro system, Surrey, UK). The following settings were applied for measuring textural parameters (hardness and fructurability): load cell of 5 kg, test speed was 1 mm/s, compression was made up to 70% and an auto trigger type was used. The peak force (N) was measured immediately prior to breakage and fructurability was also calculated. For each treatment, six replicates from each cooking method were used and values were reported in means and standard deviation.

2.5. Pasting Properties

Pasting properties of fish meat snacks were measured by following the study of Yildiz, Yurt [21]. Regarding pasting properties, snacks were ground into powder form and used for pasting properties using Rapid Visco Analyzer (RVA)-4 series (Newport Scientific Pvt., Ltd., Narrabeen, NSW, Australia). Cooked snack powder (3 g) was added in 25 mL of distilled water, which was placed in RVA cylinder and stirred manually in order to make a slurry. This slurry was fixed in RVA equipment and stirred at 960 rpm for 10 s subsequently 160 rpm during analysis. After that, paste was kept at 50 °C for 60 s and then heated up to 95 °C in 3.7 min and hold at 95 °C for 2.5 min. Hereafter, temperature was dropped to 50 °C within 3.8 min and hold at 50 °C for 2 min. Meanwhile, peak, trough, breakdown, final and setback viscosities along with pasting temperature were measured. Among these, setback viscosities were calculated by subtracting the final viscosity from trough viscosity, whereas breakdown viscosities were measured by the difference between peak viscosity and trough viscosity. All calculations were done in triplicates for each treatment in order to calculate the mean values, which were further used for the statistical study.

2.6. Measurement of Protein Oxidation

Protein oxidation in terms of protein carbonyl contents was measured through following the methodology of Hu, Ren [6]. A spectrophotometer (Model F-4600, Hitachi, Tokyo, Japan) was used to measure the absorbance at 365 nm and carbonyl content was calculated using the absorption coefficient ($22,000\ (\text{mol}/\text{L})^{-1}\ \text{cm}^{-1}$) and expressed as nmol/mg of protein.

2.7. Microstructure of Snacks

The microstructure of snacks was observed using a scanning electron microscope (NTC, JSM-6390LV; Tokyo, Japan) following the study of Nawaz, Xiong [17]. Briefly, cooked snacks (cross-sectional pieces) were freeze dried and then fixed into a bronze slide using double-edge sticky tape and then sputter-coated with platinum. The obtained images were observed at a magnification of 100 using an accelerated voltage of 15 kV.

2.8. Low-Field Nuclear Magnetic Resonance (LF-NMR) Analysis

Proton mobility distribution in cooked snacks was measured through low-field nuclear magnetic resonance using Minspec mq 20 low-field pulsed NMR spectrometer (Bruker, Ettlingen, Germany) following the previous methodology [22]. Briefly, transverse relaxation time (T2) was calculated at 20 MHz and at room temperature (25 °C) using a sequence based on Carr–Purcell–Meiboom–Gill (CPMG). The amount of proton in the population was proportional to the acquired peaks. Measurements were performed in triplicates for each treatment while the graph was plotted from the mean value of the triplicates.

2.9. Statistical Analysis

All analyses were performed in triplicates for obtaining the mean values and standard deviation. The obtained mean values were subjected for statistical analysis using one-way analysis of variance (ANOVA) and Duncan's multiple range test using SPSS version 21.0 (SPSS statistics for windows, Armonk, IBM, Carp., New York, NY, USA). The statistically significant difference was set a level of $p < 0.05$. All graphs were drawn using origin pro 9 Software.

3. Results

3.1. Physiochemical Properties of Snacks

The expansion of snacks is an important parameter regarding consumer's acceptability and appearance. Usually, expanded snacks foods are preferred, but more than 70% is unacceptable, which causes irregular volume and deshaped snacks [5]. The results of expansion are shown in Figure 1, and ranged from 1.25 to 3.75 mm, which revealed that there was a significant ($p < 0.05$) difference in frying and baking methods, while no statistically significant differences were observed between the baking and microwave methods. Although the values of expansion of baking and microwave were different, they were not statistically significant by means of the one-way ANOVA and DMR test. Overall expansion, which was almost 50% in all treatments, was attributed to the presence of starchy material and gluten protein, which both exhibit viscoelastic properties upon heating [14]. However, this expansion cannot be neglected owing to the addition of fish meat that has also been reported for the increased expansion in snack food. However, regarding the cooking methods, it was revealed that frying resulted in the highest expansion even it was more than 70% of the original volume. This is especially attributed to frying, in which high-temperature oil interacts with water, which raptures the surface and decreases the bulk density resulting in high expansion. Similar results about expansion were reported by in a previous study [20] when fish meat -based snacks were fried; however, the addition of >10% caused less expansion compared to control (devoid of meat). On the other hand, microwave heating is fast and uniform heating and based on the transition of glassy matrix to rubbery sate [23]. Moreover, during microwave heating, a sudden increase in temperature produces superheated steam and rapid escape of water molecules, which resulted in the transition to rubbery state of starch and started deformation and hindering of the proper volume. Therefore, the results of expansion in terms of baking and microwaving are acceptable in order to commercialize the product.

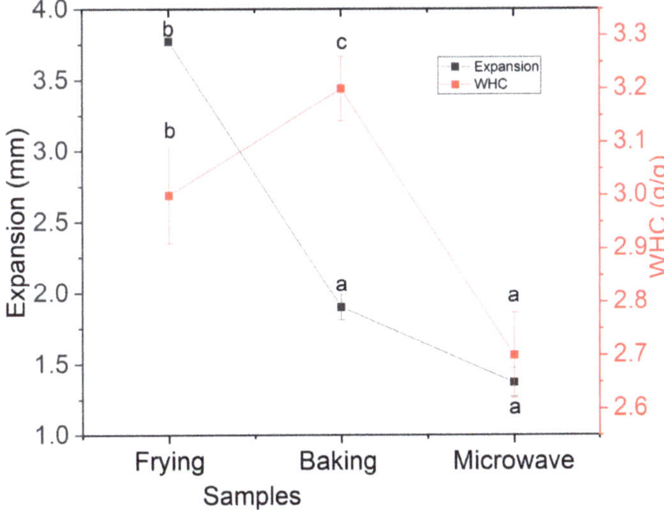

Figure 1. Effect of processing methods on expansion and WHC of fish meat-based snacks. The results of expansion and WHC are the mean and standard deviation of six and three snacks, respectively. Different letters (abc) over the error bars indicate the significant difference between processing methods at $p < 0.05$ using one-way ANOVA and Duncan's multiple range test.

Water hydration capacity (WHC) predicts the behavior of processed foods in terms of extent of gelatinization and dextrinization, which are greatly affected by various thermal processing methods and types of ingredients within the matrix. The results shown in Figure 1 reveal that all treatments showed a significant ($p < 0.05$) difference and it was highest in baked snacks, while it was found to be lowest in the microwave followed by frying. The increased WHC in baking indicated the sufficient swelling of starch granules, which absorbed more water, while less WHC in the microwave might be due to the fact that during microwave heating, protein unfolds and expands more than starch, which wraps the starch granules, resulting in insufficient gelatinization. In case of frying, oil interacts with the matrix and creates hydrophobic conditions, which may result in less WHC. The improved values of WHC in case of baking method indicates the sufficient swelling of starch granules, which endorse the feasibility of the baking method compared to frying and the microwave.

The effect of processing method on color parameters is shown in Table 1. It was revealed that color parameters, especially lightness, were significantly ($p < 0.05$) increased in baking compared to microwave and frying treatments. Actually, the lower lightness in frying might be due to the oil, which led to the formation of crust and caused browning. Another reason could be the Maillard reaction between sugars and protein. Similar results were found by another study [24], which reported the level of lightness in various cooking methods as follows: baking > microwave > grilled > frying. Regarding redness, it was found that redness was significantly higher in those treatments having lower lightness as in the case of frying. Actually, frying takes places at higher temperature and oil makes the crust darker as compared to baking and microwave heating. On the other hand, microwave heating is a transfer of heat by means of radiation, which also prevents the excessive browning; however, the high temperatures caused by microwave heating resulted in low lightness compared to baking in our study, which was also evidenced by a previous study [25], in which yellowness was higher in frying followed by microwave and baking techniques. Our findings are in agreement with data reported by Cho, Lee [26] who showed that microwave cooking results in slightly lower lightness as compared to gas stove cooking of instant noodles. Since color is very important regarding sensory appearance, the findings regarding high lightness and low redness in the baking method have the potential to be considered for a good appearance.

Table 1. Physiochemical properties of fish meat-based snacks prepared by various processing methods.

Samples	Frying	Baking	Microwave
Lightness (L *)	79.40 ± 0.50 [a]	85.69 ± 0.43 [c]	83.40 ± 0.32 [b]
Redness (a *)	5.93 ± 0.17 [c]	3.32 ± 0.21 [a]	3.97 ± 0.25 [b]
Yellowness (b *)	17.38 ± 0.16 [b]	13.85 ± 0.34 [a]	16.83 ± 0.19 [a,b]

The results of color and texture parameters are the mean and standard deviation of three and six snacks for each treatment, respectively. Different letters (a,b,c) in columns indicate the significant difference between processing methods at $p < 0.05$ using one-way ANOVA and Duncan's multiple range test.

3.2. Textural Properties of Snacks

Textural properties are of high interest regarding the consumer's acceptability. Regarding snacks, hardness and fructurability are considered as important parameters. The results of textural properties are shown in Figure 2. It was disclosed that there was statistically significant ($p < 0.05$) difference in all treatments in terms of hardness, while it was highest in microwave cooking and lowest in frying treatments. On the other hand, regarding the results of fructurability, there was a significant ($p < 0.05$) difference between frying and baking or frying and microwave heating, but no difference was found in baking and microwave heating. Low hardness in frying could be attributed to that cooking method in which high-temperature oil created high expansion (more intercellular spaces) by rupturing the surface. In addition, as reported earlier [5], higher expansion causes lower hardness and decreased bulk density. The higher hardness in microwave treatment might be due to the rapid escape of water molecules due to the superheated steam production, which makes the compact structure of snacks. This was also evidenced in terms of low expansion for microwave-baked snacks. Another reason for the increased hardness in microwave heating might be due to the denaturation of myofibrillar proteins (actually actomyosin complex), which resulted in the shrinkage of muscle fibers, as compared to the modification of the connective tissues component during cooking. Due to rapid increase in temperature, water escaped from muscle fiber more quickly, which resulted in the less tender of meat as reported by a previous study [24]. Our results regarding the hardness of snacks are in agreement with the previous study, in which microwave heating resulted in the abnormal increase in the hardness of the meat compared to other cooking methods, especially grilling [24]. Low hardness results in soft matrix while high hardness makes meat difficult to chew; thus, the values for baking were within the range and can be considered as a reference for commercialization.

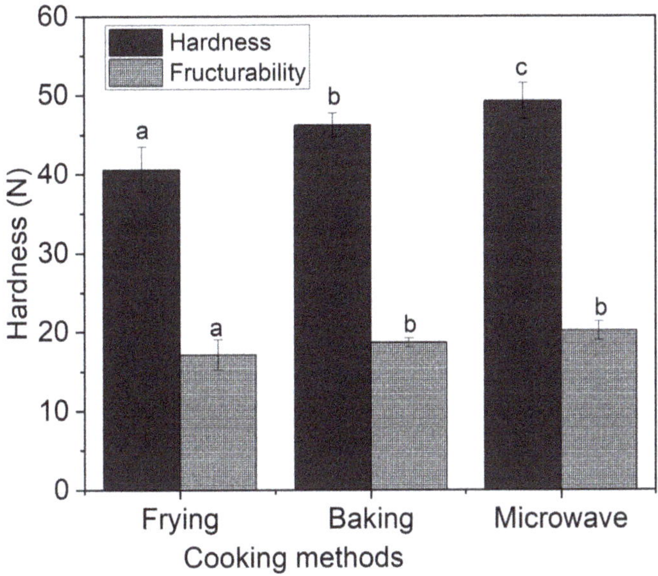

Figure 2. Effect of processing methods on textural characteristics of fish meat-based snacks. The results of texture parameters are the mean and standard deviation of six snacks for each treatment. Different letters (a,b,c) over the error bars indicate the significant difference between processing methods at $p < 0.05$ using one-way ANOVA and Duncan's multiple range test.

3.3. Pasting Properties of Snacks

Pasting properties are of extreme importance in order to predict the stability of products. These are usually determined by a combination of a complex process that occurs after the gelatinization of starch granules, which alter from swelling to rupture, amylose leaching and gel formation with the input of energy [21]. Thus, pasting properties greatly depend on processing methods along with the presence of other ingredients in a matrix, especially the type of starch [3], starch–protein interaction [5] and chemical reagents and additives [14]. The pasting properties of meat-based snacks prepared by various cooking methods are shown in Figure 3 and Table 2. Among these, peak viscosity, which is an indicative of highest viscosity acquired during gelatinization in the presence of water prior to its physical breakdown, was found to be significantly ($p < 0.05$) higher in baking, while it was lowest in frying. The increased peak viscosity in baking could be due to long time and gradual heating of starch granules that attained sufficient swelling time. On the other hand, the rapid heating or instant heating in microwave heating may partially inhibit the starch gelatinization or be due to the proteins, which showed intensive viscoelastic properties upon microwave heating. The decreased peak viscosity in frying could be due to the oil droplets that behaved as barrier and prevented the swelling of starch granules. The increased in pasting properties of baked snacks could be due to the proper gelatinization of starch, which was inhibited in microwave cooking due to sudden increase in temperature. Thus, the starch granules sufficiently swelled in baking, and pasting properties were increased.

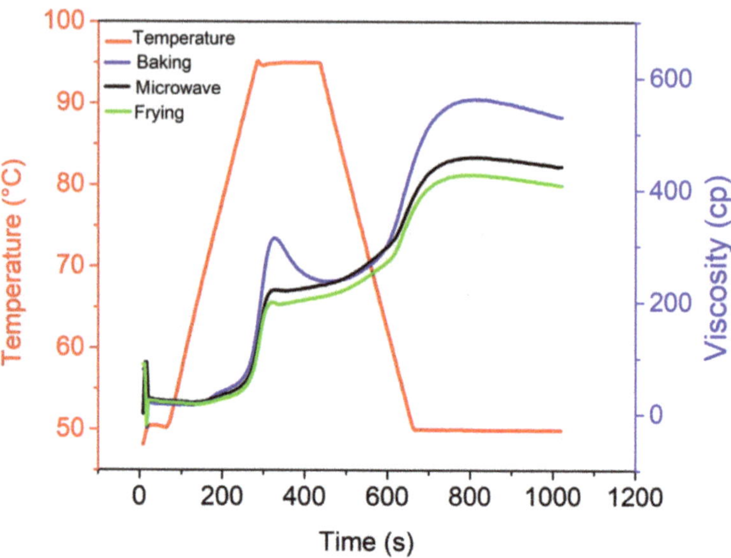

Figure 3. Pasting properties of fish meat-based snacks prepared by various processing methods.

Table 2. Pasting properties of fish meat-based snacks prepared by various processing methods.

Samples	Peak Viscosity (RVU)	Trough Viscosity (RVU)	Breakdown Viscosity (RVU)	Final Viscosity (RVU)	Setback (RVU)	Temperature (°C)
Baking	328 ± 30.50 [b]	245 ± 20.50 [b]	83 ± 22.50 [b]	549 ± 30.22 [c]	304 ± 25.50 [b]	95.25 ± 0.09 [a,b]
Microwave	237 ± 10.61 [a,b]	221 ± 0.50 [a]	16 ± 9.50 [a]	442 ± 14.50 [b]	221 ± 09.69 [a]	95.00 ± 0.04 [b]
Frying	217 ± 12.80 [a]	200 ± 0.50 [a]	17 ± 5.50 [a]	408 ± 14.50 [a]	208 ± 10.22 [a]	94.65 ± 0.50 [a]

The results of pasting properties are the mean and standard deviation of triplicates of each treatment, respectively. Different letters (a,b,c) in rows indicate the significant difference between processing methods at $p < 0.05$ using one-way ANOVA and Duncan's multiple range test.

Tough viscosity signifies the ability of paste to withstand breakdown after gelatinization when subjected to cooling and was found to be higher again in baking followed by microwave and frying treatments. This indicates that snacks cooked by baking and microwave heating were able to form a stable paste compared to frying, which was also evidenced by the results of WHC. Similarly, the results of final viscosity are also in line with peak and final viscosity, indicating the importance of cooking methods especially for baking, in which a slow heating rate provided ample time for the gradual swelling of starch granules, which were easily disrupted by sharing action of paddle [27]. The reason for the low pasting properties in microwave heating might be due to the instant heating of microwave, which degraded the starch granules leading to the disruption of hydrogen bonds and glycosidic linkage, resulting in the weakening of the granules' integrity [28]. Breakdown values represent the strength of granules, whereas setback values suggest the degree of rearrangement of starch molecules after gelatinization and determine the transformation of viscous liquid to gel. From the results, it was obvious that final and setback viscosities were significantly higher in baking, which again endorses the slow heating for a proper starch–protein network development. On the other hand, there was no significant difference in the values of setback viscosities among frying and microwave heating. Therefore, higher values for pasting properties especially in baking and microwave cooking indicate the stability of snack food.

3.4. Protein Carbonyl Content in Cooked Snacks

Protein oxidation has remained the critical challenge for meat industry, which is initiated by various factors, such as types and composition of meat, processing methods, intensity of cooking temperature and time and the presence of non-proteinous material. The results of protein carbonyl content are shown in Figure 4. The results revealed that protein carbonyl content, an important biomarker for measuring protein oxidation, was found to be significantly ($p < 0.05$) highest in frying followed by baking and microwave oven heating. The highest oxidation induced by frying could be due the lipids that are also a leading cause of protein oxidation as reported by previous study [13]. Generally, during heat treatment, meat loses its oxidative potential and cellular damage of cells occurs, which makes it more expose to oxygen, subsequently trigging the reactive oxygen species (ROS) that attack the lipid and protein molecules. Moreover, cooking promotes the oxidative cleavage of porphyrin ring, releasing the heme iron (a pro-oxidant), which accelerate oxidative deterioration [6,29].

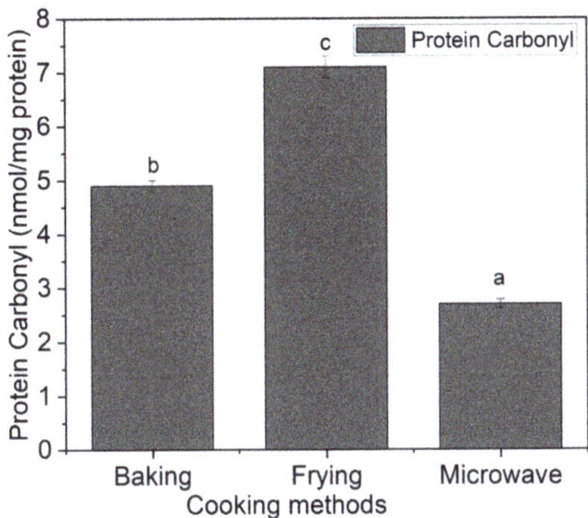

Figure 4. Protein carbonyl content of fish meat-based snacks prepared by different cooking methods. Reported values are mean of triplicate of each treatment and small letter over bar represent significant difference using one-way ANOVA and DMR's test at a significance level of $p < 0.05$.

The increased in protein oxidation in frying could be attributed to oil absorption in fillets, which exposed the proteins to oxidative cleavage. However, the increased protein oxidation in baking after frying could be attributed to long cooking time along with higher temperature, which are responsible for protein oxidation, as reported previously [6]. The increased oxidation by frying has also been reported by previous studies [9]. On the other hand, the lowest protein carbonyl content in microwave heating endorses the suitability of microwave heating, which was for less time, in terms of safety point of view.

3.5. Microstructure of Snacks

The micrographs of snacks made from different cooking methods are shown in Figure 5. The observation revealed that frying method had void holes and an irregular matrix, which was also evidenced in the form of higher expansion. Some starch granules were of small size indicating the insufficient swelling of granules, which resulted in decreased pasting properties of that method. Regarding baking method, a compact structure was found with swelled and gelatinized starch granules indicating a uniform matrix. Moreover, the starch granules were embedded within the protein matrix, suggesting a uniform starch–protein network. The spaces between starch and protein were less visible. On the other hand, the image for microwave baking was quite different from those of frying and baking with an uneven structure and non-uniform matrix. Irregular blocks were visible, which may indicate the weak interaction between starch and protein. However, the morphology was much better than frying in terms of irregular spaces and starch granules. The photos presented in Figure 5 are in accordance with previous study with same structure [12].

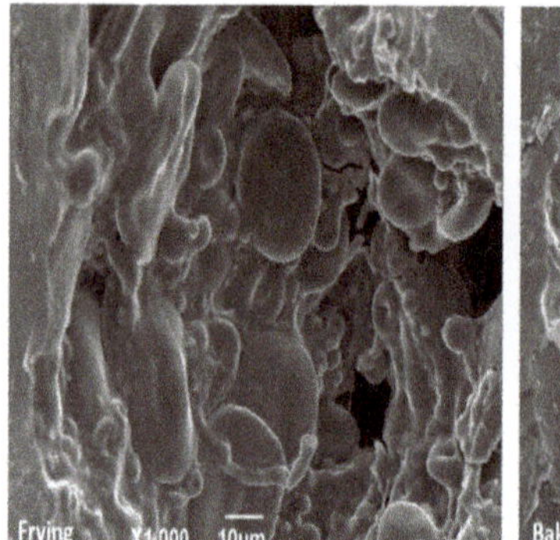

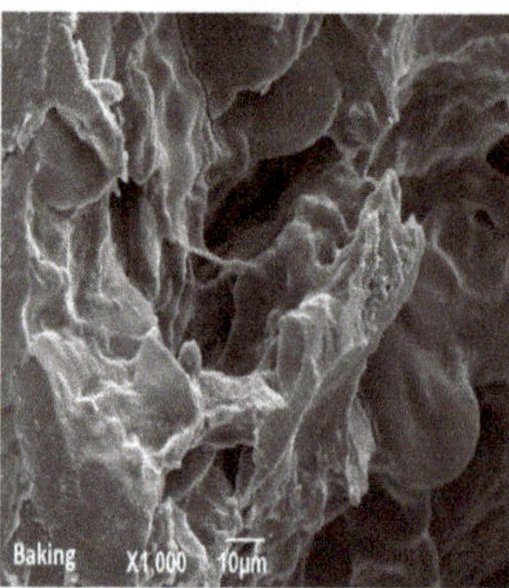

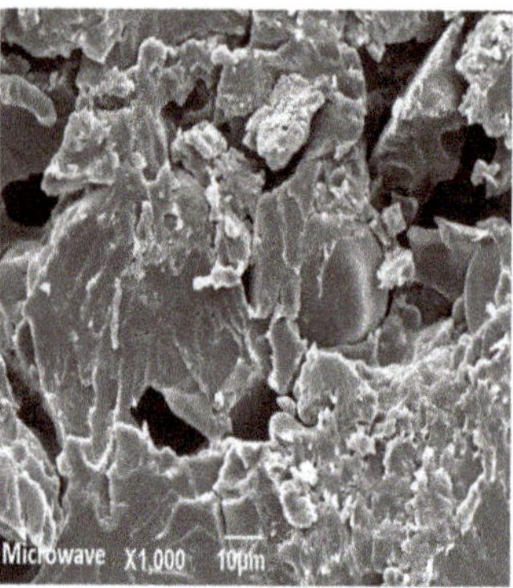

Figure 5. Microstructural analysis of fish meat-based snacks prepared by various processing methods.

3.6. LF-NMR Analysis of Snacks

The changes in water distribution and state of water during the thermal processing are essential parameters that predict the stability, physiochemical changes and shelf-life of a product. Water distribution or water mobility within the cooked matrix are usually determined by the relaxation time (T_2) by means of LF-NMR analysis [22]. The T_2 relaxation time is further categorized into three subsections: T_{21} ranges from 0.1 to 1 ms and indicates the tightly bound water, T_{22} (1–100 ms) signifies the immobilized water, whereas T_{23} represents the unbound water [14]. Generally, shorter relaxation time and shorter amplitude disclose a strong association between solids and water within a matrix, whereas long amplitude shows weak interaction of water in a heterogenous system. The results of LF-NNMR analysis are shown in Figure 6. The findings regarding T_{21} showed that relaxation time and amplitude both were highest in frying, where they were lowest in baking followed by microwave heating, suggesting that more tightly bound water was present in baked snacks compared to frying. Again, the peak for the amplitude of immobilized water T_{22} (1–100 ms) was highest in frying, which indicates the less-immobilized water in fried snacks compared to microwave and baking treatments. It was noteworthy to observe that peak for microwave cooking was very low, which suggest the strong interaction of water with other ingredients, and this indicates the proper starch–protein network development in microwave and baking methods. A similar trend was observed for T_{23} relaxation time in which the frying method showed the highest amount of unbound water. Kexin, Xu [22] also found that the frying process of turbot flesh resulted in high amount of unbound water as compared to stewing. This could be due to the oil that created a hydrophobic environment within the matrix resulting in insufficient swelling and gelatinization, which was also evidenced in pasting properties. On the other hand, the amplitude peaks for baking and microwave were almost similar in this region, which suggests that snacks prepared with these methods revealed the least amount of unbound water. This suggests that water was tightly bound within quaternary and tertiary structures of proteins during thermal processing. The high amount of unbound water causes loosening of the structure and texture, and it was leached out during heating and especially in frying methods. This unbound water was retained in the matrix owing to the crust formation by the frying method and was observed in the form of unbound water. In addition, with slow cooking for a long time, such as in baking, the starch granules absorbed sufficient amount of water with the full capacity of gelatinization, resulting in the tightly bound water within the matrix [5].

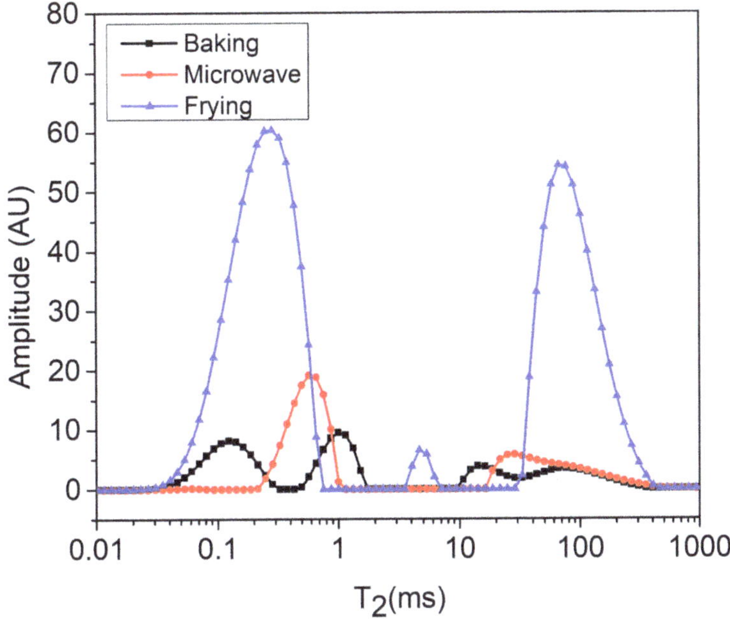

Figure 6. LF-NMR analysis of fish meat-based snacks prepared by various processing methods.

The role of water is of high importance during thermal processing, as water acts a plasticizer and promotes interaction between the molecules of starch and protein. In addition, the role of the cooking medium is very important. Once possibility in fried snacks (less immobilized and tight water) is that the frying method might breakdown the fiber bundles' network, resulting in the leaching of free water and less-immobilized water. The results are in accordance with the study of Liu, Dong [13]. Moreover, frying caused more protein aggregation and confirmation, which alter the tissue gap in meat muscle, leading to availability of free water, which was also evidenced in the form of higher protein carbonyl content in frying [6]. Similar results were reported by another study [12] in which the microwave method resulted in a higher amount of immobilized water compared to other cooking methods.

4. Conclusions

The present study reported three most used methods for fish meat-based snacks and evaluated quality, safety and structural parameters. Due to high oil uptake in frying, which causes serious diseases such as cardiovascular diseases, consumer's choice is moving toward non-oil-based methods, such as baking and microwave heating. The major findings revealed that baking and microwave cooking could be the alternative choices instead of frying, which not only improved the quality (color and texture), but also enhanced the physiochemical (WHC, pasting properties), safety (protein oxidation) and water distribution properties. The baking can be considered as an optimum method for meat-based snacks, especially for proper starch–protein interaction, which improved texture, WHC and pasting properties. The microwave method can be considered as the best choice in terms of decreased protein oxidation and water mobility, especially the presence of tightly and immobilized water, which all highly endorse the appropriateness of microwave cooking. However, in case of microwave cooking, as protein behaved extensively in response of microwaves, there is a need to further investigate the mechanism and optimum conditions for meat-based snacks. Beside this, there is need to conduct storage stability of fish meat-based snacks for these cooking methods. With all these investigations, the appropriateness of processing methods can be enhanced and commercialized for fish meat-based snacks.

Author Contributions: Conceptualization, I.K. and J.L.; methodology, A.N. and N.W.; software, S.A. and A.Z.; validation, J.M.L., M.P. and M.J.S.; formal analysis, A.N. and S.I.; investigation, A.N.; resources, E.L.; data curation, I.K.; writing—original draft preparation, A.N.; writing—review and editing, J.M.L.; supervision, E.L.; funding acquisition, E.L. All authors have read and agreed to the published version of the manuscript.

Funding: The authors acknowledge the financial support of a National Natural Science Foundation of China grant C1304013151101138—A project funded by the Priority Academic Program of Jiangsu Higher Education Institutions.

Institutional Review Board Statement: Not applicable.

Informed Consent Statement: Not applicable.

Data Availability Statement: Data are openly available for all readers after publication.

Acknowledgments: We greatly acknowledge the financial support of Postdoctoral office of Yangzhou University.

Conflicts of Interest: The authors declare no conflict of interest.

References

1. Adeleke, B.S.; Babalola, O.O. Oilseed crop sunflower (Helianthus annuus) as a source of food: Nutritional and health benefits. *Food Sci. Nutr.* **2020**, *8*, 4666–4684. [CrossRef] [PubMed]
2. Nawaz, A.; Xiong, Z.; Xiong, H.; Chen, L. The effects of fish meat and fish bone addition on nutritional value, texture and microstructure of optimised fried snacks. *Int. J. Food Sci. Technol.* **2019**, *54*, 1045–1053. [CrossRef]
3. Nawaz, A.; Alhilali, A.; Li, E.; Khalifa, I.; Irshad, S.; Walayat, N.; Chen, L.; Wang, P.; Tan, Z. The effects of gluten protein substation on chemical structure, crystallinity, and Ca in vitro digestibility of wheat-cassava snacks. *Food Chem.* **2021**, *339*, 127875. [CrossRef] [PubMed]
4. Thammahiwes, S.; Riyajan, S.-A.; Kaewtatip, K. Preparation and properties of wheat gluten based bioplastics with fish scale. *J. Cereal Sci.* **2017**, *75*, 186–191. [CrossRef]
5. Nawaz, A.; Khalifa, I.; Walayat, N.; Lorenzo, J.M.; Irshad, S.; Abdullah; Ahmed, S.; Simirgiotis, M.J.; Ali, M.; Li, E. Whole Fish Powder Snacks: Evaluation of Structural, Textural, Pasting, and Water Distribution Properties. *Sustainability* **2021**, *13*, 6010. [CrossRef]
6. Hu, L.; Ren, S.; Shen, Q.; Chen, J.; Ye, X.; Ling, J. Proteomic study of the effect of different cooking methods on protein oxidation in fish fillets. *RSC Adv.* **2017**, *7*, 27496–27505. [CrossRef]
7. Xia, C.; Wen, P.; Yuan, Y.; Yu, X.; Chen, Y.; Xu, H.; Cui, G.; Wang, J. Effect of roasting temperature on lipid and protein oxidation and amino acid residue side chain modification of beef patties. *RSC Adv.* **2021**, *11*, 21629–21641. [CrossRef]
8. Yu, T.; Morton, J.; Clerens, S.; Dyer, J. Cooking-induced protein modifications in meat. *Compr. Rev. Food Sci. Food Saf.* **2017**, *16*, 141–159. [CrossRef]
9. Yin, Y.; Zhou, L.; Pereira, J.; Zhang, J.; Zhang, W. Insights into Digestibility and Peptide Profiling of Beef Muscle Proteins with Different Cooking Methods. *J. Agric. Food Chem.* **2020**, *68*, 14243–14251. [CrossRef]
10. Echegaray, N.; Paterio, M.; Domínguez, R.; Purriños, L.; Bermúdez, R.; Carballo, J.; Lorenzo, J. Effects of different cooking methods and of the inclusion of chestnut (Castanea sativa Miller) in the finishing diet of Celta pig breed on the physicochemical parameters and volatile profile of Longissimus thoracis et lumborum muscle. *Food Res. Int.* **2020**, *137*, 109407. [CrossRef]
11. Wan, J.; Cao, A.; Cai, L. Effects of vacuum or sous-vide cooking methods on the quality of largemouth bass (Micropterus salmoides). *Int. J. Gastron. Food Sci.* **2019**, *18*, 100181. [CrossRef]
12. Wang, X.; Muhoza, B.; Wang, X.; Feng, T.; Xia, S.; Zhang, X. Comparison between microwave and traditional water bath cooking on saltiness perception, water distribution and microstructure of grass crap meat. *Food Res. Int.* **2019**, *125*, 108521. [CrossRef]
13. Liu, F.; Dong, X.; Shen, S.; Shi, Y.; OU, Y.; Cai, W.; Chen, Y.; Zhu, B. Changes in the digestion properties and protein conformation of sturgeon myofibrillar protein treated by low temperature vacuum heating during in vitro digestion. *Food Funct.* **2021**, *12*, 6981–6991. [CrossRef]
14. Nawaz, A.; Xiong, Z.; Xiong, H.; Irshad, S.; Chen, L.; Wang, P.; Ahsan, H.M.; Walayat, N.; Qamar, S.H. The impact of hydrophilic emulsifiers on the physico-chemical properties, microstructure, water distribution and in vitro digestibility of proteins in fried snacks based on fish meat. *Food Funct.* **2019**, *10*, 6927–6935. [CrossRef]

15. Nawaz, A.; Li, E.; Irshad, S.; Hammad, H.H.M.; Liu, J.; Shahbaz, H.M.; Waqas, A.; Regenstein, J.M. Improved effect of autoclave processing on size reduction, chemical structure, nutritional, mechanical and in vitro digestibility properties of fish bone powder. *Adv Powder Tech.* **2020**, *31*, 2513–2520. [CrossRef]

16. Nawaz, A.; Li, E.; Irshad, S.; Xiong, Z.; Xiong, H.; Shahbaz, H.; Siddique, F. Valorization of fisheries by-products: Challenges and technical concerns to food industry. *Trends Food Sci. Technol.* **2020**, *99*, 34–43. [CrossRef]

17. Nawaz, A.; Xiong, Z.; Li, Q.; Xiong, H.; Liu, J.; Chen, L.; Wang, P.; Walayat, N.; Irshad, S.; Regenstein, J. Effect of wheat flour replacement with potato powder on dough rheology, physiochemical and microstructural properties of instant noodles. *J. Food Process. Preserv.* **2019**, *43*, e13995. [CrossRef]

18. Jiang, S.; Zhao, S.; Jia, X.; Wang, H.; Zhang, H.; Liu, Q.; Kong, B. Thermal gelling properties and structural properties of myofibrillar protein including thermo-reversible and thermo-irreversible curdlan gels. *Food Chem.* **2020**, *311*, 126018. [CrossRef]

19. de Lima, M.M.; Tribuzi, G.; de Souza, J.A.R.; de Souza, I.G.; Laurindo, J.B.; Carciofi, B.A.M. Vacuum impregnation and drying of calcium-fortified pineapple snacks. *LWT - Food Sci. Technol.* **2016**, *72*, 501–509. [CrossRef]

20. Nawaz, A.; Xiong, Z.; Li, Q.; Xiong, H.; Irshad, S.; Chen, L.; Wang, P.; Zhang, M.; Hina, S.; Regenstein, J.M. Evaluation of physicochemical, textural and sensory quality characteristics of red fish meat-based fried snacks. *J. Sci. Food Agric.* **2019**, *99*, 5771–5777. [CrossRef]

21. Yildiz, Ö.; Yurt, B.; Baştürk, A.; Toker, Ö.; Yilmaz, M.; Karaman, S.; Dağlıoğlu, O. Pasting properties, texture profile and stress–relaxation behavior of wheat starch/dietary fiber systems. *Food Res. Int.* **2013**, *53*, 278–290. [CrossRef]

22. Xia, K.; Xu, W.; Huang, L.; Song, Y.; Zhu, B.; Tian, M. Water dynamics of turbot flesh during frying, boiling, and stewing processes and its relationship with color and texture properties: Low-field NMR and MRI studies. *J. Food Process. Preserv.* **2018**, *42*, e13338. [CrossRef]

23. Gutiérrez-Cano, J.D.; Hamilton, I.; Catala-Civera, J.; Bows, J.; Penaranda-Foix, F. Effect of water content on the dynamic measurement of dielectric properties of food snack pellets during microwave expansion. *J. Food Eng.* **2018**, *232*, 21–28. [CrossRef]

24. Lorenzo, J.M.; Cittadini, A.; Munekata, P.E.; Domínguez, R. Physicochemical properties of foal meat as affected by cooking methods. *Meat Sci.* **2015**, *108*, 50–54. [CrossRef]

25. Jogihalli, P.; Singh, L.; Sharanagat, V.S. Effect of microwave roasting parameters on functional and antioxidant properties of chickpea (Cicer arietinum). *LWT-Food Sci. Technol.* **2017**, *79*, 223–233. [CrossRef]

26. Cho, S.Y.; Lee, J.W.; Rhee, C. The cooking qualities of microwave oven cooked instant noodles. *Int. J. Food Sci. Technol.* **2010**, *45*, 1042–1049. [CrossRef]

27. Lu, X.; Xu, R.; Zhan, J.; Chen, L. Pasting, rheology, and fine structure of starch for waxy rice powder with high-temperature baking. *Int. J. Biol. Macromol.* **2020**, *146*, 620–626. [CrossRef] [PubMed]

28. Susanna, S.; Prabhasankar, P. A comparative study of different bio-processing methods for reduction in wheat flour allergens. *Eur. Food Res. Technol.* **2011**, *233*, 999–1006. [CrossRef]

29. Soladoye, O.P.; Shand, P.; Dugan, M.E.R.; Gariépy, C.; Aalhus, J.L.; Estévez, M.; Juárez, M. Influence of cooking methods and storage time on lipid and protein oxidation and heterocyclic aromatic amines production in bacon. *Food Res. Int.* **2017**, *99*, 660–669. [CrossRef]

 *foods*

Article

Evolution of Extra Virgin Olive Oil Quality under Different Storage Conditions

Soraya Mousavi [1], Roberto Mariotti [1,*], Vitale Stanzione [2], Saverio Pandolfi [1], Valerio Mastio [3], Luciana Baldoni [1] and Nicolò G. M. Cultrera [1]

[1]   Institute of Biosciences and Bioresources, National Research Council, 06128 Perugia, Italy; soraya.mousavi@ibbr.cnr.it (S.M.); saverio.pandolfi@ibbr.cnr.it (S.P.); luciana.baldoni@ibbr.cnr.it (L.B.); niccolo.cultrera@ibbr.cnr.it (N.G.M.C.)
[2]   Institute for Agricultural and Forest Systems of the Mediterranean, National Research Council, 06128 Perugia, Italy; vitale.stanzione@cnr.it
[3]   Estación Experimental Agropecuaria San Juan, Instituto Nacional de Tecnología Agropecuaria (INTA), Consejo Nacional de Investigaciones Científicas y Técnicas (CONICET), Ing. Marcos Zalazar (Calle 11) y Vidart. Villa Aberastain, Pocito, San Juan 5427, Argentina; mastio.valerio@inta.gob.ar
*   Correspondence: roberto.mariotti@ibbr.cnr.it; Tel.: +39-075-5014809

Abstract: The extent and conditions of storage may affect the stability and quality of extra virgin olive oil (EVOO). This study aimed at evaluating the effects of different storage conditions (ambient, 4 °C and −18 °C temperatures, and argon headspace) on three EVOOs (low, medium, and high phenols) over 18 and 36 months, analyzing the main metabolites at six time points. The results showed that low temperatures are able to maintain all three EVOOs within the legal limits established by the current EU regulations for most compounds up to 36 months. Oleocanthal, squalene, and total phenols were affected by storage temperatures more than other compounds and degradation of squalene and α-tocopherol was inhibited only by low temperatures. The best temperature for 3-year conservation was 4 °C, but −18 °C represented the optimum temperature to preserve the organoleptic properties. The present study provided new insights that should guide EVOO manufacturers and traders to apply the most efficient storage methods to maintain the characteristics of the freshly extracted oils for a long conservation time.

Keywords: extra virgin olive oil; storage; oxidation; phenols; sterols; tocopherols; temperature; argon

Citation: Mousavi, S.; Mariotti, R.; Stanzione, V.; Pandolfi, S.; Mastio, V.; Baldoni, L.; Cultrera, N.G.M. Evolution of Extra Virgin Olive Oil Quality under Different Storage Conditions. *Foods* **2021**, *10*, 1945. https://doi.org/10.3390/foods10081945

Academic Editor: Theodoros Varzakas

Received: 9 July 2021
Accepted: 18 August 2021
Published: 21 August 2021

Publisher's Note: MDPI stays neutral with regard to jurisdictional claims in published maps and institutional affiliations.

1. Introduction

Extra virgin olive oil (EVOO) is one of the most valuable vegetable oils and its consumption has now expanded worldwide owing to its unique flavor and richness in bioactive compounds, such as phenols, tocopherols, squalene, and sterols [1–3].

Various factors may act on the characteristics and chemical composition of EVOOs, including varieties, orchard management, harvesting time, oil extraction, and storage conditions [4,5]. In particular, storing can greatly affect the oxidation stability of EVOO, one of the primary causes of its degradation, which can progress in the dark (autoxidation), and can be accelerated in the presence of light (photo-oxidation) and enzymes (enzymatic oxidation) [6]. The high oleic acid content and the natural antioxidants present in EVOOs, including phenolic compounds, tocopherols (mainly α-tocopherol), squalene, chlorophyll, carotenoids, and vitamins, are tied to oil stability by providing an effective defense against free radicals by different mechanisms [7–11]. Minor polar phenolic compounds are constituted by simple phenols, such as tyrosol and hydroxytyrosol, and by their combination with other moieties to form oleuropein and ligstroside and their derivatives, cinnamic acids, as well as lignans and flavonoids [12]. It was demonstrated that the degradation rate of phenolic compounds during oil aging is strongly related to their initial concentration [13]. Some of the simple phenols, such as tyrosol and hydroxytyrosol, increase over

235

time, likely owing to hydrolytic processes of secoiridoid derivatives representing their linked forms [14]. The percentage of hydrolysis typically increases in aged EVOO; therefore, the EVOO quality should be correlated to lower and not to higher values of tyrosol and hydroxytyrosol [15]. Tyrosol and hydroxytyrosol increase has been observed in EVOOs stored for three months at room temperature [16], and a high decrease in secoiridoids (close to 50%) was observed in EVOOs with low initial minor polar compounds after 18 months of conservation at room temperature, while in oils with a high initial amount, the decrease was close to 20% [17]. A decrease in phenol content was reported in EVOOs stored at 1 °C for 12 months, although lignans were more stable in the same conditions [15].

The effects of the storage conditions on acidity, peroxide value, and fatty acid profile as criteria of the EVOO quality have been studied in different types of containers and at different temperatures, from 5 to 25 °C [13,15,16,18]. High-quality EVOOs stored at 12−15 °C showed a reduced lipid oxidation [16,19] and peroxide value and acidity, and delta-K did not change in EVOOs conserved at 0–8 °C for 24 months [13]. The autoxidation stability of several EVOOs in 21 months of storage in dark and at room temperature showed that the antioxidant concentration directly affected the extinction coefficient K_{232}, correlated to product oxidation [20]. In a recent study, it was reported that K_{232} could be used as an effective and cheap measure to monitor the quality evolution during storage [18]. Fadda et al. [21] found only 13% α-tocopherol degradation in EVOOs over 18 months of storage in the dark, while Okogeri and Tasioula-Margari [22] reported up to 79% loss of α-tocopherol after 4 months of storage under light, and nearly complete loss over 12–24 months of storage [23].

Squalene is a rather stable molecule in the absence of oxygen and has a weak antioxidant activity, owing to competitive oxidation with the lipid substrate [24]. In a study on degradation of squalene during EVOO storage, oil samples were stored in dark and colorless bottles, filled completely or halfway, in order to simulate the domestic storage conditions. The content of squalene decreased significantly only after 6 months in half-empty bottles, and diffused lighting did not play a significant role in squalene degradation after 12 months of storage in filled colorless bottles under lighting [25]. Nevertheless, the contribution of squalene to olive oil stability under light exposure or in the dark, scarcely investigated up to now, would deserve further attention to obtain conclusive results [8].

Phytosterol isomers decrease during EVOO storage and sterol oxidation increase through their reaction with light, high temperature, and oxygen, and this increment depends on their unsaturation level [26]. A high concentration of chlorophylls increases the stability of EVOO exposed to light and oil photo-oxidation in the presence of chlorophylls, leading to the formation of highly unstable and reactive singlet oxygen that tends to react with the unsaturated fatty acids, leading to the formation of hydroperoxides. Carotenoids are effective inhibitors of photo-oxidation by quenching singlet oxygen and triplet excited states of photosensitizers [7].

Most EVOO producers and researchers consider 12−18 months as the maximum commercial storage period from bottling to consumption and current guides on best practices delivered by the International Olive Council (COI/BPS/Doc. No 1—Best practice guidelines for the storage of olive oils and olive-pomace oils for human consumption, guide No 1/2018) suggest the commercialization within 24 months, but 18 months is still commonly reported as the "best before" date since bottling.

While extensive information is available on EVOO storage at high and ambient temperatures and light exposure, few works have been performed on the effects of low and freezing temperatures [6,15,16] in which the effects of low temperatures were studied on few oil compounds, mostly just phenols, and the extent of the experiment was limited to 18 months. The present study aims to evaluate how different storage conditions (ambient temperature, with and without argon, and cold treatments at +4 °C and −18 °C) on three EVOOs with different initial phenol content may preserve chemical and organoleptic characteristics over a long and very long time of conservation (18 and 36 months).

2. Materials and Methods

2.1. Materials and Storage Conditions

EVOOs with low phenols (LP, up to 200 mg kg^{-1}), medium phenols (MP, up to 400 mg kg^{-1}), and high phenols (HP, more than 400 mg kg^{-1}) were provided by a quality-assured industrial oil mill. After filtration, oils were stored in 250 mL amber glass bottles, closed with hermetic caps and maintained in the dark for all treatments and along the entire storage duration.

Four different treatments were settled for storing the bottles (12 bottles for each oil in each storage condition): (1) ambient temperature (AT); (2) ambient temperature + argon saturated headspace (AT + Ar) (excepting for HP oil); (3) refrigerated at 4 ± 1 °C; and (4) freezed at −18 ± 1 °C. In treatments 1, 3, and 4, the headspace was occupied by unmodified atmosphere. Two different storage durations were considered: 18 months (long-term storage, T18) and 36 months (2× long-term storage, T36) (Figure S1).

The effects of the storage conditions on oil quality were evaluated at the following time points: at the beginning of treatment (T0) and after 18 months storage (T18). To analyze the effects of 18 months of storage on the posterior oil stability, oil samples were evaluated at 72 h (T18-72 h), one month (T18-1 m), and eight months (T18-8 m) after treatment completion and bottle opening. After each time point, the oil bottles were closed and placed in the dark condition at ambient temperature for the next time point. Bottles were closed as in everyday use; no special procedure was done to remove oxygen. Through these analyses, we tried to simulate the gradual usage of oil and its progressive oxidation that usually happen after a long storage period and its usage by the consumers.

To study the effects of 2× long-term storage (T36), EVOO bottles of each treatment were kept closed for the entire storage period (Figure S1).

Six bottles for each oil, storage condition, and time point were used, three for chemical analysis and three for organoleptic evaluation. Samples stored at 4 °C and −18 °C were taken out from refrigerators and maintained at room temperature for 6 h, in order to thaw oil at liquid status.

2.2. Quality and Oxidation Indices

Acidity was measured and expressed as a percentage of free oleic acid. Peroxide value (PV) was stated as milliequivalents of active oxygen per kilogram of oil (meq O_2 kg^{-1}). K_{232} and K_{270} extinction coefficients (UV absorbance) were calculated from absorption at the exact λ wavelengths in nm, following the analytical methods described in the European Commission Regulation EEC 2568/91 (Regulation, 1991) and later amendments (the latest being EU 1348/2013).

2.3. Analysis of Total Phenolic Compounds

The content of total phenols was determined by the Folin–Ciocalteu (FC) method, according to the analytical protocol described by Singleton et al. [27]. The method was adapted for oils as follows: 5 g of oil was extracted with 5 mL of methanol/water (80:20 *v/v*) by 30 min shaking and 5 min centrifugation (1700× *g*). A 1 mL extract was added to 0.25 mL of FC reagent and 1.5 mL of Na_2CO_3 (20% *w/v*), in a 10 mL volumetric flask, reaching the final volume with purified water. Each sample was stored for 90 min at a controlled temperature of 25 °C in dark conditions, and the spectrophotometric analysis was performed at λ = 725 nm. The results, expressed in mg kg^{-1} of gallic acid (GA), were obtained through a calibration curve ranging from 1 to 15 µg mL^{-1} (R^2 = 0.9985).

2.4. Extraction of the Phenolic Fraction

Following the method described by the International Olive Council (Method COI/T.20/Doc. No 29, 2009), 5 g of olive oil was added to 5 mL methanol/water (80/20 *v/v*) and 100 µL of syringic acid as an internal standard (IS) was added to each sample. Samples were shaken for 1 min to homogenize the mixture and then centrifuged at 2200× *g* for 25 min at 4 °C. Finally, the supernatant was injected into the HPLC-DAD system. An external standard method was used in quantifying the phenolic compounds using their related calibration curves and the contribution of the internal standard.

2.5. Analysis of Phenolic Compounds through Reverse-Phase HPLC

The HPLC analyses of the phenolic extracts were conducted according to Selvaggini et al. [28], with a reversed-phase column using a HPLC-DAD, Varian ProStar-Diode Array Detector 330. The machinery was composed of a vacuum degasser, a quaternary pump, an autosampler, a thermostatic column compartment, and a DAD, using a 250 × 4.6 mm column 5 µm Kinetex EVO C18 100A (Phenomenex, Torrance, CA, USA). Eluent "A" was made with water and phosphoric acid 0.2% (Carlo Erba, Milano, Italy) and eluent "B" was methanol/acetonitrile (Carlo Erba) 50:50 (v/v). The elution gradient started from 4% eluent B and reached 100% B after 55 min for 15 min at a flow rate of 1.2 mL min^{-1}. Phenolic compounds were quantified at three wavelengths: 280, 310, and 360 nm using an authentic external standard. Secoiridoid derivatives, oleacein (3,4-DHPEA-EDA), and oleocanthal (p-HPEA-EDA) with 95% purity were provided by Prof. P. Magiatis (University of Athens, Greece). All other identified phenols were purchased from Sigma-Aldrich (St. Louis, MO, USA).

2.6. Analysis of Tocopherols

Tocopherol composition was determined by modifying the HPLC procedure described in [29]; 0.15 g of olive oil was dissolved in 5 mL hexane and homogenized by stirring. Samples were analyzed using HPLC-DAD 330 and the same column as for the phenolic compounds. The calibration curve was obtained by injecting standard solutions of (±)-α-Tocopherol-synthetic, ≥96% (Sigma-Aldrich, St. Louis, MO, USA) at different concentrations. The HPLC analysis was performed using a mobile phase composed by wluent "A" water with phosphoric acid 0.2% (Carlo Erba) and eluent "B" was methanol (Carlo Erba), at a ratio A/B 10:90. The flow rate was 1.2 mL/min, the injection volume was 30 µL, and the time of analysis was set for 20 min. Detection and quantification were performed at 290 nm.

2.7. Sterol and Squalene Analyses

Around 200 mg of each sampled oil was placed in a 10 mL propylene tube. Then, 200 µL of an internal standard solution (5α-cholestan-3β-ol, Sigma-Aldrich) in hexane was added. The analysis of the unsaponifiable fraction was performed by GC, without a preliminary thin-layer chromatography fractionation. This approach was used for the analysis of sterols and squalene in olive fruit and olive oil by many authors [30,31]. Alkaline hydrolysis was performed by adding 2 mL of KOH 2%, then tubes were soaked in a water bath at 80 °C for 15 min, and the unsaponifiable fraction was extracted by vortexing with 1 mL hexane and 1.5 mL NaCl 1%. The upper hexane layer was transferred to 2 mL glass vials. Samples were conserved at −20 °C until analysis, within 24 h from preparation.

Analyses were performed on a GC-FID, using a ZB-5HT Inferno capillary column (15 m × 0.32 mm × 0.10 µm film thickness, Phenomenex, Torrance, CA, USA). The carrier gas was helium (column flow 1.5 mL/min), and the split ratio was 1:100. The oven temperature was programmed as follows: 0.5 min at 150 °C, from 150 °C to 240 °C at 8 °C/min and from 240 °C to 370 °C at 25 °C/min held for 5 min, at 370 °C, followed by 320 °C for the injector and 350 °C for the detector (FID). The quantification was performed by external standards of squalene and sterols purchased from Sigma-Aldrich (St. Louis, MO, USA).

2.8. Analysis of Chlorophylls and Carotenoids

Chlorophylls and carotenoids were determined at 670 nm and 470 nm, respectively, following Minguez-Mosquera et al. [32] protocol. The oil samples were dissolved in cyclohexane (1.5:5 *w/v*) and absorbance was measured using a Perkin Helmer Lambda 10 UV–vis spectrophotometer.

$$\text{chlorophylls} = (A670 \times 106)/(613 \times 100 \times d) \tag{1}$$

$$\text{carotenoids} = (A470 \times 106)/(2000 \times 100 \times d) \tag{2}$$

where A is the absorbance and d is the path length of the cell (1 cm).

2.9. Fatty acid Methyl Ester (FAME) Analysis

Approximately 150 μL of oil in 2 mL of hexane was trans-methylated with 200 μL of a cold solution of KOH in methanol (2 M), according to the European Standard NF EN ISO 12966-2 [33]. Fatty acid methyl esters (FAMEs) were analyzed in accordance with the European Standard NF EN ISO 5508. Analyses were performed on a Varian Gas Chromatograph CP3800 equipped with the flame ionization detector (GC-FID) (T = 320 °C), using a capillary column (60 m × 0.25 mm i.d., 0.25 μm film thickness) coated with polyethylene glycol (Zebron, ZB-WAX, Phenomenex, Torrance, CA, USA). The carrier gas was helium (column flow 1.5 mL/min) and the split ratio was 1:100. The oven temperature was programmed as follows: 2 min at 140 °C, increased from 140 °C to 240 °C at 4 °C/min, held for 15 min, then 42 min at 240 °C. FAMEs were identified by comparing the retention times with the standard solution of Supelco 37 Component FAME Mix (Sigma-Aldrich, St. Louis, MO, USA).

2.10. Organoleptic Evaluation

The organoleptic profile of the oil samples was evaluated at all time points (T0, T18, T18-72 h, T18-1 m, T18-8 m, and T36) by the panel taste of the Slow Food organization, following the IOC method for the organoleptic assessment of virgin olive oil (COI/T.20/Doc. No 15/Rev. 10 2018) with some modification. The most important differences concerned the number of tasters, which was five plus a leader, and the number of analyzed oil per section, which was five out of four, as reported in the IOC method, and all the other recommendation were applied. The scale used for each olfactory and gustatory sensation was from 0 to 10 for each perception. The presentation of each sample to the panelist was blind. The tasting session begins with a calibration between the members, with the panel leader, who provided different oils and evaluated the response of each member. After the calibration, the panel leader, in a separate room, numbered every replica of oil and then offered to the panel members one glass with the same code at a time with a maximum of five oils for each section. Moreover, the panel leader checked the results, performing the average of them, and controlled if reported results were out of range. If this last case occurred, the panel leader started a new calibration test and the sample was numbered and tasted again. Olfactory and gustatory sensations were evaluated considering positive attributes, i.e., fruitiness, bitterness, pungency and persistence, and negative traits, as the presence of defects (i.e., fusty/muddy sediment, musty/humid/earthy, winey/vinegary/acid/sour, rancid, and others). The panel leader compiled the notes given by each taster and the statistical evaluation was carried out by the median of each parameter. This test provided sequential information about the sensory characteristics of the samples and allowed identifying changes in the organoleptic profile of oils under different treatments and time points.

2.11. Statistical Analysis

Data were analyzed by DAASTAT [34] using one-way ANOVA ($p < 0.01$ = ** and $p < 0.05$ = *, $n = 3$), separately for the three oils and among different time points. Tukey test was used to compare mean values. A principal component analysis (PCA) was applied for a total of 22 chemical variables using the percentage of reduction in content of chemical parameters after long term storage (from T18 to T18-8 m) and among T18 and T36 conditions, using PAST software version 4.03 [35]. A boxplot was made by the same software to show the variability of each chemical parameters for the studied samples and treatments.

3. Results and Discussion

3.1. Free Acidity, Oxidation Indices, and Total Phenolic Content

Analyses were performed to monitor the evolution of EVOOs' composition stored under four different treatments (temperatures/atmospheres) along time, measuring the oxidative stability and total phenols (Figure 1).

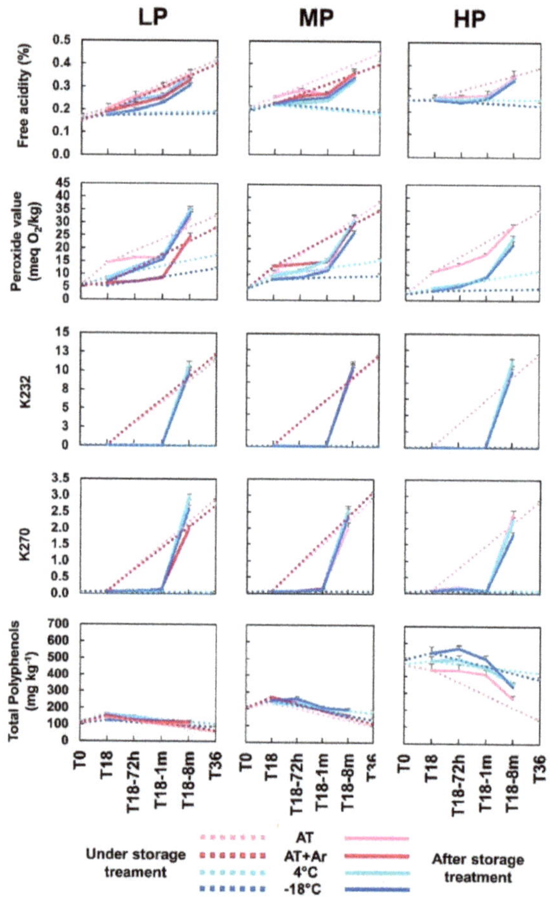

Figure 1. Free acidity (percentage of oleic acid), peroxide value (meq O_2/kg), K_{232} and K_{270} extinction coefficients, and total polyphenols (mg kg^{-1} of oil) of low, medium, and high phenolic (LP, MP, and HP, respectively) EVOOs under different storage conditions and time points. Dot lines: T0, T18, and T36 (under storage treatment); solid lines: T18-72 h, T18-1 m, and T18-8 m (after storage treatment). Bars represent means ± SD of three replicates.

From an initial level of free acidity (FA) ranging between 0.15 and 0.25 for the three oils, a slight increase was detected at T18 in LP and MP oils. The same trend was observed after one month of bottle opening (T18-1 m), and at T18-8 m, values rose rapidly above the maximum values of EVOO quality parameter (EU Reg. 2015/1830) (Figure 1, Table S1). It was evident that storage at low temperatures has guaranteed the maintenance of low levels of FA in all oils, but after opening bottles, the FA level increases as in the oils stored at room temperature.

Peroxide values rose during storage at AT in all three EVOOs, but argon headspace allowed to maintain very low levels, similar to frozen oils, during the first month after treatment completion, and then increased as in bottles without argon. Thus, cold temperatures maintained PV at low levels, with $-18\,°C$ being more effective than $4\,°C$. Very interestingly, also after 36 months of storage (T36), low temperatures allowed to keep PV levels lower than 20 meq O_2 kg^{-1} and, in the case of HP oil, the level remained confined to around 11.6 meq O_2 kg^{-1} for cooled oil and 4.7 meq O_2 kg^{-1} for frozen oil. In general, only storage at low $(4\,°C)$ or very low $(-18\,°C)$ temperatures avoided the increase of PV, but at T18-8 m, all oil samples exceeded the permitted limits of PV (Figure 1, Table S1). These results are in accordance with the observations of Mulinacci et al. [16], who reported that PV remained below 9 meq O_2 kg^{-1} in frozen oils, while in samples stored at room temperature after nine months, its vaue rose to 21.2 meq O_2 kg^{-1}.

At T0, all EVOOs showed almost undetectable values of oxidation indices K_{232} and K_{270}. These levels remained very low over 18 months of storage under any condition, but later, at T18-1 m, a sharp increase was detected in all cases, until they reached values around 10 for K_{232} and from 2 to 3 for K_{270} at T18-8 m, exceeding the legal limits for both parameters (Table S2). Moreover, oils stored at AT or AT + Ar deteriorated progressively; meanwhile, storage at $4\,°C$ or $-18\,°C$ was very effective in preserving the oils, which showed very low levels of both parameters (Figure 1).

Similar results were obtained in previous studies, showing an increase of FA and PV along with storage at room temperature [9,18,36], but on the contrary, it has been observed that most of stability parameters, notably K_{270}, are directly affected by ambient temperatures [37,38].

The use of argon in the bottle headspace under AT condition was effective in maintaining low FA and PV, as previously shown [39], whereas no effects were observed on other stability parameters, such as K_{232} and K_{270}.

Oil samples maintained under AT and AT + Ar storage conditions for 36 months were excluded from further analyses because their quality and oxidation indices exceeded the legal limits of EVOOs (Tables S1 and S2).

The EVOOs used for the analyses were distinguished based on their content in total phenols before storage: ~450 mg kg^{-1} for HP oil, ~200 mg kg^{-1} for MP oil, and ~100 mg kg^{-1} for LP oil. After 18 months of storage, total phenols showed a slight, but not significant increase for all treatments and oils. These values did not change during the 8 months after treatment completion for LP and MP oils, whereas in HP oil, phenols significantly decreased to 270–310 mg kg^{-1} for all treatments, even though their total content remained higher than that of the other two oils (Figure 1, Table S1).

After 36 months storage, LP and MP oils stored at ambient temperature, either with or without controlled atmosphere, lost about half of their phenolic content, whereas the HP oil lost more than 70%, whereas at $4\,°C$ and, to a lesser extent, at $-18\,°C$, phenols did not undergo any significant changes in all EVOOs. Thus, phenol content can be well preserved at low temperatures such as $4\,°C$ and $-18\,°C$ even after T36, regardless of their initial content (Figure 1, Table S1). Similar results have already been obtained for storage periods up to 18 months [15], but, to the best of our knowledge, we show the first evidence that oil samples stored for 3 years can retain a good phenolic profile if kept at low temperatures, and that there is no need to use lower temperatures below $4\,°C$.

3.2. Phenolic Compounds

In order to evaluate if and how storage conditions can modify the content of main phenolic compounds and the evolution of their derivatives, detailed analyses were performed on the EVOOs for all storage treatments (Figure 2, Table S3).

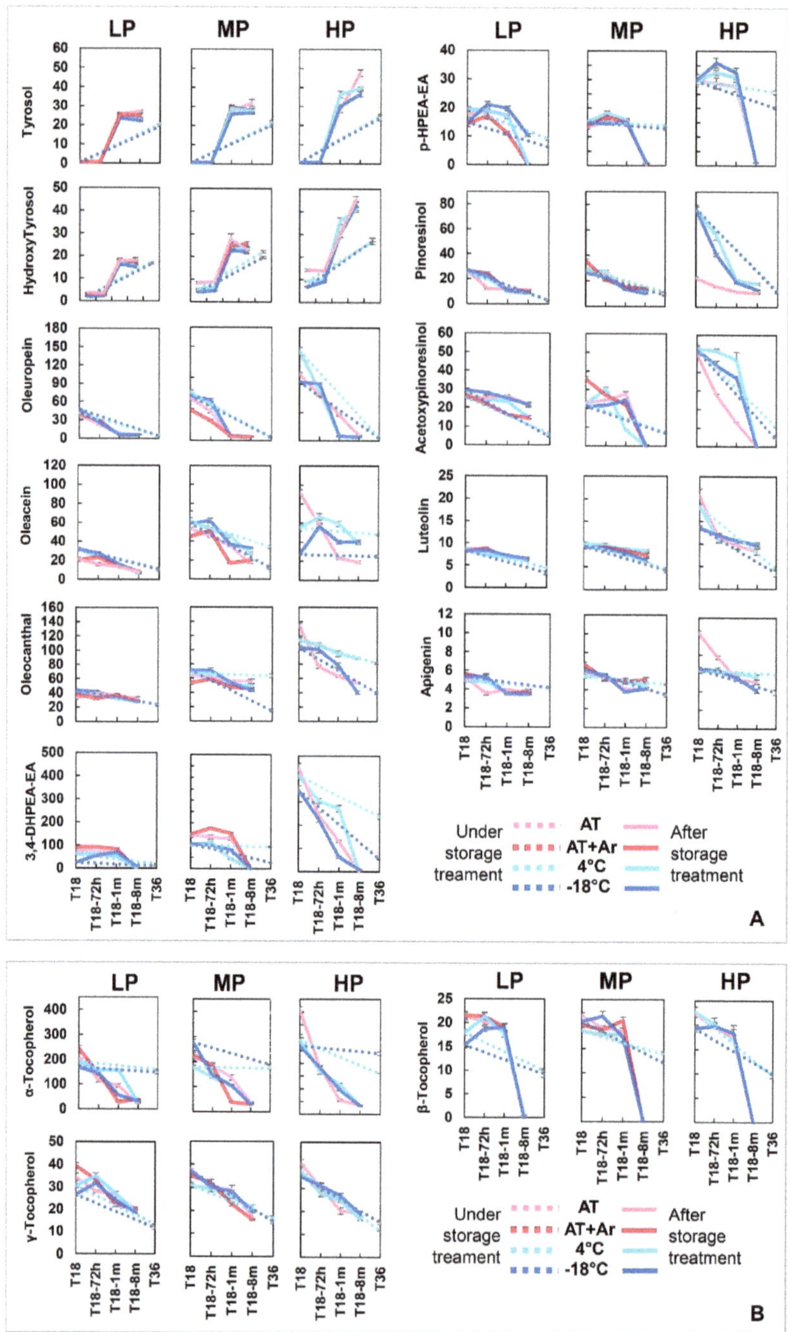

Figure 2. Evolution of phenolic compounds (**A**) and tocopherols (**B**) in LP, MP, and HP EVOOs under different storage conditions and time points. Dot lines: T18 and T36 (under storage treatment); solid lines: T18-72 h, T18-1 m, and T18-8 m (after storage treatment). All values are expressed in mg kg^{-1}. Bars represent means ± SD of three replicates (*n* = 3).

Simple phenols, such as tyrosol and hydroxytyrosol, remained very low at 18 months of conservation under all storage treatments, and significantly increased after storage completion and bottle opening. The same increase was observed in closed bottles conserved for 36 months (Figure 2, Table S3). This trend is probably due to the hydrolytic and oxidative degradation processes of main secoiridois (oleuropein, oleacin, oleocanthal, 3,4-DHPEA-EA, and p-HPEA-EA) [40,41], which in fact decreased significantly under all conditions. It is widely recognized that tyrosol and hydroxytyrosol typically increase in aged EVOOs [6,9,15,16,42] and, considering that hydroxytyrosol represents an important bioactive compound for human health, with a strong antioxidant activity [43,44], a low concentration of this molecule, when balanced by a high concentration of complex phenols, may represent a good compromise for EVOOs from a quality standpoint. The relationship between complex and simple phenolic compounds also depends on the initial content and processing conditions [5,45,46]. In fact, different contents of hydroxytyrosol were observed at T18, depending on the oil and storage conditions, showing the highest values in HP oil stored in AT. Eight months after bottles' opening, the highest tyrosol and hydroxytyrosol content was measured in HP under AT condition, the same oil and condition in which the PV and FA were higher than others (Figure 2, Table S1).

The two main families of complex phenolic compounds of EVOOs are 3,4-DHPEA-EDA (oleacein) and 3,4-DHPEA-EA (oleuropein aglycon), representing the dialdehydic and aldehydic forms of elenolic acid linked to hydroxytyrosol, and p-HPEA-EDA (oleocanthal) and p-HPEA-EA (ligstroside aglycon), the dialdehydic and aldehydic forms of elenolic acid linked to tyrosol, respectively. In general, phenol reduction was more evident within secoiridoid derivatives, indicating a more active participation in the oxidative processes. On average, the best 18 months storage condition for secoiridoids' conservation was 4 °C in all three EVOOs, especially for oleuropein, while at −18 °C, the two phenols oleacein and 3,4-DHPEA-EA showed a severe decay (25% and 48%, respectively). Oleuropein showed a significant decrease in all EVOOs and conditions, while oleocanthal was the most stable secoiridoid. The aldehydic forms of secoiridoids, such as oleuropein aglycone (3,4-DHPEA-EA) and ligstroside aglycone (p-HPEA-EA), almost disappeared eight months after treatment completion, independently from storage conditions. After 36 months, these two forms did not show a drastic reduction, but the reduction was significant under −18 °C treatment (Figure 2, Table S3); no significant differences were observed for the analyzed secoiridoids between AT and AT + Ar. The highest decrease was observed in LP oil at −18 °C followed by AT for 3,4-DHPEA-EA, p-HPEA-EA, p-HPEA-EDA, and oleacein.

Freezing temperatures led to changes in the oil physical state that passed from liquid into a solid state with a porous structure [47]. The presence of air and water through the whole volume of frozen oil could lead to a greater oxidative deterioration, and thus to a high loss of phenols [16,23]. It is important to notice that almost all secoiridoids, especially in MP and HP oils, had an increase after 72 h from 18 months storage, and the rise was higher for storage at low temperatures. It can be hypothesized that the equilibrium among secoiridoids within the oil is temperature-dependent, and the relative amounts of the isobaric forms are different in frozen and unfrozen oils. This different equilibrium can affect the measured absorbance values at 280 nm, the wavelength widely used in all HPLC/DAD methods for their determination [16]. Moreover, it is well known that the presence of a high amount of oxygen can accelerate the oxidation of phenolic compounds in the EVOOs and, consequently, it seems impossible to obtain a real increase in secoiridoids under these experimental conditions. On the other hand, the statistically significant differences between values immediately after 18 months storage stop (T18) and after 72 h (T18-72 h), observed for the peroxide values in both frozen and unfrozen oils, as expected, confirmed an increase of the primary products of autoxidation, being well known that a high amount of oxygen in the bottle headspace accelerates this process [16].

Lignan phenols (pinoresinol and acetoxypinoresinol) decreased in all EVOOs under any storage condition, as reported in other studies performed at room temperature [5,18,48]. MP and HP oils 18 months after treatment completion had a continuous loss of acetoxypinoresinol, at a maximum after eight months (93%), while storage at 4 °C was the best to conserve pinoresinol, especially when compared with the AT condition. The highest decrease of lignans was observed in HP oil under any storage condition (Figure 2).

Flavonoids (apigenin and luteolin) showed a slight decrease in LP and MP oils from T18 to T18-8 m, whereas both had a sharp decrease in HP oil. In oils stored for 36 months, luteolin decreased more at low temperatures than apigenin (Figure 2). Other authors reported the reduction of flavonoids after 12 months of storage at room temperature [5].

In order to evaluate the percentage of reduction in the content of all eleven phenolic compounds from T18 to T18-8 m and from T18 to T36, a PCA was performed on all phenolic compounds, oil samples, and storage conditions (Figure S2). The three oil samples stored at 4 °C, especially HP and MP oils, showed the lowest phenol oxidation. These results are in accordance with recent studies reporting that EVOOs with high oleacein and oleocanthal content have undergone a greater decrease in phenolic content than oils rich in other phenols [9,49]. Furthermore, after 18 months of storage at 4 °C, great stability of phenols was observed in all three EVOOs. Indeed, storage at AT + Ar in LP and MP oils was similar to that at 4 °C, in terms of stability of total phenols. It is interesting to notice that, even for 36 months of storage, HP and MP oils had the lowest phenol oxidation under the 4 °C storage condition. These results have definitely stated how storage at low temperature, when joined with low oxygen and low light availability, may allow for good preservation of the most important phenolic compounds, regardless of their initial content.

3.3. Tocopherols

Tocopherols were well preserved in all EVOOs after 18 months of storage at AT + Ar and at AT, but they declined after bottle opening. Among them, α-tocopherol decreased harshly at T18-8 m (89–91% in all oils), whereas β-tocopherol, albeit present in small quantities, had a significant decrease in all samples and storage conditions, and at T18-8, m dropped to zero, while the reduction of γ-tocopherol was lower than that of the other ones, with a very similar pattern for all EVOOs. In contrast, for 36 months of storage, α-tocopherol remained approximately the same, whereas β and γ-tocopherols decreased slightly (Figure 2, Table S3). Indeed, tocopherol loss was higher than that observed for polar phenols, in agreement with the results of Fregapane and Salvador [12]. The importance of tocopherols as antioxidants, by the induction period of oxidation, was revealed by a 9-month study conducted by Samaniego-Sanchez et al. [19], showing that tocopherols fell around 70–90% in all oils after storage, with higher losses in oils stored at 20 °C over those stored at 4 °C. In another study, an almost total loss of α-tocopherol in oils stored for 12 months in dark glass bottles at ambient temperature was reported [50], attributing the substantial loss of tocopherols during the storage to their function as hydrogen donors or oxygen quenchers, to stop the chain mechanism of auto- and photo-oxidation, respectively.

The good stability of α-tocopherol at low temperatures allows to hypothesize that freezing could decrease the oxidation process of tocopherols [16]. In general, the reduction in α-tocopherol suggested a protective effect of hydrophilic phenols probably based on their ability to reduce its oxidized forms [51]. In fact, the α-tocopherol value was lower than oleuropein and the aldehydic forms of elenolic acid linked to tyrosol and hydroxytyrosol (Figure 2).

3.4. Squalene and Sterols

After 18 months of storage, squalene was better preserved in HP oil under AT storage. After stopping the storage and bottles' opening, a sharp decrease in squalene was reported in all EVOOs for all treatments (Figure 3, Table S4). The decrease was from a minimum of 76% in LP oil to a maximum of 91% for HP oil at ambient temperature. The very fast decline occurred 72 h after treatment completion, and then the squalene content remained approximately constant for one month, followed by a second decline to the minimum levels after 8 months. Naziri et al. [24] and Rastrelli et al. [25] reported squalene degradation along time under room temperature. On the contrary, storage at 4 °C and −18 °C preserved the level of squalene over 36 months.

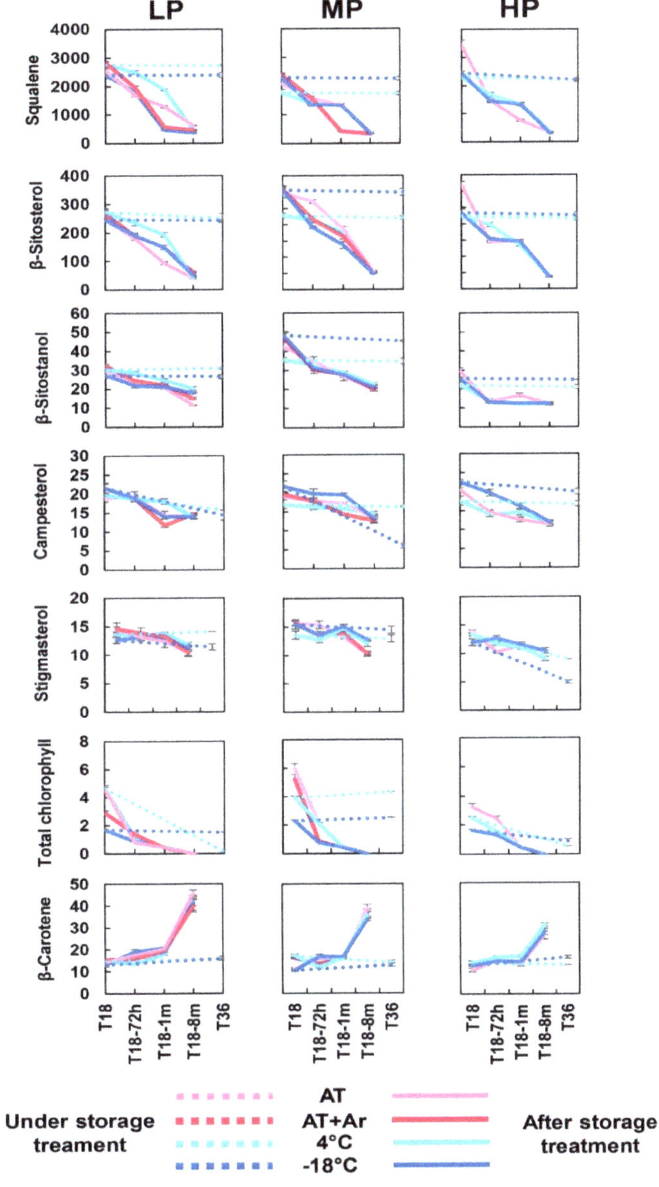

Figure 3. Changes of squalene, sterols, chlorophyll, and β-carotene in LP, MP, and HP EVOOs under different storage conditions and time points. Dot lines: T18 and T36 (under storage treatment); solid lines: T18-72 h, T18-1 m, and T18-8 m (after storage treatment). All values are expressed in mg kg^{-1}. Bars represent means ± SD of three replicates.

Squalene loss during oil storage was similar to that of α-tocopherol and, to a lower extent, to that of phenolic compounds. Squalene and tocopherols are considered important contributors to oil oxidative stability in the dark (autoxidation), under light exposure (photo-oxidation), and upon heating [7,8].

After 18 months of storage, sterol content was similar for all three oils under any storage condition (Figure 3), but after treatment completion, β-sitosterol had the highest decrease, up to 89%, while stigmasterol had the minimum loss among sterols under −18 °C storage condition (Figure 3, Table S4).

All analyzed sterols were quite well preserved after 36 months of storage treatments. Similar results were reported by Woranik and Rekas [52], showing that the loss of total sterols in oils kept closed over the entire storage period was negligible. Our results confirmed that cold temperatures are particularly effective for preserving the sterols present in the oil.

3.5. Pigments (Chlorophylls and Carotenoids)

The highest chlorophyll degradation at T18 and after bottle opening was observed under the −18 °C storage condition for all three EVOOs (Figure 3, Table S4). Total chlorophyll decreased sharply immediately 72 h after treatment completion, with a lighter reduction only in HP oil, until its total degradation. A slight antioxidant activity of chlorophylls in the dark was reported, which was attributed to the possible donation of a hydrogen radical to break free-radical chain reactions [53]. Even if the impact of chlorophylls on the oxidative stability of VOO, i.e., in the presence of other antioxidants, seems to be rather limited, our results demonstrated that the degradation caused by the oxygen is highly comparable to that reported for more known antioxidants studied in the present work.

From 18 to 36 months of storage, total chlorophyll content remained almost the same at −18 °C in all three EVOOs, while at 4 °C, the trend was negative in HP and mostly in LP oils. The evolution of chlorophyll content under low temperature storage was not previously reported and the present study allowed to demonstrate how oil refrigeration, in the absence of oxygen, can restrain chlorophyll degradation.

After 18 months of storage, as well as after 36 months of storage, the β-carotene content remained almost unchanged in all EVOOs and storage conditions. Carotenoids (and especially β-carotene, as the most studied representative) have been generally recognized as inhibitors of photooxidation, thanks to their ability to quench singlet oxygens [7]. However, the antioxidative effect of carotenoids in EVOOs under conditions of autoxidation seems to be very limited or even negative, owing to their oxidation products, which may possibly react with the lipid substrate, and thus accelerate oxidation [54]. In the present study, β-carotene increase was correlated with the increase of oxidation after treatment completion, from 1 to 8 months, mainly in LP oil.

3.6. Fatty Acid Profiles

No significant changes in monounsaturated and polyunsaturated fatty acid percentages under different treatments and after 18 and 36 months storage were observed (Table S5). Oleic, palmitic, and linoleic acids remained in the normal range for the three EVOOs (~73, 12, and 6%, respectively), as well as the polyunsaturated linoleic and linolenic acids, that play a very important role on oil stability [54].

The present results are in agreement with studies reporting minor changes in fatty acids of oils stored at room temperature, under light, or at 2 °C in the dark [12,55,56], and no changes in polyunsaturated fatty acids for olive oils stored for 18 months in dark glass at room temperature [5]. No significant changes in both unsaturated and saturated fatty acids were reported either for low temperature storage conditions in the dark [20].

3.7. The Overall Analysis of EVOOs under Different Storage Conditions

To explore the distribution pattern and visualize the effects of storage conditions on EVOOs, twenty-two variables, based on their percentage of reduction/increase during the time, were used to perform a PCA analysis (Figure 4). An evident and negative separation of samples according to their exposure to oxygen was observed. In PC1, explaining 65.40% of total variance, 14 out of 22 parameters were positive and, among them, the highest values were for 3,4-DHPEA-EA and p-HPEA-EA, squalene, β-sitosterol, β-sitostanol, α- and β-tocopherol, and total chlorophyll. PC2, with 11.20% of total variance, had 13 positive parameters, especially for oleacein, oleocanthal, 3,4-DHPEA-EA, acetoxypinoresinol, and campesterol. Eight months after treatment completion, PCA analysis grouped together all oil samples (with three out-grouping) in the negative plot of PC1 axis, independently from the storage condition, especially in MP oil. These results confirmed that autoxidation started after conservation and bottle opening, even for low temperature storage. After 18 months at AT, HP oil showed the highest degradation; noteworthy, in this oil, despite the highest concentration of antioxidants, after treatment completion and bottle opening, oxidation started very quickly. For what concerns the effect of argon on oil stability, it was observed that, after 18 months of conservation with and without modified atmosphere, only slight differences among oils were observed.

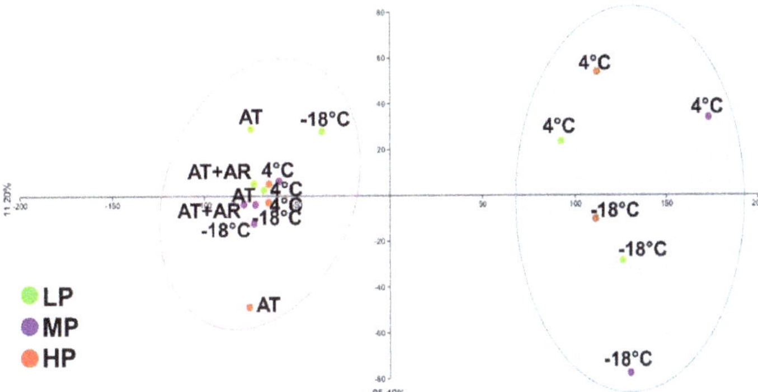

Figure 4. The scatter plot of the principal component analysis (PCA) using the percentage of reduction in content of twenty-two chemical parameters from T18 to T18-8 m (pink area) and between T18 and T36 (blue area) conditions. Colored circles indicate the type of oil, green: LP oil; purple: MP oil; orange: HP oil.

All three oils under the 4 °C storage condition were placed in the positive PCA plot area, showing high stability for all analyzed parameters, owing to a more suitable temperature and to oxygen limitation. These results are in accordance with [15], who reported that, by storing EVOOs at different temperatures (5, 15, 25, and 50 °C), their shelf-life was considerably extended at lower temperatures. Li et al. [7], evaluating the effects of different storage conditions (25, 4.5, and −27 °C) on EVOOs with different composition, showed that cold storage conditions (either 4.5 or −27 °C) were successful in retarding oxidation and hydrolysis level during storage, with no significant changes in the flavor over 4.5 months of storage. Mulinacci et al. [16] also reported that the phenolic fraction had lower oxidation after conservation at −23 °C, indicating a far superior quality of frozen EVOOs.

The box plot analysis allowed identifying, among 22 variables, those that determined the response of oils to the storage treatments from T18 to T18-8 m (Figure 5). In general, chemical parameters more prone to change under different storage conditions were almost the same; however, the analyzed oils showed different responses at various temperatures. In LP oil, differences among storage conditions derived mainly from squalene, total phenols, and β-sitosterol. Squalene showed the highest amount at AT, followed by AT + Ar and 4 °C, while the lowest value was detected at −18 °C. In MP oil, the main variables among different storage conditions again were squalene and total phenols, where the highest amounts for both components were observed at −18 °C. Four chemical parameters played the main effect on HP oil stability under three storage conditions: squalene, total phenols, oleacein, and oleocanthal, and for all of them, the highest value of positive attributes was recorded for storage at 4 °C (Figure 5). The overall analysis on all oil samples and storage conditions showed 4 °C as the best temperature to preserve the oil for 36 months and, even after this long storage, all three oils were at the standard level for EVOO, as well as a good level for all 22 chemical components.

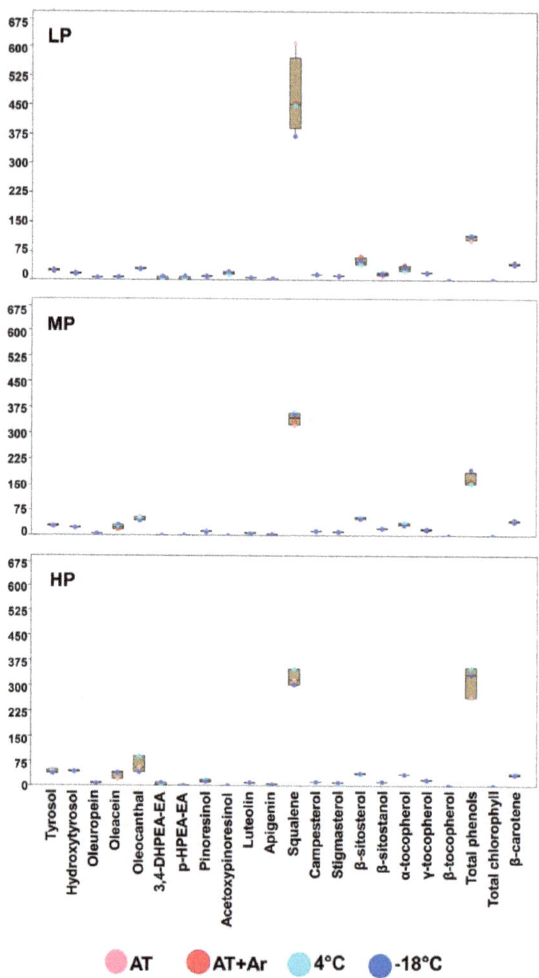

Figure 5. Box and jitter plot representing significant differences in the concentration of twenty-two chemical parameters at 8 months after bottles' opening (T18-8 m) under different conditions. Colored circles indicate the different storage conditions; pink: AT; dark pink: AT + Ar; light blue: 4 °C; and dark blue: −18 °C.

These results show that squalene, sterols, and tocopherols play a more important role as antioxidants in low phenolic EVOOs; instead, in high phenolic oil, phenols and especially secoiridoids are the main antioxidants.

Storage at −18 °C for 36 months showed that the three oils substantially maintained their characteristics, but, considering the energy and structural costs to maintain the cold chain, it can be assumed that 4 °C should be considered as the best condition for EVOO long storage.

3.8. Organoleptic Evaluation

An organoleptic evaluation was carried out at every time point (T0, T18, T18-72 h, T18-1 m, T18-8 m, T36) (Figure 6) (owing to the lack of differences between T18 and T18-72 h, this latter time point was not reported in the figure).

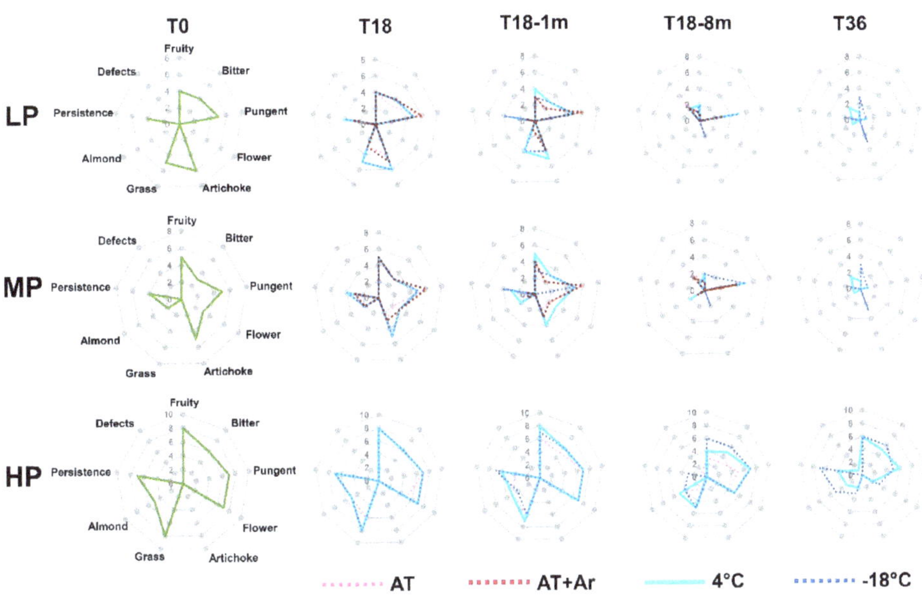

Figure 6. Radar chart reporting the median of fifteen organoleptic evaluations (three replicas X oil X panel members) carried out on EVOOs at different time points (T0, T18, T18-1 m, T18-8 m, and T36).

Before storage, LP and MP oils showed good levels of fruitiness, bitterness, and pungency, but a general low level of flavors, while HP oil showed very good taste and flavor. After 18 months of conservation at 4 °C and −18 °C, flavors remained almost unchanged, while a reduction in flavor and start of grassy occurred at AT. The same was observed at 1 month after storage completion, when oils stored at 4 °C and at −18 °C again did not have considerable changes, while oils stored at AT degraded significantly. The decrease in fruitiness, grassy, and almond at T18-8 m was perceived both under 4 °C and −18 °C storage conditions. In fact, at T18-8 m, defects were perceived at AT, AT + Ar, and 4 °C for LP and MP oils, while at −18 °C, even if bitterness and flavors almost disappeared, no defects were perceived in all analyzed EVOOs. It was interesting to note that, in HP oil, even 8 months after LTS, no defects were noticed and this oil still had pungency, fruitiness, flavor, and bitterness, especially for that treated at −18 °C.

The oil samples at −18 °C did not have any defect even after 36 months of conservation; moreover, in HP oil, artichoke flavor was detected, with a good perception of fruitiness, pungency, and persistence.

In the present study, only HP oil eight months after 18 months of storage completion and even after 36 months of storage under a low temperature preserved the sensorial attributes. The impact of storage conditions on the sensory attributes seems to be strictly connected to the initial content of hydrophilic phenols in the EVOO and only HP oil may avoid defects and maintain gustatory and olfactory sensations.

Among low temperature treatments, −18 °C was a bit more effective than 4 °C in preserving the persistency, fruitiness, bitterness, and some olfactory sensations.

The hydrophilic phenols greatly influence the taste quality (bitter and pungency attributes), as well as the beneficial biological activity and oxidative stability of EVOOs [57,58]. Among these phenols, oleacin and oleuropein aglycone are mainly responsible for the bitter taste, whereas oleocanthal is the main agent responsible for pungency. In fact, the significant reduction in oleacin and oleuropein affected the bitterness during conservation, especially in LP and MP oils, while oleocanthal had the minimum reduction among secoiridoids, allowing to perceive the pungency even 8 months after storage end, especially in HP oil. Some flavor attributes (i.e., fresh fruity and artichoke) may be related to the decrease of some secoiridoids during storage [59].

4. Conclusions

Long (18 months) and very-long (36 months) storage under different conditions (AT, AT + Ar, 4 °C, and −18 °C) was evaluated for low, medium, and high-phenolic EVOOs, in order to verify their effects on oil preservation at different time points (0 and 72 h, 1 and 8 months) after treatment completion.

The decay along all storage treatments was mostly evident for phenolic compounds, while tocopherols, squalene, and sterols have undergone a low reduction and fatty acids showed negligible alteration. The use of argon in the bottle headspace at an ambient temperature could be recommended only for 18 months of storage, whereas, later on, the use of modified atmosphere did not allow significantly reducing EVOO degradation. On the contrary, cold storage was successful in retarding the oxidation and hydrolysis at least until the oxygen availability was limited. After 18 months of storage and bottle opening, in fact, degradation rates increased in all treatments and most metabolites declined, independently from the previous storage treatment. The 36 months of storage definitively showed the effect of low temperatures to avoid oxidation and preserve the metabolite profile of EVOOs, while the organoleptic profile was preserved only for oils with a high initial content of hydrophilic phenols. The conservation of EVOOs at −18 °C could be applied for high-quality oils (with a high level of phenols and a good organoleptic profile) to avoid their degradation during storage and delivery chain.

To maintain the nutraceutical properties of EVOOs, mild cold at 4 °C was the best storage temperature, especially during the first 18 months of conservation, in comparison with the conventional storage or even at −18 °C, also considering the elevated costs and energy use for maintaining freezing temperatures during long-term storage and delivery.

These results demonstrated that it is possible to preserve EVOO quality for a very long time, up to 3 years, if proper storage technologies are applied. The present study has provided important knowledge on the change in key chemical and organoleptic parameters in extra virgin olive oils during the storage. A new strategy to better preserve the characteristics of EVOOs at industrial and commercial level has been provided.

Supplementary Materials: The following are available online at https://www.mdpi.com/article/10.3390/foods10081945/s1: Figure S1: Schematic view of the experimental design. At each time point, three bottles were used for the analyses, Figure S2: The scatter plot (A) and biplot (B) of the principal component analysis (PCA), using the percentage of reduction in content of 11 phenolic compounds in EVOOs from T18 to T18-8 m (pink area) and between T18 and T36 (blue area) conditions. Colored circles indicate the type of oil, green: LP oil; purple: MP oil; orange: HP oil, Table S1: Free acidity (% of oleic acid), peroxide value (meq O_2 kg^{-1}), and total phenols (mg kg^{-1} of oil) of three kinds of oil in different storage conditions at each time point. Different letters ($p < 0.01$ and $p < 0.05$) show significant differences in each single column, Table S2: K_{232} and K_{270} extinction coefficients of three kinds of oils in different storage conditions at each time point. Different letters ($p < 0.01$ and $p < 0.05$) show significant differences in each single column, Table S3: The p-value of ANOVAs among different time points for three kinds of oil. The non-significant and $p < 0.05$ were shown by bold superscript ns and *, respectively, and all other values were significant at $p < 0.01$, Table S4: The p-value of ANOVAs among different time points for three kinds of oil. The non-significant and

$p < 0.05$ were shown by bold superscript ns and *, respectively, and all other values were significant at $p < 0.01$, Table S5: Mean value of the main fatty acids in three different oil samples, under four storage conditions and in six time points.

Author Contributions: Conceptualization, S.M., R.M., and N.G.M.C.; methodology, S.M., R.M., V.S., V.M., S.P., and N.G.M.C.; software, S.M. and R.M.; validation, S.M., S.P., R.M., and L.B.; formal analysis, S.M., R.M., S.P., V.S., V.M., and N.G.M.C.; investigation, S.M., R.M., V.S., V.M., S.P., and N.G.M.C.; resources, S.M., R.M., and N.G.M.C.; data curation, S.M., R.M., N.G.M.C., and L.B.; writing—original draft preparation, S.M. and R.M; writing—review and editing, S.M., R.M., N.G.M.C., and L.B.; visualization, S.M., R.M., and L.B.; supervision, S.M. and R.M.; project administration, S.M. and R.M.; funding acquisition, N.G.M.C. All authors have read and agreed to the published version of the manuscript.

Funding: This research was funded by a Contract on technological development of Superior Quality Extra Virgin Olive Oil between CNR—Institute of Biosciences and Bioresources and Castello Monte Vibiano Vecchio s.r.l (Year 2015, PROT. N.0006201).

Data Availability Statement: The data presented in this study are available here and in the Supplementary Material.

Conflicts of Interest: The authors declare no conflict of interest. The funders had no role in the design of the study; in the collection, analyses, or interpretation of data; in the writing of the manuscript; or in the decision to publish the results.

References

1. Tresserra-Rimbau, A.; Lamuela-Raventos, R.M. Olives and olive oil: A Mediterranean source of polyphenols. In *Olives and Olive Oil as Functional Foods: Bioactivity, Chemistry and Processing*; Shahidi, F., Kiritsakis, A., Eds.; John Wiley & Sons, Ltd.: Chichester, UK, 2017; pp. 417–428. [CrossRef]
2. Martinez-Gonzalez, M.A.; Martin-Calvo, N. Mediterranean diet and life expectancy; beyond olive oil, fruits and vegetables. *Curr. Opin. Clin. Nutr. Metab. Care* **2016**, *19*, 401. [CrossRef]
3. Piroddi, M.; Albini, A.; Fabiani, R.; Giovannelli, L.; Luceri, C.; Natella, F.; Rosignoli, P.; Rossi, T.; Taticchi, A.; Servili, M.; et al. Nutrigenomics of extra-virgin olive oil: A review. *Biofactors* **2017**, *43*, 17–41. [CrossRef]
4. Borges, T.H.; Pereira, J.A.; Cabrera-Vique, C.; Lara, L.; Oliveira, A.F.; Seiquer, I. Characterization of Arbequina virgin olive oils produced in different regions of Brazil and Spain: Physicochemical properties, oxidative stability and fatty acid profile. *Food Chem.* **2017**, *215*, 454–462. [CrossRef]
5. Di Stefano, V.; Melilli, M.G. Effect of storage on quality parameters and phenolic content of Italian extra-virgin olive oils. *Nat. Prod. Res.* **2020**, *34*, 78–86. [CrossRef]
6. Li, X.; Zhu, H.; Shoemaker, C.F.; Wang, S.C. The effect of different cold storage conditions on the compositions of extra virgin olive oil. *J. Am. Oil Chem. Soc.* **2014**, *91*, 1559–1570. [CrossRef]
7. Psomiadou, E.; Tsimidou, M. Stability of virgin olive oil. 2. Photo-oxidation studies. *J. Agric. Food Chem.* **2002**, *50*, 722–727. [CrossRef] [PubMed]
8. Tsimidou, M.Z. Squalene and tocopherols in olive oil: Importance and methods of analysis. In *Olives and Olive Oil in Health and Disease Prevention*; Elsevier: San Diego, CA, USA, 2010; pp. 561–567. [CrossRef]
9. Shendi, E.G.; Ozay, D.S.; Ozkaya, M.T.; Ustunel, N.F. Changes occurring in chemical composition and oxidative stability of virgin olive oil during storage. *OCL* **2018**, *25*, A602. [CrossRef]
10. Redondo-Cuevas, L.; Castellano, G.; Torrens, F.; Raikos, V. Revealing the relationship between vegetable oil composition and oxidative stability: A multifactorial approach. *J. Food Compos. Anal.* **2018**, *66*, 221–229. [CrossRef]
11. Tsimidou, M.Z.; Nenadis, N.; Servili, M.; García-González, D.L.; Gallina Toschi, T. Why tyrosol derivatives have to be quantified in the calculation of "olive oil polyphenols" content to support the health claim provisioned in the EC Reg. 432/2012. *Eur. J. Lipid Sci. Technol.* **2018**, *120*, 1800098. [CrossRef]
12. Fregapane, G.; Salvador, M.D. Oxidative stability and the role of minor and functional components of olive oil. Olives and Olive Oil as Functional Foods: Bioactivity, Chemistry and Processing. In *Olives and Olive Oil as Functional Foods: Bioactivity, Chemistry and Processing*; Shahidi, F., Kiritsakis, A., Eds.; John Wiley & Sons, Ltd.: Hoboken, NJ, USA, 2017; pp. 249–265.
13. Bosque-Sendra, J.M.; De La Mata-Espinosa, P.; Cuadros-Rodríguez, L.; González-Casado, A.; Rodríguez-García, F.P.; García-Toledo, H. Stability for olive oil control materials. *Food Chem.* **2011**, *125*, 1418–1422. [CrossRef]
14. Abbattista, R.; Losito, I.; Castellaneta, A.; De Ceglie, C.; Calvano, C.D.; Cataldi, T.R.I. Insight into the storage-related oxidative/hydrolytic degradation of olive oil secoiridoids by liquid chromatography and high-resolution Fourier transform mass spectrometry. *J. Agric. Food Chem.* **2020**, *68*, 12310–12325. [CrossRef] [PubMed]
15. Krichene, D.; Salvador, M.D.; Fregapane, G. Stability of virgin olive oil phenolic compounds during long-term storage (18 months) at temperatures of 5–50 C. *J. Agric. Food Chem.* **2015**, *63*, 6779–6786. [CrossRef] [PubMed]

16. Mulinacci, N.; Ieri, F.; Ignesti, G.; Romani, A.; Michelozzi, M.; Creti, D.; Innocenti, M.; Calamai, L. The freezing process helps to preserve the quality of extra virgin olive oil over time: A case study up to 18 months. *Int. Food Res. J.* **2013**, *54*, 2008–2015. [CrossRef]

17. Romani, A.; Lapucci, C.; Cantini, C.; Ieri, F.; Mulinacci, N.; Visioli, F. Evolution of minor polar compounds and antioxidant capacity during storage of bottled extra virgin olive oil. *J. Agric. Food Chem.* **2007**, *55*, 1315–1320. [CrossRef] [PubMed]

18. Esposto, S.; Selvaggini, R.; Taticchi, A.; Veneziani, G.; Sordini, B.; Servili, M. Quality evolution of extra-virgin olive oils according to their chemical composition during 22 months of storage under dark conditions. *Food Chem.* **2020**, *311*, 126044. [CrossRef] [PubMed]

19. Samaniego-Sánchez, C.; Oliveras-López, M.J.; Quesada-Granados, J.J.; Villalón-Mir, M.; Serrana, H.L.G. Alterations in Picual extra virgin olive oils under different storage conditions. *Eur. J. Lipid Sci. Technol.* **2012**, *114*, 194–204. [CrossRef]

20. Gómez-Alonso, S.; Mancebo-Campos, V.; Salvador, M.D.; Fregapane, G. Evolution of major and minor components and oxidation indices of virgin olive oil during 21 months storage at room temperature. *Food Chem.* **2007**, *100*, 36–42. [CrossRef]

21. Fadda, C.; Del Caro, A.; Sanguinetti, A.M.; Urgeghe, P.P.; Vacca, V.; Arca, P.P.; Piga, A. Changes during storage of quality parameters and in vitro antioxidant activity of extra virgin monovarietal oils obtained with two extraction technologies. *Food Chem.* **2012**, *134*, 1542–1548. [CrossRef]

22. Okogeri, O.; Tasioula-Margari, M. Changes occurring in phenolic compounds and α-tocopherol of virgin olive oil during storage. *J. Agric. Food Chem.* **2002**, *50*, 1077–1080. [CrossRef]

23. Spika, M.J.; Kraljic, K.; Koprivnjak, O.; Skevin, D.; Anetic, M.; Katalinic, M. Effect of agronomical factors and storage conditions on the tocopherol content of Oblica and Leccino virgin olive oils. *J. Am. Oil Chem. Soc.* **2015**, *92*, 1293. [CrossRef]

24. Naziri, E.; Consonni, R.; Tsimidou, M.Z. Squalene oxidation products: Monitoring the formation, characterisation and pro-oxidant activity. *Eur. J. Lipid Sci. Technol.* **2014**, *116*, 1400–1411. [CrossRef]

25. Rastrelli, L.; Passi, S.; Ippolito, F.; Vacca, G.; De Simone, F. Rate of degradation of α-tocopherol, squalene, phenolics, and polyunsaturated fatty acids in olive oil during different storage conditions. *J. Agric. Food Chem.* **2002**, *50*, 5566–5570. [CrossRef] [PubMed]

26. El Bernoussi, S.; Boujemaa, I.; Harhar, H.; Belmaghraoui, W.; Matthäus, B.; Tabyaoui, M. Evaluation of oxidative stability of sweet and bitter almond oils under accelerated storage conditions. *J. Stored Prod. Res.* **2020**, *88*, 101662. [CrossRef]

27. Singleton, V.L.; Orthofer, R.; Lamuela-Raventós, R.M. Analysis of total phenols and other oxidation substrates and antioxidants by means of folin-ciocalteu reagent. In *Methods in Enzymology*; Elsevier: San Diego, CA, USA, 1999; pp. 152–178. [CrossRef]

28. Selvaggini, R.; Esposto, S.; Taticchi, A.; Urbani, S.; Veneziani, G.; Di Maio, I.; Sordini, B.; Servili, M. Optimization of the temperature and oxygen concentration conditions in the malaxation during the oil mechanical extraction process of four Italian olive cultivars. *J. Agric. Food Chem.* **2014**, *62*, 3813–3822. [CrossRef] [PubMed]

29. Tura, D.; Gigliotti, C.; Pedò, S.; Failla, O.; Bassi, D.; Serraiocco, A. Influence of cultivar and site of cultivation on levels of lipophilic and hydrophilic antioxidants in virgin olive oils (*Olea europaea* L.) and correlations with oxidative stability. *Sci. Hortic.* **2007**, *112*, 108–119. [CrossRef]

30. Fernández-Cuesta, A.; León, L.; Velasco, L.; De la Rosa, R. Changes in squalene and sterols associated with olive maturation. *Food Res. Int.* **2013**, *54*, 1885–1889. [CrossRef]

31. Mousavi, S.; Stanzione, V.; Mencuccini, M.; Baldoni, L.; Bufacchi, M.; Mariotti, R. Biochemical and molecular profiling of unknown olive genotypes from central Italy: Determination of major and minor components. *Eur. Food Res. Technol.* **2019**, *245*, 83–94. [CrossRef]

32. Minguez-Mosquera, M.I.; Gandul-Rojas, B.; Gallardo-Guerrero, M.L. Rapid method of quantification of chlorophylls and carotenoids in virgin olive oil by high-performance liquid chromatography. *J. Agric. Food Chem.* **1992**, *40*, 60–63. [CrossRef]

33. UNI ISO 12966-2. *Animal and Vegetables Fat and Oils. Gas Chromatography of Fatty acid Methyl Esters. Part2: Preparation of Methyl Esters of Fatty Acids*; International Organization for Standardization: Geneva, Switzerland, 2011; Available online: https://www.iso.org/standard/72142.html (accessed on 5 March 2019).

34. Onofri, A. *DSAASTAT a New Excel VBA Macro to Perform Basic Statistical Analyses of Field Trials*; Department of Agriculture and Environmental Sciences, University of Perugia: Perugia, Italy, 2010.

35. Hammer, Ø. *Past 4—The Past of the Future*; Natural History Museum, University of Oslo: Problemveien, Norway, 2020.

36. Baiano, A.; Terracone, C.; Viggiani, I.; Del Nobile, M.A. Changes produced in extra-virgin olive oils from cv. Coratina during a prolonged storage treatment. *Czech J. Food Sci.* **2014**, *32*, 1–9. [CrossRef]

37. Alvarruiz, A.; Pardo, J.E.; Copete, M.E.; de Miguel, C.; Rabadán, A.; López, E.; Álvarez-Ortí, M. Evolution of virgin olive oil during long-term storage. *J. Oleo Sci.* **2020**, *69*, 809–814. [CrossRef]

38. Conte, L.; Milani, A.; Calligaris, S.; Rovellini, P.; Lucci, P.; Nicoli, M.C. Temperature dependence of oxidation kinetics of extra virgin olive oil (EVOO) and shelf-life prediction. *Foods* **2020**, *9*, 295. [CrossRef]

39. Sanmartin, C.; Venturi, F.; Macaluso, M.; Nari, A.; Quartacci, M.F.; Sgherri, C.; Flamini, G.; Taglieri, I.; Ascrizzi, R.; Andrich, G.; et al. Preliminary results about the use of argon and carbon dioxide in the extra virgin olive oil (EVOO) storage to extend oil shelf life: Chemical and sensorial point of view. *Eur. J. Lipid Sci. Technol.* **2018**, *120*, 1800156. [CrossRef]

40. Obied, H.K.; Prenzler, P.D.; Ryan, D.; Servili, M.; Taticchi, A.; Esposto, S.; Robards, K. Biosynthesis and biotransformations of phenol-conjugated oleosidic secoiridoids from *Olea europaea* L. *Nat. Prod. Rep.* **2008**, *25*, 1167–1179. [CrossRef] [PubMed]

41. Di Maio, I.; Esposto, S.; Taticchi, A.; Selvaggini, R.; Veneziani, G.; Urbani, S.; Servili, M. HPLC-ESI-MS investigation of tyrosol and hydroxytyrosol oxidation products in virgin olive oil. *Food Chem.* **2011**, *125*, 21–28. [CrossRef]
42. Krichene, D.; Allalout, A.; Mancebo-Campos, V.; Salvador, M.D.; Zarrouk, M.; Fregapane, G. Stability of virgin olive oil and behaviour of its natural antioxidants under medium temperature accelerated storage conditions. *Food Chem.* **2010**, *121*, 171–177. [CrossRef]
43. Giannakopoulos, E.; Salachas, G.; Zisimopoulos, D.; Barla, S.A.; Kalaitzopoulou, E.; Papadea, P.; Skipitari, M.; Georgiou, C.D. Long-term preservation of total phenolic content and antioxidant activity in extra virgin olive oil: A physico-biochemical approach. *Free Radicals Antiox.* **2020**, *10*, 4–9. [CrossRef]
44. Olajide, T.M.; Liu, T.; Liu, H.; Weng, X. Antioxidant properties of two novel lipophilic derivatives of hydroxytyrosol. *Food Chem.* **2020**, *315*, 126197. [CrossRef] [PubMed]
45. Lozano-Sánchez, J.; Bendini, A.; Quirantes-Piné, R.; Cerretani, L.; Segura-Carretero, A.; Fernández-Gutiérrez, A. Monitoring the bioactive compounds status of extra-virgin olive oil and storage by-products over the shelf life. *Food Control* **2013**, *30*, 606–615. [CrossRef]
46. Ramos-Escudero, F.; Morales, M.T.; Asuero, A.G. Characterization of bioactive compounds from monovarietal virgin olive oils: Relationship between phenolic compounds-antioxidant capacities. *Int. J. Food Prop.* **2015**, *18*, 348–358. [CrossRef]
47. Mallamace, D.; Longo, S.; Corsaro, C. Proton NMR study of extra virgin olive oil with temperature: Freezing and melting kinetics. *Phys. A Stat. Mech. Appl.* **2018**, *499*, 20–27. [CrossRef]
48. Lanza, B.; Di Serio, M.G.; Giansante, L.; Di Loreto, G.; Di Giacinto, L. Effect of shelf conditions on the phenolic fraction and oxidation indices of monovarietal extra virgin olive oil from cv. 'Taggiasca'. *Acta Aliment.* **2015**, *44*, 585–592. [CrossRef]
49. Castillo-Luna, A.; Criado-Navarro, I.; Ledesma-Escobar, C.A.; Bascón, M.L.; Capote, F.P. The decrease in the health benefits of extra virgin olive oil during storage is conditioned by the initial phenolic profile. *Food Chem.* **2020**, *336*, 127730. [CrossRef] [PubMed]
50. Morelló, J.R.; Motilva, M.J.; Tovar, M.J.; Romero, M.P. Changes in commercial virgin olive oil (cv Arbequina) during storage, with special emphasis on the phenolic fraction. *Food Chem.* **2004**, *85*, 357–364. [CrossRef]
51. Boskou, D. Other important minor constituents. In *Olive Oil: Minor Constituents and Health*; CRC Press: Boca Raton, FL, USA, 2009; pp. 45–54.
52. Wroniak, M.; Rękas, A. Nutritional value of cold-pressed rapeseed oil during long term storage as influenced by the type of packaging material, exposure to light & oxygen and storage temperature. *J. Food Sci. Technol.* **2016**, *53*, 1338–1347. [CrossRef]
53. Hrncirik, K.; Fritsche, S. Relation between the endogenous antioxidant system and the quality of extra virgin olive oil under accelerated storage conditions. *J. Agric. Food Chem.* **2005**, *53*, 2103–2110. [CrossRef]
54. Lolis, A.; Badeka, A.V.; Kontominas, M.G. Effect of bag-in-box packaging material on quality characteristics of extra virgin olive oil stored under household and abuse temperature conditions. *Food Packag. Shelf Life* **2019**, *21*, 100368. [CrossRef]
55. Gutiérrez, F.; Fernández, J.L. Determinant parameters and components in the storage of virgin olive oil. Prediction of storage time beyond which the oil is no longer of "extra" quality. *J. Agric. Food Chem.* **2002**, *50*, 571–577. [CrossRef] [PubMed]
56. Gargouri, B.; Zribi, A.; Bouaziz, M. Effect of containers on the quality of Chemlali olive oil during storage. *J. Food Sci. Technol.* **2015**, *52*, 1948–1959. [CrossRef]
57. Servili, M. The phenolic compounds: A commercial argument in the economic war to come on the quality of olive oil? *OCL* **2014**, *21*, D509. [CrossRef]
58. Bendini, A.; Cerretani, L.; Salvador, M.D.; Fregapane, G.; Lercker, G. Stability of the sensory quality of virgin olive oil during storage: An overview. *Ital. J. Food Sci.* **2009**, *21*, 389–406.
59. Sinesio, F.; Moneta, E.; Raffo, A.; Lucchetti, S.; Peparaio, M.; D'Aloise, A.; Pastore, G. Effect of extraction conditions and storage time on the sensory profile of monovarietal extra virgin olive oil (cv Carboncella) and chemical drivers of sensory changes. *LWT-Food Sci. Technol.* **2015**, *63*, 281–288. [CrossRef]

 *foods*

Article

Changes of the Aroma Composition and Other Quality Traits of Blueberry 'Garden Blue' during the Cold Storage and Subsequent Shelf Life

Xiaoxue Yan [1], Jun Yan [1], Siyi Pan [1,2] and Fang Yuan [1,2,*]

[1]   College of Food Science and Technology, Huazhong Agricultural University, Wuhan 430070, China; yxx1647883233@163.com (X.Y.); juneyan19@163.com (J.Y.); pansiyi@mail.hzau.edu.cn (S.P.)
[2]   Key Laboratory of Environment Correlative Dietology, Huazhong Agricultural University, Wuhan 430070, China
*   Correspondence: fyuan@mail.hzau.edu.cn

Received: 10 August 2020; Accepted: 31 August 2020; Published: 2 September 2020

Abstract: The changes of volatile composition and other quality traits of blueberry during postharvest storage were investigated. Blueberries were packaged in vented clam-shell containers, and stored at 0 °C for 0, 15 and 60 days, followed by storage at room temperature (25 °C) for up to 8 days for quality evaluation. The firmness, pH, and total soluble solids increased by 8.42%, 8.92% and 42.9%, respectively, after 60 days of storage at 0 °C. Titratable acidity decreased 18.1% after 60 days of storage at 0 °C. The volatile change was monitored using headspace–solid-phase microextraction–gas chromatography–quadrupole time-of-flight–mass spectrometry (HS-SPME-TOF-MS) and off-odor was evaluated by sensory panel. Volatile compounds generally showed a downward trend during cold storage. However, the subsequent shelf life was the most remarkable period of volatile change, and was represented by the strong fluctuation of ethyl acetate and the rapid decrease of terpenoids. Extending storage from 15 to 60 days under cold condition still resulted in an acceptable odor. However, subsequent storage at higher temperature resulted in a quick deterioration in sensory acceptability. The results proved that cold storage was a reliable way to maintain the quality of blueberry, and flavor deterioration during subsequent shelf life was more fatal to the blueberry flavor.

Keywords: rabbiteye blueberry; postharvest storage; firmness; aroma compounds; off-odor

---

1. Introduction

Due to their unique flavor and nutritional value, blueberry fruit and products are sold at a high price in the international market, and the production is growing every year [1,2]. As the production increases, blueberry deterioration is often a problem for the industry. Like other soft berries, blueberries can be processed into blueberry juice, jam and other products, but the market price is several times lower than that of fresh fruit. Therefore, the sale of fresh fruit is important for the blueberry industry. Thanks to rapidly developed preservation technology, the shelf life of blueberries can be extended to up to one month [3–6]. Cold storage and transportation are most the widely used means of blueberry preservation all over the world [7]. However, for countries with an undeveloped cold chain system, room temperature is still a common condition for short-distance transportation and retail display, which may lead to problems such as the loss of aroma and the generation of off-odor, which can greatly affect the flavor perception of consumers.

The aroma of small berries is more complicated than other fruit and is hard to manipulate. Unlike the climacteric fruit, ethylene does not stimulate the ripening of small berries such as strawberries [8]. Blueberries are climacteric fruits and will respond to ethylene. However, blueberry flavor does not improve by ethylene after harvest [9]. Therefore, a comprehensive understanding of the berry aroma change is essential for possible flavor manipulation during postharvest storage. Among the small berries, strawberry and raspberry are the most widely studied in aroma research [10,11]. More than 360 and 200 volatile compounds have been reported in strawberries and raspberries, respectively [12]. Relatively speaking, the research on the aroma of the blueberry is lagging. There have been fewer than 100 volatile compounds reported in blueberry [13]. The main aroma substances found in rabbiteye blueberry are ethyl acetate, limonene, hexanol, (Z)-3-hexenol, heptanol, β-ionone, terpinene-4-ol, α-terpineol, vanillin, nerol and eugenol [14,15]. Farneti et al. [16] analyzed the volatile composition of eleven different blueberry cultivars and found that for most cultivars, aldehydes, alcohols, terpenoids, and esters can be used as putative biomarkers to evaluate the blueberry aroma variations. Recent reports on blueberry wine and vinegar also showed that terpenes such as linalool and α-terpineol were important berry-derived aroma compounds in blueberries [17,18]. Zhu et al. [19] reported that esters, aldehydes, C13-norisoprenoids, as well as several terpenes and eugenol, were the aroma-active compounds in freshly pressed blueberry juice. From the existing reports, the main aroma substances of fresh blueberry fruits are alcohols, esters, terpenes and aldehydes, and the aroma contents are affected by cultivar and environmental factors [16,20–22].

The methods used for blueberry volatile analysis were thoroughly reviewed by Sater et al. [2]. Static headspace sampling using solid phase microextraction (SPME) is the most commonly used extraction method in the blueberry volatile studies. The separation of compounds in blueberry studies is most commonly accomplished by a gas chromatograph (GC) directly connected to a mass spectrometer (MS). Older GCs often use a flame ionized detector (FID) to quantify the relative abundance of each compound. In recent years, more sensitive methods such as headspace–solid-phase microextraction–gas chromatography–quadrupole time-of-flight–mass spectrometry (HS-SPME-TOF-MS) and proton transfer reaction–time-of-flight-mass spectrometry (PTR-TOF-MS) technology have been applied to blueberry volatile analysis [16,23].

In the postharvest chain, flavor maintenance is an important aspect for ensuring the quality of fresh fruit. Many studies have been carried out in blueberry for understanding the impact of different storage conditions and treatments with regard to phenolics, anthocyanins and flavonoids [24,25]. However, there are fewer reports about ways to preserve the aroma/flavor quality of blueberry. Based on our literature search, only one study showed that the content of volatile terpenes, phenols and anthocyanins in blueberry fruits was increased by UV-B irradiation during postharvest storage [26], proving that it is feasible to manipulate the aroma quality of blueberry by appropriate means. Therefore, in this study, we investigated the quality changes of blueberries during different durations of cold storage conditions and subsequent room temperature storage. The volatile composition change was monitored using HS-SPME-GC-QTOF-MS and off-odor was evaluated. The aim of this study was to have a better understanding of the changes occurring in blueberry during postharvest storage.

2. Materials and Methods

2.1. Plant Material and Postharvest Storage

Blueberries were hand harvested from a local blueberry orchard in Huangpi, Hubei, China. The fruits of rabbiteye blueberry (*Vaccinium virgatum*) cultivar 'Garden blue' were randomly harvested in July of 2018. Berries that were at commercial maturity (a completely blue color) were selected. The blueberries were precooled in an air-conditioned room (~20 °C) for half an hour to remove the field heat, divided into PET boxes (classic commercial "clam-shells" container with openings for ventilation, ~125 g per box) and transported to the lab. To simulate commercial storage conditions, fruits were stored in a cooler at 0 ± 0.5 °C and 85% relative humidity (RH) in the dark for up to 60 days, and were then removed and placed at room temperature (to simulate retail display conditions) at 25 ± 0.5 °C and 60% RH under normal room light for up to 8 d. At the 0th, 15th and 60th day of cold storage, 15 boxes were picked from the top, middle and bottom of the cooler and used as samples for the subsequent

25 °C storage. At the 0th, 2nd, 4th, 6th and 8th day of 25 °C storage, three boxes were picked as three biological replicates for the following analysis.

2.2. Measurements of Weight Loss and Decay Index

At sampling date, each box of samples was weighed to calculate the weight loss. Decay refers to berry juice leakage, visible microbial growth, or pronounced rot on berry surface. Decay index was used for decay evaluation according to Cao et al. [27] with minor modification, calculated as follows: no decay = 1; decay berries < 5% = 2; decay berries < 10% = 3; decay berries < 20% = 4; decay berries > 20% = 5.

2.3. Measurements of Total Soluble Solids, pH and Titratable Acidity

For each biological replicate, approximately 30 g of berries was randomly selected and crushed in liquid nitrogen. The powdered sample was thoroughly homogenized and put in a 30 °C water bath to thaw, and centrifuged at 12,000 rpm for 10 min. The clear juice obtained was used for total soluble solids (TSS) and pH measurements. TSS was measured at room temperature using a PAL-1 pocket refractometer (Atago USA, Inc., Bellevue, WA, USA). The pH of the juice was measured using a pH meter (Rex PHS-2F, Shanghai, China). Five mL of juice was diluted with 50 mL of distilled water, and the titratable acidity (TA), as g/100 g of citric acid, was measured using burette with phenolphthalein as indicator. Three measurements were made for each sample and means were used for each biological replicate.

2.4. Measurement of Firmness

Ten berries without decay were randomly selected from each replicate for firmness analysis. Berry firmness was assessed using a texture analyzer (TA. XT plus, Godalming, UK), with 8 mm probe diameter, across a 7 mm distance, and a measuring speed of 0.5 mm/s [28]. Each berry was measured once. The peak force (firmness, expressed as N) was calculated by the integrated software.

2.5. Analysis of Volatile Compounds

The analysis of volatile metabolites was previously described by Cheng et al. [23]. Briefly, for each biological replicate, 30 g of berries were randomly selected and crushed in liquid nitrogen. 10 g of blended blueberry powder was weighed and mixed with 10 mL of propyl gallate (10 mM). The sample was homogenized at 4 °C for 24 h and centrifuged (10,000 rpm, 30 min, 4 °C). The clear supernate was used for volatile compound analysis.

The volatile compound analysis was performed using the solid phase microextraction (SPME) technique coupled with gas chromatography–quadrupole time-of-flight–mass spectrometry (GC-QTOF-MS) (7200 accurate-mass, Agilent Technologies, Santa Clara, CA, USA). Two mL of juice and 8 mL of saturated saltwater were mixed in a headspace sampling vial. Ten μL internal standard (50 mg/L 4-octanol in methanol) was added. A small magnetic stir bar was added to the vial and equilibrated at 50 °C in a water bath for 15 min with stirring. After equilibration, headspace volatiles were collected on a SPME fiber (2 cm, DVB/CAR/PDMS, 50/30 μm, Supelco, Bellefonte, PA, USA). The triple phase fiber was selected because it can cover more volatiles with different molecular weight. The adsorption time was 45 min. Headspace temperature was set at 50 °C. Desorption temperature was 250 °C, and desorption time was 5 min with split ratio of 1:10.

The GC-QTOF-MS was equipped with a HP-5MS (30 m × 250 μm × 0.25 μm) column. Oven temperature program setting: 40 °C for 5 min, increased to 180 °C at 3 °C/min, hold for 1 min, increased to 250 °C at 20 °C/min, hold for 2 min. The helium flow rate was 1.2 mL/min. The interface temperature was 300 °C and ion source temperature was 250 °C. Mass spectrum data from m/z 25 to 300 were collected. The ionization voltage was 70 eV. To ensure the accuracy of the instrument, mass calibration was performed daily.

Identification of volatile compounds were performed by comparing the mass spectra with records from external databases such as NIST, HMDB, MassBank and an internal database for the wine volatiles based on the literature, and by comparing the Kovats retention indices (RI) in NIST database and published literatures. Calibration plots were constructed using authentic standards (Table S1 (Supplementary Materials)). Ten microliters of 50 mg/L 4-octanol were added to each calibration mixture (10 mL of saturated saltwater plus authentic standards). The volatile extraction and detection methods were the same as for sample analysis. Seven-level calibration plots for each volatile compound were built using the MassHunter software to quantify the volatile compounds in the blueberry samples. Peak area of target ion for each compound was plot against the peak area of the target ion of internal standard. The GC-QTOF-MS data processing was performed with MassHunter B.06.00 software (Agilent Technologies). A representative chromatogram could be found in Supplementary Materials (Figure S1 (Supplementary Materials)). The average value of three measurements was used for each biological replicate.

2.6. Sensory Evaluation

The sensory test was conducted prior to storage, after storage at 0 °C for 15 and 60 days, and after post-storage conditioning at 25 °C for 4 more days, respectively. Six samples (prior to storage, 0 d at 0 °C + 4 d at 25 °C, 15 d at 0 °C, 15 d at 0 °C + 4 d at 25 °C, 60 d at 0 °C, and 60 d at 0 °C + 4 d at 25 °C) were evaluated in six different sensory sections. Eighteen panelists were recruited from the campus (10 female and 8 male). Panelists' ages ranged from 20 to 50 years. The panelists attended three half-hour training sessions prior to the test. During the training session, the panelists got familiar with the five-point scale, and the difference between fresh and stored blueberries through discussion. During the formal sensory evaluation, a set of three sensory replications, defined as "samples" for the panelists, were presented to each panelist. Each sensory replication (or "sample") consisted of a set of 10 berries, which were placed into one 120 mL plastic soufflé cup with lids and labeled with a random three-digit code, which corresponded to the code presented in the evaluation score sheet. The panelists were asked to take at least five berries each time, chew thoroughly and rate. Off-odor was assessed according to the 1–5 scale, in which 1 was used to refer to excellent odor; 2 for good freshness odor; 3 for neither good nor bad odor; 4 for unpleasant off-odor; and 5 for extensively severe off-odor.

2.7. Statistical Analysis

To determine the changes of blueberry quality parameters, a one-way ANOVA was performed for each storage day, using the days of shelf-life as treatments. For volatile compounds, a one-way ANOVA was also performed for each storage day, using the days of shelf-life as treatments. For data on sensory evaluation and decay index, a non-parametric analysis was conducted. All statistical analyses were carried out using SPSS 22.0 (SPSS, Chicago, IL, USA). A post hoc range test of Tukey's HSD (honestly significant difference) was used to identify homogeneous subsets of means that are not different from each other.

3. Results

3.1. Weight Loss and Decay Index

As expected, weight loss increased during storage, with longer storage time resulting in increased percentage of weight loss (Figure 1). It was unexpected that the percentage of weight loss was the highest (22.4%) after 60 d at 0 °C, followed by 2 d at 25 °C, possibly due to the abnormally high percentage of decayed berries in two of the three replicates (5.7%, 14.3% and 13.8%, respectively). We also observed a remarkably higher variation of weight loss for the samples stored at 0 °C for 60 days. Weight loss values observed at the completion of the 0 °C storage (2.4% for 15 d and 10.3% for 60 d) were similar to the values found by previous research work. Paniagua et al. [29] reported weight loss of approximately 10% after 21 d at 0 °C. Similarly, Sanford et al. [30] reported a 5.3% weight loss after storage at 0 °C for 14 d. Accordingly, the weight loss values generated in this experiment represent the range of weight loss values that are representative of possible postharvest systems.

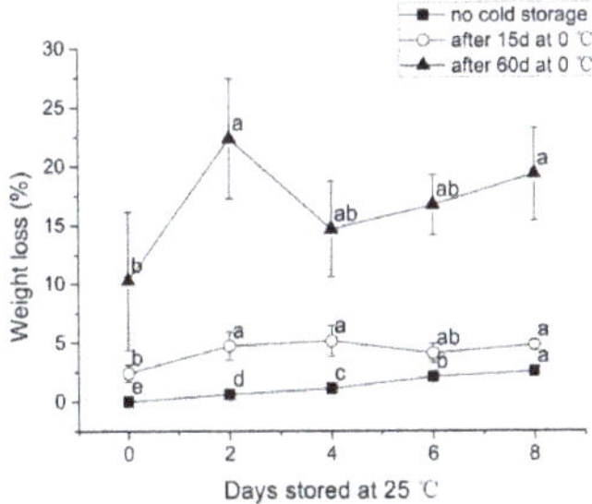

Figure 1. Berry weight loss during 25 °C storage period after different 0 °C storage lengths. Different letters indicate statistically significant difference (Tukey's HSD, $p < 0.05$) between samples.

Decay index was calculated during the 25 °C storage period after different times of cold storage, respectively (Figure 2). The decay incidence significantly increased during the 25 °C storage period, mainly due to the juice leakage, according to our observations. Without cold storage, about 10% of the berries decayed after 8 days of storage at 25 °C. Cold storage resulted in more severe decay (about 20% of the berries decayed) at the end of the experimental period compared to the samples without cold storage. A fluctuation was observed for decay index after 60 d of cold storage due to the unexpectedly severe deterioration at the second day, possibly due to the berries being damaged during handling or temperature change. Apart from the second day, the samples stored at 25 °C for 4 and 6 days exhibited a low decay index (<3) compared to other samples (no cold and after 15 days of storage at 0 °C).

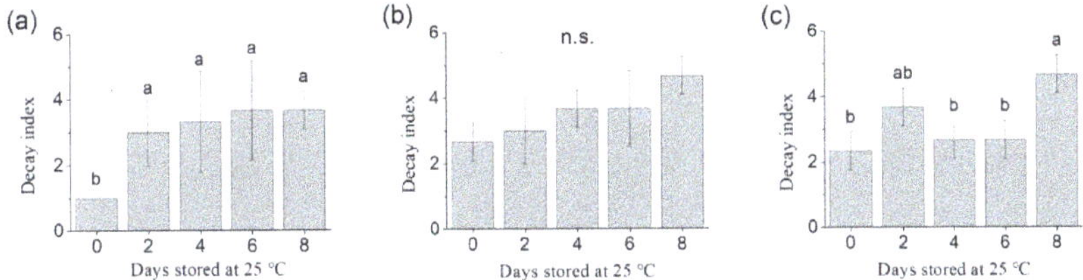

Figure 2. Decay index during 25 °C storage periods after (**a**) no cold storage, (**b**) 15 days of 0 °C storage and (**c**) 60 days of 0 °C storage. Different letters indicate statistically significant difference (Tukey HSD, $p < 0.05$) between samples. n.s. No statistically significant difference between samples. Decay index was calculated as follows: no decay = 1; decay berries < 5% = 2; decay berries < 10% = 3; decay berries < 20% = 4; decay berries > 20% = 5.

3.2. TSS, pH and TA

TSS of the 'Garden blue' was 14.3% at the time of harvest (Table 1). After 15 d and 60 d of cold storage (0 °C), the TSS increased to 14.8% and 15.7%, respectively, which is common in fruits because of the sugar concentration due to the moisture loss during the cold storage. During the 25 °C storage period, the TSS of the blueberry fruits generally showed a declining trend, indicating the sugar metabolism was activated at higher storage temperature. The pH of the berry did not change after 15 d of 0 °C storage, compared to the fresh sample at time of harvest (pH = 2.94). However, the pH increased to 3.21 after 60 d of 0 °C storage. After different cold storage times, the pH of the berry slightly increased during the 25 °C storage period for all samples, while the TA decreased.

Table 1. Changes of pH, total soluble solids and firmness of blueberry stored at different temperatures.

0 °C Storage Time	Subsequent 25 °C Storage Time	pH	Total Soluble Solids (Brix)	TA (meq/L)	Firmness (N)
	0 d	2.94 ± 0.2 [b]	14.3 ± 0.6	6.52 ± 0.10 [a]	4.08 ± 1.16 [b]
	2 d	3.01 ± 0.1 [ab]	13.9 ± 0.6	6.40 ± 0.22 [ab]	4.45 ± 0.72 [b]
0 d	4 d	3.06 ± 0.2 [ab]	14.0 ± 0.3	6.35 ± 0.16 [ab]	4.68 ± 1.22 [ab]
	6 d	3.07 ± 0.2 [ab]	13.2 ± 0.8	6.37 ± 0.35 [ab]	5.06 ± 1.13 [ab]
	8 d	3.17 ± 0.1 [a]	13.7 ± 0.3	6.21 ± 0.12 [b]	5.58 ± 1.26 [a]
	0 d	3.04 ± 0.09 [c]	13.5 ± 1.0 [ab]	6.44 ± 0.23	4.00 ± 1.00 [c]
	2 d	3.14 ± 0.04 [abc]	14.5 ± 0.3 [a]	6.47 ± 0.19	4.86 ± 1.42 [b]
15 d	4 d	3.20 ± 0.03 [ab]	14.6 ± 0.3 [a]	6.30 ± 0.22	7.06 ± 1.34 [a]
	6 d	3.21 ± 0.05 [a]	13.0 ± 0.8 [b]	6.33 ± 0.14	5.96 ± 1.82 [ab]
	8 d	3.14 ± 0.03 [bc]	12.5 ± 0.4 [b]	6.21 ± 0.54	6.95 ± 0.99 [a]
	0 d	3.21 ± 0.09 [bc]	15.7 ± 0.5 [a]	5.34 ± 0.24	5.83 ± 2.31 [a]
	2 d	3.27 ± 0.03 [abc]	15.0 ± 0.6 [ab]	5.26 ± 0.33	3.79 ± 1.53 [b]
60 d	4 d	3.31 ± 0.05 [ab]	13.7 ± 0.3 [c]	5.39 ± 0.45	6.53 ± 2.82 [a]
	6 d	3.36 ± 0.03 [a]	13.6 ± 0.6 [c]	5.45 ± 0.39	5.35 ± 1.88 [ab]
	8 d	3.19 ± 0.03 [c]	14.2 ± 0.3 [bc]	5.54 ± 0.47	6.10 ± 1.30 [a]

Different letters in same column means there is statistically significant difference (Tukey's HSD, $p < 0.05$) between samples.

3.3. Berry Firmness

For no cold storage treatment, fruit firmness significantly increased after 8 d of storage at 25 °C (approximately 29.5%) in comparison to initial firmness (4.08 N) and reached a maximum value of 5.58 N (Table 1). Firmness also showed a significant increase during the 25 °C storage period after 15 d of cold storage. For the 60 d cold storage treatment, firmness fluctuated during the 25 °C storage period due to the large sample variation, which was consistent with the weight loss and decay index data.

3.4. Aroma Compounds

The volatile profile of 'Garden blue' blueberry has been investigated previously, and the results showed that only part of the compounds played important roles in the blueberry aroma [23]. In fact, in this study, we also observed that most of the volatile compounds in blueberry were present at levels far below their sensory threshold. We did not observe any new compounds generated during the experimental period compared to the samples at harvest. Therefore, in this study, we mainly focused on the change of the volatile compounds in blueberry at the level higher or closer to their sensory thresholds (Figure 3). Nine volatile compounds were selected, including three esters (ethyl acetate, methyl isovalerate and ethyl 2-methylbutanoate), two C6 aldehydes (hexanal and E-2-hexenal), three monoterpenes (linalool, eucalyptol and α-terpineol), and one volatile phenol (eugenol).

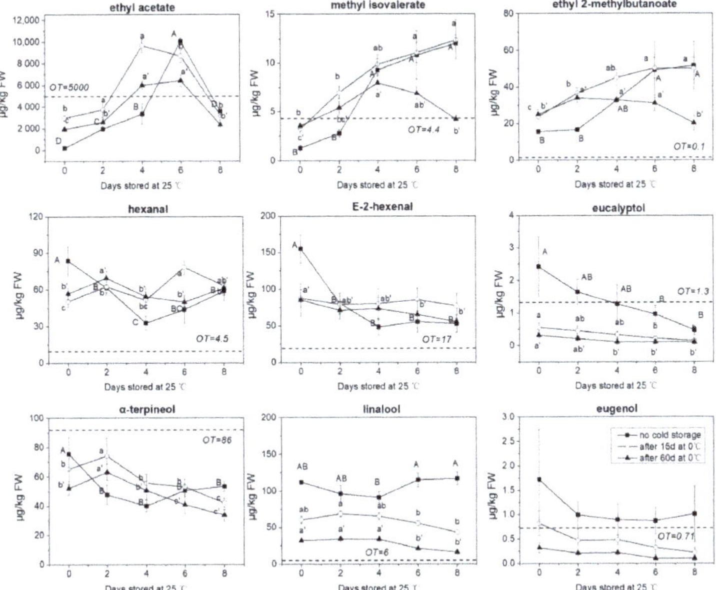

Figure 3. The changes of key aroma compounds in blueberry during 25 °C storage period after different lengths of 0 °C storage. Volatile concentrations are presented as µg/kg fresh weight (FW). Different letters indicate statistically significant difference (Tukey HSD, $p < 0.05$) between samples. The dashed line indicates the odor threshold (OT, µg/L in water) of the compound. The odor thresholds (µg/L in water) were obtained from [29]. Odor descriptions for each compound can be found in Table S2 (Supplementary Materials).

The results show that ethyl acetate is the most abundant ester in 'Garden blue' blueberry. However, ethyl 2-methylbutanoate probably has higher aroma potency due to its low odor threshold. For the room temperature (no cold storage) treatment, the concentration of ethyl acetate significantly increased on the 4th day of storage at 25 °C, reached 10,064 µg/L on the 6th day, and then decreased quickly. Similar trends were also observed for ethyl acetate during the shelf life for the 15 and 60 d of cold storage treatments. Ethyl acetate reached its maximum level two days earlier than with no cold storage treatment. Without cold storage, methyl isovalerate content increased continually, with a large increment during the first two days. After 15 days of cold storage, the methyl isovalerate level also increased during the 25 °C storage period. However, after 60 days of cold storage, the methyl isovalerate level firstly increased then decreased during the subsequent 25 °C storage. A similar phenomenon was also observed for ethyl 2-methylbutanoate.

Hexanal and E-2-hexenal are two C6 aldehydes in the blueberry that mainly contribute to the green or leafy odor [21]. For the no cold storage treatment, the concentration of hexanal and E-2-hexenal decreased fast in the first 4 days, then hexanal started to increase, but the E-2-hexenal level became stable. After 15 and 60 days of cold storage, the concentrations of hexanal and E-2-hexenal were lower than in the fresh sample. The concentration of hexanal fluctuated, while E-2-hexenal decreased slowly during shelf life for both 15 and 60 d cold storage treatments.

Linalool and α-terpineol contribute to the floral notes of blueberry [23]. In this study, the concentration of α-terpineol was slightly below its sensory threshold (Figure 3). Considering that an aroma compound close to its threshold could also have sensory impact [31], we included α-terpineol in the quantification. Without cold storage, linalool content decreased from 0 to 4 d and increased from 4 to 8 d after harvest. After 15 and 60 days of cold storage, linalool content was significantly lower than the fresh sample. After cold storage, the linalool content in blueberry was relatively stable from

0 d to 4 d of subsequent shelf life and started to decrease afterward. Similar to linalool, α-terpineol content decreased from 0 d to 4 d and increased from 4 d to 8 d at 25 °C after harvest. For 15 d and 60 d cold storage treatments, α-terpineol showed a short increase during the first 2 days and decreased afterward during the subsequent shelf life. The aroma of eucalyptol has been described as 'eucalyptus', 'fresh', 'cool', 'medicinal', and 'camphoraceous' [32], and has been found to be closely related to the minty odor of blueberry [23]. The concentration of eucalyptol decreased quickly during the 25 °C storage and also decreased fast during the cold storage period. After 15 days and 60 days of cold storage, eucalyptol content in the blueberry decreased 76.9% and 86.8% respectively compared to its initial content at harvest (2.42 µg/kg).

3.5. Off-Odor

Off-flavor is defined as unpleasant odors or flavors imparted to food through internal deteriorative change [33]. In this study, off-flavor was defined as unpleasant odor that differed from the fresh samples. The results of sensory evaluation showed that the score of off-odor increased with increasing storage time (Figure 4). Off-odor developed faster under higher temperature storage conditions, and unpleasant odor became noticeable only after 4 days of storage at 25 °C. Cold storage could keep the blueberry odor at acceptable level (off-flavor score < 3) for over 60 days.

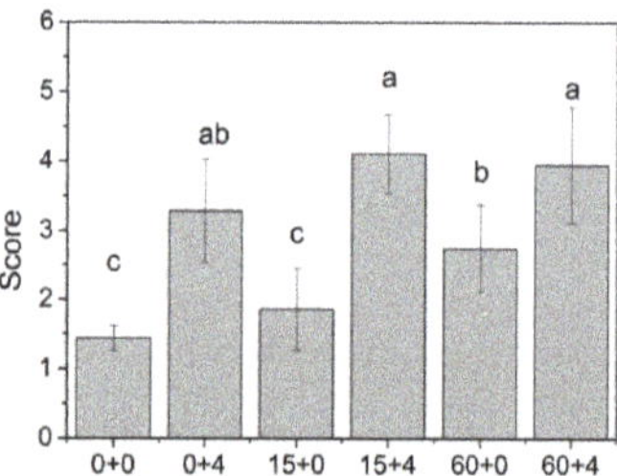

Figure 4. Sensory evaluation of off-flavor. 0 + 0: the blueberries prior to storage; 0 + 4: after storage at 25 °C for 4 days; 15 + 0: after storage at 0 °C for 15 days; 15 + 4: after storage at 0 °C for 15 days and subsequent storage at 25 °C for 4 days; 60 + 0: after storage at 0 °C for 60 days; 60 + 4: after storage at 0 °C for 60 days and subsequent storage at 25 °C for 4 days. Different letters indicate statistically significant difference (Tukey's HSD, $p < 0.05$) between samples.

4. Discussion

'Garden blue' is a rabbiteye blueberry cultivar with high total soluble solids and good hardiness [34]. It is generally accepted that blueberry tends to soften during transportation and postharvest storage. However, in our study, increase of firmness was observed during storage. Berry firmness continually increased when the weight loss was below 2.55%. After 15 days of cold storage, the berry firmness still increased for 4 days until the weight loss increased to 5%, then the berry firmness started to decrease. Our results suggested that if the weight loss were well controlled, the firmness of blueberries could be well maintained at low temperature, and slightly increasing the storage temperature could increase the firmness of blueberries in the short term. Since the room temperature storage period was short compared to other studies, the expected berry softening was not obvious. Nevertheless, blueberry firming during storage has been observed by many researchers. For example, Chen et al. [23] reported that, irrespective of the storage temperatures, firmness of blueberries increased during the first 7 days of storage, and declined gradually afterwards. Chiabrando et al. [35] also reported that for blueberry cultivars 'Bluecrop' and 'Coville', berry firmness increased over 36% after storage at 0 °C for 5 weeks. Unfortunately, unlike the berry softening, the mechanism of berry firming has not been well explained yet. Paniagua et al. [36] pointed out that berry firming occurred consistently with low levels of weight loss (0.22–1.34%), whereas softening occurred with higher weight loss (3.47–15.06%). It was suggested that the firming of the blueberry during storage may be related to the parenchyma cell wall thickening, stone cell arrangement, and thickening of stone cell walls [37].

In recent decades, it has been generally accepted that the drop-off in flavor quality is one of the major sources of consumer dissatisfaction with fresh fruit products [38,39]. However, berry fruit flavor is delicate and hard to control during transportation and storage. Cold storage and transportation are the most widely used techniques for maintaining blueberry quality. Indeed, our results showed that esters in blueberries were relatively stable during cold storage for up to 60 days, but monoterpenes such as eucalyptol and linalool continually decreased. We also observed that the aroma compounds changed much more quickly under the subsequent higher temperature condition, indicating the flavor change could occur not only during the commercial storage and transportation process, but also during retail display or home storage. In fact, the aroma change under these conditions could be more noteworthy, since the changes occurring in just a few days depend on the temperature.

The change of blueberry odor during postharvest storage was probably due to the changes of esters, C6 aldehydes and terpenes. The most remarkable change was the concentration fluctuation of ethyl acetate. To better understand the possible sensory effect of these changes, the odor threshold of each compound was plotted with the concentration change in Figure 3. It has to be mentioned that the odor threshold of ethyl acetate was different in different studies. Therefore, we chose the most frequently reported threshold: 5000 μg/L in water [40]. In the fresh berries, the concentration of ethyl acetate was lower than its odor threshold. However, during 25 °C storage, the concentration of ethyl acetate increased to the level well above its threshold in just 4~6 days, then dropped quickly. Our observation was consistent with a previous report that the concentrations of esters were higher in overripe blueberries compared to ripe ones [16]. A similar phenomenon has also been observed in other fruits, such as over-ripened bananas and apples [41,42], and was associated with increased pyruvate decarboxylase activity [43]. The accumulated ethyl acetate during storage might impart an 'over-ripe' off-odor, and could result in a suppressive effect on the perception of other esters [44].

It has been reported that monoterpene formed in blueberry during the berry ripening stage [15], but little information was found regarding the monoterpene change during postharvest storage. In 'Garden blue', the linalool content was much higher than that of other monoterpenes, which was consistent with a previous report that linalool was an important compound contributing to the flavor of 'Garden blue' [23]. It has been reported in grapes that during storage, linalool can be lost by emission from the berries' surfaces [45], which probably also occurs in blueberry, as the linalool content successively decreased after 15 d and 60 d of cold storage. Interestingly, the linalool content, which decreased during low-temperature storage, slightly increased during the subsequent storage at 25 °C. On the contrary, the eucalyptol content decreased rapidly during postharvest storage despite the temperature. Eucalyptol, which imparts a subtle minty aroma to blueberry, dropped to a level lower than its odor threshold in just four days. Minty aroma is a pleasant and distinct flavor for some blueberry cultivars, but seems hard to maintain during storage according to our observation.

The C6 aldehydes are generally considered to be products of fatty acid oxidation through the lipoxygenase pathway [46]. Like in cucumber and tomato, C6 aldehydes are important flavor compounds in blueberries, regardless of whether they are generated by oxidation [21,47]. Increase of C6 aldehydes is often observed in fruit during the postharvest storage [48], and has been found to be related to chilling injuries in some fruits such as in peaches and bananas [41,49]. However, in our study, cold storage did not result in increment of C6 aldehydes in "Garden blue" blueberry. This proves that cold storage is an effective way of preserving the quality of blueberries.

According to the sensory evaluation, prolonged storage did not negatively affect the flavor of 'Garden blue' blueberries. Extending postharvest storage from 15 to 60 days still resulted in an acceptable odor. However, subsequent storage at higher temperature resulted in a quick deterioration in sensory acceptability from "good" and "excellent" at harvest to "unpleasant off-odor" after 4 days. Due to the experimental design, we could not perform comparative descriptive analysis, since the samples had to be evaluated at different times. However, according to the panelists, the observed deterioration in blueberry flavor during storage was attributed to a slight decrease in sourness, a gradual decrease in typical blueberry flavor such as minty and floral odors, and enhanced accumulation of off-flavors described as 'over-ripe', 'stale', and 'solvent-like', which was consistent with the instrumental data.

5. Conclusions

Blueberry aroma is one of the most important factors affecting its quality. To simulate the storage conditions of blueberries in real production, we studied the changes of different quality parameters of blueberries during different length of cold storage conditions and subsequent room temperature storage, with special regard to volatile changes and 'off-odor' generation. The results proved that cold storage was a reliable way to preserve most of the quality traits of blueberry, as the firmness, pH, TSS and TA of blueberry changed slowly during cold storage. Aroma compounds generally showed a downward trend during the cold storage, but the decline was relatively slow. In contrast, the shelf life after cold storage is the most remarkable period of aroma change, which is represented by the strong fluctuation of ethyl acetate and the rapid decrease of terpenoids. Our results showed that aroma preservation during shelf life may be more important in the postharvest chain, and more research is still needed to better understand the mechanism behind the aroma change, and to explore ways to regulate the flavor of blueberries during shelf life.

Supplementary Materials: The following are available online at http://www.mdpi.com/2304-8158/9/9/1223/s1, Figure S1: A representative chromatograph of 'Garden blue' blueberry volatiles detected by SPME-GC-QTOF-MS. The peak number is corresponding to the compound list in Table S1, Table S1: Quantification information for GC-QTOF-MS analysis, Table S2: Odor threshold and description of volatile compounds.

Author Contributions: Conceptualization, F.Y., investigation, X.Y. and J.Y.; writing—original draft preparation, X.Y. and J.Y.; writing—review and editing, F.Y.; supervision, S.P.; project administration, S.P.; funding acquisition, F.Y. All authors have read and agreed to the published version of the manuscript.

Funding: This research was funded by the National Natural Science Foundation of China, grant number 31701561; Students' Innovation and Entrepreneurship Training Program of Hubei Province [grant number S201910504067].

Conflicts of Interest: The authors declare no conflict of interest.

References

1. Howard, L.R.; Prior, R.L.; Liyanage, R.; Lay, J.O. Processing and Storage Effect on Berry Polyphenols: Challenges and Implications for Bioactive Properties. *J. Agric. Food Chem.* **2012**, *60*, 6678–6693. [CrossRef]
2. Sater, H.M.; Bizzio, L.N.; Tieman, D.M.; Muñoz, P.D. A Review of the Fruit Volatiles Found in Blueberry and Other Vaccinium Species. *J. Agric. Food Chem.* **2020**, *68*, 5777–5786. [CrossRef] [PubMed]
3. Caleb, O.J.; Mahajan, P.V.; Al-Said, A.J.; Opara, U.L. Modified Atmosphere Packaging Technology of Fresh and Fresh-cut Produce and the Microbial Consequences—A Review. *Food Bioprocess Technol.* **2013**, *6*, 303–329. [CrossRef]
4. Wang, C.; Gao, Y.; Tao, Y.; Wu, X. Influence of γ-irradiation on the reactive-oxygen metabolism of blueberry fruit during cold storage. *Innov. Food Sci. Emerg.* **2017**, *41*, 397–403. [CrossRef]
5. Bounous, G.; Giacalone, G.; Guarinoni, A.; Peano, C. Modified atmosphere storage of high bush blueberry. *Acta Hortic.* **1997**, *446*, 197–203. [CrossRef]
6. Almenar, E.; Samsudin, H.; Auras, R.; Harte, B.; Rubino, M. Postharvest shelf life extension of blueberries using a biodegradable package. *Food Chem.* **2008**, *110*, 120–127. [CrossRef] [PubMed]
7. Reque, P.M.; Steffens, R.S.; Jablonski, A.; Flôres, S.H.; Rios, A.D.O.; De Jong, E.V. Cold storage of blueberry (*Vaccinium* spp.) fruits and juice: Anthocyanin stability and antioxidant activity. *J. Food Compos. Anal.* **2014**, *33*, 111–116. [CrossRef]
8. Villarreal, N.M.; Bustamante, C.A.; Civello, P.M.; Martinez, G.A. Effect of ethylene and 1-MCP treatments on strawberry fruit ripening. *J. Sci. Food Agric.* **2010**, *90*, 683–689. [CrossRef] [PubMed]
9. Duan, W.; Sun, P.; Chen, L.; Gao, S.; Shao, W.; Li, J. Comparative analysis of fruit volatiles and related gene expression between the wild strawberry *Fragaria pentaphylla* and cultivated *Fragaria × ananassa*. *Eur. Food Res. Technol.* **2018**, *244*, 57–72. [CrossRef]

10. Maclean, D.; Nesmith, D.S. Rabbiteye Blueberry Postharvest Fruit Quality and Stimulation of Ethylene Production by 1-Methylcyclopropene. *Hortscience* **2011**, *46*, 1278–1281. [CrossRef]

11. Fu, X.M.; Cheng, S.H.; Zhang, Y.Q.; Du, B.; Feng, C.; Zhou, Y.; Mei, X.; Jiang, Y.M.; Duan, X.W.; Yang, Z.Y. Differential responses of four biosynthetic pathways of aroma compounds in postharvest strawberry (*Fragaria × ananassa* Duch.) under interaction of light and temperature. *Food Chem.* **2017**, *221*, 356–364. [CrossRef] [PubMed]

12. Forney, C.F. Horticultural and other Factors Affecting Aroma Volatile Composition of Small Fruit. *HortTechnology* **2001**, *11*, 529–538. [CrossRef]

13. Dymerski, T.; Namieśnik, J.; Vearasilp, K.; Arancibia-Avila, P.; Toledo, F.; Weisz, M.; Katrich, E.; Gorinstein, S. Comprehensive two-dimensional gas chromatography and three-dimensional fluorometry for detection of volatile and bioactive substances in some berries. *Talanta* **2015**, *134*, 460–467. [CrossRef] [PubMed]

14. Horvat, R.J.; Senter, S.D.; Dekazos, E.D. GLC-MS Analysis of Volatile Constituents in Rabbiteye Blueberries. *J. Food Sci.* **1983**, *48*, 278–279. [CrossRef]

15. Horvat, R.J.; Schlotzhauer, W.S.; Chortyk, O.T.; Nottingham, S.F.; Payne, J.A. Comparison of Volatile Compounds from Rabbiteye Blueberry (*Vaccinium ashei*) and Deerberry (*V. stamineum*) during Maturation. *J. Essent. Oil Res.* **1996**, *8*, 645–648. [CrossRef]

16. Farneti, B.; Khomenko, I.; Grisenti, M.; Ajelli, M.; Betta, E.; Algarra, A.A.; Cappellin, L.; Aprea, E.; Gasperi, F.; Biasioli, F. Exploring blueberry aroma complexity by chromatographic and direct-injection spectrometric techniques. *Front. Plant Sci.* **2017**, *8*, 617. [CrossRef]

17. Liu, F.; Li, S.; Gao, J.; Cheng, K.; Yuan, F. Changes of terpenoids and other volatiles during alcoholic fermentation of blueberry wines made from two southern highbush cultivars. *LWT Food Sci. Technol.* **2019**, *109*, 233–240. [CrossRef]

18. Yuan, F.; Cheng, K.; Gao, J.; Pan, S. Characterization of Cultivar Differences of Blueberry Wines Using GC-QTOF-MS and Metabolic Profiling Methods. *Molecules* **2018**, *23*, 2376. [CrossRef]

19. Zhu, N.; Zhu, Y.; Yu, N.; Wei, Y.; Zhang, J.; Hou, Y.; Sun, A.D. Evaluation of microbial, physicochemical parameters and flavor of blueberry juice after microchip-pulsed electric field. *Food Chem.* **2018**, *274*, 146–155. [CrossRef]

20. Gilbert, J.L.; Schwieterman, M.L.; Colquhoun, T.A.; Clark, D.G.; Olmstead, J.W. Potential for Increasing Southern Highbush Blueberry Flavor Acceptance by Breeding for Major Volatile Components. *Hortscience* **2013**, *48*, 835–843. [CrossRef]

21. Du, X.; Rouseff, R. Aroma active volatiles in four southern highbush blueberry cultivars determined by gas chromatography–olfactometry (GC-O) and gas chromatography–mass spectrometry (GC-MS). *J. Agric. Food Chem.* **2014**, *62*, 4537–4543. [CrossRef] [PubMed]

22. Ferrao, L.F.V.; Johnson, T.S.; Benevenuto, J.; Edger, P.P.; Colquhoun, T.A.; Munoz, P.R. Genome-wide association of volatiles reveals candidate loci for blueberry flavor. *New Phytol.* **2020**, *226*, 1725–1737. [CrossRef]

23. Cheng, K.; Peng, B.Z.; Yuan, F. Volatile composition of eight blueberry cultivars and their relationship with sensory attributes. *Flavour Fragr. J.* **2020**, *35*, 443–453. [CrossRef]

24. Yang, J.; Shi, W.; Li, B.; Bai, Y.; Hou, Z. Preharvest and postharvest uv radiation affected flavonoid metabolism and antioxidant capacity differently in developing blueberries (*Vaccinium corymbosum* L.). *Food Chem.* **2019**, *301*, 125248. [CrossRef] [PubMed]

25. Li, X.; Jin, L.; Pan, X.; Yang, L.; Guo, W. Proteins expression and metabolite profile insight into phenolic biosynthesis during highbush blueberry fruit maturation. *Food Chem.* **2019**, *290*, 216–228. [CrossRef] [PubMed]

26. Eichholz, I.; Huyskens-Keil, S.; Keller, A.; Ulrich, D.; Kroh, L.W.; Rohn, S. UV-B-induced changes of volatile metabolites and phenolic compounds in blueberries (*Vaccinium corymbosum* L.). *Food Chem.* **2011**, *126*, 60–64. [CrossRef]

27. Cao, S.; Hu, Z.; Pang, B.; Wang, H.; Xie, H.; Wu, F. Effect of ultrasound treatment on fruit decay and quality maintenance in strawberry after harvest. *Food Control* **2010**, *21*, 529–532. [CrossRef]

28. Lin, Y.; Huang, G.; Zhang, Q.; Wang, Y.; Meng, X. Ripening affects the physicochemical properties, phytochemicals and antioxidant capacities of two blueberry cultivars. *Postharvest Biol. Technol.* **2020**, *162*, 111097. [CrossRef]

29. Paniagua, A.C.; East, A.R.; Heyes, J.A. Effects of delays in cooling on blueberry quality outcomes. *Acta Hortic.* **2013**, *1012*, 1493–1498. [CrossRef]

30. Sanford, K.A.; Lidster, P.D.; Mcrae, K.B.; Jackson, E.D.; Lawrence, R.A.; Stark, R.; Prange, R.K. Lowbush blueberry quality changes in response to mechanical damage and storage temperature. *J. Am. Soc. Hortic. Sci.* **1991**, *116*, 47–51. [CrossRef]

31. Ribereau-Gayon, P.; Boidron, J.N.; Terrier, A. Aroma of Muscat grape varieties. *J. Agric. Food Chem.* **1975**, *23*, 1042–1047. [CrossRef]

32. Capone, D.L.; Van Leeuwen, K.; Taylor, D.K.; Jeffery, D.W.; Pardon, K.H.; Elsey, G.M.; Sefton, M.A. Evolution and occurrence of 1,8-cineole (eucalyptol) in australian wine. *J. Agric. Food Chem.* **2011**, *59*, 953–959. [CrossRef] [PubMed]

33. Kilcast, D. *Taints and Off-Flavours in Foods*; Baigrie, B., Ed.; CRC Press LLC: Boca Raton, FL, USA, 2003.

34. Dozier, W.A.; Caylor, A.W.; Himelrick, D.G.; Powell, A.A.; Akridge, J.R. Rabbiteye blueberry cultivar performance. *Fruit Var. J.* **1991**, *45*, 128–134. [CrossRef]

35. Chiabrando, V.; Giacalone, G.; Rolle, L. Mechanical behaviour and quality traits of highbush blueberry during postharvest storage. *J. Sci. Food Agric.* **2009**, *89*, 989–992. [CrossRef]

36. Paniagua, A.C.; East, A.R.; Hindmarsh, J.P.; Heyes, J. Moisture loss is the major cause of firmness change during postharvest storage of blueberry. *Postharvest Biol. Technol.* **2013**, *79*, 13–19. [CrossRef]

37. Allanwojtas, P.M.; Forney, C.F.; Carbyn, S.E.; Nicholas, K.U. Microstructural Indicators of Quality-related Characteristics of Blueberries—An Integrated Approach. *LWT Food Sci. Technol.* **2001**, *34*, 23–32. [CrossRef]

38. Alvarez, M.V.; Ponce, A.G.; Moreira, M.R. Influence of polysaccharide-based edible coatings as carriers of prebiotic fibers on quality attributes of ready-to-eat fresh blueberries. *J. Sci. Food Agric.* **2018**, *98*, 2587–2597. [CrossRef]

39. Klee, H.J. Improving the flavor of fresh fruits: Genomics, biochemistry, and biotechnology. *New Phytol.* **2010**, *187*, 44–56. [CrossRef]

40. van Gemert, L.J. *Odour Thresholds. Compilations of Odour Threshold Values in Air, Water and Other Media*; Oliemans, Punter & Partners BV: Utrecht, The Netherlands, 2011.

41. Dixon, J.; Hewett, E.W. Factors affecting apple aroma/flavour volatile concentration: A review. *N. Z. J. Crop Hortic. Sci.* **2000**, *28*, 155–173. [CrossRef]

42. Zhu, X.; Luo, J.; Li, Q.; Li, J.; Liu, T.; Wang, R.; Chen, W.; Li, X. Low temperature storage reduces aroma-related volatiles production during shelf-life of banana fruit mainly by regulating key genes involved in volatile biosynthetic pathways. *Postharvest Biol. Technol.* **2018**, *146*, 68–78. [CrossRef]

43. Yuan, F.; Yan, J.; Yan, X.; Liu, H.; Pan, S. Comparative transcriptome analysis of genes involved in volatile compound synthesis in blueberries (*Vaccinium virgatum*) during postharvest storage. *Postharvest Biol. Technol.* **2020**, *170*, 111327. [CrossRef]

44. Francis, I.; Newton, J. Determining wine aroma from compositional data. *Aust. J. Grape Wine Res.* **2005**, *11*, 114–126. [CrossRef]

45. Matsumoto, H.; Ikoma, Y. Effect of postharvest temperature on the muscat flavor and aroma volatile content in the berries of 'Shine Muscat' (*Vitis labruscana* Baily × *V. vinifera* L.). *Postharvest Biol. Technol.* **2016**, *112*, 256–265. [CrossRef]

46. Siegmund, B. *Flavour Development Analysis & Perception in Food & Beverages*; Woodhead Publishing: Cambridge, UK, 2015; pp. 127–149.

47. Beaulieu, J.C.; Stein-Chisholm, R.E.; Boykin, D.L. Qualitative Analysis of Volatiles in Rabbiteye Blueberry Cultivars at Various Maturities Using Rapid Solid-phase Microextraction. *J. Am. Soc. Hortic. Sci.* **2014**, *139*, 167–177. [CrossRef]

48. Krumbein, A.; Peters, P.; Bruckner, B. Flavour compounds and a quantitative descriptive analysis of tomatoes (*Lycopersicon esculentum* Mill.) of different cultivars in short-term storage. *Postharvest Biol. Technol.* **2004**, *32*, 15–28. [CrossRef]

49. Brovelli, E.A.; Brecht, J.K.; Sherman, W.B.; Sims, C.A. Quality of Fresh-Market Melting- and Nonmelting-Flesh Peach Genotypes as Affected by Postharvest Chilling. *J. Food Sci.* **1998**, *63*, 730–733. [CrossRef]

Article

Effect of Freeze-Thaw Cycles on Juice Properties, Volatile Compounds and Hot-Air Drying Kinetics of Blueberry

Lin Zhu [1], Xianrui Liang [2], Yushuang Lu [1], Shiyi Tian [1], Jie Chen [1], Fubin Lin [1] and Sheng Fang [1,*]

[1] School of Food Science and Biotechnology, Zhejiang Gongshang University, Xuezheng Street No. 18, Hangzhou 310018, China; wishzzl@163.com (L.Z.); lyszjgsl@163.com (Y.L.); tianshiyi@zjgsu.edu.cn (S.T.); chenjie@zjgsu.edu.cn (J.C.); linfubinwin@126.com (F.L.)

[2] Collaborative Innovation Center of Yangtze River Delta Region Green Pharmaceuticals, College of Pharmaceutical Sciences, Zhejiang University of Technology, Hangzhou 310014, China; liangxrvicky@zjut.edu.cn

[*] Correspondence: fangsheng@zjgsu.edu.cn; Tel.: +86-13093752831

Abstract: This paper studied the effects of freeze-thaw (FT) cycles on the juice properties and aroma profiles, and the hot-air drying kinetics of frozen blueberry. After FT treatment, the juice yield increased while pH and total soluble solids of the juice keep unchanged. The total anthocyanins contents and DPPH antioxidant activities of the juice decreased by FT treatments. The electronic nose shows that FT treatments significantly change the aroma profiles of the juice. The four main volatile substances in the fresh juice are (E)-2-hexenal, α-terpineol, hexanal and linalyl formate, which account for 48.5 ± 0.1%, 17.6 ± 0.2%, 14.0 ± 1.5% and 7.8 ± 2.7% of relative proportions based on total ion chromatogram (TIC) peak areas. In the FT-treated samples, the amount of (E)-2-hexenal and hexanal decreased significantly while α-terpineol and linalyl formate remained almost unchanged. Repeated FT cycles increased the ethanol content and destroyed the original green leafy flavor. Finally, the drying kinetics of FT-treated blueberries was tested. One FT treatment can shorten the drying time by about 30% to achieve the same water content. The D_{eff} values of the FT-treated sample are similar, which are about twice as large as the value of the fresh sample. The results will be beneficial for the processing of frozen blueberry into juice or dried fruits.

Keywords: freeze-thaw cycles; anthocyanins; gas chromatography-mass spectrometry; aroma profiles; hot-air drying; blueberry

Citation: Zhu, L.; Liang, X.; Lu, Y.; Tian, S.; Chen, J.; Lin, F.; Fang, S. Effect of Freeze-Thaw Cycles on Juice Properties, Volatile Compounds and Hot-Air Drying Kinetics of Blueberry. *Foods* **2021**, *10*, 2362. https://doi.org/10.3390/foods10102362

Academic Editors: Sidonia Martinez, Javier Carballo and Susana Casal

Received: 17 July 2021
Accepted: 1 October 2021
Published: 4 October 2021

Publisher's Note: MDPI stays neutral with regard to jurisdictional claims in published maps and institutional affiliations.

1. Introduction

Blueberry is widely grown all over the world and its production has largely increased in recent years [1]. Due to seasonality and short shelf life, approximately 50% of blueberries are processed into food products such as juice and dry fruits [2]. Most of the blueberries used for juice processing and drying are frozen fruits. During storage and cold chain transportation, frozen blueberries might be subjected to several freeze-thaw (FT) cycles [3]. Insights into the changes of frozen blueberries during repeated FT cycles are essential for the processing of blueberry products [4].

FT treatment will affect the qualities and flavors of final food products [5,6]. The ice crystal formed in FT treatments facilitates the rupture of fruit cell walls and also changes the texture and flavor during the thawing process. For blueberries, the study has shown that FT treatment facilitated moisture transfer during drying and produced softer, less chewable and less gummy berries [7]. The juice yield of fruits and extraction efficiency of bioactive compounds were also improved by the FT treatments of blueberries. However, Nowak et al. [8] found that FT treatments induced considerable changes in the color of blueberries. Yan et al. [9] found that aroma deterioration during subsequent shelf life after cold storage was more fatal to the blueberry flavor. So, the FT treatment exerts both desirable and undesirable effects on the physicochemical properties of blueberry [10]. To

267

the best of our knowledge, there are no studies on the effects of multiple FT cycles on the physicochemical properties of frozen blueberry, especially focus on, aromatic compounds and drying kinetics.

In this study, the effects of FT cycles on the physicochemical properties and volatile compounds of blueberry juice, and hot-air drying kinetics of frozen blueberry fruits were studied. The juice yields, pH, total soluble solids, total anthocyanin contents and antioxidant properties of the juice were firstly characterized. Then, the electronic nose and gas chromatography-mass spectrometry (GC-MS) was used to study the influences on aroma profiles. Finally, the kinetic characteristics of hot-air drying of blueberry at 60 °C were studied, and the drying model and parameters were obtained. The results will be beneficial for the further processing of blueberry frozen fruit into juice and dried fruits.

2. Materials and Methods

2.1. Materials

Blueberry which belongs to a southern-highbush variety was harvested at the mature stage (fully blue and firm) from a local orchard in Zhuji, Zhejiang Province, China. Fruit without disease and wounds were harvested. Chromatography grade acetonitrile, methanol, ethanol, formic acid (purity $\geq$ 98%) and acetic acid (purity $\geq$ 99%) was purchased from Aladdin (Shanghai, China). All other reagents were of analytical grade.

2.2. Freeze-Thaw Treatments

Blueberries were washed with water and put in a sealed bag. For each bag, a total of 60 blueberries were carefully selected without physical damage and used. Three bags were set as one group for each FT treatment. The blueberry in the bag was frozen at −25 °C [11] for 6 h in a low-temperature refrigerator (MDF-330, SANYO, Japan) and then thawed at 4 °C for 3 h. FT-1, FT-2 and FT-3 represent the sample with FT treatment once, twice and third times. After FT treatments, the blueberry was juiced with a juicer. The juice was centrifuged at 10,000 rpm and 20 °C for 15 min. The supernatant was taken and analyzed.

2.3. The pH and Total Soluble Solids (TSS)

The pH values of blueberry juices were obtained using a pH meter (PHS-3G, Shanghai INESA, China) at 25 °C. TSS was determined by measurement of refractive index at 25 ± 1 °C using a digital refractometer (WAY-2S, Shanghai Shengke, Shanghai, China).

2.4. Total Anthocyanins Content by the pH Differential Method

The anthocyanin content in the juice was determined using the pH differential method [12]. The content was expressed based on cyanidin-3-O-glucoside (C3G). The absorbance (Ab) was measured using a UV spectrophotometer (UV-2600, Shimadzu, Japan) at 510 and 700 nm. The total anthocyanins (TA) are calculated as:

$$\text{TA (mg/L)} = \frac{\Delta Ab \times M \times Df}{L \times \varepsilon} \tag{1}$$

where ΔAb is calculated as $(Ab_{510}\text{-}Ab_{700})_{\text{pH}=1.0}\text{-}(Ab_{510}\text{-}Ab_{700})_{\text{pH}=4.5}$; M is the molar mass of C3G, 449.2 g/mol; Df is the dilution factor; L is the path length in cm; ε is the molar extinction coefficient of C3G, 26900 L/mol/cm.

2.5. DPPH Free Radical Scavenging Test

The free radical scavenging ability was measured according to the method of Kahkonen [13] with some modifications. 100 mL ethanol solution contains 0.1 mM DPPH was prepared and stored in a refrigerator at 4 °C. Incubate 2 mL sample solution and 2 mL ethanol DPPH solution (0.1 mM) at room temperature (25 °C) for 0.5 h. Finally, the absorbance of the mixture was measured at 517 nm using a UV spectrophotometer (UV-2600, Shimadzu, Japan). As a control, distilled water was used instead of DPPH.

2.6. Electronic Nose Test

The sensory profiles of the blueberry juice were tested by an electronic nose according to a previous study [14]. The juice sample was put into a 500 mL beaker, wrapped and sealed with tin foil, and the headspace was allowed to balance for 30 min at room temperature. The equipment was exhausted for 120 s at first. Then, the headspace gas of the beaker was pumped through the sensor array at a rate of 0.6 L/min. The measurements lasted for 120 s to ensure a stable response [15].

2.7. Volatile Compounds Profile Analyzed by GC-MS

The volatile compounds were analyzed according to a previous study [16]. An Agilent 7890B-5977B gas chromatograph-mass spectrometer equipped with an Agilent 7697A Headspace auto-sampler (Santa Clara, CA, USA) was used. A capillary column (DB-624 UI, Agilent Technologies, Santa Clara, CA, USA) of 30 m × 0.25 mm with 1.4 µm film was selected. The blueberry juice (3 mL) and saturated NaCl solution (1 mL) were mixed into a 20 mL headspace bottle for the GC-MS test. The headspace equilibrium temperature and time are 80 °C and 40 min, respectively. The quantitative loop and transmission line temperature are set at 90 °C and 100 °C, respectively. Helium (>99.9%, 1 mL/min) was used as carrier gas. The sample was injected into the GC column in a split ratio (1:5) mode. The initial column temperature is set at 35 °C and kept for 2 min. Then increase to 190 °C at a rate of 4 °C/min. Then increase to 240 °C at a rate of 8 °C/min and hold for 5 min, and finally cool to 35 °C. The MS spectrum was obtained at 70 eV in electron ionization (EI) mode with the scanning range 35–600 m/z. The MS ion source and detector temperatures are 230 °C and 260 °C, and the solvent delay time is set to 2 min. The compound is identified by comparing the mass spectrum with the reference mass spectrum in the NIST 14. The relative content of each compound was calculated based on the peak area of total ion chromatograms (TIC). Each sample was measured in duplicate, and the values were presented as mean ± standard deviation.

2.8. Hot-Air Drying Conditions and Modeling

2.8.1. Hot-Air Drying Conditions

The drying experiments were conducted by a hot-air convective dryer that is the same as a previous study [17]. The dryer was set at 60 °C with an air velocity of 1.2 m/s and was equilibrated for at least 1 h. A single layer of blueberries (about 14 g) was placed on stainless steel wire meshes in the dryer that weighted in real-time with an electronic balance. The sample weight was recorded every 5 min for 10 h. After drying, the sample continues to be dried to a constant weight at 110 °C [18] in an oven, and the final mass is taken as the absolute dry mass of the sample (W_d). The moisture ratio (*MR*) at time t in the drying process is calculated as follows:

$$MR = \frac{W_t - W_d}{W_0 - W_d} \tag{2}$$

where W_t, and W_0 represent the mass of sample at time t, and the initial mass of the sample, respectively.

2.8.2. Mathematical Models

Nine empirical equations (the Newton, Page, modified Page, Henderson and Pabis, Logarithmic, two-term, two-term exponential, Wang and Singh, and diffusion approaches) [17] used to fit the experimental data are listed in Table S1. The model parameters were obtained by fitting the experimental drying values with the nonlinear least square method. The determination coefficient (R^2) and root mean square error (*RMSE*) were calculated to test the fitting performance of each model.

The effective moisture diffusivity coefficient (D_e) is an important transport property. Based on the general solution of Fick's second law of diffusion and assumptions for spherical particles, the following equation can be obtained:

$$\ln MR = \ln\left(\frac{6}{\pi^2}\right) - \pi^2\left(\frac{D_e}{R_p^2}\right)t \tag{3}$$

where R_p is the average radius of spherical particle (m). As with most calculations of logarithmic kinetic models, a linear plot of lnMR versus time t was used to obtain the value of D_e [19].

2.9. Statistical Analysis

All experiments were conducted at least in triplicate, and data were analyzed by Origin 2018 software (Origin Lab Corporation, USA). The obtained results were statistically evaluated by ANOVA single factor analyze ($p < 0.05$) with Duncan's multiple comparisons using SPSS 19.0 software (SPSS Inc, Chicago, IL, USA).

3. Results and Discussion

3.1. Effect of FT Cycles on the Physicochemical Properties of Blueberry Juice

Table 1 shows the effects of FT cycles on the juice yield, pH and, TSS of the blueberry juice. The dry and wet water content of fresh blueberries studied are 6.3 ± 0.4 kg H_2O/kg db (dry basis) and $86.3 \pm 0.7\%$, respectively. The juice yield of fresh blueberries is approximately 50%, while the sample after FT treatment is approximately 60%. The increase in FT cycles did increase the juice yield as compared to the fresh one ($p < 0.05$). The FT treatment causes the cells to rupture and therefore makes the release of cellular juice easier [20,21]. The pH and TSS of the fresh juice are 3.64 ± 0.02 and 12.33 ± 0.33, respectively. The results are within the value range of fresh blueberry juice [22]. However, there are no significant differences in the pH and TSS of blueberry juice with FT treatments. Similar results were also found in the effects of FT treatment on carrot juice [23].

Table 1. Effect of freeze-thaw (FT) cycles on juice yield, pH and total soluble solids (TTS) of the juice. FT-1, FT-2 and FT-3 represent the sample with FT treatment once, twice and third times.

Sample	Juice Yield (g/g)	pH	TSS (°Brix)
Fresh	0.51 ± 0.00 [d]	3.64 ± 0.02 [c]	12.33 ± 0.33 [a]
FT-1	0.57 ± 0.00 [c]	3.70 ± 0.01 [b]	12.33 ± 0.38 [a]
FT-2	0.59 ± 0.00 [b]	3.67 ± 0.03 [bc]	12.73 ± 0.26 [a]
FT-3	0.61 ± 0.00 [a]	3.75 ± 0.03 [a]	12.27 ± 0.19 [a]

Values presented as mean $\pm$ standard deviation ($n = 3$). Different letters in each column denote significant differences ($p < 0.05$).

Figure 1A shows the effects of FT cycles on the total anthocyanins content of the blueberry juice. The total anthocyanins content in the initial blueberry juice is 285.0 ± 0.5 mg/L and is lower than the literature values that by ethanol extraction methods [24,25]. It might be attributed to the fact that anthocyanins were not fully extracted by the juicing process in this study and remained in the pomace [26]. The total anthocyanins decreased significantly ($p < 0.05$) after repeated FT treatment. After one, two and three FT cycles, the total anthocyanins in the juice decreased by 9.7%, 17.5% and 33.7%, respectively. Anthocyanins are unstable and are easily degraded by free radicals and enzymes [26]. Cell structure damage caused by the FT treatments leads to the leak of enzymes such as polyphenol oxidase easier, thereby improved the enzyme-substrate interaction [27]. Similar results were also found in the effects of FT treatment on cherry anthocyanins [27].

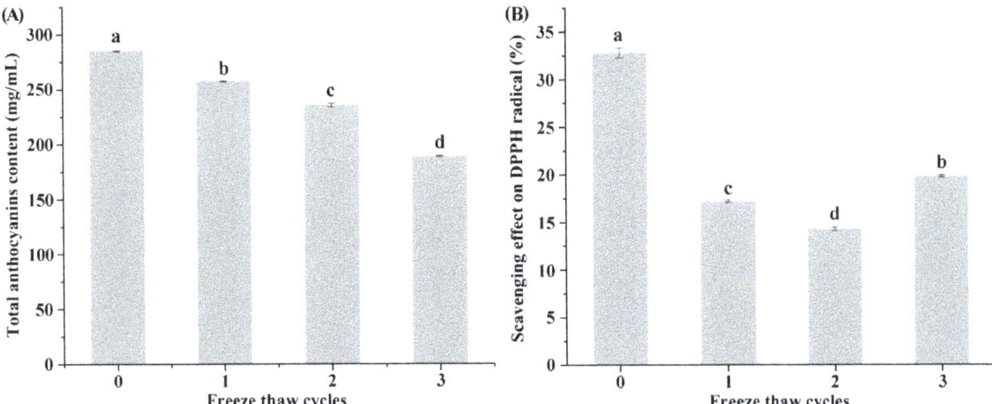

Figure 1. The effects of freeze thaw cycles on (**A**) the total anthocyanins content (mg/mL), and (**B**) the DPPH radical scavenging (%). Values are presented as mean ± standard deviation (*n* = 3). Different letters indicate significant differences (*p* < 0.05).

Figure 1B shows the effects of FT cycles on the antioxidant capacity of the juice. Fresh blueberry juice has the strongest antioxidant capacity. As the number of FT cycles increases, the antioxidant capacity of the samples first decreased and then slightly increased. However, after three FT cycles, the antioxidant capacity is only about 50% of that for the fresh sample. It is known that the anthocyanins content is positively related to DPPH antioxidant capacity [28]. The degradation of anthocyanins in the blueberry may result in the reduction of antioxidant activity. However, the DPPH antioxidant activity is not only determined by anthocyanins, but also by many other compounds including the products of degradation. The formed compounds might have higher activities than the original ones. The study also found that the antioxidant effect of juices containing anthocyanins and other phenolic compounds is increased after biotransformation [29].

3.2. Effect of FT Cycles on the Aroma Profiles of Blueberry Juice

The aroma profiles of the juice samples were first evaluated by electronic nose. PCA visualization for the electronic nose was shown in Figure 2. The first and second principal components accounted for 97.25% and 1.99%, respectively, which explained 99.24% of the variance. Fresh samples and freeze-thaw processed samples can be distinguished well. The abscissas of fresh samples and FT processed samples are far apart, which shows the good discrimination of different samples. Samples by FT once and 2 times can also be distinguished to a certain extent, but the difference is small. Nevertheless, the overlap observed at samples of 2 and 3 FT cycles indicated that these samples might have similar aroma profiles. Overall, the electronic nose shows differences for different FT treated samples [30].

The typical GC-MS total ion chromatograms (TIC) of fresh and frozen blueberry juice are shown in Figure S1 (supporting information). Most volatile compounds can be separated under the chromatographic conditions in this study. Many gas chromatographic conditions including the headspace parameters were compared. It is worth pointing out, without adding saturated NaCl, the peaks in TIC are relatively weak. When 1 mL of saturated NaCl solution is added to 3 mL of blueberry juice, the number of peaks and signal intensity are greatly enhanced. The chemical structure of most of the peaks in TIC can be obtained by matching the mass spectrum with the NIST database, and the reported flavor substances of blueberry in literatures [31,32].

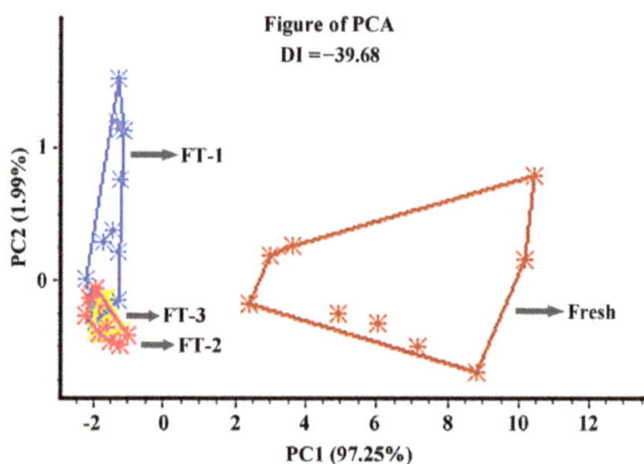

Figure 2. The principal component analysis (PCA) results of freeze-thaw (FT) treated samples. PC1 and PC2 represent the first and second principal component, respectively; FT-1, FT-2 and FT-3 represent the sample with FT treatment once, twice and third times.

A total of 28 main volatiles are identified and grouped into five groups, including 10 alcohols, 4 aldehydes, 3 esters, 2 hydrocarbons and 9 monoterpenes as shown in Table 2. Figure 3 shows the average profiles of peak area abundance on TIC and the distribution of each volatile class. Three mainly volatile substance classes in the fresh juice are aldehydes, alcohols and esters. The relative proportions of aldehydes, alcohols and esters calculated from TIC peak areas are approximately 62.7 ± 0.4%, 21.7 ± 1.0% and 10.1 ± 1.8%, respectively. The peak area values of volatile compounds in the fresh juice and FT treated samples are shown in Figure S2. The four main volatile substances in the fresh juice are (E)-2-hexenal, α-terpineol, hexanal and linalyl formate, which account for 48.5 ± 0.1%, 17.6 ± 0.2%, 14.0 ± 1.5% and 7.8 ± 2.7%, respectively. The result is similar to the literature that the aroma profiles of blueberry are mainly aldehydes including (E)-2-hexenal and hexanal, and additional relatively high amounts of terpene compounds [31,32]. It is known that the (E)-2-hexenal has a relatively low odor threshold of 17 ppb with a green leafy aroma descriptor [33]. Esters of linalool such as linalyl formate are responsible for the fruity aroma descriptor [34]. The results are consistent with the sensory evaluation of the fresh blueberry sample.

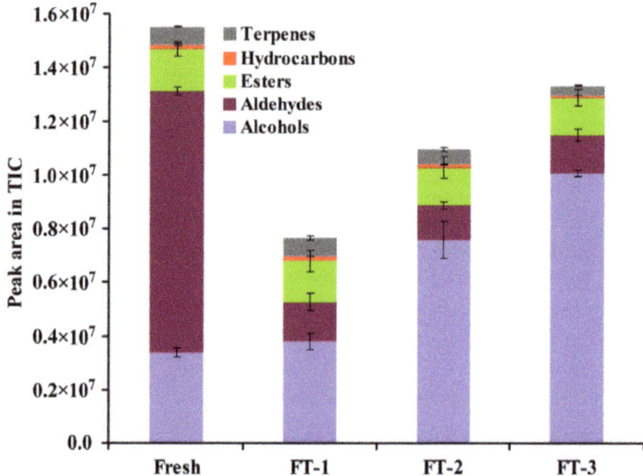

Figure 3. Abundance of peak area in TIC assigned to different class of volatiles for different samples. FT-1, FT-2 and FT-3 represent the sample with FT treatment once, twice and third times.

Table 2. Volatile organic compounds found in different freeze-thaw (FT) treated samples and its relative proportion (%) based on total ion chromatogram (TIC) peak areas in headspace gas chromatography-mass spectrometry. FT-1, FT-2 and FT-3 represent the sample with FT treatment once, twice and third times.

NAME	CAS	Relative Proportion (%)			
		Fresh	FT-1	FT-2	FT-3
ALCOHOLS					
Ethanol	64-17-5	0.81 ± 0.31 [b]	5.48 ± 0.49 [b]	38.61 ± 4.09 [a]	43.46 ± 2.34 [a]
(Z)-2-Hexen-1-ol	111-27-3	0.33 ± 0.06 [c]	0.96 ± 0.09 [c]	2.33 ± 0.12 [b]	6.56 ± 0.24 [a]
Eucalyptol	470-82-6	0.11 ± 0.08 [a]	0.16 ± 0.02 [a]	0.16 ± 0.09 [a]	0.03 ± 0.04 [a]
cis-Linalool oxide	5989-33-3	0.98 ± 0.28 [b]	1.64 ± 0.29 [a]	1.33 ± 0.33 [a,b]	1.70 ± 0.07 [a]
Hotrienol	20053-88-7	0.37 ± 0.02 [a]	0.74 ± 0.29 [a]	0.54 ± 0.11 [a]	0.53 ± 0.15 [a]
cis-Ocimenol	7643-59-6	0.47 ± 0.20 [b]	0.73 ± 0.03 [a]	0.69 ± 0.23 [a,b]	0.87 ± 0.07 [a]
Ocimenol	5986-38-9	0.26 ± 0.11 [a]	0.38 ± 0.01 [a]	0.40 ± 0.17 [a]	0.38 ± 0.01 [a]
trans-4-Thujanol	17699-16-0	0.33 ± 0.10 [a]	0.51 ± 0.15 [a]	0.37 ± 0.04 [a]	0.26 ± 0.03 [b]
α-Terpineol	98-55-5	17.56 ± 0.24 [c]	36.70 ± 0.42 [a]	23.90 ± 1.70 [b]	20.67 ± 2.21 [b]
p-Mentha-1(7),8-dien-2-ol	35907-10-9	0.11 ± 0.09 [a]	0.23 ± 0.02 [a]	0.14 ± 0.04 [a]	0.20 ± 0.08 [a]
ALDEHYDES					
Hexanal	66-25-1	13.99 ± 1.55 [a]	1.88 ± 0.25 [b]	2.37 ± 0.22 [b]	2.22 ± 0.36 [b]
(Z)-3-Hexenal	6789-80-6	0.86 ± 0.26 [a]	0.15 ± 0.01 [b]	0.08 ± 0.00 [c]	0.07 ± 0.01 [c]
(E)-2-Hexenal	6728-26-3	48.47 ± 0.10 [a]	15.43 ± 2.13 [b]	8.84 ± 0.38 [c]	6.97 ± 0.58 [c]
Carvomenthenal	29548-14-9	0.39 ± 0.08 [a]	1.66 ± 1.55 [a]	0.60 ± 0.24 [a]	1.33 ± 0.76 [a]
ESTERS					
Ethyl Acetate	141-78-6	0.37 ± 0.27 [a]	0.44 ± 0.12 [a]	0.23 ± 0.33 [a]	0.54 ± 0.50 [a]
Linalyl formate	115-99-1	7.81 ± 2.70 [b]	18.96 ± 4.11 [a]	12.00 ± 3.04 [a]	9.43 ± 0.83 [b]
3-Cyclohexen-1-ol, 5-methylene-6-(1-methylethenyl)-, acetate	54832-23-4	1.32 ± 0.32 [ab]	2.68 ± 1.12 [a]	1.25 ± 0.10 [b]	1.48 ± 0.06 [a]
HYDROCARBONS					
o-Cymene	527-84-4	0.35 ± 0.06 [b]	0.52 ± 0.02 [a]	0.22 ± 0.04 [c]	0.10 ± 0.01 [d]
p-Cymenene	1195-32-0	0.83 ± 0.17 [b]	1.47 ± 0.10 [a]	0.96 ± 0.12 [b]	0.60 ± 0.00 [c]
TERPENES					
γ-Ionone	79-76-5	0.14 ± 0.01 [b]	0.39 ± 0.03 [a]	0.11 ± 0.05 [bc]	0.07 ± 0.00 [c]
β-Myrcene	123-35-3	0.07 ± 0.03 [b]	0.22 ± 0.03 [a]	0.10 ± 0.01 [b]	0.11 ± 0.09 [ab]
α-Terpinene	99-86-5	0.57 ± 0.13 [bc]	1.19 ± 0.07 [a]	0.71 ± 0.03 [b]	0.41 ± 0.0 [c]
D-Limonene	5989-27-5	0.76 ± 0.03 [b]	1.64 ± 0.29 [a]	0.91 ± 0.13 [b]	0.35 ± 0.00 [c]
1S-α-Pinene	7785-26-4	0.22 ± 0.01 [b]	0.49 ± 0.19 [a]	0.25 ± 0.03 [b]	0.05 ± 0.01 [c]
3-Carene	13466-78-9	0.39 ± 0.00 [b]	0.87 ± 0.29 [a]	0.36 ± 0.21 [bc]	0.18 ± 0.01 [c]
γ-Terpinene	99-85-4	0.15 ± 0.04 [b]	0.32 ± 0.09 [a]	0.14 ± 0.03 [b]	0.03 ± 0.05 [c]
Terpinolene	586-62-9	1.30 ± 0.10 [b]	3.23 ± 0.42 [a]	1.75 ± 0.24 [b]	0.79 ± 0.00 [c]
L-β-Pinene	18172-67-3	0.68 ± 0.07 [a]	0.93 ± 0.33 [a]	0.64 ± 0.07 [a]	0.62 ± 0.17 [a]

Values presented as mean ± standard deviation ($n = 2$). Different letters in each row denote significant differences ($p < 0.05$).

Figure 3 shows that the total amount of volatile substances in the juice is reduced overall after FT treatments. This is also consistent with the result of the weakening of the flavor after FT treatments in the sensory evaluation. Figure 4 shows the changes in the peak area abundance of 10 main aroma compounds in different samples. In the FT treated samples, even if only once (FT-1), the amount of C6 aldehydes including (E)-2-hexenal ($p < 0.001$) and hexanal ($p < 0.01$) decreased significantly as compared to that in the fresh juice. The amount of hexanal and (E)-2-hexenal in juice decreased approximately 93.2 ± 0.4% and 84.0 ± 2.8%, respectively, by only one FT treatment. Even though not very obvious, an increase in (Z)-2-hexen-1-ol was observed at the same time. Interestingly, the amount of α-terpineol and linalyl formate remained unchanged after FT cycles and become the main aroma compounds other than C6 aldehydes. Therefore, we speculate that the weakening and changes in the flavor of FT treated blueberries are mainly due to the degradation of C6 aldehydes. It is supposed that these C6 aldehydes might be converted into C6 alcohols during the thawing process by enzymatic reduction. During the process, the cell structure changes which might facilitate the release of enzymes such as alcohol dehydrogenase and lipoxygenase [35]. In the fermentation process of some grape wines, the study found that C6 aldehydes (hexanal and trans-2-hexenal) decreased significantly during the process [36].

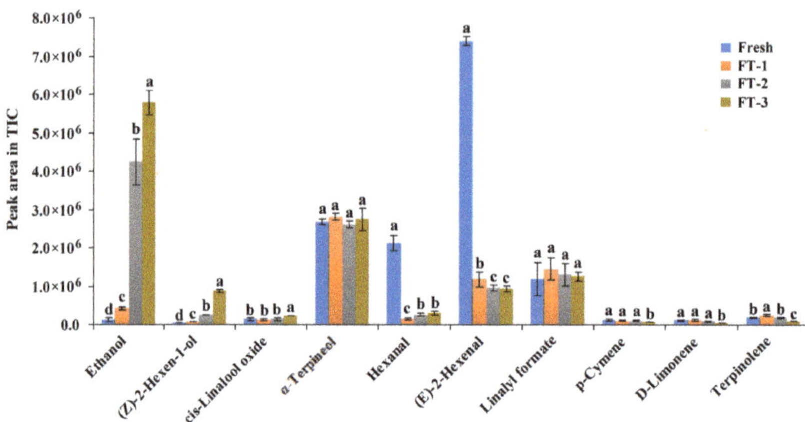

Figure 4. The peak area in total ion chromatogram (TIC) assigned to 10 main compounds for freeze-thaw (FT) treated samples. FT-1, FT-2 and FT-3 represent the sample with FT treatment once, twice and third times. Values are presented as mean $\pm$ standard deviation (n = 2). Different letters indicate significant differences (p < 0.05).

In the samples after multiple FT cycles, the ethanol content increased significantly. Moreover, the increase in ethanol content has little correlation with the decrease in other flavor substances. On the other hand, the samples that were FT treated 2 and 3 times showed little change in other flavor substances except for ethanol and (Z)-2-hexen-1-ol. This is consistent with the increase in alcohol taste in the samples after multiple FT cycles. It is known that ethanol is relatively low in most fresh fruits, but it may accumulate in senescent fruits [37]. We speculate that the FT cycle destroys the blueberry tissue structure and promotes anaerobic respiration thereby accumulating ethanol in the juice [38]. In general, repeated FT cycles will increase the ethanol content of the blueberry and dilute or destroy the original green leafy flavor.

3.3. Effect of FT Cycles on Hot-Air Drying Kinetics of Blueberry Fruit

Early results show that blueberry anthocyanins can remain stable below 60 °C [26,39]. Therefore, the hot-air temperature at 60 °C was selected to study the effect of freezing and thawing on the drying kinetics of blueberry. As predicted, freezing and thawing can shorten the hot-air drying time of blueberries. Figure 5 shows that the total time required to dry blueberry to 50% MR at 60 °C is about 290 min, while it only takes about 200 min for the FT-1 samples. So, one FT treatment can shorten the drying time by about 30% to achieve the same water content. Similar trends are found in the drying process of other agricultural products [7,40,41]. For example, the ultrasound freeze-thawing pretreatment reduced the drying time and also energy consumption in the drying of okra [40]. However, the samples after multiple FT cycles gave similar hot-air drying curves, which did not further improve the drying efficiency. From the perspective of energy consumption and costs, it is appropriate to choose one cycle of FT treatment for the drying of blueberry. The fitting results of nine models on the hot-air drying curve of blueberry at 60 °C are listed in Table S2 and Figure S3. The regression drying constants a and b of the Wang and Singh equation which gives the best results are shown in Table S3. The model and parameters will facilitate the calculation of hot-air drying of blueberry for further studies.

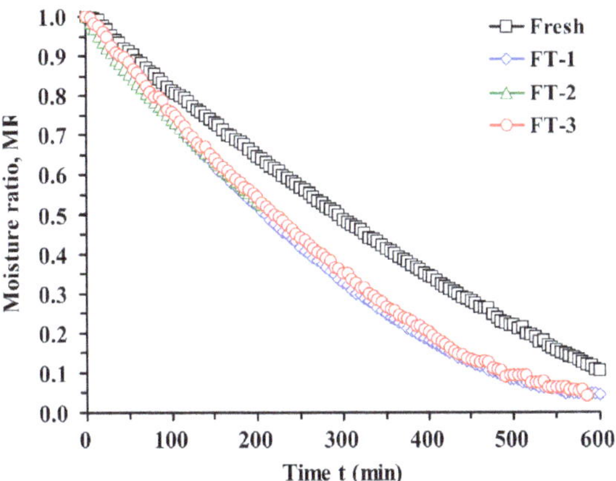

Figure 5. The hot-air drying curves of blueberry at 60 °C subjected to freeze-thaw (FT) treatments. FT-1, FT-2 and FT-3 represent the sample with FT treatment once, twice and third times.

The D_e value of the fresh blueberry sample is 2.69×10^{-10} m^2/s, while the FT treated samples range between 4.43×10^{-10} and 4.59×10^{-10} m^2/s, as shown in Table 3. The values are in the range of those reported for hot-air drying of different materials (10^{-8} to 10^{-10} m^2/s) [41,42]. The D_e values of the FT-treated sample are similar, which are about twice as large as the value of the fresh sample. This again means that the freeze-thawed promotes the diffusion rate of water in blueberries. It is known that freeze-thawing cycles can rupture cell walls and membrane, and change osmotic pressure, thereby promoted water transfer rate in the drying process [40]. Studies also found that the high D_e values shown by FT treated samples were due to the formation of large ice crystals inside cells [40,41].

Table 3. The effective moisture diffusivity (D_e, m^2/s) of raw blueberry and freeze-thaw (FT) treated blueberry fruit at 60 °C. FT-1, FT-2 and FT-3 represent the sample with FT treatment once, twice and third times.

Samples	Effective Diffusivity, D_e (m^2/s)	R^2
Fresh	2.69×10^{-10}	0.962
FT-1	4.43×10^{-10}	0.973
FT-2	4.68×10^{-10}	0.955
FT-3	4.59×10^{-10}	0.964

4. Conclusions

The effects of FT cycles on juice physicochemical properties, juice aroma profiles and hot-air drying kinetics of blueberry were studied. We found that the FT cycles will increase the juice yield of blueberries and reduce the anthocyanin content, while the pH and solids contents keep unchanged. (E)-2-hexenol, ethanol and 1-hexanol are characteristic substances of blueberry passing through FT cycles. The former gradually decreases, while the latter two gradually increase. The electronic nose shows the different aroma parents of FT-treated blueberries. The main volatile substances in the fresh juice are (E)-2-hexenal, α-terpineol, hexanal and linalyl formate, respectively. In the FT-treated samples, the amount of C6 aldehydes including (E)-2-hexenal and hexanal decreased significantly while the amount of α-terpineol and linalyl formate remained unchanged. Repeated FT cycles will increase the ethanol content of the blueberry. Finally, the drying kinetics of FT-treated blueberries was studied. FT treatment can shorten the drying time by about 30% to achieve the same water content. The D_e values of the FT-treated sample are similar, which are about twice as large as the value of the fresh sample. The results will be beneficial for the processing of blueberry into juice or dried fruits.

Supplementary Materials: The following are available online at https://www.mdpi.com/article/10.3390/foods10102362/s1, Figure S1: The top-down pictures represent the GC-MS base peak chromatogram of the total ion chromatogram (TIC) for the fresh, FT-1, FT-2 and FT-3 samples. FT-1, FT-2 and FT-3 represent the blueberry with freeze-thaw (FT) treatment once, twice and third times; Figure S2: Peak area in TIC assigned to each compound for (A) fresh juice, (B) freeze-thaw (FT) once, (C) FT twice and (D) FT third times; Figure S3: The correlation coefficient of a different model for the fitting of the hot-air drying curve of frozen blueberry. FT-1, FT-2 and FT-3 represent the blueberry with freeze-thaw (FT) treatment once, twice and third times; Table S1: Mathematical models applied to fit the drying curves; Table S2: The fitting models and statistical results for models. FT-1, FT-2 and FT-3 represent the blueberry with freeze-thaw (FT) treatment once, twice and third times; Table S3: Parameter of the Wang and Singh model for the correlation of hot-air drying curve of freeze-thaw (FT) treated blueberry fruit at 60 °C. FT-1, FT-2 and FT-3 represent the sample with FT treatment once, twice and third times.

Author Contributions: Formal analysis, L.Z. and S.F.; investigation, L.Z. and X.L.; methodology, Y.L., S.T., J.C. and F.L.; resources, S.F.; supervision, S.F.; writing—original draft, L.Z. and X.L.; writing—review and editing, X.L. and S.F. All authors have read and agreed to the published version of the manuscript.

Funding: This research was supported by the Cooperative project with Zhejiang Hisoar Pharmaceutical Co., Ltd. (KYY-HX-20180525), the Key Research and Development Project of Zhejiang Province, China, grant number 2019C02088.

Data Availability Statement: Data sharing is not applicable to this article.

Conflicts of Interest: The authors declare no conflict of interest.

References

1. Strik, B.C. Perspective on the U.S. and global blueberry industry. *J. Am. Pomol. Soc.* **2007**, *61*, 144.
2. Nindo, C.I.; Tang, J.; Powers, J.R.; Singh, P. Viscosity of blueberry and raspberry juices for processing applications. *J. Food Eng.* **2005**, *69*, 343–350. [CrossRef]
3. Cao, X.; Zhang, F.; Zhao, D.; Zhu, D.; Li, Y.; Liu, L. Effects of Immersion Freezing Methods on Freezing Characteristics of Blueberry. *J. Chin. Inst. Food Sci. Technol.* **2018**, *18*, 184–190.
4. Munzenmayer, P.; Ulloa, J.; Pinto, M.; Ramirez, C.; Valencia, P.; Simpson, R.; Almonacid, S. Freeze-Drying of Blueberries, Effects of Carbon Dioxide (CO_2) Laser Perforation as Skin Pretreatment to Improve Mass Transfer, Primary Drying Time, and Quality. *Foods* **2020**, *9*, 211. [CrossRef]
5. Ali, S.; Zhang, W.; Rajput, N.; Khan, M.A.; Li, C.; Zhou, G. Effect of multiple freeze–thaw cycles on the quality of chicken breast meat. *Food Chem.* **2015**, *173*, 808–814. [CrossRef]
6. Hajji, W.; Bellagha, S.; Allaf, K. Effect of partial drying intensity, frozen storage and repeated freeze-thaw cycles on some quality attributes of dehydrofrozen quince fruit. *J. Food Meas. Charact.* **2020**, *14*, 353–365. [CrossRef]
7. Zielinska, M.; Sadowski, P.; Błaszczak, W. Freezing/thawing and microwave-assisted drying of blueberries (*Vaccinium corymbosum* L.). *LWT-Food Sci. Technol.* **2015**, *62*, 555–563. [CrossRef]
8. Nowak, K.W.; Zielinska, M.; Waszkieli, K.M. The effect of ultrasound and freezing/thawing treatment on the physical properties of blueberries. *Food Sci. Biotechnol.* **2018**, *28*, 741–749. [CrossRef]
9. Yan, X.; Yan, J.; Pan, S.; Yuan, F. Changes of the Aroma Composition and Other Quality Traits of Blueberry 'Garden Blue' during the Cold Storage and Subsequent Shelf Life. *Foods* **2020**, *9*, 1223. [CrossRef]
10. Vollmannová, A.; Tóth, T.; Urminská, D.; Poláková, Z.; Timoracká, M.; Margitanová, E. Anthocyanins Content in Blueberries (*Vaccinium corymbosum* L.) in Relation to Freesing Duration. *Czech J. Food Sci.* **2009**, *27*, S204–S206. [CrossRef]
11. González, E.M.; de Ancos, B.; Cano, M.P. Relation between bioactive compounds and free radical-scavenging capacity in berry fruits during frozen storage. *J. Sci. Food Agric.* **2003**, *83*, 722–726. [CrossRef]
12. Jeong, J.Y.; Kim, G.D.; Yang, H.S.; Joo, S.T. Effect of freeze–thaw cycles on physicochemical properties and color stability of beef semimembranosus muscle. *Food Res. Int.* **2011**, *44*, 3222–3228. [CrossRef]
13. Kähkönen, M.P.; Heinonen, M. Antioxidant activity of anthocyanins and their aglycons. *J. Agric. Food Chem.* **2003**, *51*, 628–633. [CrossRef]
14. Chen, J.; Gu, J.; Zhang, R.; Mao, Y.; Tian, S. Freshness Evaluation of Three Kinds of Meats Based on the Electronic Nose. *Sensors* **2019**, *19*, 605. [CrossRef] [PubMed]

15. Xie, H.J.; Zhang, Y.T.; Cao, M.N.; Liu, C.Z.; Mao, Y.Z.; Ren, G.R.; Wu, Z.Y.; Fang, S.; Tian, S.Y.; Wu, D. Fabrication of PGFE/CN-stabilized β-carotene-loaded peppermint oil nanoemulsions: Storage stability, rheological behavior and intelligent sensory analyses. *LWT Food Sci. Technol.* **2021**, *138*, 110688. [CrossRef]
16. Xia, J.; Guo, Z.; Fang, S.; Gu, J.; Liang, X. Effect of Drying Methods on Volatile Compounds of Burdock (*Arctium lappa* L.) Root Tea as Revealed by Gas Chromatography Mass Spectrometry-Based Metabolomics. *Foods* **2021**, *10*, 868. [CrossRef] [PubMed]
17. Cheng, L.S.; Fang, S.; Ruan, M.L. Influence of Blanching Pretreatment on the Drying Characteristics of Cherry Tomato and Mathematical Modeling. *Int. J. Food Eng.* **2015**, *11*, 265–274. [CrossRef]
18. Doymaz, İ. Drying Kinetics and Rehydration Characteristics of Convective Hot-Air Dried White Button Mushroom Slices. *J. Chem.* **2014**, 453175. [CrossRef]
19. Mo, X.; Peng, X.; Liang, X.; Fang, S.; Xie, H.; Chen, J.; Meng, Y. Development of antifungal gelatin-based nanocomposite films functionalized with natamycin-loaded zein/casein nanoparticles. *Food Hydrocoll.* **2021**, *113*, 106506. [CrossRef]
20. Vallespir, F.; Rodríguez, Ó.; Eim, V.S.; Rosselló, C.; Simal, S. Effects of freezing treatments before convective drying on quality parameters, Vegetables with different microstructures. *J. Food Eng.* **2019**, *249*, 15–24. [CrossRef]
21. Celli, G.B.; Ghanem, A.; Brooks, S.L. Influence of freezing process and frozen storage on the quality of fruits and fruit products. *Food Rev. Int.* **2016**, *32*, 280–304. [CrossRef]
22. Magro, R.P.; Stroschoen, S.R.; Da, S.; André, J.; Hickmann, F.S.; Oliveira, R.A.D. Characterization of blueberry fruits (*vaccinium* spp.) and derived products. *Food Sci. Technol.* **2014**, *34*, 773–779.
23. Nadulski, R.; Grochowicz, J.Z.; Sobczak, P.; Kobus, Z.; Panasiewica, M. Application of Freezing and Thawing to Carrot (*Daucus carota* L.) Juice Extraction. *Food Bioprocess Technol.* **2015**, *8*, 218–227. [CrossRef]
24. Ferreira, L.F.; Minuzzi, N.M.; Rodrigues, R.F.; Pauletto, R.; Rodrigues, E.; Emanuelli, T.; Bochi, V.C. Citric acid water-based solution for blueberry bagasse anthocyanins recovery, Optimization and comparisons with microwave-assisted extraction (MAE). *LWT-Food Sci. Technol.* **2020**, *133*, 110064. [CrossRef]
25. Mu, C.; Yuan, Z.; Ouyang, X.; Sun, P.; Wang, B. Non-destructive detection of blueberry skin pigments and intrinsic fruit qualities based on deep learning. *J. Sci. Food Agric.* **2021**, *101*, 3165–3175. [CrossRef]
26. Lu, Y.; Liang, X.; Cheng, L.; Fang, S. Microencapsulation of Pigments by Directly Spray-Drying of Anthocyanins Extracts from Blueberry Pomace, Chemical Characterization and Extraction Modeling. *Int. J. Food Eng.* **2020**, *16*, 20190247. [CrossRef]
27. Kong, C.H.Z.; Hamid, N.; Ma, Q.; Lu, J.; Wang, B.-G.; Sarojini, V. Antifreeze peptide pretreatment minimizes freeze-thaw damage to cherries, An in-depth investigation. *LWT-Food Sci. Technol.* **2017**, *84*, 441–448. [CrossRef]
28. Abdel-Aal, E.S.M.; Hucl, P.; Rabalski, I. Compositional and antioxidant properties of anthocyanin-rich products prepared from purple wheat. *Food Chem.* **2018**, *254*, 13. [CrossRef]
29. Suthanthangjai, W.; Davies, K.; Phillips, A.; Ansell, J.; Kilmartin, P. Biotransformation of blueberry anthocyanins to bioavailable phenolic compounds by Lactobacillus. *Planta Med.* **2013**, *79*, PJ46. [CrossRef]
30. Adelina, N.M.; Wang, H.; Zhang, L.; Zhao, Y. Comparative analysis of volatile profiles in two grafted pine nuts by headspace-SPME/GC–MS and electronic nose as responses to different roasting conditions. *Food Res. Int.* **2021**, *140*, 110026. [CrossRef] [PubMed]
31. Qian, Y.L.; Zhang, D.; An, Y.; Zhou, Q.; Qian, M.C. Characterization of Aroma-Active Compounds in Northern Highbush Blueberries "Bluecrop" (Vaccinium corymbosum "Bluecrop") and "Elliott" (Vaccinium corymbosum "Elliott") by Gas Chromatography–Olfactometry Dilution Analysis and Odor Activity Value. *J. Agric. Food Chem.* **2021**, *69*, 5691–5701. [CrossRef] [PubMed]
32. Cheng, K.; Peng, B.; Yuan, F. Volatile composition of eight blueberry cultivars and their relationship with sensory attributes. *Flavour. Frag. J.* **2020**, *35*, 443–453. [CrossRef]
33. Czerny, M.; Christlbauer, M.; Christlbauer, M.; Fischer, A.; Granvogl, M.; Hammer, M.; Hartl, C.; Hernandez, N.M.; Schieberle, P. Re-investigation on odour thresholds of key food aroma compounds and development of an aroma language based on odour qualities of defined aqueous odorant solutions. *Eur. Food Res. Technol.* **2008**, *228*, 265–273. [CrossRef]
34. Elsharif, S.A.; Banerjee, A.; Buettner, A. Structure-odor relationships of linalool, linalyl acetate and their corresponding oxygenated derivatives. *Front. Chem.* **2015**, *3*, 1–10. [CrossRef] [PubMed]
35. Campestre, C.; Angelini, G.; Gasbarri, C.; Angerosa, F. The Compounds Responsible for the Sensory Profile in Monovarietal Virgin Olive Oils. *Molecules* **2017**, *22*, 1833. [CrossRef]
36. Noguerol-Pato, R.; González-Álvarez, M.; González-Barreiro, C.; Cancho-Grande, B.; Simal-Gándara, J. Evolution of the aromatic profile in Garnacha Tintorera grapes during raisining and comparison with that of the naturally sweet wine obtained. *Food Chem.* **2013**, *139*, 1052–1061. [CrossRef] [PubMed]
37. Nichols, W.C.; Patterson, M.E. Ethanol accumulation and poststorage quality of 'Delicious' apples during short-term, low-O$_2$, CA storage. *HortScience* **1987**, *22*, 89–92.
38. Pesis, E. The role of the anaerobic metabolites, acetaldehyde and ethanol, in fruit ripening, enhancement of fruit quality and fruit deterioration. *Postharvest Biol. Technol.* **2005**, *37*, 1–19. [CrossRef]
39. Wang, T.; Zhong, J.; Fang, S.; Chen, J.; Li, Y.; Meng, Y. Response surface optimization and kinetics model of osmotic dehydration blueberry in hot-air drying. *Food Sci. Technol.* **2016**, *41*, 52–57.
40. Xu, X.; Zhang, L.; Feng, Y.; Zhou, C.; Yagoub, A.E.A.; Wahia, H.; Ma, H.; Zhang, J.; Sun, Y. Ultrasound freeze-thawing style pretreatment to improve the efficiency of the vacuum freeze-drying of okra (*Abelmoschus esculentus* (L.) Moench) and the quality characteristics of the dried product. *Ultrason. Sonochem.* **2021**, *70*, 105300. [CrossRef]

41. Feng, Y.; Tan, C.P.; Zhou, C.; Yagoub, A.E.A.; Xu, B.; Sun, Y.; Ma, H.; Xu, X.; Yu, X. Effect of freeze-thaw cycles pretreatment on the vacuum freeze-drying process and physicochemical properties of the dried garlic slices. *Food Chem.* **2020**, *324*, 126883. [CrossRef] [PubMed]
42. Jayatunga, G.K.; Amarasinghe, B.M.W.P.K. Drying kinetics, quality and moisture diffusivity of spouted bed dried Sri Lankan black pepper. *J. Food Eng.* **2019**, *263*, 38–45. [CrossRef]

 *foods*

Article

Impact of Cold Storage on Bioactive Compounds and Their Stability of 36 Organically Grown Beetroot Genotypes

Khadijeh Yasaminshirazi [1,*], Jens Hartung [2], Michael Fleck [3] and Simone Graeff-Hönninger [1]

[1] Group of Cropping Systems and Modelling, Institute of Crop Science, University of Hohenheim, Fruwirthstr. 23, 70599 Stuttgart, Germany; simone.graeff@uni-hohenheim.de
[2] Department of Biostatistics, Institute of Crop Science, University of Hohenheim, Fruwirthstr. 23, 70599 Stuttgart, Germany; jens.hartung@uni-hohenheim.de
[3] Kultursaat e.V., Kronstraße 24, 61209 Echzell, Germany; michael.fleck@kultursaat.org
* Correspondence: khadijeh.yasaminshirazi@uni-hohenheim.de; Tel.: +49-711-459-24186

Abstract: In order to exploit the functional properties of fresh beetroot all year round, maintaining the health-benefiting compounds is the key factor. Thirty-six beetroot genotypes were evaluated regarding their content of total dry matter, total phenolic compounds, betalain, nitrate, and total soluble sugars directly after harvest and after cold storage periods of one and four months. Samples were collected from two field experiments, which were conducted under organic conditions in Southwestern Germany in 2017 and 2018. The outcome of this study revealed a significant influence of genotype ($p < 0.05$) on all measured compounds. Furthermore, significant impacts were shown for storage period on total dry matter content, nitrate, and total phenolic compounds. The medians of nitrate content based on the genotypes studied within the experiment ranged between 4179 ± 1267–$20,489 \pm 2988$ mg kg^{-1} DW (dry weight), and that for the total phenolic compounds varied between 201.45 ± 13.13 mg GAE 100 g^{-1} DW and 612.39 ± 40.58 mg GAE 100 g^{-1} DW (milligrams of gallic acid equivalents per 100 g of dry weight). According to the significant influence of the interactions of storage period and genotype on total soluble sugars and betalain, the decrease or increase in the content of the assessed compounds during the cold storage noted to be genotype-specific. Therefore, to benefit beetroots with retained quality for an extended time after harvest, selection of the suitable genotype based on the intended final use is recommended.

Keywords: beetroot; organic farming; storage; bioactive compounds; betalain; nitrate; sugar; phenolic compounds; total dry matter

Citation: Yasaminshirazi, K.; Hartung, J.; Fleck, M.; Graeff-Hönninger, S. Impact of Cold Storage on Bioactive Compounds and Their Stability of 36 Organically Grown Beetroot Genotypes. *Foods* **2021**, *10*, 1281. https://doi.org/10.3390/foods10061281

Academic Editors: Sidonia Martinez and Javier Carballo

Received: 11 May 2021
Accepted: 2 June 2021
Published: 4 June 2021

Publisher's Note: MDPI stays neutral with regard to jurisdictional claims in published maps and institutional affiliations.

1. Introduction

Containing a high amount of health-promoting compounds [1], suitable cultivation and storage [2] brought attention to the use of beetroot (*Beta vulgaris* subsp. *vulgaris* L.) and its products. Classification of beetroot as a super food has increased the importance of this vegetable [3]. Beetroot is included in the ten vegetables with the highest antioxidant capacity due to its promising amount of betalain and phenolic compounds [4] as well as the presence of carotenoid and ascorbic acid [5,6].

Betalains are nitrogen-containing, water-soluble plant secondary metabolites, which are derivatives of betalamic acid [7] and can be found in different parts of plants, but are restricted to the *Caryophyllales* order [8]. Red beetroot, as one of the chief and the most commercially sources [9] of betalain, has been approved by European Union to be used as a natural colorant (E162) [8] in dairy, confectionery, beverages, and meat products [10]. Based on the nature of the substituent of betalamic acid residue, betalains can be divided into two major groups: betacyanin which is responsible for red-violet color, and betaxanthin representing the yellow-orange color [11]. Investigating different beetroot genotypes indicated that the main betacyanins in beetroot are betanin, isobetanin, betanidin, and isobetanidin, and the chief betaxanthin components are vulgaxanthin I and vulgaxanthin

II [12]. In spite of the fact that betalains show low stability at higher temperatures [13], due to its high stability in the wide range of pH values (between three and seven [9]), betacyanins are considered to be a better choice providing a red-violet color range for coloring foods with a low acid content, compared to anthocyanins. Anthocyanins are the most common pigments for this color range [14], despite showing instability at pH values above three [15]. Indicating anti-lipidemic, anti-cancer, and antimicrobial activities made betalain a beneficial component for human health [16]. Moreover, flavonoids, phenolic acids, and various organic and inorganic acids, which are the major phenolic compounds of beetroot [5], further increase the anti-radical activity of this crop, which leads to preventing cancer and cardiovascular disease [17,18]. Total phenolics are part of fruits and vegetables' bioactive compounds, which not only benefit plants with assisting the growth and protective mechanisms under abiotic stress conditions [19], but also promote human health with influencing the functional qualities of plant products [20].

In addition, the potential profiting effects of nitrate in beetroot on human health have drawn a lot of attention. Green leafy vegetables including beetroot are considered as major dietary sources of nitrate [21], which is the chief contributor in nitric oxide production [19,22]. Nitric oxide demonstrated an essential role in the gastric [23] and cardiovascular [24] regulations. Latterly, preventing ischemia-reperfusion damages, regulating the blood pressure [25], enhancing muscle efficiency and endurance [1], were reported as the potential positive effects of dietary nitrate. Therefore, beetroot has been recently used as a dietary supplement for patients with hypertension or cardiovascular diseases [23] or as powder formulation in different products, such as yogurt, drinks, and snacks, for consumption by athletes before physical exercises [5]. Moreover, the contribution of nitric oxide in the improvement of seed germination, growth performance, and mineral adjustment in plants under different abiotic stress conditions has been reported [26–28]. Furthermore, the sugar composition of beetroot was reported as a dominant proportion of sucrose (91.6%) [29], with a small and relatively similar proportion of glucose and fructose [5]. Information on the amount and composition of carbohydrates in vegetables is essential when considering the importance of sugar content in different controlled diets (such as for diabetic patients, athletes, and vegetarians) as well as in the food industry for optimization of processing practices.

Due to the rise in consumers' awareness of the advantages of organic products [30], the demand for such products is steadily growing [29]. Comparing the quality of organic and conventional cultivation, previous studies reported higher contents of total phenolic compounds, betalain, and antioxidant capacities in organically grown beetroots. However, the extent of difference between the cultivation methods highly depends on the genotype [31]. Considering the functional characteristics of beetroot, the stability of its heath beneficial compounds plays an important role [32]. In order to have fresh beetroot throughout the year and be able to use them for the processed products with promising quality, preserving the nutritional properties of beetroot during storage is crucial. Previous studies reported the impact of genotypes, fertilization, and the storage environment on the storability of beetroot [2]. Nevertheless, an evaluation of various biologically active compounds considering a broad range of beetroot genotypes, which were cultivated organically was missing. Cold or refrigerated storage is one of the prevalent postharvest practices, which is applied for extending the shelf-life of vegetables. Due to their perishable nature, vegetables and fruits often need to be stored at low temperatures to minimize their physiological and chemical changes [33], and to extend their marketing after the harvest season [34]. Different studies claimed that advanced storage technologies, such as a controlled or modified atmosphere, are ideal options to preserve bioactive compounds in vegetables [2,35]. However, due to the high prices of such technologies, small-scale producers cannot afford them and cheap and locally-available technologies are demanded [36]. Therefore, using the genetic potential of forgotten varieties or breeding new and promising genotypes with high storability and evaluating their performances under organic farming conditions could be one of the most reasonable solutions.

The present study aimed to determine the impact of genotype and cold storage period on the stability of different bioactive compounds of 36 beetroot genotypes, grown under organic farming conditions in Southwestern Germany. The outcomes of this study can be beneficial for household consumers, who tend to favor fresh beetroot for extended time rather than the processed ones, for farmers to profit their grown beetroots with maintained quality and less storage loss, and for food industries accessing beetroot with retained health-promoting compounds.

2. Materials and Methods

2.1. Chemicals and Reagents

For quantification of nitrate, sulfanilamide (AppliChem GmbH, Darmstadt, Germany), ammonium chloride and hydrochloric (Th. Geyer, Renningen, Germany), sodium nitrite and ammonia solution 25% (Merck, Darmstadt, Germany), and N-(1-naphthyl)-ethylene diamine dihydrochloride (Carl Roth GmbH, Karlsruhe, Germany) were used. Regarding total phenolic content measurement, Folin–Ciocalteu reagent and gallic acid were provided by Merck (Darmstadt, Germany). Na_2CO_3 and methanol were purchased from AppliChem GmbH (Darmstadt, Germany) and Carl Roth GmbH (Karlsruhe, Germany), respectively. Ethanol needed for betalain analysis was purchased from Th. Geyer (Renningen, Germany).

2.2. Plant Materials and Sample Preparation

The beetroots analyzed in the present study were grown under organic conditions at the research station for organic farming Kleinhohenheim, University of Hohenheim, Stuttgart, Baden-Wuerttemberg, Germany (48°44'14 N, 9°12'01 E, 430 m above the sea level). Two field experiments were conducted in which, in 2017, 40 genotypes, and in 2018, 36 genotypes were cultivated. In 2017, the field experiment was conducted as row-column design with three replicates. In 2018, the experiment was carried out as non-resolvable block design, in which a block size of ten and a treatment number of 36, with six replicates for four genotypes and three replicates for 32 genotypes were applied. Although the data from all genotypes were statistically analyzed together, the results shown in the present study were limited to the 36 genotypes, which occurred in both years. Table 1 presents the detailed information on beet color, beet shape, and seed origin of the studied beetroot genotypes.

During the growth period in 2017, the mean precipitation and temperature were 77.42 mm and 17.6 °C, respectively. In 2018, the mean precipitation reached 38.2 mm and the mean temperature was 19.0 °C during the growth period. Detailed information on monthly precipitation and mean temperature, fertilization, sowing and harvest dates, and soil management practices can be found in Yasaminshirazi et al. [37].

Three randomly selected beetroots per plot were collected each year for analysis of the bioactive compounds in freshly harvested beetroots. Additionally, beetroots from each plot were stored in vegetable net sacks in a cooling chamber at 6 °C, directly after harvest. All stored beetroots met the marketability criteria (including beet diameter of 5–13 cm, no deformation or remarkable damages, diseases, etc.). Beetroots for analysis of bioactive compounds were taken from the cooling chamber one and four months after the storage. After washing and cutting the leaves-growth-base and root tail, a sectional cut of each beet (flesh including peel) was diced and mixed in order to have a homogenous sample from each plot. After collecting the diced beetroots in a plastic flask, to prevent any further enzymatic processes, samples were immediately frozen by liquid nitrogen, kept at −18 °C, and then followed by lyophilization using the Dieter Piatkowski–Forschungsgeraete freeze-dryer (Munich, Germany). The dried samples were milled using GRINDOMIX GM 200 (Retsch GmbH, Haan, Germany) up until a fine powder texture was reached. Until analysis, the powdered samples were stored in closed plastic bottles in a dark and dry box at ambient temperature. Total dry matter content (TDMC), total phenolic compounds, betalain, nitrate, and total soluble sugars of beetroots of freshly harvested samples were determined and compared with those of samples taken after one and four months of cold storage.

Table 1. List of the 36 investigated beetroot genotypes indicating the beet color, shape, and seed origin.

Genotype	Beet Color	Shape	Seed Origin
Ägyptische Plattrunde (Ä. P.)	red	flat-spherical	Sativa (DE)
Akela Rijk Zwaan (RZ)	red	spherical	Rijk Zwaan (NL)
Alvro Mono	red	spherical	Vitalis (US)
Betina	red	spherical	Moravo Seeds (CZ)
Bolivar	red	spherical	Hild (DE)
Bona	red	spherical	Moravo Seeds (CZ)
Bordo	red	spherical	Seklos (LT)
Boro F1	red	spherical	Bejo (DE)
BoRu1	red	spherical	Kultursaat e.V. (DE)
Borus	red	spherical	Spójnia (PL)
Burpees Golden (Burpees G.)	yellow	spherical	Bingenheimer S. AG (DE)
Carillon RZ	red	cylindrical	Rijk Zwaan (NL)
Cervena Kulata (Cervena K.)	red	spherical	Moravo Seeds (CZ)
Ceryl	red	spherical	Spójnia (PL)
Chrobry	red	spherical	Spójnia (PL)
Czerwona Kula 2 (Czerwona K. 2)	red	spherical	Spójnia (PL)
Detroit 2 Dark Red (Detroit 2 D. R.)	red	spherical	Samen Schenker (DE)
Detroit 3	red	spherical	Caillard (FR)
Detroit Globe (Detroit G.)	red	spherical	King Seed (UK)
Formanova	red	cylindrical	Sativa (DE)
Forono	red	cylindrical	Bingenheimer S. AG (DE)
Gesche SG	red	spherical	Christiansens Biolandhof (DE)
Jannis	red	spherical	Bingenheimer S. AG (DE)
Jawor	red	spherical	Snówidza (PL)
Libero RZ	red	spherical	Rijk Zwaan (NL)
Monty RZ F1	red	spherical	Rijk Zwaan (NL)
Nobol	red	spherical	Vilmorin (PL)
Nochowski	red	spherical	Spójnia (PL)
Pablo F1	red	spherical	Bejo (DE)
Regulski Okragly (Regulski O.)	red	spherical	Pnos (PL)
Robuschka	red	spherical	Bingenheimer S. AG (DE)
Ronjana	red	spherical	Bingenheimer S. AG (DE)
Hilmar	red	spherical	Hild (DE)
Sniezna Kula	white	spherical	Torseed (PL)
Tondo de Chioggia (Tondo d. Ch.)	red-white	spherical	Bingenheimer S. AG (DE)
UB-E3	red	spherical	U.Behrendt (DE)

2.3. Total Dry Matter Content

Before and after freeze-drying, the diced beetroot samples (i) were weighed. Equation (1) was used to calculate the TDMC:

$$\text{TDMC}_i \, [\%] = \left(\frac{\text{weight after drying}_i}{\text{weight before drying}_i} \right) \times 100, \tag{1}$$

2.4. Total Phenolic Content (TPC)

The TPC quantification was conducted according to the methodology of Folin–Ciocalteu [38].

Briefly, the extraction was prepared by mixing 10 mL of methanol with approximately 0.5 g dried beetroot sample in a falcon tube. After shaking the mixture for 30 min, the tubes were placed in a centrifuge (Centrifuge 5810 R, Eppendorf AG, Hamburg, Germany) at 4000 rounds per minute (rpm) for 20 min (20 °C) for separation of supernatant from the solid phase. Later, 0.6 mL of the prepared extract was mixed with 60 mL of distilled water and 5 mL of Folin–Ciocalteu's reagent in a 100 mL volumetric flask. After two to six minutes, with adding 25 mL of sodium carbonate (15%) and adjusting the final volume with distilled water to 100 mL, the mixture was left for two hours at room temperature. The absorbance at 760 nm was measured spectrophotometrically (Ultrospec 3100 Pro, Amersham Bioscience, Buckinghamshire, UK) and TPC was reported as mg GAE 100 g^{-1} DW. In order to draw a standard curve, six different concentrations of gallic acid solution (0.03–1.5 g L^{-1} gallic acid in distilled water) were used.

2.5. Betalain Content

Determination of two chief subgroups of betalains, namely betacyanin and betaxanthin, was conducted spectrophotometrically (Ultrospec 3100 Pro, Amersham Bioscience, Buckinghamshire, UK) in accordance with the method used by Koubaier et al. [39] and Sawicki et al. [40].

A mixture of roughly 0.04 g of the dried beetroot samples and 30 mL of 50% (*v/v*) ethanol was shaken for two hours with the speed of 100 rpm and followed by centrifuging the samples at 20 °C for 10 min with the speed of 400 rpm (Centrifuge 5810 R, Eppendorf AG, Hamburg, Germany). The absorption of betaxanthin and betacyanin was measured at 480 nm and 538 nm, respectively.

2.6. Nitrate Content Determination

The nitrate content was determined according to the flow injection analysis method (FIA) [21] using FIASTAR 5000 (FOSS Analytical AB, Hilleroed, Denmark). The detailed extract preparation can be found in Yasaminshirazi et al. [41].

2.7. Total Soluble Sugar Content

The degree of Brix corresponding to the percentage of total soluble sugar content was measured utilizing a digital handheld refractometer (Kruess, Hamburg, Germany). After measuring each sample in duplicate, their mean value was calculated directly.

2.8. Statistical Analysis

Data from both years and experiments were jointly analyzed using a mixed model approach. The model can be described by:

$$y_{ijklmn} = \mu + \tau_i + \varphi_j + a_k + (\tau\varphi)_{ij} + (\tau a)_{ik} + (\varphi a)_{jk} + (\tau\varphi a)_{ijk} + b_{kl} + r_{klm} + c_{kln} + e_{ijklmn}, \qquad (2)$$

where y_{ijklmn} is the observation of genotype i after storage period j in the m row, nth column of block l in year k, μ is the intercept, τ_i is the fixed effect of genotype i, φ_j is the fixed effect of the jth storage period, a_k is the fixed effect of the kth year, $(\tau\varphi)_{ij}$, $(\tau a)_{ik}$, and $(\tau\varphi a)_{ijk}$ are the random interaction effects of the corresponding main effects, $(\varphi a)_{jk}$ is the fixed effect of storage period j and year k, and b_{kl}, r_{klm}, and c_{kln} are the random effects of block, row and column within block. e_{ijklmn} is the error of y_{ijklmn}. Note that the effect $(\varphi a)_{jk}$ is the confounded with the effect of sampling day and the interaction effect of year k and storage period j, as thus was taken as fixed in the model. Due to this confounding, the interpretation of common or marginal means should be made with caution, too. Further note that data of the same sample (beetroots form the same plot) were taken for different storage periods within one experiment. Thus, repeated data were taken on each sample. The model accounted for this repeated measures structure by allowing a first order autoregressive variance-covariance structure for random effects including the year and the residual error. Year specific variances and covariances were fitted to all random effects and the error except for year-by-genotype effects. As row and columns existed only in the first year, effects for both were fitted only for first year data. Pre-requirements of normally distributed residuals and homogeneous variance (despite the year specific variances) were checked graphically. In case of nitrate content, data were square-root transformed prior to analysis. For total phenolic compounds, betacyanin, and betaxanthin, data were logarithmically transformed prior to analysis. In both cases, means were back-transformed for presentation purpose only. Back-transformed values were denoted as medians. Standard errors were back-transformed using the delta method. In case of finding significant differences, Tukey test was used for multiple comparisons. A letter display was used to present their results [42].

Additionally, genotype-by-storage period means were calculated for all six traits. These simple means were standardized to have a mean of zero and a variance of one. A Principal Component Analysis (PCA) was applied on these standardized means. The two first components were presented via biplot using the default setting of the %biplot macro for SAS (factype = SYM). Thus, scaling of score and loading plot was done with $US^{\frac{1}{2}}$ and $VS^{\frac{1}{2}}$, respectively, where USV' is the single value decomposition of the two-dimensional approximation of the data matrix.

3. Results and Discussion

3.1. Total Dry Matter Content

The outcome of analysis of variance (ANOVA) demonstrated a significant effect of genotype on the content of total dry matter ($p < 0.0001$) (Table 2). This is in line with the findings of Kosson et al. [43] who claimed a significant influence of genotype on TDMC. Furthermore, it was revealed that the TDMC changed significantly during the cold storage ($p = 0.0005$). Likewise, significant influence of the interactions of storage period and year on the TDMC was noted (Table 2). Hagen et al. [44] stated no significant influence of cold storage for six weeks on the percentage of total dry matter in curly kale. Nonetheless, a strong influence of the storage on transpiration and consequently on the weight loss of the stored beetroot has been reported [45]. In this regard, Gawęda [45] assessed the difference of two storage methods on two beetroot cultivars and reported not only a significant difference between the two cultivars, but also a two-times-decline in dry weight of beetroots stored in polyethylene film bags than those in traditional plastic boxes, after 6 months of cold storage.

Based on the significant impacts of genotype and the interactions of storage period and year, Table 3 presents the means of TDMC of the investigated genotypes within the trial (A) and based on three storage periods in each year (B). The TDMC of the tested genotypes ranged between $11.49 \pm 0.55\%$ and $18.15 \pm 0.55\%$. The highest contents of total dry matter were noted in the genotypes Nochowski ($18.15 \pm 0.55\%$), Chrobry ($18.14 \pm 0.55\%$), and Betina ($16.94 \pm 0.56\%$) and the lowest in the genotypes Alvro Mono ($11.49 \pm 0.55\%$), Libero RZ ($12.31 \pm 0.56\%$), and Tondo d. Ch. ($13.37 \pm 0.58\%$) (Table 3).

Considering the means of different storage periods in the first year, the TDMC directly after the harvest was $15.19 \pm 0.14\%$. After one month of cold storage, the TDMC increased to $15.63 \pm 0.14\%$. After a storage period of four months, the TDMC significantly decreased compared to the samples taken one month after the harvest and reached $14.95 \pm 0.14\%$, which was not significantly different from the mean of freshly harvested beetroots. Therefore, the overall change in the TDMC was minor. In the second year, the TDMC directly after the harvest was $14.51 \pm 0.21\%$. After one month of cold storage, the content rose significantly to $15.34 \pm 0.21\%$. Afterwards, up to a storage period of four months, the change was not significant (Table 3). A corresponding outcome was noted by Jakopic et al. [46], who reported a higher TDMC in rutabaga turnips after a cold storage of four months. In contrast, a slight decrease in the TDMC of ten organically grown onion genotypes after five months of cold storage was noted [47]. Regarding the influence of the year, the amount of total dry matter did not differ significantly between the values of the samples from the freshly harvested beetroots in each year. Likewise, the TDMC contents after cold storage periods of one and four months did not differ significantly between both years. According to the significant increase in the TDMC within the first one month of cold storage in both years, it can be concluded that the highest amount of water loss could occur at the first four weeks of cold storage.

Table 2. ANOVA of results of total dry matter, total phenolic compounds, betaxanthin, betacyanin, nitrate, and total soluble sugars of 36 beetroot genotypes grown in research station Kleinhohenheim within the years 2017 and 2018 for three storage periods (storage period expressed as cold storage durations of zero (directly after harvest), one, and four months).

Effect	Total Dry Matter Content	Total Phenolic Compounds	Betaxanthin	Betacyanin	Nitrate	Total Soluble Sugars
Genotype	<0.0001	<0.0001	<0.0001	<0.0001	0.0005	<0.0001
Storage period	0.0005	<0.0001	<0.0001	<0.0001	<0.0001	n.s. [1]
Genotype × Storage period	n.s.	n.s.	<0.0001	<0.0001	n.s.	0.0121
Year	n.s.	<0.0001	0.0001	<0.0001	n.s.	0.0008
Storage period × Year	0.0009	0.0001	n.s.	n.s.	n.s.	<0.0001

[1] not significant.

3.2. Total Phenolic Content

In accordance with the results of ANOVA, genotypes can significantly impact the amount of total phenolic compounds ($p < 0.0001$). This is in agreement with Lattanzio et al. [48], who stated the key effect of genotype on the content of phenolic contents in fresh fruit and vegetables. Furthermore, a significant influence of the storage period ($p < 0.0001$), year ($p < 0.0001$), and the interactions between storage period and year were noted (Table 2).

According to the significant impacts of genotype and the interactions of storage period and year, Table 3 exhibits the medians of TPC of the examined genotypes within the trial (A) and based on three storage periods in each year (B). The TPC in investigated red-colored beetroot genotypes varied from 322.93 ± 21.04 mg GAE 100 g^{-1} DW to 612.39 ± 40.58 mg GAE 100 g^{-1} DW measured in genotypes Robuschka and Alvro Mono, respectively. Following Alvro Mono, the cylindrical-shaped genotype Forono with 561.74 ± 37.37 mg GAE 100 g^{-1} DW and Monty RZ F1 with 519.91 ± 33.89 mg GAE 100 g^{-1} DW indicated the highest TPC. Taking all the genotypes into account, the lowest TPC possessed by the yellow-colored genotype Burpees G. (201.45 ± 13.13 mg GAE 100 g^{-1} DW), the red-white-colored Tondo d. Ch. (241.16 ± 20.35 mg GAE 100 g^{-1} DW), and white-colored Sniezna Kula (242.55 ± 16.07 mg GAE 100 g^{-1} DW). Based on the median values of different storage periods in the first year, the TPC directly after the harvest was 294.55 ± 8.57 mg GAE 100 g^{-1} DW. After one month of cold storage, a significant increase up to 341.18 ± 9.93 mg GAE 100 g^{-1} DW was noted.

Table 3. Mean and median values of total dry matter content (%) and total phenolic content (mg GAE 100 g^{-1} DW) of 36 beetroot genotypes grown at the research station Kleinhohenheim within the years 2017 and 2018. Results represent the mean (median) values ± (approximate) standard error. In section (A), means (medians) followed by at least one identical lower-case letter in one column did not differ significantly between genotypes at experiment-wise Type 1 error α = 0.05. In section (B), means (medians) followed by at least one identical lower-case letter in one column did not differ significantly within different storage periods at experiment-wise Type 1 error α = 0.05 and means (medians) followed by at least one identical upper-case letter in one column did not differ significantly within year at experiment-wise Type 1 error α = 0.05.

(A) Means (Medians) Based on Genotype		
Genotype	**TDMC (%)**	**Total Phenolic Content (mg GAE 100 g^{-1} DW)**
Ä. P.	14.77 [bcde] ± 0.55	417.20 [ac] ± 27.18
Akela RZ	14.81 [bcde] ± 0.55	413.30 [ac] ± 26.94
Alvro Mono	11.49 [e] ± 0.55	612.39 [a] ± 40.58
Betina	16.94 [ac] ± 0.56	413.71 [ac] ± 27.41
Bolivar	14.85 [ad] ± 0.55	364.91 [cdef] ± 23.79
Bona	14.78 [bcde] ± 0.55	405.26 [ac] ± 26.41
Bordo	16.79 [ac] ± 0.55	449.00 [ac] ± 29.26
Boro F1	13.84 [bcde] ± 0.55	414.61 [ac] ± 27.03
BoRu1	14.85 [ad] ± 0.55	417.89 [ac] ± 27.25
Borus	16.39 [ab] ± 0.56	376.35 [bcde] ± 25.94
Burpees G.	14.87 [ad] ± 0.55	201.45 [i] ± 13.13
Carillon RZ	13.81 [bcde] ± 0.55	432.35 [ac] ± 28.19
Cervena K.	16.15 [ab] ± 0.56	465.96 [ac] ± 31.57
Ceryl	16.25 [ab] ± 0.55	426.45 [ac] ± 27.80
Chrobry	18.14 [a] ± 0.55	449.16 [ac] ± 29.26
Czerwona K. 2	15.92 [ab] ± 0.55	355.25 [cdefg] ± 24.04
Detroit 2 D. R.	15.64 [ad] ± 0.56	418.26 [ac] ± 28.83
Detroit 3	14.30 [bcde] ± 0.55	403.59 [ac] ± 26.73
Detroit G.	14.89 [ad] ± 0.56	350.31 [cdefg] ± 24.15
Formanova	14.46 [bcde] ± 0.55	397.19 [bcd] ± 26.31
Forono	15.21 [ad] ± 0.55	561.74 [ab] ± 37.37
Gesche SG	15.25 [ad] ± 0.55	484.20 [ac] ± 31.56
Jannis	14.68 [bcde] ± 0.55	414.71 [ac] ± 27.02
Jawor	15.17 [ad] ± 0.55	430.75 [ac] ± 28.07
Libero RZ	12.31 [de] ± 0.56	408.49 [ac] ± 29.38
Monty RZ F1	15.65 [ad] ± 0.55	519.91 [ad] ± 33.89
Nobol	14.20 [bcde] ± 0.55	456.04 [ac] ± 29.74
Nochowski	18.15 [a] ± 0.55	486.33 [ac] ± 31.69
Pablo F1	14.58 [bcde] ± 0.55	440.53 [ac] ± 29.33
Regulski O.	15.89 [ab] ± 0.55	381.10 [bcde] ± 24.84
Robuschka	15.57 [ad] ± 0.55	322.93 [cefgh] ± 21.04
Ronjana	15.39 [ad] ± 0.55	508.49 [ad] ± 33.16
Hilmar	15.01 [ad] ± 0.55	420.64 [ac] ± 27.41
Sniezna Kula	15.65 [ad] ± 0.55	242.55 [fi] ± 16.07
Tondo d. Ch.	13.37 [bde] ± 0.58	241.16 [ei] ± 20.35
UB-E3	16.06 [ab] ± 0.55	393.07 [bcd] ± 26.04

(B) Means (Medians) Based on Storage Period			
	Storage Period		
Year 1	Directly after harvest	15.19 [abA] ± 0.14	294.55 [cB] ± 8.57
	1 month after cold storage	15.63 [aA] ± 0.14	341.18 [bA] ± 9.93
	4 months after cold storage	14.95 [bA] ± 0.14	437.83 [aA] ± 12.78
Year 2	Directly after harvest	14.51 [bA] ± 0.21	395.91 [aA] ± 12.59
	1 month after cold storage	15.34 [aA] ± 0.21	379.09 [aA] ± 12.17
	4 months after cold storage	15.46 [aA] ± 0.22	425.18 [aA] ± 14.47

Likewise, after a storage period of four months the TPC further increased and reached 437.83 ± 12.78 mg GAE 100 g^{-1} DW (Table 3). In the second year, the TPC directly after the harvest was 395.91 ± 12.59 mg GAE 100 g^{-1} DW. After one month of cold storage, it slightly decreased, however, the change was not significant. Afterwards, up to a storage period of four months, an increasing trend in the TPC was noticed (Table 3). The median values of TPC after a cold storage period of four months in both years were higher than those at harvest time, which indicated the potential of the investigated genotypes in providing a promising content of phenolic compounds for an extended time after the harvest. This is in agreement with the study of Jakopic et al. [46], who reported an increase of TPC of rutabaga root after four months of cold storage. Regarding the influence of the year, the amount of total phenolic compounds differed significantly between the values of the samples from the freshly harvested beetroots in each year. Nevertheless, the TPC after cold storage periods of one and four months did not differ significantly between both years.

Evaluation of the stability of TPC of red beetroot (var. Little Ball) during the storage in 5 °C for 196 days indicated a slight decrease in the first 63 days of storage and afterwards the change was minor [49]. In line with the results of the present study, in which both decreasing and increasing trends in the TPC during the storage period was noted, both decreases (in broccoli [50] and pomegranate [51]) and increases (in pigmented potato tuber [52]) in TPC have been reported in the previous studies. This may result from the differences between the impact's extent of cold storage on individual constituents of phenolic compounds [49].

Corleto et al. [32] investigated the stability of TPC in beetroot juice, which were stored for 32 days at four different temperatures and significant differences were noted during the storage at refrigeration temperature (4 °C). Nevertheless, it was revealed that under refrigeration and freezing conditions, antioxidant activity and TPC remain more stable in comparison to storage at room temperature. Assessing the impact of temperature and period of storage on beetroot snack bars indicated that TPC decreased constantly during six months storage at all studied temperatures (6 °C, 22–32 °C, and 37 °C). However, the TPC loss at 6 °C was less than at higher temperatures [53]. High temperature is reported as the main factor causing the reduction of TPC in vegetables due to change in the phenolic profiles [54].

The cylindrical-shaped genotype, Forono, was noted to be among the genotypes with the highest TPC (Table 3), betacyanin, and betaxanthin (Table 4) after the cold storage, which can be correlated to the high antioxidant activity of this genotype. Furthermore, another cylindrical-shaped beetroot, Carillon RZ, was noted as a genotype with an average TPC and betalain content. Additionally, this genotype indicated the highest total and marketable yield as well as high resistance against the common beet disease, scab, among 15 investigated beetroot genotypes in our previous study [37] which further reveals its latent promising characteristics. In contradiction to Forono and Carillon RZ, the other studied cylindrical-shaped beetroot, Formanova, was among the red-colored genotypes with the lowest betalain content and an average TPC. This may explain the impact of genotype rather than the shape on the content of the discussed compounds.

Table 4. Median values of betaxanthin (mg g^{-1} DW) and betacyanin (mg g^{-1} DW) of 36 beetroot genotypes grown at the research station Kleinhohenheim within the years 2017 and 2018, directly after harvest, and after the cold storage periods of one and four months. Results represent the median values $\pm$ approximate standard error. Medians followed by at least one identical lower-case letter in one column did not differ significantly between genotypes at experiment-wise Type 1 error α = 0.05. Medians followed by at least one identical upper-case letter in one row did not differ significantly between storage periods at experiment-wise Type 1 error α = 0.05.

Genotype	Betaxanthin (mg g^{-1} DW)			Betacyanin (mg g^{-1} DW)		
	Directly after Harvest	1 Month after Cold Storage	4 Months after Cold Storage	Directly after Harvest	1 Month after Cold Storage	4 Months after Cold Storage
Ä. P.	4.41 [aA] $\pm$ 0.84	4.62 [abA] $\pm$ 0.88	4.40 [abA] $\pm$ 0.84	6.12 [abcA] $\pm$ 1.10	6.17 [aA] $\pm$ 1.11	5.42 [aA] $\pm$ 0.97
Akela RZ	4.44 [aA] $\pm$ 0.85	5.70 [aA] $\pm$ 1.09	5.04 [abA] $\pm$ 0.96	6.65 [abA] $\pm$ 1.19	8.02 [aA] $\pm$ 1.44	6.77 [aA] $\pm$ 1.22
Alvro Mono	2.69 [aA] $\pm$ 0.51	3.08 [adA] $\pm$ 0.59	3.05 [bcdA] $\pm$ 0.60	3.61 [abcdA] $\pm$ 0.65	4.13 [aA] $\pm$ 0.74	4.28 [aA] $\pm$ 0.79
Betina	4.31 [aA] $\pm$ 0.82	4.73 [abA] $\pm$ 0.90	5.13 [abA] $\pm$ 1.00	5.47 [abcdA] $\pm$ 0.98	7.20 [aA] $\pm$ 1.29	6.83 [aA] $\pm$ 1.26
Bolivar	4.65 [aA] $\pm$ 0.89	4.55 [abA] $\pm$ 0.87	5.28 [abA] $\pm$ 1.01	6.30 [abcA] $\pm$ 1.13	5.93 [aA] $\pm$ 1.07	6.83 [aA] $\pm$ 1.23
Bona	4.07 [aA] $\pm$ 0.78	4.76 [abA] $\pm$ 0.91	3.71 [beA] $\pm$ 0.71	5.79 [abcA] $\pm$ 1.04	6.39 [aA] $\pm$ 1.15	4.94 [aA] $\pm$ 0.89
Bordo	5.03 [aA] $\pm$ 0.96	4.48 [abA] $\pm$ 0.86	5.23 [abA] $\pm$ 1.00	6.99 [abA] $\pm$ 1.25	7.05 [aA] $\pm$ 1.27	7.28 [aA] $\pm$ 1.31
Boro F1	4.75 [aA] $\pm$ 0.91	4.88 [aA] $\pm$ 0.93	4.58 [abA] $\pm$ 0.87	6.50 [abA] $\pm$ 1.17	6.17 [aA] $\pm$ 1.11	5.74 [aA] $\pm$ 1.03
BoRu1	5.38 [aA] $\pm$ 1.03	5.88 [aA] $\pm$ 1.12	4.99 [abA] $\pm$ 0.95	7.23 [aA] $\pm$ 1.30	7.56 [aA] $\pm$ 1.36	6.34 [aA] $\pm$ 1.14
Borus	4.10 [aA] $\pm$ 0.78	4.63 [abA] $\pm$ 0.88	4.28 [bcA] $\pm$ 0.88	5.75 [abcA] $\pm$ 1.03	6.40 [aA] $\pm$ 1.15	5.63 [aA] $\pm$ 1.10
Burpees G.	0.20 [bB] $\pm$ 0.04	1.15 [dA] $\pm$ 0.22	1.03 [deA] $\pm$ 0.20	0.43 [egA] $\pm$ 0.08	0.09 [bB] $\pm$ 0.02	0.11 [bB] $\pm$ 0.02
Carillon RZ	4.83 [aA] $\pm$ 0.92	5.17 [aA] $\pm$ 0.99	4.72 [abA] $\pm$ 0.90	6.65 [abA] $\pm$ 1.20	7.47 [aA] $\pm$ 1.34	6.51 [aA] $\pm$ 1.17
Cervena K.	4.11 [aA] $\pm$ 0.79	4.49 [abA] $\pm$ 0.86	4.54 [abA] $\pm$ 0.90	5.75 [abcA] $\pm$ 1.03	6.30 [aA] $\pm$ 1.13	5.93 [aA] $\pm$ 1.11
Ceryl	5.57 [aA] $\pm$ 1.06	6.40 [aA] $\pm$ 1.22	5.48 [abA] $\pm$ 1.05	7.91 [aA] $\pm$ 1.42	8.55 [aA] $\pm$ 1.54	7.22 [aA] $\pm$ 1.30
Chrobry	5.93 [aA] $\pm$ 1.13	4.58 [abA] $\pm$ 0.88	4.71 [abA] $\pm$ 0.90	7.84 [aA] $\pm$ 1.41	7.69 [aA] $\pm$ 1.38	7.73 [aA] $\pm$ 1.39
Czerwona K. 2	4.51 [aA] $\pm$ 0.88	4.41 [abcA] $\pm$ 0.84	3.91 [bcA] $\pm$ 0.76	6.67 [abA] $\pm$ 1.22	6.17 [aA] $\pm$ 1.11	5.48 [aA] $\pm$ 1.01
Detroit 2 D. R.	4.67 [aA] $\pm$ 0.89	4.40 [abcA] $\pm$ 0.84	4.63 [abA] $\pm$ 0.95	6.65 [abA] $\pm$ 1.19	6.22 [aA] $\pm$ 1.12	6.20 [aA] $\pm$ 1.21
Detroit 3	5.46 [aA] $\pm$ 1.04	4.58 [abA] $\pm$ 0.87	5.41 [abA] $\pm$ 1.05	7.73 [aA] $\pm$ 1.39	6.15 [aA] $\pm$ 1.10	6.72 [aA] $\pm$ 1.24
Detroit G.	4.67 [aA] $\pm$ 0.89	4.08 [adA] $\pm$ 0.78	4.45 [bcA] $\pm$ 0.90	7.02 [abA] $\pm$ 1.26	5.59 [aA] $\pm$ 1.00	5.63 [aA] $\pm$ 1.10
Formanova	3.63 [aA] $\pm$ 0.69	3.54 [adA] $\pm$ 0.69	3.28 [bcdA] $\pm$ 0.63	5.01 [abcdA] $\pm$ 0.90	5.08 [aA] $\pm$ 0.94	5.00 [aA] $\pm$ 0.90
Forono	5.76 [aA] $\pm$ 1.10	6.39 [aA] $\pm$ 1.25	6.70 [bA] $\pm$ 1.28	8.39 [aA] $\pm$ 1.51	7.99 [aA] $\pm$ 1.46	8.97 [aA] $\pm$ 1.61
Gesche SG	6.60 [aA] $\pm$ 1.26	5.74 [aA] $\pm$ 1.10	5.38 [abA] $\pm$ 1.03	7.79 [aA] $\pm$ 1.40	7.73 [aA] $\pm$ 1.42	7.11 [aA] $\pm$ 1.28
Jannis	4.17 [aA] $\pm$ 0.80	4.81 [abA] $\pm$ 0.92	4.90 [abA] $\pm$ 0.94	5.85 [abcA] $\pm$ 1.05	6.49 [aA] $\pm$ 1.16	6.17 [aA] $\pm$ 1.11
Jawor	4.24 [aA] $\pm$ 0.81	4.23 [adA] $\pm$ 0.81	3.85 [bcA] $\pm$ 0.74	5.86 [abcA] $\pm$ 1.05	6.18 [aA] $\pm$ 1.11	5.84 [aA] $\pm$ 1.05
Libero RZ	6.25 [aA] $\pm$ 1.19	6.73 [aA] $\pm$ 1.38	5.24 [abA] $\pm$ 1.07	8.63 [aA] $\pm$ 1.55	8.70 [aA] $\pm$ 1.70	6.51 [aA] $\pm$ 1.27
Monty RZ F1	5.87 [aA] $\pm$ 1.12	5.40 [aA] $\pm$ 1.03	5.29 [abA] $\pm$ 1.01	8.70 [aA] $\pm$ 1.56	8.20 [aA] $\pm$ 1.47	7.32 [aA] $\pm$ 1.32
Nobol	5.72 [aA] $\pm$ 1.09	5.38 [aA] $\pm$ 1.03	5.83 [abA] $\pm$ 1.11	8.18 [aA] $\pm$ 1.47	7.64 [aA] $\pm$ 1.37	8.16 [aA] $\pm$ 1.47
Nochowski	6.08 [aA] $\pm$ 1.16	5.05 [aA] $\pm$ 0.97	5.95 [abA] $\pm$ 1.14	7.77 [aA] $\pm$ 1.39	7.57 [aA] $\pm$ 1.36	8.49 [aA] $\pm$ 1.53
Pablo F1	4.35 [aA] $\pm$ 0.83	5.00 [aA] $\pm$ 0.96	4.72 [abA] $\pm$ 0.92	6.13 [abcA] $\pm$ 1.10	6.36 [aA] $\pm$ 1.14	5.71 [aA] $\pm$ 1.04
Regulski O.	3.91 [aA] $\pm$ 0.75	4.96 [aA] $\pm$ 0.95	4.48 [abA] $\pm$ 0.86	5.93 [abcA] $\pm$ 1.06	6.88 [aA] $\pm$ 1.24	6.23 [aA] $\pm$ 1.12
Robuschka	3.85 [aA] $\pm$ 0.74	4.60 [abA] $\pm$ 0.88	5.34 [abA] $\pm$ 1.02	5.50 [abcA] $\pm$ 0.99	6.22 [aA] $\pm$ 1.12	6.70 [aA] $\pm$ 1.20
Ronjana	6.16 [aA] $\pm$ 1.18	4.68 [abA] $\pm$ 0.89	6.08 [abA] $\pm$ 1.16	8.07 [aA] $\pm$ 1.45	7.42 [aA] $\pm$ 1.33	9.05 [aA] $\pm$ 1.63
Hilmar	4.42 [aA] $\pm$ 0.84	4.73 [abA] $\pm$ 0.90	5.05 [abA] $\pm$ 0.97	5.89 [abcA] $\pm$ 1.06	6.54 [aA] $\pm$ 1.17	6.63 [aA] $\pm$ 1.19
Sniezna Kula	0.13 [bA] $\pm$ 0.03	0.18 [eA] $\pm$ 0.05	0.16 [gA] $\pm$ 0.05	0.12 [gA] $\pm$ 0.03	0.17 [bA] $\pm$ 0.04	0.15 [bA] $\pm$ 0.04
Tondo d. Ch.	0.22 [bA] $\pm$ 0.04	0.26 [eA] $\pm$ 0.05	0.19 [fgA] $\pm$ 0.05	0.30 [fgA] $\pm$ 0.05	0.25 [bA] $\pm$ 0.05	0.20 [bA] $\pm$ 0.05
UB-E3	3.40 [aA] $\pm$ 0.65	4.21 [adA] $\pm$ 0.80	3.50 [bcdA] $\pm$ 0.68	5.31 [abcdA] $\pm$ 0.95	5.75 [aA] $\pm$ 1.03	4.29 [aA] $\pm$ 0.79

3.3. Betalain Content

The findings of this study revealed a significant effect of the interaction between genotype and storage period on both betacyanin and betaxanthin content ($p < 0.0001$). Moreover, it was noted that the year impacted the betaxanthin and betacyanin contents significantly (Table 2). Kujala et al. [6] claimed that beetroot genotype, cultivation, and storage conditions can affect the content of betanin and isobetanin (the main betacyanins found in beetroot). Cejudo-Bastante et al. [55] reported the influence of storage duration and temperature on the betalain content of fruits. Generally, temperature has been reported as a key factor for the stability of the betalain [56].

Two major subgroups of betalain, betaxanthin and betacyanin, were measured in this study. Respecting the significant impact of the interactions between genotype and storage period on both betaxanthin and betacyanin contents, Table 4 demonstrates the contents based on different storage periods for each genotype separately. Considering all studied genotypes, the highest betaxanthin contents in beetroots directly after harvest were measured in the genotypes Gesche SG (6.60 ± 1.26 mg g^{-1} DW), Libero RZ (6.25 ± 1.19 mg g^{-1} DW), and Ronjana (6.16 ± 1.18 mg g^{-1} DW) and the lowest values were noted in the white-colored genotype Sniezna Kula, yellow-colored Burpees G., and red-white Tondo d. Ch., with 0.13 ± 0.03 mg g^{-1} DW, 0.20 ± 0.04 mg g^{-1} DW, and 0.22 ± 0.4 mg g^{-1} DW, respectively (Table 4). Comparing the red-colored genotypes, the betaxanthin content after one month of cold storage varied between 3.08 ± 0.59 mg g^{-1} DW and 6.73 ± 1.38 mg g^{-1} DW belonging to the genotypes Alvro Mono and Libero RZ, respectively. Following Alvro Mono, the genotypes Formanova (3.54 ± 0.69 mg g^{-1} DW), Detroit G. (4.08 ± 0.78 mg g^{-1} DW), and UB-E3 (4.21 ± 0.80 mg g^{-1} DW) possessed the lowest betaxanthin values (Table 4). However, the white-colored genotype Sniezna Kula with 0.18 ± 0.05 mg g^{-1} DW, the red-white genotype Tondo d. Ch. with 0.26 ± 0.05 mg g^{-1} DW, and yellow-colored genotype Burpees G. with 1.15 ± 0.22 mg g^{-1} DW indicated the lowest betaxanthin values when taking all studied genotypes into account (Table 4). After the storage period of four months, the betaxanthin content of the red-colored genotypes ranged between 3.05 ± 0.60 mg g^{-1} DW and 6.70 ± 1.28 mg g^{-1} DW belonging to the genotypes Alvro Mono and the cylindrical-shaped Forono, respectively. However, considering all studied genotypes, the lowest betaxanthin contents were observed in the white-colored genotype Sniezna Kula with 0.16 ± 0.05 mg g^{-1} DW, the red-white genotype Tondo d. Ch. with 0.19 ± 0.05 mg g^{-1} DW, and yellow-colored genotype Burpees G. with 1.03 ± 0.20 mg g^{-1} DW (Table 4). Regarding the influence of cold storage, in the yellow-colored genotype Burpees G., a significant increase in the betaxanthin content after one month of cold storage was observed and afterwards, up to a storage period of four months, no significant change occurred (Table 4).

The betacyanin content of the red-colored genotypes directly after the harvest ranged from 3.61 ± 0.65 mg g^{-1} DW to 8.70 ± 1.56 mg g^{-1} DW. The three highest betacyanin contents were noted in genotypes Monty RZ F1, Libero RZ, and Forono, respectively (Table 4). Taking all the examined genotypes into account, the white-colored genotype Sniezna Kula with 0.12 ± 0.03 mg g^{-1} DW, the red-white genotype Tondo d. Ch. with 0.30 ± 0.05 mg g^{-1} DW, and yellow-colored genotype Burpees G. with 0.43 ± 0.08 mg g^{-1} DW possessed the lowest betacyanin contents. After one month of cold storage, the betacyanin content of the red-colored genotypes varied between 4.13 ± 0.74 mg g^{-1} DW and 8.70 ± 1.70 mg g^{-1} DW belonging to Alvro Mono and Libero RZ, respectively, and that after a cold storage of four months, ranged between 4.28 ± 0.79 mg g^{-1} DW and 9.05 ± 1.63 mg g^{-1} DW found in the genotypes Alvro Mono and Ronjana, respectively. After cold storage periods of one and four months, the lowest betacyanin contents were measured in Burpees G., Sniezna Kula, and Tondo d. Ch., respectively (Table 4). Among all investigated genotypes, the betacyanin content in yellow-colored genotype, Burpees G., was significantly influenced by the duration of the cold storage. In this regard, a significant decrease in the betacyanin content after one month of cold storage was noted and afterwards up to a storage period of four months, no significant change was observed (Table 4).

Kujala et al. [49] studied the effect of cold storage at 5 °C on betanin and isobetanin content of the red beetroots (var. Little Ball) grown in Finland and significant differences in the amounts of betanin and isobetanin during the cold storage (0–196 days) were noted. Moreover, it was reported that the content of betanin in red beetroot peel decreased in the first 140 days of cold storage and then slightly increased. In terms of isobetanin, until 98 days, an increasing trend and afterward, up to 140 days of storage, a light decrease were noticed [49]. Maity et al. [53] investigated the effect of storage temperature and duration on betacyanin and betaxanthin contents of compressed beetroot snack bars and the maximum retention was noted in those stored at 6 °C and the content did not change significantly after four months of storage.

Moreover, storage of beetroot powder in three different temperatures (namely 10, 25, and 40 °C) indicated the minimum loss in the content of betacyanin and betaxanthin at the lowest temperature up to five weeks [56]. Yong et al. [57] investigated the effect of seven days of cold storage on betacyanin content of red pitahaya at 4 °C and reported a significant increase after six days of storage and then a slight decrease on day seven. In contrast, Obenland et al. [58] noted no significant impact of two weeks storage of red pitahaya at 5 and 10 °C on the betacyanin content.

3.4. Nitrate Content

The average nitrate content differed significantly between the genotypes (p = 0.0005) (Table 2). A corresponding result was found by Kosson et al. [43], who reported a significant influence of variety on the nitrate content in two beetroot genotypes. Moreover, a significant impact of the storage period on the nitrate content was noted (p < 0.0001). On the other hand, the interactions between storage period and genotype, year, and interactions between storage period and year on the nitrate content were not significant (Table 2).

Corleto et al. [32] reported a significant impact of different refrigeration temperatures and periods on the nitrate level of freshly pressed beetroot juice. Moreover, their study revealed that in the first eight days of storage at 4 °C, the nitrate value did not change significantly, while between day 8 and 32 of the storage, the nitrate level decreased drastically. In contrast, Chung et al. [59] studied the impact of cold storage on four different leafy vegetables, including spinach, crown daisy, organic Chinese spinach, and organic non-heading Chinese cabbage, and no significant changes in the nitrate content could be proven during the storage period.

According to the significant effect of genotype and storage period on the content of nitrate, Figure 1 demonstrates the medians of nitrate content of the studied genotypes as well as medians based on three storage periods studied within the experiment. Moreover, the precise median and asymptotic standard error values of nitrate content are available in Table S1 in the supplementary material. The nitrate values exhibited in Figure 1 were reported on a dry weight basis, while nitrate content in literature is often also presented on a fresh weight (FW) basis. Consequently, to compare the findings of this work with other studies, the TDMC should be considered. The amount of nitrate of the studied genotypes ranged between 4179 ± 1267 mg kg^{-1} DW and 20,489 ± 2988 mg kg^{-1} DW found in Chrobry and Libero RZ, respectively (Figure 1). Following Libero RZ, the genotypes Bona (16,794 ± 2539 mg kg^{-1} DW) and Alvro Mono (16,438 ± 2539 mg kg^{-1} DW) indicated the highest nitrate contents within the trial. The second and third lowest nitrate content were measured in genotypes Nochowski with 4602 ± 1329 mg kg^{-1} DW, and Hilmar with 5190 ± 1411 mg kg^{-1} DW, respectively.

With regards to the medians based on the storage period, the nitrate content of 9095 ± 414 mg kg^{-1} DW was calculated for all genotypes in both years for samples collected directly after the harvest. After one month of cold storage, the median nitrate content increased significantly and reached a median of 10,977 ± 459 mg kg^{-1} DW. No further significant increase was observed after a cold storage period of four months (Figure 1). Consequently, based on the outcome of this study, the main change in the nitrate content of the investigated genotypes arose within the first four weeks of storage.

According to the point that the nitrate content may not only vary between the genotypes but also among the cultivars of the same species and plant tissues [60–62], assessing a greater number of plants per genotype for having a better evaluation can be recommended.

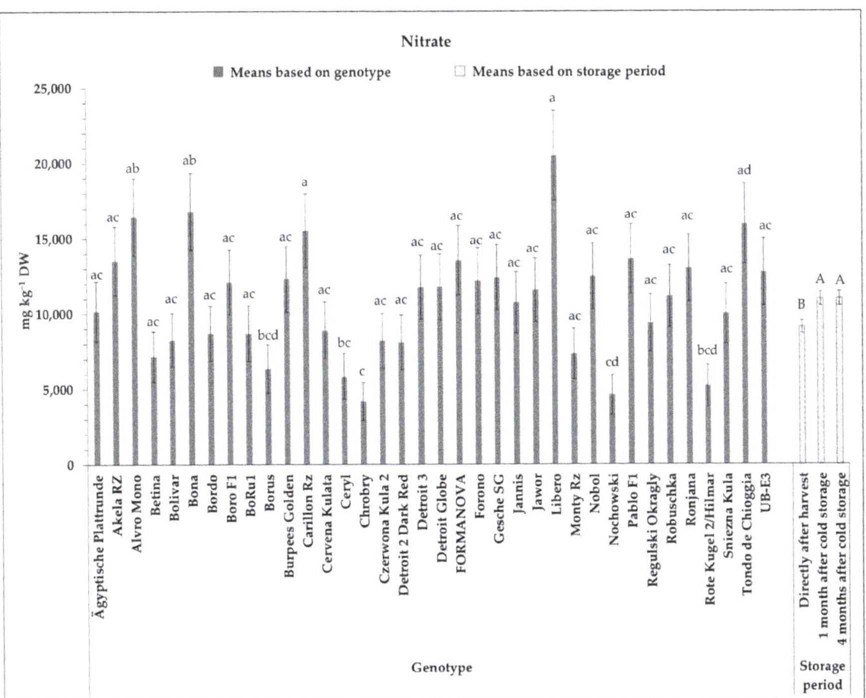

Figure 1. Median values of nitrate content (mg kg^{-1} DW) of 36 beetroot genotypes grown at the research station Kleinhohenheim within the years 2017 and 2018, based on genotype and storage period. Results represent the median values ± asymptotic standard error. Medians covered by at least one identical lower-case letter did not differ significantly between genotypes at experiment-wise Type 1 error α = 0.05. Medians covered by at least one identical upper-case letter did not differ significantly between storage periods at experiment-wise Type 1 error α = 0.05.

3.5. Total Soluble Sugar Content

According to the statistical analysis, interactions between storage period and genotype were significant (p = 0.0121). Viskelis et al. [2] reported no significant change of total sugar content of 11 beetroot genotypes, grown in Lithuania, during the storage at 1 ± 1 °C and relative humidity of 90–95% for the storage period of seven months. Nonetheless, in the case of four common genotypes in both studies, namely Bona, Boro F1, Detroit 2 D. R., and Pablo F1, despite the lower values in our study, the storage period did not impact the content of total soluble sugars significantly. Thus, it disclosed that the influence of cold storage might be genotype-dependent.

To better appraise the impact of cold storage on sugar content, it is noteworthy to know the amount of sugar in freshly harvest beetroot. In this regards, the freshly harvested beetroots in this study contained the total soluble sugar content in the range of 8.55 ± 0.67 °Bx to 15.43 ± 0.67 °Bx possessing by the genotypes Alvro Mono and Nochowski, respectively. After one month of cold storage, the three highest total soluble sugar contents belonged to the genotypes Nochowski, Chrobry, and Cervena K., with 14.88 ± 0.67 °Bx, 13.96 ± 0.67 °Bx, and 13.89 ± 0.67 °Bx, respectively (Table 5). The genotypes Alvro Mono (9.00 ± 0.46 °Bx), Bolivar (9.64 ± 0.48 °Bx), and Libero RZ (9.64 ± 0.67 °Bx) exhibited the lowest total soluble sugar contents. After a cold storage period of four months, the total soluble sugar content ranged between 8.41 ± 0.71 °Bx and 14.92 ± 0.67 °Bx, belonging to the genotypes Alvro Mono and Chrobry, respectively. Following the genotype Alvro Mono, the red-colored Libero RZ, red-white-colored Tondo d. Ch., and yellow-colored Burpees G. indicated the lowest total soluble sugar contents.

Table 5. Mean values of total soluble sugars (°Bx) of 36 beetroot genotypes grown at the research station Kleinhohenheim within the years 2017 and 2018, directly after harvest, and after the cold storage periods of one and four months. Results represent the mean values ± standard error. Means followed by at least one identical lower-case letter in one column did not differ significantly between genotypes at experiment-wise Type 1 error α = 0.05. Means followed by at least one identical upper-case letter in one row did not differ significantly between storage periods at experiment-wise Type 1 error α = 0.05.

Genotype	Total Soluble Sugars (°Bx)		
	Directly after Harvest	1 Month after Cold Storage	4 Months after Cold Storage
Ä. P.	11.08 [adA] ± 0.67	11.77 [adA] ± 0.67	12.31 [bcA] ± 0.67
Akela RZ	11.26 [adA] ± 0.67	11.58 [adA] ± 0.67	12.10 [bcA] ± 0.67
Alvro Mono	8.55 [dA] ± 0.67	9.49 [dA] ± 0.67	8.41 [cA] ± 0.71
Betina	13.40 [abA] ± 0.67	13.55 [adA] ± 0.67	13.74 [abA] ± 0.71
Bolivar	11.14 [adA] ± 0.67	9.53 [cdA] ± 0.67	12.37 [bcA] ± 0.67
Bona	11.18 [adA] ± 0.67	11.14 [adA] ± 0.67	10.88 [bcA] ± 0.67
Bordo	12.04 [adA] ± 0.67	13.50 [adA] ± 0.67	12.47 [bcA] ± 0.67
Boro F1	10.99 [bcdA] ± 0.67	11.13 [adA] ± 0.67	11.16 [bcA] ± 0.67
BoRu1	11.79 [adA] ± 0.67	10.60 [adA] ± 0.67	11.53 [bcA] ± 0.67
Borus	12.40 [adA] ± 0.67	13.53 [adA] ± 0.67	13.38 [abA] ± 0.76
Burpees G.	10.66 [bcdA] ± 0.67	10.59 [adA] ± 0.67	10.29 [acA] ± 0.67
Carillon RZ	9.76 [bdA] ± 0.67	11.87 [adA] ± 0.67	11.15 [bcA] ± 0.67
Cervena K.	13.85 [abA] ± 0.67	13.89 [abcA] ± 0.67	11.88 [bcA] ± 0.77
Ceryl	13.47 [abA] ± 0.67	12.97 [adA] ± 0.67	13.70 [abA] ± 0.71
Chrobry	14.40 [acA] ± 0.67	13.96 [abA] ± 0.67	14.92 [bA] ± 0.67
Czerwona K. 2	12.21 [adA] ± 0.67	12.14 [adA] ± 0.67	13.44 [abA] ± 0.71
Detroit 2 D. R.	12.28 [adA] ± 0.67	12.91 [adA] ± 0.67	11.29 [bcA] ± 0.84
Detroit 3	10.98 [bcdA] ± 0.67	11.34 [adA] ± 0.67	11.22 [bcA] ± 0.71
Detroit G.	10.53 [bcdA] ± 0.67	11.47 [adA] ± 0.67	11.72 [bcA] ± 0.76
Formanova	11.15 [adA] ± 0.67	11.11 [adjA] ± 0.67	11.34 [bcA] ± 0.67
Forono	11.75 [adA] ± 0.67	12.05 [adA] ± 0.67	11.87 [bcA] ± 0.67
Gesche SG	13.07 [abeA] ± 0.67	12.63 [adA] ± 0.67	11.80 [bcA] ± 0.67
Jannis	11.19 [adA] ± 0.67	11.72 [adA] ± 0.67	11.87 [bcA] ± 0.67
Jawor	11.71 [adA] ± 0.67	12.22 [adA] ± 0.67	12.93 [abA] ± 0.67
Libero RZ	8.97 [deA] ± 0.67	9.64 [bdA] ± 0.67	9.60 [acA] ± 0.76
Monty RZ F1	11.36 [adA] ± 0.67	13.17 [adA] ± 0.67	12.90 [bcA] ± 0.71
Nobol	10.61 [bcdA] ± 0.67	11.53 [adA] ± 0.67	11.17 [bcA] ± 0.67
Nochowski	15.43 [aA] ± 0.67	14.88 [aA] ± 0.67	12.20 [bcA] ± 0.67
Pablo F1	10.81 [bcdA] ± 0.67	11.11 [adA] ± 0.67	11.61 [bcA] ± 0.81
Regulski O.	11.76 [adA] ± 0.67	12.26 [adA] ± 0.67	11.99 [bcA] ± 0.67
Robuschka	11.86 [adA] ± 0.67	12.82 [adA] ± 0.67	11.99 [bcA] ± 0.67
Ronjana	12.71 [adA] ± 0.67	11.71 [adA] ± 0.67	11.73 [bcA] ± 0.67
Hilmar	11.43 [adA] ± 0.67	12.71 [adA] ± 0.67	11.75 [bcA] ± 0.67
Sniezna Kula	11.60 [adA] ± 0.67	10.74 [adA] ± 0.67	11.50 [bcA] ± 0.71
Tondo d. Ch.	8.77 [deA] ± 0.68	9.71 [bdA] ± 0.71	10.07 [bcA] ± 0.84
UB-E3	12.70 [adA] ± 0.67	12.58 [adA] ± 0.67	13.00 [abA] ± 0.67

Depending on the stored crop, the effect of cold storage on the total soluble sugar content can be different. Hagen et al. [63] stated a significant decrease in the total content of soluble sugars in curly kale stored for six weeks at 1 °C, while an increase in sugar content was noted in potato tubers stored for six months at 2–4 °C. Jakopic et al. [46] investigated the change in the content of soluble sugars in rutabaga during the cold storage and an increase in the first month of storage and afterwards up to the cold storage for four months a slight decrease was reported. Barboni et al. [64] investigated the impact of cold storage on the total soluble sugar content of kiwi fruit and an increase in the first seven weeks and a constant amount after 21 weeks of the storage was noted. This reveals the considerable variation in the effect of cold storage on vegetables and fruits, which is highly dependent on the species [65].

3.6. Principal Component Analysis

The biplot of the genotype-by-storage period means (Figure 2) demonstrated that more than 83% of the variation is elucidated by two examined components.

It revealed that the TDMC and total soluble sugar content are highly positively correlated, whereas, high negative correlations between these two traits and nitrate content was noted. This is in agreement with Anjana and Iqbal [66], who reported a negative correlation between the sugar and nitrate contents and a positive correlation between carbohydrate concentration and TDMC. The biplot further confirmed that the genotypes Alvro Mono and Libero RZ, which indicated the highest nitrate values possessed the lowest sugar content among the studied genotypes. On the other hand, the highest total soluble sugar contents were noted in genotypes Chrobry and Nochowski, which were included in the genotypes with the lowest nitrate content. Moreover, the biplot visualized a high positive correlation between the contents of betacyanin and betaxanthin. Likewise, the contents of these two compounds were positively correlated with the TPC. Corresponding findings were reported by Kugler et al. [67] regarding a positive correlation between betacyanin and betaxanthin, by Kujala et al. [49] about a positive correlation between betacyanin and TPC, and Čanadanović- Brunet et al. [68] stated a significant correlation between the total phenolic compounds and betaxanthin. The biplot of the interactions of genotype and storage period approved that the lowest TPC and betalain content belonged to the non-red genotypes, including Tondo d. Ch., Burpees G., and Sniezna Kula.

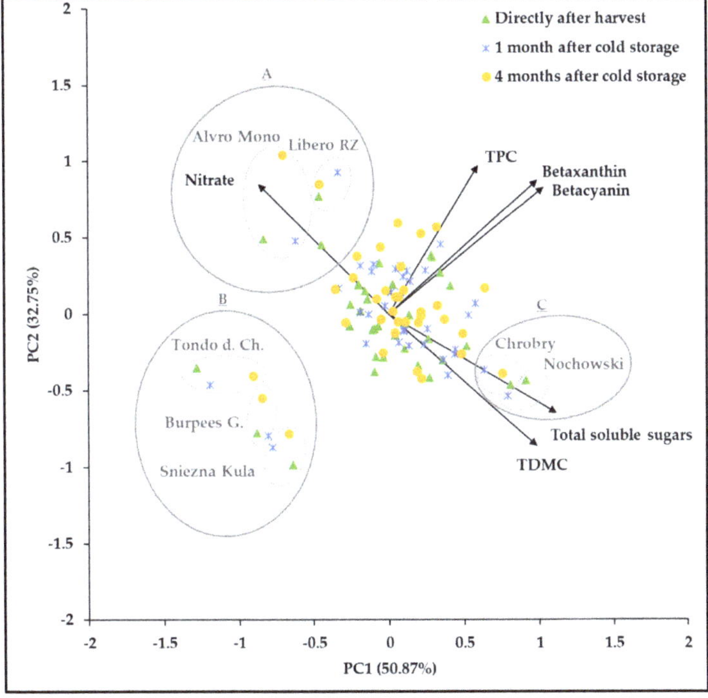

Figure 2. Biplot of the impact of the interactions of genotype-by-storage period means on the traits TDMC, TPC, betaxanthin, betacyanin, nitrate, and total soluble sugars for 36 beetroot genotypes grown at the research station Kleinhohenheim within the years 2017 and 2018, directly after harvest, and after the cold storage periods of one and four months. Group A shows the genotypes with the highest nitrate contents, group B the genotypes with the highest total soluble sugar contents, and group C, the non-red-colored genotypes. PC1 and PC2 are principal components 1 and 2.

4. Conclusions

Besides investigating a great number of beetroot genotypes and disclosing a high genetic variability regarding the content of the bioactive compounds, this study examined the storability of beetroot genotypes at a low temperature and the extent of change in their compositional quality in order to prolong the use of this vegetable for an extended time after the harvest. However, to have a thorough insight on the genetic potential of the examined beetroot genotypes for their application in various sections, the agronomic performance, and their sensory quality should be additionally considered. The genotype 'Chrobry' was characterized by the lowest nitrate content and indicated a high total soluble sugar content with no significant changes during four months of cold storage, thus, it may be of interest for beetroot juice production. 'Nochowski' and 'Cervena K.' which were among the top three genotypes with the highest total soluble sugar contents directly after the harvest and retained contents after one month of cold storage, indicated a significant decrease in the amount of total soluble sugars after four months of cold storage. Therefore, it is beneficial to use these beetroot genotypes freshly or within the first month of storage, when a high sugar content is desired. The genotypes 'Forono', 'Ronjana', 'Monty RZ F1', and 'Nobol' which were characterized by a high amount of betacyanin and betaxanthin, and their stability during a cold storage period of four months, can be of interest for the use as natural food colorants. The cylindrical-shaped genotype 'Forono' characterized by a high content of betalain and TPC, can serve as an option for value-added food products. Further studies considering beet firmness, and impact of other atmospheric conditions such as relative humidity to further improve the storability of beetroot, can be recommended.

Supplementary Materials: The following is available online at https://www.mdpi.com/article/10.3390/foods10061281/s1, Table S1: Median values of nitrate content (mg kg^{-1} DW) of 36 beetroot genotypes grown at the research station Kleinhohenheim within the years 2017 and 2018, based on genotype and storage period.

Author Contributions: Conceptualization, K.Y. and S.G.-H.; software, K.Y. and J.H.; formal analysis, K.Y. and J.H.; investigation, K.Y.; writing—original draft preparation, K.Y.; writing—review and editing, S.G.-H. and J.H.; supervision, S.G.-H.; project administration, S.G.-H. and M.F.; funding acquisition, S.G.-H. and M.F. All authors have read and agreed to the published version of the manuscript.

Funding: This research was funded by the Federal Ministry of Food and Agriculture of Germany (BMEL) based on a decision of the Parliament of the Federal Republic of Germany via the Federal Office for Agriculture and Food (BLE) under the innovation support program (Grant number: 2818201015).

Data Availability Statement: The datasets presented in this study are available on request to the corresponding author.

Acknowledgments: The authors would like to thank the members of the group of Cropping Systems and Modelling for their support in collecting the samples in the field experiments and chemical analysis. We acknowledge the technical staff of the experimental station Kleinhohenheim (University of Hohenheim) for their help during the field experiments.

Conflicts of Interest: The authors declare no conflict of interest. The funders had no role in the design of the study; in the collection, analyses, or interpretation of data; in the writing of the manuscript, or in the decision to publish the results.

References

1. Wruss, J.; Waldenberger, G.; Huemer, S.; Uygun, P.; Lanzerstorfer, P.; Müller, U.; Höglinger, O.; Weghuber, J. Compositional Characteristics of Commercial Beetroot Products and Beetroot Juice Prepared from Seven Beetroot Varieties Grown in Upper Austria. *J. Food Compos. Anal.* **2015**, *42*, 46–55. [CrossRef]
2. Viskelis, J.; Nevidomskis, S.; Bobinas, C.; Urbonaviciene, D.; Bobinaite, R.; Karkleliene, R.; Viskelis, P. Evaluation of Beetroot Quality during Various Storage Conditions. In Proceedings of the FOODBALT 2019 13th Baltic Conference on Food Science and Technology "Food, Nutrition, Well-being", Jelgava, Latvia, 2–3 May 2019; pp. 170–175. [CrossRef]

3. Clifford, T.; Howatson, G.; West, D.J.; Stevenson, E.J. The Potential Benefits of Red Beetroot Supplementation in Health and Disease. *Nutrients* **2015**, *7*, 2801–2822. [CrossRef] [PubMed]

4. Harsh, C.; Milind Parle, K.S. Beetroot: A Health Promoting Functional Food. *Inven. Rapid Nutraceuticals* **2016**, *1*, 1–5.

5. Vasconcellos, J.; Conte-Junior, C.; Silva, D.; Pierucci, A.P.; Paschoalin, V.; Alvares, T.S. Comparison of Total Antioxidant Potential, and Total Phenolic, Nitrate, Sugar, and Organic Acid Contents in Beetroot Juice, Chips, Powder, and Cooked Beetroot. *Food Sci. Biotechnol.* **2016**, *25*, 79–84. [CrossRef]

6. Kujala, T.S.; Vienola, M.S.; Klika, K.D.; Loponen, J.M.; Pihlaja, K. Betalain and Phenolic Compositions of Four Beetroot (*Beta Vulgaris*) Cultivars. *Eur. Food Res. Technol.* **2002**, *214*, 505–510. [CrossRef]

7. Lombardelli, C.; Benucci, I.; Mazzocchi, C.; Esti, M. A Novel Process for the Recovery of Betalains from Unsold Red Beets by Low-Temperature Enzyme-Assisted Extraction. *Foods* **2021**, *10*, 236. [CrossRef]

8. Martins, N.; Roriz, C.L.; Morales, P.; Barros, L.; Ferreira, I.C.F.R. Coloring Attributes of Betalains: A Key Emphasis on Stability and Future Applications. *Food Funct.* **2017**, *8*, 1357–1372. [CrossRef] [PubMed]

9. Strack, D.; Vogt, T.; Schliemann, W. Recent Advances in Betalain Research. *Phytochemistry* **2003**, *62*, 247–269. [CrossRef]

10. Georgiev, V.G.; Weber, J.; Kneschke, E.M.; Denev, P.N.; Bley, T.; Pavlov, A.I. Antioxidant Activity and Phenolic Content of Betalain Extracts from Intact Plants and Hairy Root Cultures of the Red Beetroot Beta Vulgaris Cv. Detroit Dark Red. *Plant Foods Hum. Nutr.* **2010**, *65*, 105–111. [CrossRef] [PubMed]

11. Skalicky, M.; Kubes, J.; Shokoofeh, H.; Tahjib-Ul-Arif, M.; Vachova, P.; Hejnak, V. Betacyanins and Betaxanthins in Cultivated Varieties of *Beta Vulgaris* L. Compared to Weed Beets. *Molecules* **2020**, *25*, 5395. [CrossRef]

12. Gasztonyi, M.N.; Daood, H.; Hájos, M.T.; Biacs, P. Comparison of Red Beet (*Beta Vulgaris* Var *Conditiva*) Varieties on the Basis of Their Pigment Components. *J. Sci. Food Agric.* **2001**, *81*, 932–933. [CrossRef]

13. Mohammed, A.N.; Ishwarya, S.P.; Nisha, P. Nanoemulsion Versus Microemulsion Systems for the Encapsulation of Beetroot Extract: Comparison of Physicochemical Characteristics and Betalain Stability. *Food Bioprocess Technol.* **2021**, *14*, 133–150. [CrossRef]

14. Azeredo, H.M.C. Betalains: Properties, Sources, Applications, and Stability—A Review. *Int. J. Food Sci. Technol.* **2009**, *44*, 2365–2376. [CrossRef]

15. Stintzing, F.C.; Carle, R. Functional Properties of Anthocyanins and Betalains in Plants, Food, and in Human Nutrition. *Trends Food Sci. Technol.* **2004**, *15*, 19–38. [CrossRef]

16. Gengatharan, A.; Dykes, G.A.; Choo, W.S. Betalains: Natural Plant Pigments with Potential Application in Functional Foods. *LWT* **2015**, *64*, 645–649. [CrossRef]

17. Delgado-Vargas, F.; Jiménez, A.R.; Paredes-López, O.; Francis, F.J. Natural Pigments: Carotenoids, Anthocyanins, and Betalains—Characteristics, Biosynthesis, Processing, and Stability. *Crit. Rev. Food Sci. Nutr.* **2000**, *40*, 173–289. [CrossRef] [PubMed]

18. Singh, B.; Hathan, B.S.; Devi, C.; Memorial, L.; Longowal, S. Chemical Composition, Functional Properties and Processing of Beetroot—A Review. *Int. J. Sci. Eng. Res.* **2014**, *5*, 679–684.

19. Blekkenhorst, L.C.; Bondonno, N.P.; Liu, A.H.; Ward, N.C.; Prince, R.L.; Lewis, J.R.; Devine, A.; Croft, K.D.; Hodgson, J.M.; Bondonno, C.P. Nitrate, the Oral Microbiome, and Cardiovascular Health: A Systematic Literature Review of Human and Animal Studies. *Am. J. Clin. Nutr.* **2018**, *107*, 504–522. [CrossRef] [PubMed]

20. Rocchetti, G.; Tomas, M.; Zhang, L.; Zengin, G.; Lucini, L.; Capanoglu, E. Red Beet (*Beta Vulgaris*) and Amaranth (*Amaranthus* Sp.) Microgreens: Effect of Storage and in Vitro Gastrointestinal Digestion on the Untargeted Metabolomic Profile. *Food Chem.* **2020**, *332*, 127415. [CrossRef] [PubMed]

21. Andac, M.; Eren, H.; Coldur, F. Determination of Nitrate in Leafy Vegetables by Flow Injection Analysis with Potentiometric Detection. *J. Food Drug Anal.* **2011**, *19*, 457–462. [CrossRef]

22. Babateen, A.M.; Fornelli, G.; Donini, L.M.; Mathers, J.C.; Siervo, M. Assessment of Dietary Nitrate Intake in Humans: A Systematic Review. *Am. J. Clin. Nutr.* **2018**, *108*, 878–888. [CrossRef] [PubMed]

23. Lundberg, J.O.; Carlström, M.; Larsen, F.J.; Weitzberg, E. Roles of Dietary Inorganic Nitrate in Cardiovascular Health and Disease. *Cardiovasc. Res.* **2011**, *89*, 525–532. [CrossRef] [PubMed]

24. Baião, D.d.S.; da Silva, D.V.T.; Paschoalin, V.M.F. Beetroot, a Remarkable Vegetable: Its Nitrate and Phytochemical Contents Can Be Adjusted in Novel Formulations to Benefit Health and Support Cardiovascular Disease Therapies. *Antioxidants* **2020**, *9*, 960. [CrossRef]

25. Chhikara, N.; Kushwaha, K.; Sharma, P.; Gat, Y.; Panghal, A. Bioactive Compounds of Beetroot and Utilization in Food Processing Industry: A Critical Review. *Food Chem.* **2019**, *272*, 192–200. [CrossRef]

26. Fan, H.; Du, C.; Xu, Y.; Wu, X. Exogenous Nitric Oxide Improves Chilling Tolerance of Chinese Cabbage Seedlings by Affecting Antioxidant Enzymes in Leaves. *Hortic. Environ. Biotechnol.* **2014**, *55*, 159–165. [CrossRef]

27. Fancy, N.N.; Bahlmann, A.K.; Loake, G.J. Nitric Oxide Function in Plant Abiotic Stress. *Plant Cell Environ.* **2017**, *40*, 462–472. [CrossRef] [PubMed]

28. Sohag, A.A.M.; Tahjib-Ul-Arif, M.; Afrin, S.; Khan, M.K.; Hannan, M.A.; Skalicky, M.; Mortuza, M.G.; Brestic, M.; Hossain, M.A.; Murata, Y. Insights into Nitric Oxide-Mediated Water Balance, Antioxidant Defence and Mineral Homeostasis in Rice (*Oryza Sativa* L.) under Chilling Stress. *Nitric. Oxide Biol. Chem.* **2020**, *100–101*, 7–16. [CrossRef] [PubMed]

29. Bavec, M.; Turinek, M.; Grobelnik-Mlakar, S.; Slatnar, A.; Bavec, F. Influence of Industrial and Alternative Farming Systems on Contents of Sugars, Organic Acids, Total Phenolic Content, and the Antioxidant Activity of Red Beet (*Beta Vulgaris* L. Ssp. *Vulgaris* Rote Kugel). *J. Agric. Food Chem.* **2010**, *58*, 11825–11831. [CrossRef]

30. Cicia, G.; Del Giudice, T.; Scarpa, R. Consumers' Perception of Quality in Organic Food: A Random Utility Model under Preference Heterogeneity and Choice Correlation from Rank-Orderings. *Br. Food J.* **2002**, *104*, 200–213. [CrossRef]

31. Carrillo, C.; Wilches-Pérez, D.; Hallmann, E.; Kazimierczak, R.; Rembiałkowska, E. Organic versus Conventional Beetroot. Bioactive Compounds and Antioxidant Properties. *LWT* **2019**, *116*, 108552. [CrossRef]

32. Corleto, K.A.; Singh, J.; Jayaprakasha, G.K.; Patil, B.S. Storage Stability of Dietary Nitrate and Phenolic Compounds in Beetroot (*Beta Vulgaris*) and Arugula (*Eruca Sativa*) Juices. *J. Food Sci.* **2018**, *83*, 1237–1248. [CrossRef] [PubMed]

33. Lal Basediya, A.; Samuel, D.V.K.; Beera, V. Evaporative Cooling System for Storage of Fruits and Vegetables—A Review. *J. Food Sci. Technol.* **2013**, *50*, 429–442. [CrossRef] [PubMed]

34. Gopala Rao, C. *Engineering for Storage of Fruits and Vegetables: Cold Storage, Controlled Atmosphere Storage, Modified Atmosphere Storage*; Academic Press: Cambridge, MA, USA, 2015. [CrossRef]

35. Lepse, L.; Viskelis, P.; Lepsis, J.; Bimsteine, G. Influence of Controlled Atmosphere on the Carrot Storage Quality. *ISHS Acta Hortic.* **2012**, *1033*, 59–64. [CrossRef]

36. El-Ramady, H.R.; Domokos-Szabolcsy, É.; Abdalla, N.A.; Taha, H.S.; Fári, M. *Postharvest Management of Fruits and Vegetables Storage. In Sustainable Agriculture Reviews*; Lichtfouse, E., Ed.; Springer: Cham, Switzerland, 2015; Volume 15, pp. 65–152. [CrossRef]

37. Yasaminshirazi, K.; Hartung, J.; Groenen, R.; Heinze, T.; Fleck, M.; Zikeli, S.; Graeff-Hoenninger, S. Agronomic Performance of Different Open-Pollinated Beetroot Genotypes Grown Under Organic Farming Conditions. *Agronomy* **2020**, *10*, 812. [CrossRef]

38. Singleton, V.L.; Orthofer, R.; Lamuela-Raventós, R. Analysis of Total Phenols and Other Oxidation Substrates and Antioxidants by Means of Folin-Ciocalteu Reagent. *Methods Enzym.* **1999**, *299*, 152–178. [CrossRef]

39. Koubaier, H.B.H.; Snoussi, A.; Essaidi, I.; Chaabouni, M.M.; Thonart, P.; Bouzouita, N. Betalain and Phenolic Compositions, Antioxidant Activity of Tunisian Red Beet (*Beta Vulgaris* L. Conditiva) Roots and Stems Extracts. *Int. J. Food Prop.* **2014**, *17*, 1934–1945. [CrossRef]

40. Sawicki, T.; Bączek, N.; Wiczkowski, W. Betalain Profile, Content and Antioxidant Capacity of Red Beetroot Dependent on the Genotype and Root Part. *J. Funct. Foods* **2016**, *27*, 249–261. [CrossRef]

41. Yasaminshirazi, K.; Hartung, J.; Fleck, M.; Graeff-Hoenninger, S. Bioactive Compounds and Total Sugar Contents of Different Open-Pollinated Beetroot Genotypes Grown Organically. *Molecules* **2020**, *25*, 4884. [CrossRef]

42. Piepho, H.P. An Algorithm for a Letter-Based Representation of All-Pairwise Comparisons. *J. Comput. Graph. Stat.* **2004**, *13*, 456–466. [CrossRef]

43. Kosson, R.; Elkner, K.; Szafirowska, A. Quality of Fresh and Processed Red Beet from Organic and Conventionl Cultivation. *Veg. Crop. Res. Bull.* **2011**, *75*, 125–132. [CrossRef]

44. Hagen, S.F.; Borge, G.I.A.; Solhaug, K.A.; Bengtsson, G.B. Effect of Cold Storage and Harvest Date on Bioactive Compounds in Curly Kale (*Brassica oleracea* L. Var. Acephala). *Postharvest Biol. Technol.* **2009**, *51*, 36–42. [CrossRef]

45. Gawęda, M. The Effect of Storage Conditions on Red Beetroot Quality. *Veg. Crop. Res. Bull.* **2006**, *65*, 85–94.

46. Jakopic, J.; Veberic, R.; Slatnar, A. Changes in Quality Parameters in Rutabaga (*Brassica Napus* Var. Napobrassica) Roots during Long Term Storage. *LWT* **2021**, *147*, 111587. [CrossRef]

47. Romo-Pérez, M.L.; Weinert, C.H.; Häußler, M.; Egert, B.; Frechen, M.A.; Trierweiler, B.; Kulling, S.E.; Zörb, C. Metabolite Profiling of Onion Landraces and the Cold Storage Effect. *Plant Physiol. Biochem.* **2020**, *146*, 428–437. [CrossRef]

48. Lattanzio, V.; Lattanzio, V.M.; Cardinali, A. Role of Phenolics in the Resistance Mechanisms of Plants against Fungal Pathogens and Insects. *Phytochemistry* **2006**, *661*, 23–67.

49. Kujala, T.S.; Loponen, J.M.; Klika, K.D.; Pihlaja, K. Phenolics and Betacyanins in Red Beetroot (*Beta Vulgaris*) Root: Distribution and Effect of Cold Storage on the Content of Total Phenolics and Three Individual Compounds. *J. Agric. Food Chem.* **2000**, *48*, 5338–5342. [CrossRef] [PubMed]

50. Serrano, M.; Martinez-romero, D.; Guill, F. Maintenance of Broccoli Quality and Functional Properties during Cold Storage as Affected by Modified Atmosphere Packaging. *Postharvest Biol. Technol.* **2006**, *39*, 61–68. [CrossRef]

51. Sayyari, M.; Valero, D.; Babalar, M.; Kalantari, S.; Zapata, P.J.; Serrano, M. Prestorage Oxalic Acid Treatment Maintained Visual Quality, Bioactive Compounds, and Antioxidant Potential of Pomegranate after Long-Term Storage at 2 °C. *J. Agric. Food Chem.* **2010**, *58*, 6804–6808. [CrossRef]

52. Külen, O.; Stushnoff, C.; Holm, D.G. Effect of Cold Storage on Total Phenolics Content, Antioxidant Activity and Vitamin C Level of Selected Potato Clones. *J. Sci. Food Agric.* **2013**, *93*, 2437–2444. [CrossRef] [PubMed]

53. Maity, T.; Bawa, A.S.; Raju, P.S. Optimization and Quality Assessment of Ready-to-Eat Intermediate Moisture Compressed Beetroot Bar. *J. Food Sci. Technol.* **2016**, *53*, 3233–3243. [CrossRef]

54. Ismail, A.; Marjan, Z.M.; Foong, C.W. Total Antioxidant Activity and Phenolic Content in Selected Vegetables. *Food Chem.* **2004**, *87*, 581–586. [CrossRef]

55. Cejudo-Bastante, M.J.; Hurtado, N.; Delgado, A.; Heredia, F.J. Impact of PH and Temperature on the Colour and Betalain Content of Colombian Yellow Pitaya Peel (*Selenicereus Megalanthus*). *J. Food Sci. Technol.* **2016**, *53*, 2405–2413. [CrossRef]

56. Pandey, G.; Pandey, V.; Pandey, P.R.; Thomas, G. Effect of Extraction Solvent Temperature on Betalain Content, Phenolic Content, Antioxidant Activity and Stability of Beetroot (*Beta Vulgaris* L.) Powder under Different Storage Conditions. *Plant Arch.* **2018**, *18*, 1623–1627.

57. Yong, Y.Y.; Dykes, G.; Lee, S.M.; Choo, W.S. Effect of Refrigerated Storage on Betacyanin Composition, Antibacterial Activity of Red Pitahaya (*Hylocereus Polyrhizus*) and Cytotoxicity Evaluation of Betacyanin Rich Extract on Normal Human Cell Lines. *LWT Food Sci. Technol.* **2018**, *91*, 491–497. [CrossRef]

58. Obenland, D.; Cantwell, M.; Lobo, R.; Collin, S.; Sievert, J.; Arpaia, M.L. Impact of Storage Conditions and Variety on Quality Attributes and Aroma Volatiles of Pitahaya (*Hylocereus* Spp.). *Sci. Hortic. (Amst.)* **2016**, *199*, 15–22. [CrossRef]

59. Chung, J.C.; Chou, S.S.; Hwang, D.F. Changes in Nitrate and Nitrite Content of Four Vegetables during Storage at Refrigerated and Ambient Temperatures. *Food Addit. Contam.* **2004**, *21*, 317–322. [CrossRef]

60. Felczynski, K.; Elkner, K. Effect of Long-Term Organic and Mineral Fertilization on the Yield and Quality of Red Beet (*Beta Vulgaris* L.). *Veg. Crop. Res. Bull.* **2008**, *68*, 111–125. [CrossRef]

61. Grzebelus, D.; Baranski, R. Identification of Accessions Showing Low Nitrate Accumulation in a Germplasm Collection of Garden Beet. *Acta Hortic.* **2001**, *563*, 253–257. [CrossRef]

62. Wang, X.; Cai, X.; Xu, C.; Wang, S.; Dai, S.; Wang, Q. Nitrate Accumulation and Expression Patterns of Genes Involved in Nitrate Transport and Assimilation in Spinach. *Molecules* **2018**, *23*, 2231. [CrossRef]

63. Pal, S.; Bhattacharya, A.; Konar, A.; Mazumdar, D.; Das, A.K. Chemical Composition of Potato at Harvest and after Cold Storage. *Int. J. Veg. Sci.* **2008**, *14*, 162–176. [CrossRef]

64. Barboni, T.; Cannac, M.; Chiaramonti, N. Effect of Cold Storage and Ozone Treatment on Physicochemical Parameters, Soluble Sugars and Organic Acids in *Actinidia Deliciosa*. *Food Chem.* **2010**, *121*, 946–951. [CrossRef]

65. Galani, J.H.Y.; Patel, J.S.; Patel, N.J.; Talati, J.G. Storage of Fruits and Vegetables in Refrigerator Increases Their Phenolic Acids but Decreases the Total Phenolics, Anthocyanins and Vitamin C with Subsequent Loss of Their Antioxidant Capacity. *Antioxidants* **2017**, *6*, 59. [CrossRef] [PubMed]

66. Anjana, S.U.; Iqbal, M. Nitrate Accumulation in Plants, Factors Affecting the Process, and Human Health Implications. A Review. *Agron. Sustain. Dev.* **2007**, *27*, 45–57. [CrossRef]

67. Kugler, F.; Graneis, S.; Stintzing, F.C.; Carle, R. Studies on Betaxanthin Profiles of Vegetables and Fruits from the Chenopodiaceae and Cactaceae. *Z. Nat. Sect. C J. Biosci.* **2007**, *62*, 311–318. [CrossRef] [PubMed]

68. Vulić, J.; Čanadanović-Brunet, J.; Ćetković, G.; Tumbas, V.; Djilas, S.; Četojević-Simin, D.; Čanadanović, V. Antioxidant and Cell Growth Activities of Beet Root Pomace Extracts. *J. Funct. Foods* **2012**, *4*, 670–678. [CrossRef]

 *foods*

Article

Influence of Refrigerated Storage on Water Status, Protein Oxidation, Microstructure, and Physicochemical Qualities of Atlantic Mackerel (*Scomber scombrus*)

Rong Lin [1,2,3], Shasha Cheng [1,2,3,*], Siqi Wang [1,2,3], Mingqian Tan [1,2,3] and Beiwei Zhu [1,2,3]

[1] National Engineering Research Center of Seafood, School of Food Science and Technology, Dalian Polytechnic University, Dalian 116034, China; 18308320017021@xy.dlpu.edu.cn (R.L.); 17308320017012@xy.dlpu.edu.cn (S.W.); mqtan@dlpu.edu.cn (M.T.); zhubw@dlpu.edu.cn (B.Z.)

[2] Collaborative Innovation Center of Seafood Deep Processing, Dalian 116034, China

[3] National Engineering Research Center of Seafood, Dalian 116034, China

* Correspondence: chengss@dlpu.edu.cn; Tel.: +86-411-86318657

Abstract: Moisture migration, protein oxidation, microstructure, and the physicochemical qualities of Atlantic mackerel during storage at 4 °C and 0 °C were explored in this study. Three proton components were observed in mackerel muscle using low-field nuclear magnetic resonance relaxation, which were characterized as bound water, immobilized water, and lipid. The relaxation peak of immobilized water shifted to a shorter relaxation time and its intensity decreased with the proceeding of the storage process. T_1 and T_2 weighted images obtained by magnetic resonance imaging showed a slightly continuous decrease in the intensity of water. The significant decrease in sulfhydryl (SH) content and the increase in carbonyl group (CP) content, disulfide bond content, and hydrophobicity revealed the oxidation of protein during storage. The contents of α-helixes in proteins decreased while that of random coils increased during storage, which suggested changes in the secondary structure of mackerel protein. The storage process also caused the contraction and fracture of myofibrils, and the granulation of endolysin protein. In addition, the drip loss, total volatile basic nitrogen (TVB-N), thiobarbituric acid-reactive substances (TBARS) value, and *b** value increased significantly with the storage time.

Keywords: water status and distribution; microstructure; secondary structure of protein; Atlantic mackerel

Citation: Lin, R.; Cheng, S.; Wang, S.; Tan, M.; Zhu, B. Influence of Refrigerated Storage on Water Status, Protein Oxidation, Microstructure, and Physicochemical Qualities of Atlantic Mackerel (*Scomber scombrus*). *Foods* **2021**, *10*, 214. https://doi.org/10.3390/foods10020214

Academic Editor: Sidonia Martinez
Received: 27 December 2020
Accepted: 19 January 2021
Published: 21 January 2021

Publisher's Note: MDPI stays neutral with regard to jurisdictional claims in published maps and institutional affiliations.

1. Introduction

Atlantic mackerel (*Scomber scombrus*) is an abundant pelagic marine fish with various nutrients such as polyunsaturated fatty acids, high quality protein, and vitamin D, which is consumed by consumers worldwide [1]. The total protein content in white muscle of Atlantic mackerel is between 18.5% and 20.8% [2]. The lipid content of Atlantic mackerel can be as high as 29.7% at the optimal fishing location and season [3]. However, Atlantic mackerel easily decomposes within a short period after fishing because of its high water activity, high enzymatic activity, high protein content, and high lipid content, which leads to a series of quality deterioration such as drip loss, lipid oxidation, protein decomposition, microbial growth, and pH change [4–6]. In recent years, low-temperature preservation has become an effective method to maintain the quality of aquatic products by reducing the activities of endogenous enzymes and spoilage microorganisms [7].

The total moisture content of the Atlantic mackerel muscle is up to 70% [2]. The water status and distribution are critical factors affecting the physical and chemical qualities of fish and fish products including sensory quality, water holding capacity (WHC), drip loss, and texture [8]. Low-field nuclear magnetic resonance (LF-NMR) and magnetic resonance imaging (MRI) have successfully realized the monitoring of water states and distribution in food during storage and processing as non-destructive and rapid technologies [9]. LF-NMR is commonly used to evaluate the proton relaxation behavior including longitudinal (T_1) relaxation and transverse (T_2) relaxation, in which T_2 relaxation could provide information about the interactions between water molecules and macromolecules (proteins) in muscle tissues [10]. The water within aquatic products are usually characterized as three populations, which are ascribed to bound water closely associated with macromolecules, immovable water located within organized protein structures and free water within the extra-myofibrillar [8]. Meanwhile, MRI provides additional spatial information about nuclear spins in the sample, which can visualize the spatial distribution of water. Andersen et al. [11] observed the close correlation between the full NMR-signal decays and WHC of cod subjected to various storages of frozen and/or chilled conditions up to 24 months. Sánchez-Alonso et al. observed three populations of water in hake using LF-NMR relaxometry and investigated the effects of different freezing mode and frozen conditions on the water status and distribution [12,13]. Obvious decreases in the relaxation time and abundance of immobilized water were observed in hake during frozen storage, and the obtained LF-NMR parameters could realize the prediction of frozen storage time and quality changes [12,13]. Sánchez-Valencia et al. also found that the frozen storage could cause decreases in relaxation time and abundance of immobilized water in hake, and the higher frozen temperature accelerated the decrease [14]. The increase in relaxation time of immobilized and free water was observed in channel catfish fillets during frozen storage, accompanied by the decrease in the abundance of immobilized water and the increase in the abundance of free water [15]. Ultrasound-assisted immersion freezing could reduce the increase in the freedom of immobilized and free water in common carp during frozen storage at −18 °C [16]. Three water components were observed in grass carp, in which the immobilized water was identified as the major water component and became more mobile during super-chilled storage with the addition of salt and sugar [17]. Sun et al. observed obvious moisture migration between bound, immobilized, and free water in mandarin fish in the early stage of 4 °C storage due to the change in protein–water interaction [18]. Wang et al. investigated the changes of water distribution in salmon [19], yellow tuna, and bigeye tuna [20] during storage at 0 and 4 °C using LF-NMR and MRI and found that the trapped water decreased while free water increased with the prolongation of storage time and the relaxation parameters had significant correlations with the protein and quality changes of fish muscle. Salting methods and modified atmosphere packaging could affect the water dynamic parameters of the muscle and muscle structure during superchilled storage [21]. Previous research found three proton components in Atlantic mackerel, and the catching season, frozen storage temperature, and duration affected the relaxation time and abundance of protons [22]. However, there is limited information about the moisture migration of Atlantic mackerel during refrigerating storage.

The quality deterioration of fish during preservation is also associated with protein degradation, protein aggregation, or protein crosslinking, which will lead to changes in the structure and conformation of the protein [23]. Previous studies have found that the protein oxidation occurred in fish under different storage conditions, which affected the elasticity of the protein gel, the total sulfhydryl (SH) and carbonyl group (CP) content of protein as well as the solubility and surface hydrophobicity of the protein [24,25]. Refrigeration storage at 4 °C caused the decrease in the contents of salt soluble protein, water soluble protein, and total protein of hybrid catfish fillets [26]. The softening of fish fillets during cold storage was attributed to the collagen solubilization and myofibrillar protein hydrolysis [27]. Detachment between myofibers and myocommata was concomitant with the loss of dystrophin, which corresponded to the reduction in flesh hardness of sea beam during ice storage [28]. Protein degradation was promoted by protein oxidation, which increased protein surface hydrophobicity and changed the secondary structure during refrigerated storage [29].

The aim of this study was to explore the changes in moisture migration, protein oxidation, microstructure, and the physicochemical quality of Atlantic mackerel in refrigerated conditions with and without ice (Scheme 1). The moisture distribution and migration of the Atlantic mackerel during the refrigeration process were monitored by LF-NMR and MRI. Changes in the secondary structure of protein were determined by Fourier transform infrared spectra (FTIR) and CP content, total SH content, disulfide bond content, and surface hydrophobicity were analyzed to evaluate the protein oxidation of mackerel muscle during refrigerated storage. The microstructure of mackerel muscle was observed using cryo-scanning electron microscopy. In addition, physicochemical properties including pH, total volatile basic nitrogen (TVB-N), thiobarbituric acid-reactive substances (TBARS), color, and other indicators were also analyzed to evaluate the quality changes.

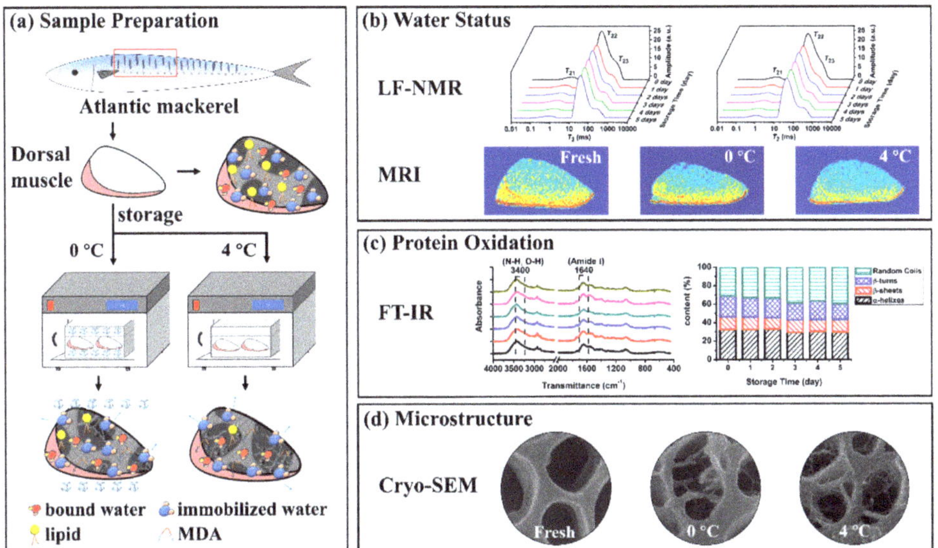

Scheme 1. Schematic diagram of this paper. (**a**) Sample preparation, (**b**) Water status, (**c**) Protein oxidation, (**d**) Microstructure. (Low-field nuclear magnetic resonance (LF-NMR), magnetic resonance imaging (MRI), fourier transform infrared spectra (FT-IR), scanning electron microscopy (SEM) and malondialdehyde (MDA)).

2. Materials and Methods

2.1. Sample Preparation

Vacuum-packed Atlantic mackerel (*Scomber scombrus*) fillets were purchased from Qingdao Yihexing Foods Co. Ltd. (Qingdao, Shandong province, China) and expressed to the laboratory under freezing conditions. Average weight of the fillets was 150 ± 10 g. Before the experiment, the frozen fillets were thawed under flowing water for half an hour. The dorsal muscles of Atlantic mackerel with 3.0 cm width and 2.0 cm thickness were cut into 3.0 cm length pieces. The mackerel pieces were placed in polythene bags, and randomly divided into two groups, which were stored in a refrigerator without ice (4 °C) and with ice (0 °C, ice replaced every day). The thawed fillets were defined as the sample stored for 0 day in this experiment.

2.2. Low-Field Nuclear Magnetic Resonance (LF-NMR) and Magnetic Resonance Imaging (MRI) Measurements

T_2 transverse relaxation data were measured by a MesoQMR23-060H NMR analyzer (Suzhou Niumag Analytical Instrument Co., Suzhou, China). The permanent magnet was kept at 32 °C with a 0.5 T magnetic field strength. A Carr–Purcell–Meiboom–Gill (CPMG) sequence with parameters including 90° pulses, 180° pulses, and π-value of 21.00 µs, 42.00 µs, and 200 µs was used to collect the relaxation decay data. After removing the surface water, the sample was placed in the 60-mm-diameter tube NMR probe and measured under 10,000 echoes with four scan repetitions. The inversion of the CPMG decay data was performed using MultiExp Inv analysis software (Suzhou Niumag Analytical Instrument Co., Suzhou, China) to obtain the T_2 distributed curve as well as the relaxation parameters of different proton components.

MRI experiments were performed by the same MesoQMR23-060H NMR analyzer with spin echo (SE) imaging sequence. The scanning parameters of matrix size, field of view, and slice width were 256 mm × 196 mm, 100 mm × 100 mm, and 3 mm, respectively. The echo time (T_E) and repetition time (T_R) of the T_1 weighted image were 40 ms and 660 ms, respectively, while that of the T_2 weighted image were 40 ms and 3300 ms. The raw MRI images were processed using OsiriXLite software (v7.0.4, Pixmeo, Geneva, Switzerland) to obtain pseudo-color images and the corresponding intensities.

2.3. Protein Oxidation

2.3.1. Protein Extraction

Water- and salt-soluble proteins were extracted from the mackerel muscle according to Yang et al. [23]. First, 5 g of mackerel muscle was mixed with 20 mL of precooled 20 mmol/L Tris-maleate solution (50 mmol/L NaCl, pH 7.0). The mixture was homogenized for 45 s on ice and centrifuged for 10 min at 4 °C with a centrifugal force of 8000× g. The supernatant was used for the measurement of water-soluble sarcoplasmic protein. Afterward, 20 mL of precooled 20 mmol/L Tris-maleate solution (0.6 mol/L NaCl, pH 7.0) was added to the precipitate, and the mixture was homogenized for 45 s on ice. After 15 min centrifugation at 8000× g at 4 °C, the supernatant was collected as the salt-soluble myofibrillar protein solution for subsequent experiments. The contents of water- and salt-soluble protein in the supernatant were measured using the method of Bradford, and bovine serum albumin (BSA) was used as the standard.

2.3.2. Analysis of Carbonyl Group (CP) Content, Total Sulfhydryl (SH) Content, Disulfide Bond Content, and Surface Hydrophobicity

According to the previous method of Ganhao et al. [30], 1 mL of 2.4-dinitrophenol hydrazine (DNPH, 10 mmol/L in 2 N HCl) was mixed with 1 mL of the salt-soluble protein solution, which was incubated in a 30 °C water bath for 1 h in darkness. Then, the protein in the mixture was precipitated by the addition of 1 mL of trichloroacetic acid. After 5 min centrifugation at 8000× g, the precipitate was washed three times using 1 mL ethanol–ethyl acetate (1:1, v/v) solution to remove the redundant DNPH. Finally, the precipitate was suspended in 3.5 mL guanidine hydrochloride (6 mol/L) and kept at 37 °C for 15 min. After 3 min centrifugation at 10,000× g, the absorbance of the supernatant was measured at 370 nm to calculate the content of carbonyl groups expressed as µmol of per gram of protein.

The total SH content was determined according to Shi's method [25] with some modifications. First, 9 mL tris-HCl buffer (0.05 mol/L with 8 mol/L urea, 10 mmol/L ethylenediamine tetra acetic acid (EDTA, pH 6.8) was added to 1 mL myofibrillar protein solution, which was incubated at 4 °C for 1 h. Then, 4 mL of the mixed solution was sampled and mixed with 400 µL Ellman's reagent (DTNB). After incubating for 25 min at 40 °C, the absorbance of the solution was measured with a wavelength of 412 nm, and the total SH content was calculated as µmol per gram protein with 13,600 M^{-1} cm^{-1} as the molar extinction coefficient.

The content of the disulfide bond was calculated by the total SH and free SH content. The free SH content was measured using the same method of total SH except for the tris-HCl buffer, which was substituted by an equal volume of buffer without urea (0.05 mol/L tris-HCl with 10 mM EDTA, pH 6.8). The disulfide bond was calculated with the following equation:

$$c_{\text{disulfide bond}} = \frac{\left(c_{\text{total SH}} - c_{\text{free SH}}\right)}{2} \tag{1}$$

Changes of surface hydrophobicity were determined according to the methodology of Kobayashi et al. [31]. For this experiment, 8-Anilinonaphthalene-1-sulfonic acid (ANS, 8 mmol/L in 0.1 mol/L PBS buffer, pH 7.0) was used as the fluorescent probe. Myofibrillar protein solutions with gradient concentrations of 0.05, 0.10, 0.15, and 0.2 mg/mL were prepared by the dilution of stock solution. Then, 4 mL of the diluted myofibrillar solution was mixed with 10 μL ANS. The fluorescence intensity of the mixture was determined by an F-2700 Fluorescence Spectrophotometer (Hitachi, Tokyo, Japan) using a 390 nm excitation wavelength and 470 nm emission wavelength. S_0-ANS was calculated based on the slope of the curve between fluorescence intensity and protein concentration.

2.3.3. Fourier Transform Infrared (FTIR) Spectra

Freeze-dried myofibrillar protein was used for the FTIR analysis (Frontier, PerkinElmer, Norwalk, CT, USA). The sample was mixed with KBr (1:100), and the mixture was ground into a uniform powder and pressed into a transparent sheet. Under transmittance mode, all spectra were collected in the wavelength range of 4000 to 400 cm^{-1} with the resolution of 4 cm^{-1} at room temperature. Omnic software was used for basic spectrum analysis. The data of the amide I bands were fitted to a Gaussian curve by Peakfit 4.12 software (Seasolve Software, Inc. Framingham, LA, USA). The secondary structure contents of each component were calculated from the peak area.

2.4. Microstructure

Samples with different storage times were cut into 2.0 mm × 2.0 mm × 5.0 mm pieces for microstructural analysis using cryo-scanning electron microscopy (SU8010/PP3010T, Hitachi, Tokyo, Japan). The pieces were placed on a sample tray coated with carbon conductive adhesive and immersed into slush nitrogen (−210 °C) to cryo-fix. The sample tray was transferred into a vacuum cryo-preparation chamber, and fractured on a preparation stage at −140 °C to expose the internal structure. After sublimating at −90 °C for 10 min, the sample was sputtered for 60 s at a 10 mA current, and transferred into the scanning electron microscopy (SEM) chamber for observation. The accelerating voltage of the electron beam was set at 1.0 kV.

2.5. Physicochemical Parameters

2.5.1. Drip Loss

The samples were taken out of the polythene bags and blotted with filter paper before weighing. Drip loss was calculated based on the weight difference of the mackerel samples before and after storage.

2.5.2. pH Value

The mixture of mackerel muscle and distilled water in aa ratio of 1:5 (w/v) was homogenized, and the suspension was used for pH value measurement with a digital pH meter (S210 Seven compact, Mettler-Toledo Instrument Co., Ltd., Shanghai, China).

2.5.3. Total Volatile Basic Nitrogen (TVB-N)

TVB-N was measured according to Conway's micro-diffusion method of association of official analytical chemists (AOAC) [32]. Briefly, the mixture of 5 g of the homogenized mackerel muscle and 50 mL distilled water was placed in a conical flask at ambient temperature for 30 min and filtered. Then, 1 mL of 2% boric acid was added in the inner ring of the Conway dishes, while 1 mL of the filtrate and 1 mL of saturated K_2CO_3 solution were placed in the outer ring. Each dish was covered by a piece of glass and sealed with gum arabic. The filtrate and saturated K_2CO_3 solution in the outer ring were mixed quickly. The Conway dishes were kept at 37 °C for 2 h, in which the formed volatile basic compounds were absorbed by the boric acid in the inner ring. The boric acid absorption solution of each Conway dish was titrated by H_2SO_4 with a micro-burette until it turned from a green color to pink. The TVB-N value was expressed as mg/100 g fish muscle according to the consumption of H_2SO_4:

$$TVB - N \ (mg/100 \ g) = \frac{V_{H_2SO_4} \times c_{H_2SO_4} \times 14}{m_{sample}} \times 50 \times 100 \tag{2}$$

where V_{H2SO4} an c_{H2SO4} are the volume and concentration of H_2SO_4 solotion, and m_{sample} is the mass of mackerel muscle.

2.5.4. Thiobarbituric Acid-Reactive Substances (TBARS)

TBARS values were measured according to John et al. [33] to estimate the level of lipid oxidation of Atlantic mackerel. First, 5 mL mixed solution (15% trichloro-acetic acid, 0.375% thiobarbituric acid, and 0.25 N HCl) was added to 1 g minced fish. Then, the mixture was incubated for 20 min in boiling water bath and centrifuged for 15 min at $8000 \times g$ by a centrifuge (CF-RXII Hitachi centrifuge, Hitachi, Japan). The supernatant was used to measure the absorbance at 532 nm by a spectrophotometer (Infinite F200 PRO, Tecan, Mannedorf, Switzerland). The content of malondialdehyde (MDA) was calculated based on absorbance and converted to a TBARS value as follows: TBARS (mg/kg) = $A_{532 \ nm} \times 2.77$.

2.6. Color Measurement

The surface color of mackerel were determined in CIE L^* (lightness) a^* (redness) b^* (yellowness) by an USP1792 UltraScan PRO colorimeter (Hunter Associates Lab., Inc., Reston, VA, USA). The colorimeter was equipped with a 10° standard observer, a D65 light source, and an aperture with 0.19 inch diameter. Three samples for each storage condition were prepared for color analysis every day, and three different surface areas of each sample were selected to measure the color value.

2.7. Statistical Analysis

All experimental data were displayed as mean values ± standard deviation. A commercial SPSS 19.0 software (SPSS Inc., Chicago, IL, USA) was used for the significance analysis of experimental data with one-way ANOVA followed by Tukey's test. Parts of the figures were plotted using Origin 8.5 software (OriginLab Corporation, Northampton, MA, USA).

3. Results

3.1. LF-NMR and MRI Analysis

The proton distribution and characteristics of Atlantic mackerel during storage at 4 °C and 0 °C were analyzed by LF-NMR proton transversal relaxation measurements, and T_2 distributed curves are shown in Figure 1a,b, respectively. Three proton peaks labeled as T_{21}, T_{22}, and T_{23} with different relaxation times were observed in the Atlantic mackerel matrix. T_{21} with a relaxation time less than 10 ms was described as bound water closely attached on the polar groups of macromolecules. The strongest T_{22} peak was ascribed to immobilized water, which was entrapped within the myofibrillar network based on its relaxation time in the range of 10–100 ms. T_{23} with a relaxation time between 100–1000 ms appeared as a shoulder peak of T_{22}, which was obviously different from the free water peak reported in other muscle food [8]. To identify the assignment of T_{23}, the T_2 relaxation distribution of fresh and freeze-dried Atlantic mackerel samples were compared, as shown in Figure S1. After the freeze-drying, the bound water T_{21} and the immobilized water T_{22} disappeared. Only one predominant proton peak with relaxation time around 100 ms was observed in the freeze-dried sample, which was ascribed to lipid based on the high lipid content of Atlantic mackerel. Therefore, the T_{23} peak of Atlantic mackerel sample was assigned to the protons of lipid.

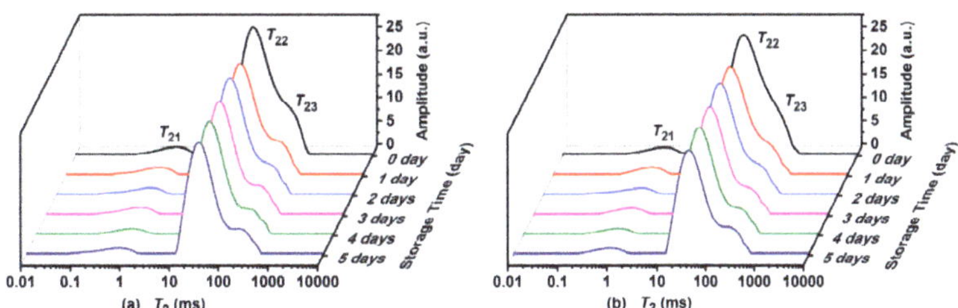

Figure 1. T_2 curves of Atlantic mackerel during five days of storage at (**a**) 4 °C and (**b**) 0 °C.

The changes in relaxation parameters for Atlantic mackerel during two storage conditions are shown in Figure 2. With the proceeding of refrigeration storage at 4 and 0 °C, the relaxation peaks of T_{21} and T_{22} shifted to shorter relaxation time, suggesting the decrease in water mobility of Atlantic mackerel during storage. The decrease of T_{22} might be caused by the denaturation and aggregation of protein, which led to changes in chemical exchange between water and protein protons [13,20]. As shown in Figure 2c,d, the change in the peak area of A_{21} was nonsignificant, while A_{22} dropped rapidly on the first day, and then decreased slightly during subsequent storage. The status and distribution of water could influence other qualities such as WHC and textural properties.

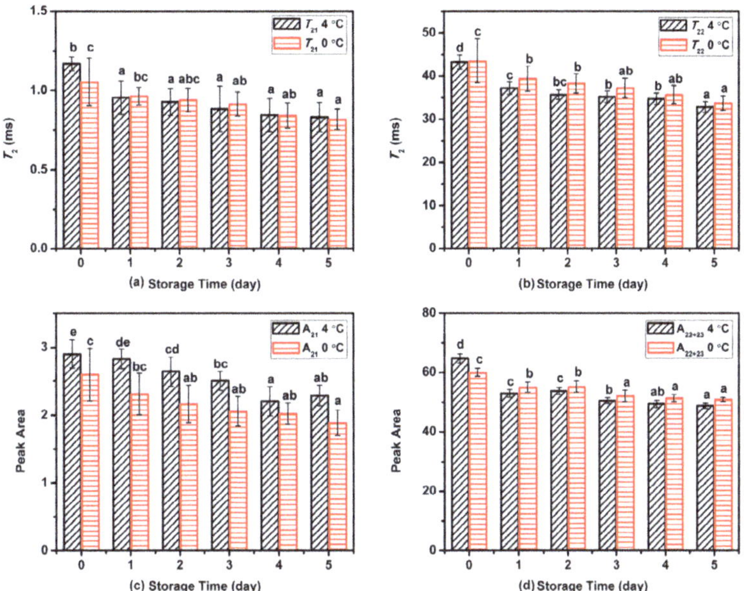

Figure 2. Variation in (**a**) T_{21}, (**b**) T_{22}, (**c**) A_{21}, and (**d**) A_{22} of water components in Atlantic mackerel during storage at 4 °C and 0 °C for five days. Different letters a–d mean significant difference ($p < 0.05$).

Figure 3a,b shows the T_1 and T_2 weighted pseudo-color images of Atlantic mackerel under different refrigeration conditions, in which the red color represents high proton density and the blue color represents low proton density. Protons with low and high mobility are highlighted in the T_1 and T_2 images, respectively. As shown in the pseudo-color image, the signal intensity was mainly distributed in the skin and the dark meat near the skin. During storage of Atlantic mackerel, the MRI intensity was weakened at both 4 °C and 0 °C, suggesting that the water content of the fish decreased during refrigeration storage. The quantitative signal intensity of the T_1 and T_2 weighted images are shown in Figure 3c,d. Significant decreases in the intensities of the T_1 and T_2 images under two storage conditions with the storage time were observed, which was consistent with the decrease in the T_{21} and $T_{22} + T_{23}$ peak area observed in T_2 relaxation. The decrease in the signal intensity of T_1 and T_2 images demonstrated the decrease in the moisture content of mackerel muscle, which could be ascribed to the degradation and denaturation of proteins during refrigeration.

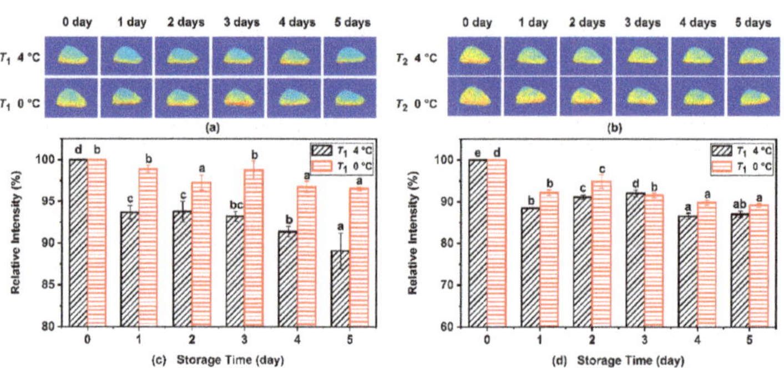

Figure 3. (**a**) T_1 and (**b**) T_2 weighted magnetic resonance imaging (MRI) images of Atlantic mackerel during storage at 4 °C and 0 °C, and the relative intensities of (**c**) T_1 and (**d**) T_2 images. Different letters a–e mean significant difference ($p < 0.05$).

3.2. Protein Oxidation Analysis

The water-soluble protein in fish is mainly sarcoplasmic protein, while the salt-soluble protein is mainly myofibrillar protein. Changes in sarcoplasmic protein and myofibrillar protein of Atlantic mackerel during five days of storage at 4 °C and 0 °C are shown in Figure 4. The initial content of water-soluble protein in Atlantic mackerel stored for 0 days was 38.28 ± 2.76 mg/g, which increased to 48.52 ± 3.07 mg/g and 46.92 ± 2.06 mg/g after five days at 4 °C and 0 °C, respectively. Meanwhile, the content of myofibrillar protein in Atlantic mackerel before storage was 29.01 ± 0.72 mg/g, which decreased to 17.40 ± 0.72 mg/g and 21.67 ± 1.80 mg/g after five days at 4 °C and 0 °C, respectively. The decrease in salt-soluble protein content of Atlantic mackerel stored at 4 °C was more significant than that of 0 °C. The protein oxidation, leading to protein crosslinking and protein aggregation, could be explained by the decrease in salt-soluble protein extractability during storage [34]. The increase in the water-soluble protein content of Atlantic mackerel during storage corresponded to the decrease in salt-soluble protein content, which was in agreement with the study on bighead carp fillets under chilled storage [35].

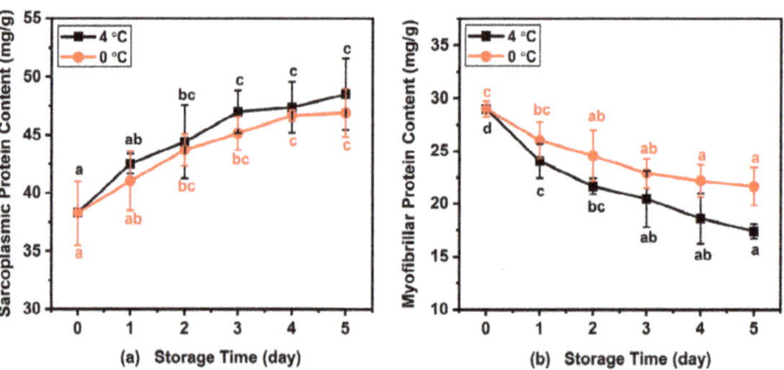

Figure 4. Variations in (**a**) sarcoplasmic protein and (**b**) myofibrillar protein of Atlantic mackerel during five days storage at 4 °C and 0 °C. Different letters a–d mean significant difference ($p < 0.05$).

Changes in CP content, sulfhydryl groups, and hydrophobicity are usually used to quantify the protein oxidation of muscle food [36]. Figure 5 shows the changes in CP content, sulfhydryl group content, disulfide bond content, and surface hydrophobicity of Atlantic mackerel during 4 °C and 0 °C storage. The carbonyl content of myofibrillar in Atlantic mackerel showed a significant increase from an initial value of 0.84 ± 0.15 µmol/g at 0 day to 1.54 ± 0.16 and 1.49 ± 0.18 µmol/g on the fifth day stored at 4 °C and 0 °C, as shown in Figure 5a. The metal-catalyzed oxidation, and reaction with lipid peroxidation products and the oxidation of the side chain or backbone of some amino acids could all produce carbonyl groups, which reduces the protein solubility [36]. Similar results were also observed in the minced mackerel during 4 °C storage, which suggested the relationship between the formation of protein carbonyl groups and lipid oxidation [37]. The total SH content showed a decrease from an initial value of 18.28 ± 0.15 µmol/g to 16.77 ± 0.09 and 17.63 ± 0.12 µmol/g protein at 4 °C and 0 °C, respectively (Figure 5b). Corresponding to the decrease in the total SH content, the disulfide bond content of myofibrillar protein from Atlantic mackerel increased significantly during the 4 °C and 0 °C storage (Figure 5c). The formation of disulfide bonds was probably ascribed to the oxidation of SH groups of the protein during the refrigeration storage [20]. The surface hydrophobicity of Atlantic mackerel protein significantly increased during storage at 4 °C and 0 °C (Figure 5d), which indicated the occurrence of protein denaturation. The increase in the surface hydrophobicity could be explained by the unfolding of the proteins during cold storage, which led to the exposure of hydrophobic aliphatic and aromatic amino acids [20]. The above results show that the storage at 0 °C could diminish the protein oxidation of Atlantic mackerel compared with storage at 4 °C.

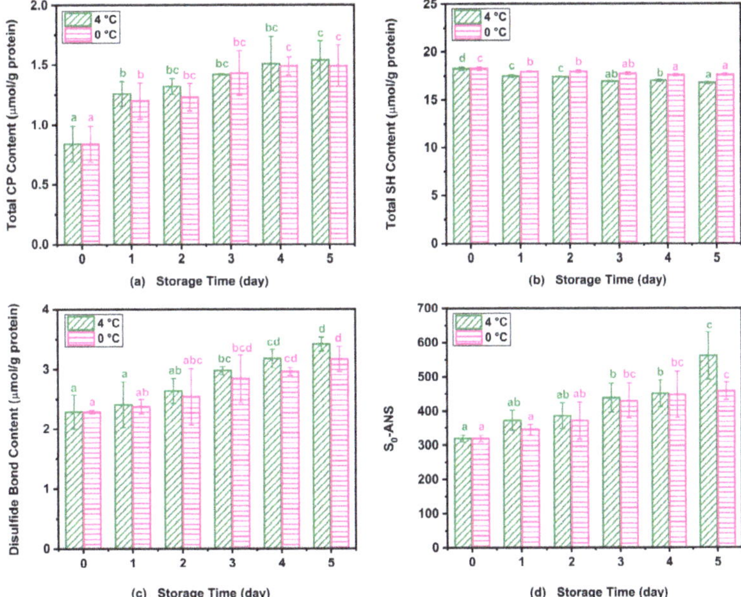

Figure 5. Variations in (**a**) carbonyl, (**b**) total sulfhydryl, (**c**) disulfide bond, and (**d**) surface hydrophobicity (S_0-ANS) of Atlantic mackerel protein during five days of storage at 4 °C and 0 °C. Different letters a–d mean significant difference ($p < 0.05$).

3.3. Secondary Structure Change of Atlantic Mackerel Protein

The degradation and oxidation of proteins during storage generally lead to the destruction of secondary structure and spatial conformation of proteins. The FTIR spectra of protein extracted from Atlantic mackerel with different storage times under two storage conditions are shown in Figure 6a,b. The amide I band (1600~1700 cm^{-1}) was the main characteristic absorption band of protein in the infrared region. The FTIR spectra of Atlantic mackerel stored at 4 °C and 0 °C both showed a strong absorption peak around 1640 cm^{-1}, and the absorption intensity increased with the prolongation of storage time, which might be due to the changes in the conformation of the polypeptide chain. The percentages of α-helices, β-turns, β-sheets, and random coils of protein extracted from Atlantic mackerel under different storage conditions are shown in Figure 6c,d. Decrease in percentage of α-helix and increase in percentage of random coils were revealed by the quantitative analysis. The network structure was damaged due to the protein aggregation by the cross-linking of disulfide bonds and carbonyls [38]. The decrease in α-helices indicated the unfolding of protein induced by oxidation, which was consistent with the increase in hydrophobicity. The changes of β-turns and β-sheets were also ascribed to the oxidation of proteins, which was promoted by the free radicals produced by the lipid oxidation [20].

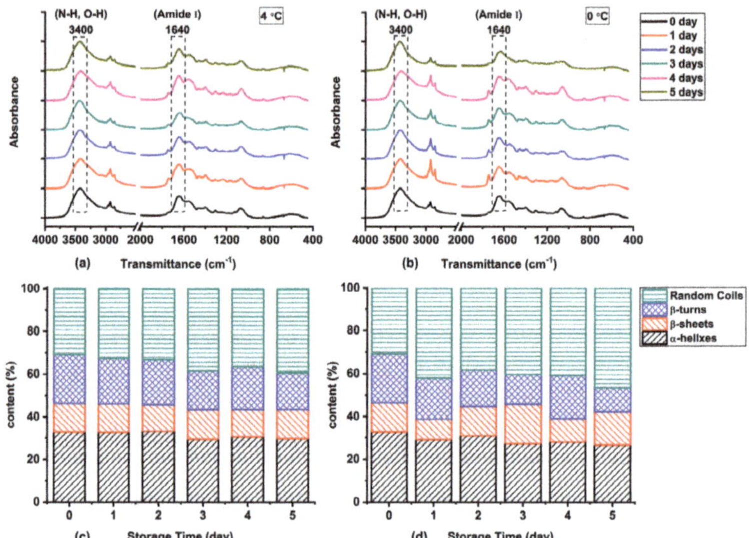

Figure 6. Fourier transform infrared (FTIR) spectra of Atlantic mackerel protein stored at (**a**) 4 °C and (**b**) 0 °C after different storage times; quantitative analysis of secondary structure changes of protein in Atlantic mackerel during storage at (**c**) 4 °C and (**d**) 0 °C.

3.4. Microstructure Analysis

Protein denaturation can be reflected in the microstructure. The qualitative changes in the microstructure of Atlantic mackerel during storage observed by cryo-scanning electron microscopy (Cryo-SEM) are shown in Figure 7. The section perpendicular to the muscle fiber direction of the sample stored for 0 days showed many holes with a smooth surface in the muscle fiber, which was formed by the ice crystals during frozen storage. With the increase in drip loss of Atlantic mackerel during storage, the myofibrils contracted and fractured, and the holes became dense and irregular. Granular materials appeared on the surface of holes after three days of storage for both 4 °C and 0 °C. A similar phenomenon also appeared in bighead carp during storage, which was explained by the ruptures of the endomysium and the perimysium and the outflow of sarcoplasm [39]. The moisture reduction and protein denaturation of Atlantic mackerel after refrigerated storage could be explained by the microstructure changes.

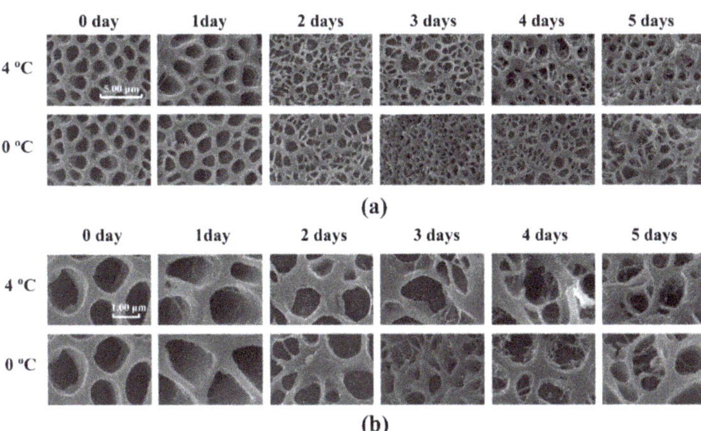

Figure 7. Microstructure of Atlantic mackerel after storage at 4 °C and 0 °C observed by cryo-scanning eletron microscopy (Cryo–SEM) with (**a**) 10 k× and (**b**) 30 k×.

3.5. Physicochemical Analysis

Variations in the drip loss of Atlantic mackerel during storage at 4 °C and 0 °C are shown in Figure 8a. The drip loss of Atlantic mackerel significantly increased to 18.73% and 14.54% after five days of storage at 4 °C and 0 °C, respectively. This trend was in agreement with the changes in drip loss in sea bream during storage under different conditions [40]. Drip loss including water and lipid loss increased rapidly in the early storage stage and slightly in subsequent storage, which was consistent with the change in A_{22}. The increase in drip loss of mackerel muscle during storage might be ascribed to the breakdown of protein, which led to the decrease of the WHC of muscle. Generally, lower storage temperature leads to lower drip loss.

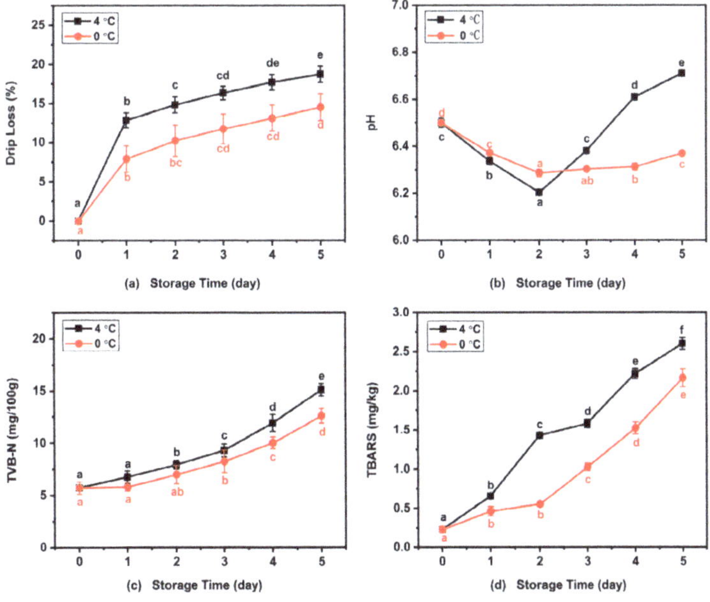

Figure 8. Variations of (**a**) drip loss, (**b**) pH, (**c**) TVB-N, and (**d**) TBARS of Atlantic mackerel during storage at 4 °C and 0 °C. Different letters a–f mean significant difference ($p < 0.05$).

Figure 8b shows the changes in the pH values of mackerel muscle with the increase in storage time under two refrigeration conditions. The pH values both declined significantly at the beginning of 4 °C and 0 °C storage, which reached the lowest values of 6.20 and 6.29 after two days of storage, respectively. With the further increase of storage time, the pH value of Atlantic mackerel samples stored at 4 °C increased significantly while that of the samples stored at 0 °C increased slightly. This was consistent with the result of the grass carp fillets during 0 °C storage, in which pH value decreased within the first three days and then increased during the following storage [41]. The pH value decrease during the first two days of storage was due to the degradation of adenosine-triphosphate (ATP) and creatine phosphate, and the subsequent increase was caused by the formation of alkaline substances related to protein degradation during refrigeration.

TVB-N mainly consists of amines such as methylamine, dimethylamine, and trimethylamine, which have been considered as important quality indicators of aquatic products. Figure 8c shows the changes in TVB-N values of Atlantic mackerel during refrigeration of 4 °C and 0 °C. Significant increases in TVB-N content from an value of 5.74 ± 0.56 mg/100 g to 15.12 ± 0.60 and 12.62 ± 0.70 mg/100 g were observed after five days of storage at 4 °C and 0 °C. The increase in TVB-N value in fish during cold storage is mainly due to the alkalinity substances produced by the protein degradation by enzymatic reactions and microbial activity [42].

The lipids in Atlantic mackerel are prone to oxidize during storage, and the degree of lipid oxidation can be evaluated by the TBARS value [7]. Figure 8d shows changes in the TBARS value of Atlantic mackerel during 4 °C and 0 °C storage. The initial TBARS value at the beginning of the storage was 0.23 ± 0.04 mg/kg, which increased to 2.60 ± 0.08 mg/kg and 2.16 ± 012 mg/kg after five days of storage at 4 °C and 0 °C, respectively. Wang et al. reported that the storage at 0 °C was more effective at inhibiting lipid oxidation of salmon than 4 °C [19]. Free radicals produced by the lipid oxidation can promote protein oxidation to form carbonyl groups and change the secondary structure. Increased lipid oxidation was associated with protein carbonyl group formation and β-turn and β-sheet changes during storage.

3.6. Color Analysis

Color is a common parameter to evaluate the quality of food. The effects of the 4 °C and 0 °C storage processes on the lightness (L^*), redness (a^*), and yellowness (b^*) values of Atlantic mackerel are shown in Table 1. With the increase of refrigeration time, the L^* values of samples stored at both 4 °C and 0 °C increased significantly, and a higher storage temperature led to higher L^* value. The changes in heme proteins (particularly the heme-ring destruction) were considered to be responsible for the increased L^* values during the frozen storage of herring fillet [43]. The increase of L^* in superchilled salmon was also thought to be due to the white spots on the surface of fillets caused by drip channels [44]. The a^* values were relatively stable, while the b^* values increased significantly during storage. The lipid oxidation was considered to be the main reason for the increase in yellowness for fish, which was also supported by the increase in TBARS value in this study.

Table 1. Variations of color parameters of Atlantic mackerel during refrigeration at 4 °C and 0 °C.

Storage Time (D)	L^*		a^*		b^*	
	4 °C	0 °C	4 °C	0 °C	4 °C	0 °C
0	45.83 ± 1.51 [a]	45.83 ± 1.51 [a]	-1.67 ± 0.47 [a]	-1.67 ± 0.47 [ab]	7.50 ± 0.50 [a]	7.50 ± 0.50 [a]
1	48.71 ± 1.11 [b]	47.70 ± 0.56 [ab]	-1.79 ± 0.47 [a]	-1.88 ± 0.55 [ab]	8.78 ± 0.38 [ab]	7.82 ± 1.25 [a]
2	53.46 ± 1.27 [c]	50.44 ± 3.18 [bc]	-1.78 ± 0.19 [a]	-2.02 ± 0.46 [a]	10.88 ± 1.67 [bc]	9.22 ± 0.58 [a]
3	54.79 ± 1.13 [cd]	51.81 ± 1.35 [bc]	-1.37 ± 0.72 [a]	-2.27 ± 0.26 [a]	11.52 ± 2.09 [cd]	9.86 ± 1.37 [ab]
4	56.57 ± 1.17 [de]	54.40 ± 3.99 [c]	-2.10 ± 0.08 [a]	-1.08 ± 0.35 [b]	11.85 ± 0.29 [cd]	11.93 ± 1.80 [b]
5	57.50 ± 1.74 [e]	54.87 ± 2.09 [c]	-1.51 ± 0.56 [a]	-2.48 ± 0.69 [a]	13.70 ± 0.87 [e]	12.21 ± 1.56 [b]

Different letters [a–e] in the same column mean significant difference ($p < 0.05$).

3.7. Correlation Analysis between Protein Oxidation, Physicochemical Qualities, and Water Status

Correlation coefficients between protein oxidation indicators, physicochemical qualities, and LF-NMR relaxation times are shown in Table 2. T_{21} showed significant correlations with the contents of carbonyl and total sulfhydryl group for the samples stored at 4 °C. T_{22} was positively correlated with the contents of myofibrillar protein and total sulfhydryl, while negatively correlated with the contents of sarcoplasmic protein, carbonyl, disulfide bond, and surface hydrophobicity as well as drip loss, TBARS, TVB-N, L^*, and b^* during 4 and 0 °C storage. During storage, changes in total sulfhydryl content, disulfide bond content, and hydrophobicity affect the water binding ability of proteins. Significant correlations between relaxation time of immobilized water and physicochemical indexes were also observed in tuna [20]. Pearson's correlation analysis proved the excellent correlations between T_{22} and the protein oxidation indicators and physicochemical parameters, which indicated that the quality changes of Atlantic mackerel during refrigerated storage could be monitored by LF-NMR.

Table 2. Correlation coefficients between sarcoplasmic protein content, myofibrillar protein content, CP content, total SH content, disulfide bond content, surface hydrophobicity, drip loss, pH, TBARS, TVB-N, L^*, a^* and b^* with T_{21} and T_{22} obtained by Pearson's correlation analysis.

Parameter	4 °C		0 °C	
	T_{21}	T_{22}	T_{21}	T_{22}
Sarcoplasmic protein content	−0.269	−0.706 **	−0.536 *	−0.899 **
Myofibrillar protein content	0.573 *	0.832 **	0.458	0.650 **
Total CP content	−0.638 **	−0.849 **	−0.446	−0.542 *
Total SH content	0.617 **	0.856 **	0.466	0.681 **
Disulfide bond content	−0.242	−0.597 *	−0.573 *	−0.892 **
Surface hydrophobicity	−0.323	−0.641 **	−0.430	−0.802 **
Drip loss	−0.496 *	−0.750 **	−0.492 *	−0.792 **
pH	−0.266	−0.292	0.348	0.536 *
TBARS	−0.427	−0.822 **	−0.533 *	−0.754 **
TVB-N	−0.402	−0.739 **	−0.535 *	−0.785 **
L^*	−0.462	−0.750 **	−0.584 *	−0.721 **
a^*	−0.203	−0.760 **	−0.198	0.142
b^*	−0.588 *	−0.790 **	−0.494 *	−0.690 **

$* p < 0.01$ and $** p < 0.05$.

4. Conclusions

In this study, the moisture migration, protein oxidation, microstructure, and physicochemical qualities of Atlantic mackerel (*Scomber scombrus*) during storage at 4 and 0 °C were investigated. Three proton components assigned to bound water, immobilized water, and lipid were observed in mackerel by LF-NMR relaxation. With the prolongation of refrigeration time, the mobility and abundance of immobilized water decreased. MRI images also displayed the decrease in water content of mackerel during storage. The refrigerated storage also caused the oxidation and the change of the secondary structure of the mackerel protein, the fracture of myofibrils, and the changes in physicochemical parameters.

Supplementary Materials: The following are available online at https://www.mdpi.com/2304-8158/10/2/214/s1, Figure S1: T_2 curves of fresh and freeze-dried Atlantic mackerel.

Author Contributions: Conceptualization, R.L., S.C., S.W., and M.T.; Data curation, R.L. and S.W.; Formal analysis, R.L. and S.W.; Funding acquisition, S.C.; Investigation, R.L.; Methodology, R.L. and S.W.; Project administration, R.L.; Resources, S.C., M.T., and B.Z.; Supervision, S.C., M.T., and B.Z.; Validation, R.L.; Writing—original draft, R.L. and S.C.; writing—review & editing, S.C. and M.T. All authors have read and agreed to the published version of the manuscript.

Funding: This work was supported by the National Natural Science Foundation of China (31972105) and the Central Funds Guiding the Local Science and Technology Development of China (2020JH6/10500002).

Institutional Review Board Statement: Not applicable.

Informed Consent Statement: Informed consent was obtained from all subjects involved in the study.

Data Availability Statement: The data that support the findings of this study are available from the corresponding author upon reasonable request.

Conflicts of Interest: The authors declare no conflict of interest. The funders had no role in the design of the study; in the collection, analyses, or interpretation of data; in the writing of the manuscript, or in the decision to publish the results.

References

1. Romotowska, P.E.; Gudjonsdottir, M.; Karlsdottir, M.G.; Kristinsson, H.G.; Arason, S. Stability of frozen atlantic mackerel (*Scomber scombrus*) as affected by temperature abuse during transportation. *LWT Food Sci. Technol.* **2017**, *83*, 275–282. [CrossRef]
2. Maestre, R.; Pazos, M.; Medina, I. Role of the raw composition of pelagic fish muscle on the development of lipid oxidation and rancidity during storage. *J. Agric. Food Chem.* **2011**, *59*, 6284–6291. [CrossRef]
3. Romotowska, P.E.; Karlsdóttir, M.G.; Gudjónsdóttir, M.; Kristinsson, H.G.; Arason, S. Seasonal and geographical variation in chemical composition and lipid stability of Atlantic mackerel (*Scomber scombrus*) caught in icelandic waters. *J. Food Compos. Anal.* **2016**, *49*, 9–18. [CrossRef]
4. Mohan, C.O.; Ravishankar, C.N.; Lalitha, K.; Gopal, T.S. Effect of chitosan edible coating on the quality of double filleted Indian oil sardine (*Sardinella longiceps*) during chilled storage. *Food Hydrocoll.* **2012**, *26*, 167–174. [CrossRef]
5. Truong, B.Q.; Buckow, R.; Nguyen, M.H.; Stathopoulos, C.E. High pressure processing of barramundi (*Lates calcarifer*) muscle before freezing: The effects on selected physicochemical properties during frozen storage. *J. Food Eng.* **2016**, *169*, 72–78. [CrossRef]
6. Bono, G.; Okpala, C.O.R.; Vitale, S.; Ferrantelli, V.; Di Noto, A.; Costa, A.; Di Bella, C.; Monaco, D.L. Effects of different ozonized slurry-ice treatments and superchilling storage (−1 °C) on microbial spoilage of two important pelagic fish species. *Food Sci. Nutr.* **2017**, *5*, 1049–1056. [CrossRef]
7. Yu, D.; Wu, L.; Regenstein, J.M.; Jiang, Q.; Yang, F.; Xu, Y.; Xia, W. Recent advances in quality retention of non-frozen fish and fishery products: A review. *Crit. Rev. Food Sci. Nutr.* **2020**, *60*, 1747–1759. [CrossRef]
8. Pearce, K.; Rosenvold, K.; Andersen, H.J.; Hopkins, D.L. Water distribution and mobility in meat during the conversion of muscle to meat and ageing and the impacts on fresh meat quality attributes—A review. *Meat Sci.* **2011**, *89*, 111–124. [CrossRef]
9. Marcone, M.F.; Wang, S.; Albabish, W.; Nie, S.; Somnarain, D.; Hill, A. Diverse food-based applications of nuclear magnetic resonance (NMR) technology. *Food Res. Int.* **2013**, *51*, 729–747. [CrossRef]
10. Belton, P. Spectroscopic Approaches to the Understanding of Water in Foods. *Food Rev. Int.* **2011**, *27*, 170–191. [CrossRef]
11. Andersen, C.M.; JØRgensen, B.M. On the Relation Between Water Pools and Water Holding Capacity in Cod Muscle. *J. Aquat. Food Prod. Technol.* **2004**, *13*, 13–23. [CrossRef]
12. Sánchez-Alonso, I.; Martinez, I.; Valencia, J.R.S.; Careche, M. Estimation of freezing storage time and quality changes in hake (*Merluccius merluccius*, L.) by low field NMR. *Food Chem.* **2012**, *135*, 1626–1634. [CrossRef]
13. Sánchez-Alonso, I.; Moreno, P.; Careche, M. Low field nuclear magnetic resonance (LF-NMR) relaxometry in hake (*Merluccius merluccius*, L.) muscle after different freezing and storage conditions. *Food Chem.* **2014**, *153*, 250–257. [CrossRef]
14. Sánchez-Valencia, J.; Sánchez-Alonso, I.; Martinez, I.; Careche, M. Low-Field Nuclear Magnetic Resonance of Proton (1H LF NMR) Relaxometry for Monitoring the Time and Temperature History of Frozen Hake (*Merluccius merluccius* L.) Muscle. *Food Bioprocess Technol.* **2015**, *8*, 2137–2145. [CrossRef]
15. Xu, Y.; Song, M.; Xia, W.; Jiang, Q. Effects of freezing method on water distribution, microstructure, and taste active compounds of frozen channel catfish (*Ictalurus punctatus*). *J. Food Process Eng.* **2018**, *42*, e12937. [CrossRef]
16. Sun, Q.; Sun, F.; Xia, X.; Xu, H.; Kong, B. The comparison of ultrasound-assisted immersion freezing, air freezing and immersion freezing on the muscle quality and physicochemical properties of common carp (*Cyprinus carpio*) during freezing storage. *Ultrason. Sonochem.* **2019**, *51*, 281–291. [CrossRef]
17. Qin, N.; Zhang, L.; Zhang, J.; Song, S.; Wang, Z.; Regenstein, J.M.; Luo, Y. Influence of lightly salting and sugaring on the quality and water distribution of grass carp (*Ctenopharyngodon idellus*) during super-chilled storage. *J. Food Eng.* **2017**, *215*, 104–112. [CrossRef]
18. Sun, Y.; Ma, L.; Ma, M.; Zheng, H.; Zhang, X.; Cai, L.; Li, J.; Zhang, Y. Texture characteristics of chilled prepared Mandarin fish (*Siniperca chuatsi*) during storage. *Int. J. Food Prop.* **2018**, *21*, 242–254. [CrossRef]
19. Wang, S.; Xiang, W.; Fan, H.; Xie, J.; Qian, Y. Study on the mobility of water and its correlation with the spoilage process of salmon (*Salmo solar*) stored at 0 and 4 °C by low-field nuclear magnetic resonance (LF NMR 1H). *J. Food Sci. Technol.* **2018**, *55*, 173–182. [CrossRef]
20. Wang, X.-Y.; Xie, J. Evaluation of water dynamics and protein changes in bigeye tuna (*Thunnus obesus*) during cold storage. *LWT Food Sci. Technol.* **2019**, *108*, 289–296. [CrossRef]
21. Gudjónsdóttir, M.; Lauzon, H.L.; Magnússon, H.; Sveinsdóttir, K.; Arason, S.; Martinsdóttir, E.; Rustad, T. Low field Nuclear Magnetic Resonance on the effect of salt and modified atmosphere packaging on cod (*Gadus morhua*) during superchilled storage. *Food Res. Int.* **2011**, *44*, 241–249. [CrossRef]

22. Gudjónsdóttir, M.; Romotowska, P.E.; Karlsdóttir, M.G.; Arason, S. Low field nuclear magnetic resonance and multivariate analysis for prediction of physicochemical characteristics of Atlantic mackerel as affected by season of catch, freezing method, and frozen storage duration. *Food Res. Int.* **2019**, *116*, 471–482. [CrossRef]

23. Yang, F.; Jia, S.; Liu, J.; Gao, P.; Yu, D.; Jiang, Q.; Xu, Y.; Yu, P.; Xia, W.; Zhan, X. The relationship between degradation of myofibrillar structural proteins and texture of superchilled grass carp (*Ctenopharyngodon idella*) fillet. *Food Chem.* **2019**, *301*, 125278. [CrossRef]

24. Barraza, F.A.A.; León, R.A.Q.; Álvarez, P.X.L. Kinetics of protein and textural changes in Atlantic salmon under frozen storage. *Food Chem.* **2015**, *182*, 120–127. [CrossRef]

25. Shi, L.; Yang, T.; Xiong, G.; Li, X.; Wang, X.; Ding, A.; Qiao, Y.; Wu, W.; Liao, L.; Wang, L. Influence of frozen storage temperature on the microstructures and physicochemical properties of pre-frozen perch (*micropterus salmoides*). *LWT Food Sci. Technol.* **2018**, *92*, 471–476. [CrossRef]

26. Chomnawang, C.; Nantachai, K.; Yongsawatdigul, J.; Thawornchinsombut, S.; Tungkawachara, S. Chemical and biochemical changes of hybrid catfish fillet stored at 4 °C and its gel properties. *Food Chem.* **2007**, *103*, 420–427. [CrossRef]

27. Sáez, M.I.; Suárez, M.D.; Cárdenas, S.; Martínez, T.F.; Martínez, T.F. Freezing and Freezing-Thawing Cycles on Textural and Biochemical Changes of Meagre (*Argyrosomus regius*, L.) Fillets During Further Cold Storage. *Int. J. Food Prop.* **2014**, *18*, 1635–1647. [CrossRef]

28. Caballero, M.J.; Betancor, M.; Escrig, J.; Montero, D.; Monteros, A.E.D.L.; Castro, P.; Ginés, R.; Izquierdo, M. Post mortem changes produced in the muscle of sea bream (*Sparus aurata*) during ice storage. *Aquaculture* **2009**, *291*, 210–216. [CrossRef]

29. Zhang, X.; Huang, W.; Xie, J. Effect of Different Packaging Methods on Protein Oxidation and Degradation of Grouper (*Epinephelus coioides*) During Refrigerated Storage. *Foods* **2019**, *8*, 325. [CrossRef]

30. Ganhão, R.; Morcuende, D.; Estévez, M. Protein oxidation in emulsified cooked burger patties with added fruit extracts: Influence on colour and texture deterioration during chill storage. *Meat Sci.* **2010**, *85*, 402–409. [CrossRef]

31. Kobayashi, N.P.; Donley, C.L.; Wang, S.-Y.; Williams, R.S. Atomic layer deposition of aluminum oxide on hydrophobic and hydrophilic surfaces. *J. Cryst. Growth* **2007**, *299*, 218–222. [CrossRef]

32. AOAC. *Associaation of Official Analytical Chemists: Official Methods of Analysis*, 17th ed.; AOAC: Washington, DC, USA; Gaithersburg, MD, USA, 2005.

33. John, L.; Cornforth, D.; Carpenter, C.E.; Sorheim, O.; Pettee, B.C.; Whittier, D.R. Color and thiobarbituric acid values of cooked top sirloin steaks packaged in modified atmospheres of 80% oxygen, or 0.4% carbon monoxide, or vacuum. *Meat Sci.* **2005**, *69*, 441–449. [CrossRef]

34. Standal, I.B.; Mozuraityte, R.; Rustad, T.; Alinasabhematabadi, L.; Carlsson, N.-G.; Undeland, I. Quality of Filleted Atlantic Mackerel (*Scomber Scombrus*) During Chilled and Frozen Storage: Changes in Lipids, Vitamin D, Proteins, and Small Metabolites, including Biogenic Amines. *J. Aquat. Food Prod. Technol.* **2018**, *27*, 338–357. [CrossRef]

35. Lu, H.; Wang, H.; Luo, Y. Changes in Protein Oxidation, Water-Holding Capacity, and Texture of Bighead Carp (*Aristichthys Nobilis*) Fillets Under Chilled and Partial Frozen Storage. *J. Aquat. Food Prod. Technol.* **2017**, *26*, 566–577. [CrossRef]

36. Hematyar, N.; Rustad, T.; Sampels, S.; Dalsgaard, T.K. Relationship between lipid and protein oxidation in fish. *Aquac. Res.* **2019**, *50*, 1393–1403. [CrossRef]

37. Babakhani, A.; Farvin, K.H.S.; Jacobsen, C. Antioxidative Effect of Seaweed Extracts in Chilled Storage of Minced Atlantic Mackerel (*Scomber scombrus*): Effect on Lipid and Protein Oxidation. *Food Bioprocess Technol.* **2016**, *9*, 352–364. [CrossRef]

38. Pan, J.; Wang, Y.; Xia, L.; Lian, H.; Jin, W.; Li, S.; Qiao, F.; Dong, X. Physiochemical and rheological properties of oxidized Japanese seerfish (*Scomberomorus niphonius*) myofibrillar protein. *J. Food Biochem.* **2019**, *43*, e13079. [CrossRef]

39. Lu, H.; Zhang, L.; Shi, J.; Wang, Z.; Luo, Y. Effects of frozen storage on physicochemical characteristics of bighead carp (*Aristichthys nobilis*) fillets. *J. Food Process. Preserv.* **2019**, *43*, e14141. [CrossRef]

40. Duran-Montgé, P.; Permanyer, M.; Belletti, N. Refrigerated or superchilled skin-packed sea bream (*Sparus aurata*) compared with traditional unpacked storage on ice with regard to physicochemical, microbial and sensory attributes. *J. Food Process. Preserv.* **2015**, *39*, 1278–1286. [CrossRef]

41. Liu, D.; Liang, L.; Xia, W.; Regenstein, J.M.; Zhou, P. Biochemical and physical changes of grass carp (*Ctenopharyngodon idella*) fillets stored at −3 and 0 °C. *Food Chem.* **2013**, *140*, 105–114. [CrossRef]

42. Otero, L.; Pérez-Mateos, M.; Holgado, F.; Márquez-Ruiz, G.; López-Caballero, M. Hyperbaric cold storage: Pressure as an effective tool for extending the shelf-life of refrigerated mackerel (*Scomber scombrus*, L.). *Innov. Food Sci. Emerg. Technol.* **2019**, *51*, 41–50. [CrossRef]

43. Cavonius, L.R.; Undeland, I. Glazing herring (*Clupea harengus*) fillets with herring muscle press juice: Effect on lipid oxidation development during frozen storage. *Int. J. Food Sci. Technol.* **2017**, *52*, 1229–1237. [CrossRef]

44. Duun, A.; Rustad, T. Quality of superchilled vacuum packed Atlantic salmon (*Salmo salar*) fillets stored at −1.4 and −3.6 °C. *Food Chem.* **2008**, *106*, 122–131. [CrossRef]

MDPI

St. Alban-Anlage 66

4052 Basel

Switzerland

Tel. +41 61 683 77 34

Fax +41 61 302 89 18

www.mdpi.com

Foods Editorial Office

E-mail: foods@mdpi.com

www.mdpi.com/journal/foods